Differential Equations
A Linear Algebra Approach

Anindya Dey

CRC Press
Taylor & Francis Group
Boca Raton London New York

CRC Press is an imprint of the
Taylor & Francis Group, an **informa** business

LEVANT
Levant Books
India

First published 2022
by CRC Press
2 Park Square, Milton Park, Abingdon, Oxon, OX14 4RN

and by CRC Press
6000 Broken Sound Parkway NW, Suite 300, Boca Raton, FL 33487-2742

British Library Cataloguing-in-Publication Data
A catalogue record for this book is available from the British Library

Library of Congress Cataloging-in-Publication Data
A catalog record has been requested

ISBN: 9781032072265 (hbk)
ISBN: 9781003205982 (ebk)

DOI: 10.1201/9781003205982

Typeset in Knuth Computer Modern 10.5pt
by Levant Books, Kolkata 700014

LEVANT

Dedicated to
My wife **Aparna** & daughters **Riddhi, Anandi**

Preface

'There is nothing so practical as a good theory'

—Kurt Lewin

Mathematics plays a pivotal role in many scientific and engineering disciplines and ordinary differential equation is certainly an important branch of it. When you publish either a text book or a reference book on such a classical topic, you have to face the obvious question : 'why yet another?'

The idea of preparing a book on this classical topic came to my mind in the process of teaching this subject at the undergraduate level for more than a decade. As experienced, our undergraduate students are by and large accustomed to mechanical workout of gritty problems but they fail to understand the legitimacy of the methods in vogue and more surprisingly cannot appreciate the geometrical arguments or physical applications of this powerful tool. However, should not the learners be solely blamed for this setback as dearth of a book presenting the rigorous theoretical aspects concisely and parallelly citing a plenty of practical applications of the differential equations is also a prime reason for that shortcoming. The existing texts that I had the opportunity to have a glance of, either concentrate on a serious note of pedantic discussion of the deep-rooted results emerging from mathematical analysis (e,g existence, uniqueness, boundedness, stability of solutions) or are very informal in the theoretical aspects while elaborative on catering various routine techniques of attacking the problems. My decade long teaching experience prompts that most of the popular books on this subject neither point out the origin of a particular method nor its scope and limitations. The symmetry aspect of an ode, the interplay of linear algebra and linear differential equations, the connection between an exact ode and a conservative vector field are not usually addressed. This leads to a delinked study of this powerful subject and an ardent reader does not get the much desired interest in the topic.

The items encompassed in this book are common to any introductory course delivered at the graduate level and partly at the postgraduate level, but it has been dished out in a new mould. This volume is a self-contained introduction to the field of ode with an emphasis on the underlying role played by linear algebra in delving into the solutions of linear odes. The reader only needs a preliminary knowledge of calculus in $\mathbb{R}$ and elementary linear algebra to pursue the book. Personally I believe that the success of a book lies not in how much topic it encompassed in a pedantic style, rather how much interest could it kindle in the learners through its presentation. This motivated me to discuss the theories and its applications side-by-side. Often one will find prolonged remarks following a particular theory or even an illustrative example since through suitable examples the scope and limitations of a method comes to surface. However, as already told, it was not possible for me to confine all the study material within the grasp of the undergraduate students for whom the venture was initially aimed [chapter 4, parts of chapter 6 and chapter 8 and chapter 9 are beyond traditional graduation courses of Indian Universities]. Even with all these drawbacks if the present work could serve the learners at least to some extent, I shall feel rewarded. I look ahead for more constructive suggestions from the students as well as teachers so that I may patch up the lacunae or incompleteness and streamline the ideas better and user-friendly.

It's a pleasure to acknowledge the help I received in bringing this project to fruition. First of all I must thank all my students who at least helped me learn the subject a bit and instigated me to go beyond the barriers of protocol teaching despite many constraints. I also thank my friends and colleagues who procured a lot of suggestions and corrections. Should I express thanks to Levant Books, specially Dr. Milinda De, Mr. Debasis Auddy and their team for taking the brunt of shaping up the book through a long stint of nearly three years. Last but not least I must express my deepest thank to my wife and children whose forebearence over the years of ups and downs has been habitual.

Kolkata   **Anindya Dey**
April 2017   *Assistant Professor,*
                                        Dept. of Mathematics
                                      St. Xavier's College Kolkata

Contents

Chapter 1

A Prelude to Differential Equations

1.1 Introduction

The very title 'Differential Equation' is self-explanatory. It is an equation that involves differentials of functions. Before getting into the groove of our main discussion let us first explain the term 'differential' of a function. If $f(x)$ be a function of the independent variable x and $f'(x)$ be its derivative, then the differential of $f(x)$, denoted by $df(x)$, is defined by the relation $df(x) = f'(x).\Delta x$, where Δx stands for the increment in x. In particular if $f(x) = x$, $f'(x) = 1$ and so $dx = \Delta x$, signifying the truth that for an independent variable x, the differential dx of x and its increment are the same. Hence if $y = f(x)$, $dy = df(x) = f'(x)\Delta x$. For dependent variable $y = f(x)$, the differential dy is proportional to the differential dx, the proportionality constant being $f'(x) = \frac{dy}{dx}$. This is one reason why in elementary calculus the derivatives are called differential co-efficients. So a differential equation might involve differentials, or because of that link, derivatives of one or more dependent variables with respect to one or more independent variables. When number of independent variables is unity, the differential equation involves only ordinary derivatives of functions (the total number of which might be one or more), and they are called **ordinary differential equations**, more succinctly, **ODE**. The partial differential equations (popularly known as **PDE**) are characterised by the fact that they involve more than one independent variables with respect to which partial derivatives of one or more dependent variables appear in the equation.

Examples of ODE :

$$\frac{d^2y}{dx^2} - 8x\frac{dy}{dx} + y = 5 \qquad (1.1a)$$

$$\left(\frac{d^2x}{dt^2}\right)^2 + \omega^2 x^3 = 0 \tag{1.1b}$$

Examples of PDE :

$$\frac{\partial u}{\partial t} = \lambda \frac{\partial^2 u}{\partial x^2} \tag{1.2a}$$

$$\frac{\partial^2 u}{\partial x^2} + \frac{\partial^2 u}{\partial y^2} = 4\pi\rho(x,y) \tag{1.2b}$$

In example (1.1a), derivatives of y with respect to x appear upto the second order. In (1.1b), powers of derivatives and that of independent variables involved. In example (1.2a), λ is a constant and as it transpires from the pde itself, u is function of two independent variables x and t. In (1.2b), two independent variables x and y and two dependent variables u and ρ are involved.

In this book we shall restrict ourselves to the ordinary differential equations. Link between differentials and derivatives at once tells us that the ordinary differential equations might be put up in either of the two forms, viz, the derivative form and the differential form. The basic difference in these two forms lies in the fact that in the derivative form, irreversibly the labels of 'dependent variables' and 'independent variables' are tagged to specific variables that appear in the equations, while in the differential form there exists a flexibility of our choice—all the variables being treated in the same footing. The obvious question that strikes us can be briefed as, 'Can one always switch over from one form to the other?' The answer is in the negative but the details of the explanation will be addressed to when we deal with solutions of ODE's.

Given any ordinary differential equation, its order is the highest order derivative that appears in the equation. Degree of the differential equation is the power of the highest order derivative term involved in the ode. The following illustrations may clarify the idea of order and degree.

$$x^2 \frac{dy}{dx} + y^2 = 0 \tag{1.3a}$$

$$\frac{d^2y}{dx^2} + y \sin x = 0 \tag{1.3b}$$

$$\left(\frac{dr}{dt}\right)^3 = \sqrt{\frac{d^2r}{dt^2} + 1} \tag{1.3c}$$

The differential equation (1.3a) is of order 1 and degree 1. The equation (1.3b) is of order 2 but of degree 1. However the order and degree of the equation (1.3c) will be 2 and 1 respectively. This conclusion can be drawn immediately after squaring both sides of the equation to clear out radicals and reducing it to the form

$$\frac{d^2r}{dt^2} - \left(\frac{dr}{dt}\right)^6 + 1 = 0 \tag{1.3d}$$

One may classify the ordinary differential equation (ODE) as 'linear' or 'nonlinear'. Infact this classification is not just confined to the difference in their physical appearance but in the basic difference of the approach of their solution-process and qualitative aspects. Later on we shall cite examples to show that most of the practical-field problems are non-linear by nature and we sometimes try to solve them after linearisation under suitable restrictions. As a precaution we remark that degree, order and linearity or non linearity of differential equations are interrelated—the celebrated Riccati substitution will illustrate this point.

Definition : Linear ODE : An ordinary differential equation of order n, say, is said to be **linear** if it assumes the general form

$$a_0(x)\frac{d^n y}{dx^n} + a_1(x)\frac{d^{n-1}y}{dx^{n-1}} + \cdots + a_{n-1}(x)\frac{dy}{dx} + a_n(x)y = b(x),$$

where $a_0(x) \neq 0$ and the following conditions are satisfied :

(a) The highest possible degree of y and its derivatives is unity.
(b) Products of y and any of its derivatives do not appear in equation.
(c) No term involving transcendental functions of y or its derivatives is included.

The following illustrations will clarify the standpoint of the above definition.

Examples :

(i) $$\frac{dy}{dx} + x^2 y = xe^x$$

Observations :

- Highest order derivative involved is $\frac{dy}{dx}$.
- Co-efficient of $\frac{dy}{dx} = 1 \neq 0$.
- No term involves higher powers of y or its derivative.
- No term involves transcendental functions of y or its derivatives.

Inference : First order linear ODE.

(ii) $$\frac{d^3y}{dx^3} + 4\frac{d^2y}{dx^2} - 5\frac{dy}{dx} + 3y = \sin x$$

.

Observations :

- Highest order derivative involved is $\frac{d^3y}{dx^3}$.
- Co-efficient of $\frac{d^3y}{dx^3} = 1 \neq 0$.
- No higher degree terms of y and its derivatives appear.
- No transcendental functions of y or its derivative appear.

Inference : Third order linear ODE.

(iii) $$\frac{d^4y}{dx^4} + 3\left(\frac{d^2y}{dx^2}\right)^5 + 5y = 0$$

Observations :

- Highest order derivative involved is $\frac{d^4y}{dx^4}$.
- The highest degree of the derivative involved is five.

Inference : Non-linear fourth order ODE.

(iv) $$\frac{d^2y}{dx^2} + x\sin y = 0$$

Observations :

- Highest order derivative involved is $\frac{d^2y}{dx^2}$.

• Trigonometric function $\sin y$ appears here.

Inference : Non-linear second order ODE.

(v)
$$\frac{d^6x}{dt^6} + \frac{d^4x}{dt^4}\frac{d^3x}{dt^3} + x = t^2$$

Observations :

• Highest order derivative involved is $\frac{d^6x}{dt^6}$

• Product of two derivatives $\frac{d^4x}{dt^4}$ and $\frac{d^3x}{dt^3}$ appear in the ODE.

Inference : Non-linear sixth order ODE.

Remark:

(a) While checking out linearity of an ode it should be brought to the form free of radicals. For instance, if one is interested in finding whether equation (1.3c) is linear or not, first of all, should it be brought to the form (1.3d). It will then at once be clear what its nature is.

When we work with a first order ode we implicitly assume one variable as independent and the other as dependent. Our conclusion regarding linearity or non-linearity of the ode hinges on that assumption. For example we take up the first order ode

$$(1 + y^2)dx + (x - \tan^{-1} y)dy = 0.$$

It is a non linear differential equation with x as independent and y as dependent variable but a linear equation with y as independent and x as dependent variable.

Sometimes simple substitutions make an apparently non-linear ode linear. To this effect we take up the ode

$$\cos y\frac{dy}{dx} + \frac{1}{x}\sin y = \cos x,$$

which is apparently non-linear no matter whether we consider x or y as independent variable. However, if we substitute $u = \sin y$ then this ode assumes the form

$$\frac{du}{dx} + \frac{u}{x} = \cos x,$$

which is linear in independent variable x and dependent variable u. Thus linearity or non-linearity of an ode is not an intrinsic feature of the equation — it may sometimes seem to be a protean depending on our outlook!

(b) In case a function $f(x)$ contains only different powers of x, no matter whether the indices involved are integral or fractional, positive or negative, we call it an 'algebraic' function. Otherwise it is called a 'transcendental' function. For examples, $x^3, x^{-3}, x^{1/3}$ are all algebraic functions but $\sin x, \cos x, e^{3x}, \ln(1+x)$ are all transcendental functions. Although some authors like to single out trigonometrical functions for special class, we shall always avoid it and stick to the term 'transcendental' to mean non-algebraic functions.

(c) Linearisation of a non-linear ode under suitable restrictions is very lucidly illustrated by the simple pendulum problem.

A bob of mass m fastened at the end of an inextensible string of length l is suspended from a point S [See fig 1.1]. O is the equilibrium position of the bob. It is now displaced to the right from its equilibrium position. If P be the position of the bob at subsequent time t and θ be the corresponding angular displacement, then the motion of the bob is governed by the ode

$$m.l.\frac{d^2\theta}{dt^2} = -mg\sin\theta \tag{1.4}$$

[$mg\cos\theta$ quantifies the tension T in the string]

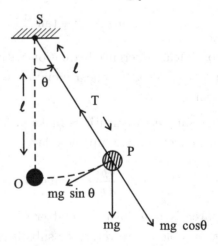

Fig 1.1 : Motion of a simple pendulum

Equation (1.4) is a non-linear ode because of the presence of the term $\sin\theta$ in it. However if θ be very small, $\sin\theta \approx \theta$ and the resulting equation is the linear SHM equation. Indeed with a view to having this linearity in our equation, in elementary physics books one assumes θ to be very small. Had we not assumed it, the solution of (1.4) could not be had without using the more complicated elliptic functions that are beyond the scope of this text.

(d) The concept of linearity or non-linearity is not limited to the case of ODE's only. It has got parallel concepts for PDE's also, only modification being the replacement of the ordinary derivatives by partial derivatives. As an example we treat the pde

$$\frac{\partial^4 u}{\partial x^2 \partial y^2} + 3\frac{\partial^2 u}{\partial x^2} - \frac{\partial^2 u}{\partial y^2} + u^2 = 0 \tag{1.5}$$

Here highest order partial derivative involved being four, the pde is of fourth order. The co-efficients of the different partial derivatives of u are constants. However, the pde is non-linear because of the presence of the term u^2 in it.

As another illustration, let us coin the Poisson's equation in 2D,viz,

$$\frac{\partial^2 u}{\partial x^2} + \frac{\partial^2 u}{\partial y^2} = 4\pi\rho(x,y) \tag{1.2b}$$

Observe that in equation (1.2b), highest order of the partial derivatives involved is two. The co-efficients of the partial derivatives of u are constants. No higher powers of the partial derivatives involved. The dependent variable appears but its power is also unity. Hence (1.2b) is linear.

1.2 Formulation of Differential Equation−Its Significance.

The equation to a curve, no matter whether it is given in the implicit or the explicit function form, gives rise to a differential equation through differentiation and elimination of arbitrary constant(s) involved. When we are given an equation to family of curves, it should involve one or more parameters, the assignment of different values to which enables us to identify the different members of the family. Elimination of family parameters gives the differential equation of the family of curves which speaks out the local characteristics common to all members of the family.

The equation to a straight line in the xy-plane can be put in one of the equivalent forms, viz,

(a) Intercept form $\left(\dfrac{x}{a} + \dfrac{y}{b} = 1 \right)$

(b) Normal form $(x \sin \alpha + y \sin \alpha = p)$

(c) Symmetrical form $\left(\dfrac{x - x_1}{\cos \theta} = \dfrac{y - y_1}{\sin \theta} \right)$

(d) Two point form $\left(\dfrac{x - x_1}{x_2 - x_1} = \dfrac{y - y_1}{y_2 - y_1} \right)$

(e) Slope $-$ intercept form $(y = mx + c)$.

Surprisingly, all the forms ultimately involve two arbitrary constants! For sake of simplificity we take up the $m - c$ given in (e) form where both m and c are parameters of the family of straight lines. Elimination of these parameters gives us the innocent looking simple ode $\frac{d^2y}{dx^2} = 0$ which effectively shows that any straight line is characterised by its zero curvature. Thus the differential equation of a family of curves always brings to light the intrinsic features of the family, either directly (as seen for straight lines) or indirectly (for circles, the fact that the curvatures at every point is same in magnitude as shown below).

Consider the family of unit circles given by $(x - a)^2 + (y - b)^2 = 1$.

We find two accessory equations:

$$(x - a) + (y - b)y_1 = 0 \quad \text{and} \quad (1 + y_1^2) + (y - b)y_2 = 0$$

through successive differentiation w.r. to x. The eliminant of parameters a and b is:

$$(1 + y_1^2)^{\frac{3}{2}} = y_2$$

which shows that the curvature of the family is $\dfrac{y_2}{(1+y_1^2)^{\frac{3}{2}}} = 1$

Hence the claim of constancy of curvature for circles.

Suppose we are to obtain the differential equation of the family of curves given by

$$y - x^2 = Ae^x + Be^{-x}, \qquad (1.6a)$$

A and B being family parameters.

On successive differentiation of the above equation we get the subsidiary relations

$$\frac{dy}{dx} - 2x = Ae^x - Be^{-x} \qquad (1.6b)$$

$$\frac{d^2y}{dx^2} - 2 = Ae^x + Be^{-x} \qquad (1.6c)$$

Substituting back the last relation to the original one we may get the required ode.

We might also get the differential equation by use of matrix and determinants. The three relations (1.6a)–(1.6c) could also be looked into as a system of three linear inhomogeneous equations in two unknowns A and B. From the theory of vector spaces it follows that this system of equations can have infinitely many solutions only if

$$\begin{vmatrix} 1 & 1 & y - x^2 \\ 1 & -1 & \frac{dy}{dx} - 2x \\ 1 & 1 & \frac{d^2y}{dx^2} - 2 \end{vmatrix} = 0,$$

which on simplification yields the required ode :

$$\frac{d^2y}{dx^2} - y + x^2 - 2 = 0 .$$

Another illustration of the matrix method is given in the problem where one has to establish that the ode of the family of circles

$$x^2 + y^2 + 2gx + 2fy + c = 0$$

$(g, f$ and c being parameters) is

$$\left\{ 1 + \left(\frac{dy}{dx}\right)^2 \right\} \frac{d^3y}{dx^3} - 3\frac{dy}{dx}\left(\frac{d^2y}{dx^2}\right)^2 = 0 . \qquad (1.7)$$

Rewriting the equation (1.7) in the form

$$c + 2gx + 2fy = -(x^2 + y^2)$$

and successively differentiating it thrice we have the relations :

$$g + f\frac{dy}{dx} = -\left(x + y\frac{dy}{dx}\right) \qquad (1.8a)$$

$$f\frac{d^2y}{dx^2} = -\left(1 + \left(\frac{dy}{dx}\right)^2 + y\frac{d^2y}{dx^2}\right) \qquad (1.8b)$$

$$f\frac{d^3y}{dx^3} = -\left(3\frac{dy}{dx}\cdot\frac{d^2y}{dx^2} + y\frac{d^3y}{dx^3}\right) \qquad (1.8c)$$

These four linear inhomogeneous equations in three unknowns g, f, c will admit infinitely many solutions provided

$$\begin{vmatrix} 1 & x & y & x^2 + y^2 \\ 0 & 1 & \frac{dy}{dx} & x + \frac{dy}{dx} \\ 0 & 0 & \frac{d^2y}{dx^2} & 1 + \left(\frac{dy}{dx}\right)^2 + y\frac{d^2y}{dx^2} \\ 0 & 0 & \frac{d^3y}{dx^3} & 3\frac{dy}{dx}\frac{d^2y}{dx^2} + y\frac{d^3y}{dx^3} \end{vmatrix} = 0,$$

which on simplification gives the required ode.

Finally we set out to determine the ode of a general family of conics in xy-plane.

The most general equation of a conic in the xy-plane is given by

$$ax^2 + 2hxy + by^2 + 2gx + 2fy + c = 0,$$

where a, h, b cannot be simultaneously zeros. In other words, at least one of these three constants must be non-zero and hence by division with that we can effectively get an equation which is second degree in x and y but involving five arbitrary constants. Without loss of generality we can start with the form that is monic in x^2 :

$$x^2 + 2Hxy + By^2 + 2Gx + 2Fy + C = 0 \qquad (1.9)$$

Now we frame the ode of this general family of conics with five parameters, viz, H, B, G, F, C involved. Without computations it is easy to predict that the desired ode must be of order five as otherwise one cannot retrieve the five arbitrary constants via successive integration.

Equation (1.9) can be regarded as a linear equation in the five unknowns H, B, G, F, C. We carry out differentiation of (1.9) successively five times and use the determinant method to eliminate these unknowns. The five new equations procured are :

$$\left. \begin{array}{l} x+ \quad H(y + xy_1) \quad + \quad Byy_1 + G \qquad\qquad + \quad Fy_1 = 0 \\ 1+ \quad H(2y_1 + xy_2) \quad + \quad B(y_1^2 + yy_2) \qquad + \quad Fy_2 = 0 \\ \quad H(3y_2 + xy_3) \quad + \quad B(3y_1y_2 + yy_3) \qquad + \quad Fy_3 = 0 \\ \quad H(4y_3 + xy_4) \quad + \quad B(3y_2^2 + 4y_1y_3 + yy_4) \quad + \quad Fy_4 = 0 \\ \quad H(5y_4 + xy_5) \quad + \quad B(10y_2y_3 + 5y_1y_4 + yy_5) + \quad Fy_5 = 0 \end{array} \right\} (1.10)$$

where $y_n \equiv \frac{d^n y}{dx^n}$ for $n = 1, 2, \cdots\cdots$

The eliminant of H, B, G, F, C from equations in (1.9) & (1.10) is:

$$
\begin{vmatrix}
x^2 & 2xy & y^2 & 2x & 2y & 1 \\
x & y + xy_1 & yy_1 & 1 & y_1 & 0 \\
1 & 2y_1 + xy_2 & (y_1^2 + yy_2) & 0 & y_2 & 0 \\
0 & 3y_2 + xy_3 & (3y_1y_2 + yy_3) & 0 & y_3 & 0 \\
0 & 4y_3 + xy_4 & (3y_2^2 + 4y_1y_3 + y_4) & 0 & y_4 & 0 \\
0 & 5y_4 + xy_5 & (10y_2y_3 + 5y_1y_4 + yy_5) & 0 & y_5 & 0
\end{vmatrix} = 0
$$

which on simplification becomes:

$$
\begin{vmatrix}
(3y_2 + xy_3) & (3y_1y_2 + yy_3) & y_3 \\
(4y_3 + xy_4) & (3y_2^2 + 4y_1y_3 + yy_4) & y_4 \\
(5y_4 + xy_5) & (10y_2y_3 + 5y_1y_4 + yy_5) & y_5
\end{vmatrix} = 0
$$

This equation represents the desired ode.

Remark: If we impose some extra conditions on the conic, for example, the conics are coaxial or confocal, or the axes of the conics being parallel to the co-ordinate axes, or the conics are a family of concentric circles, then accordingly the number of independent family parameters get reduced and this in turn, produces odes of lower order. In other words, imposition of more restrictions on the family of conics injects more symmetry into the system and thereby cutting down its degree of freedom connoted by the number of family parameters involved in the general equation.

The above matrix method exploits the basic results of the consistency of a system of inhomogeneous linear equations when number of equations exceeds the number of variables. Unfortunately for non linear equations there is no such parallel idea. This is why one cannot use the above method to construct the ode of the family of curves whose equation is given by

$$y = kx + k^3, \quad k \text{ being family parameter.} \tag{1.11}$$

Although its use is rather restricted, the matrix method has been really an eye-opener to the reader as by now he/she could sense how odes work as liasions between different branches of mathematics and so occupy a central position in its study.

Example (1) : Show that the differential equation of all parabolas with foci at the origin and axes along the x-axis is :

$$y\left(\frac{dy}{dx}\right)^2 + 2x\frac{dy}{dx} - y = 0$$

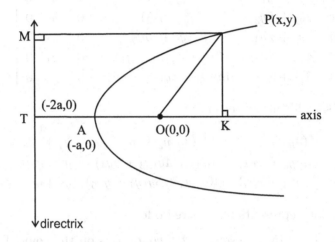

Fig 1.2 : Parabola with focus as origin and axis along x-axis.

Focus $O \equiv (0, 0)$ vertex $A \equiv (-a, 0)$

Directrix $\overleftrightarrow{TM} : x + 2a = 0$

$P(x, y)$ is arbitrary point on the parabola. By the geometric condition, the equation to parabola reads : $x + 2a = \sqrt{x^2 + y^2}$

which on simplification gives $y^2 = 4a(a+x)$, 'a' being family parameter.

Elimination of 'a' between the above equation and $y\frac{dy}{dx} = 2a$ gives the required ode of the family of parabolas.

Example (2) : Show that the differential equation of the family of circles having their centres on the y-axis is given by $xy'' - (y')^3 - y'$, where $y' \equiv \frac{dy}{dx}$

Centre $\equiv (0, k)$; radius $= r$

Equation to the family of circles : $x^2 + (y - k)^2 = r^2$, k and r being family parameters.

Differentiating w.r. to x to have

$$x + (y - k)\frac{dy}{dx} = 0,$$

Differentiating once more,

$$1 + \left(\frac{dy}{dx}\right)^2 + (y - k)\frac{d^2y}{dx^2} = 0$$

Multiplying both sides by $\frac{dy}{dx}$ and using the first relation we get,

$$-\left[1 + \left(\frac{dy}{dx}\right)^2\right]\frac{dy}{dx} + x\frac{d^2y}{dx^2} = 0$$

Note : The equation to the family of circles involved two parameters viz, k and r and so the differential equation of the family is of second order as expected.

Example (3) : Obtain the differential equation of the system of confocal conics

$$\frac{x^2}{a^2 + \lambda} + \frac{y^2}{b^2 + \lambda} = 1,$$

in which λ is the arbitrary parameter, a and b being preassigned constants. Differentiation of the given equation w.r. to x we get:

$$\frac{x}{a^2 + \lambda} + \frac{yy'}{b^2 + \lambda} = 0 \quad \left[\text{where } y' \equiv \frac{dy}{dx}\right]$$

or, $\quad \lambda = \dfrac{-(a^2yy' + b^2x)}{(x + yy')}$

Put back this λ into the original equation we get

$$\frac{x^2}{a^2 - \frac{a^2yy'+b^2x}{x+yy'}} + \frac{y^2}{b^2 - \frac{a^2yy'+b^2x}{x+yy'}} = 1$$

or, $\quad (x + yy') \cdot \left\{\dfrac{x}{a^2 - b^2} + \dfrac{y}{b^2 - a^2}\right\} = 1$

or, $\quad (xy' - y)(x + yy') = (a^2 - b^2)y',$

which is the first order ode whose complete integral or primitive is the equation of the system of confocal conics.

Remark : Although at first sight the form of the equation seems to be representing a family of confocal ellipses, it is infact a family of ellipses and hyperbolas, all confocal. This categorisation owes its origin to the fact that λ may assume both +ve and -ve values. Later on we shall see the interrelation between these two subfamilies when we discuss self-orthogonal trajectories in the next chapter.

Example (4) : Find the differential equation of the family of parabolas touching the co-ordinate axes.

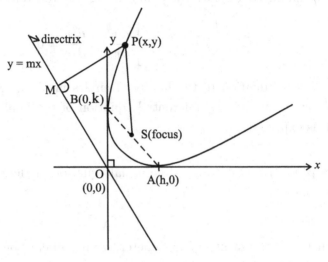

Fig 1.3 : Parabola touching the co-ordinate axes at $A(h, 0)$ and $B(0, k)$

Since tangents at the extremities of a focal chord of a parabola are mutually perpendicular and intersect on the directrix the point $A(h, 0)$ and $B(0, k)$ in the figure are the extremities of a focal chord, the equation to which in the intercept form is given by

$$\frac{x}{h} + \frac{y}{k} = 1.$$

Thus parametrically any point on the focal chord AB is given by $(h(1 - t), kt), t$ being the parameter.

By geometric conditions defining a parabola it follows that

$$\left. \begin{array}{l} k^2 = (1 - t_0)^2(1 + m^2)(h^2 + k^2) \\ m^2 h^2 = t_0^2(1 + m^2)(h^2 + k^2) \end{array} \right\} t = t_0 \text{ corresponds to focus S.}$$

where from follows that $m = -\frac{h}{k}$.

Using this result in the above relations we get $t_o = \frac{h^2}{h^2 + k^2}$.

$$\therefore \text{ Focus S} \equiv (h(1 - t_0), kt_0) = \left(\frac{hk^2}{h^2 + k^2}, \frac{h^2 k}{h^2 + k^2} \right)$$

If $P(x, y)$ be a variable point on the parabola then $PS = PM$ in the figure and this yields the equation to the parabola :

$$\frac{(hx + ky)^2}{h^2 + k^2} = \left(x - \frac{hk^2}{h^2 + k^2}\right)^2 + \left(y - \frac{h^2 k}{h^2 + k^2}\right)^2,$$

On simplification, the above equation reads

$$\left(\frac{y}{k} - \frac{x}{h}\right)^2 - 2\left(\frac{y}{k} + \frac{x}{h}\right) + 1 = 0,$$

or more compactly,

$$\pm\sqrt{\frac{x}{h}} \pm \sqrt{\frac{y}{k}} = 1, \quad h \text{ and } k \text{ being parameters.}$$

Carrying out differentiation w. r. to x we have

$$x^{1/2} y^{-1/2} \frac{dy}{dx} = \pm\sqrt{\frac{k}{h}}$$

On differentiating again, w.r. to x,

$$\frac{1}{2} x^{-1/2} y^{-1/2} \frac{dy}{dx} - \frac{1}{2} x^{1/2} y^{-3/2} \left(\frac{dy}{dx}\right)^2 + x^{1/2} y^{-1/2} \frac{d^2y}{dx^2} = 0$$

or, $$y \frac{dy}{dx} - x \left(\frac{dy}{dx}\right)^2 + 2xy \frac{d^2y}{dx^2} = 0,$$

which is the required ode of the family of parabolas.

Example (5) : Show that the differential equation

$$(4x + 3y + 1)dx + (3x + 2y + 1)dy = 0$$

represents a family of hyperbolas having as asymptotes the lines $x+y = 0$ and $2x + y + 1 = 0$.

Solution: In the derivative form the ode reads:

$$\frac{dy}{dx} = -\frac{(4x + 3y + 1)}{(3x + 2y + 1)}$$

This ode can be recast as homogeneous one by means of the transformations:

$$x = x' - 1 \quad \text{and} \quad y = y' + 1$$

so that the newlook ode is :

$$\frac{dy'}{dx'} = \frac{-(4x' + 3y')}{(3x' + 2y')}$$

Using $y' = vx'$ transformation we can make it changed to

$$x'\frac{dv}{dx'} + \frac{2v^2 + 6v + 4}{3 + 2v} = 0$$

$$\text{i, e,} \quad \frac{(3 + 2v)dv}{(v^2 + 3v + 2)} + 2\frac{dx'}{x'} = 0,$$

which on integration yields :

$$x'^2(v^2 + 3v + 2) = \text{constant}$$

$$\text{i, e,} \quad 2x'^2 + 3x'y' + y'^2 = \text{constant}$$

Above equation represents a hyperbola-family (the constant of integration serving as the family parameter) as we can verify by the matrix methods of determining the nature of quadratic forms.

One can write the above equation as :

$$\begin{pmatrix} x' & y' \end{pmatrix} \begin{pmatrix} 2 & \frac{3}{2} \\ \frac{3}{2} & 1 \end{pmatrix} \begin{pmatrix} x' \\ y' \end{pmatrix} = \text{constant}$$

and the eigenvalues of the middle-matrix are given by

$$\begin{vmatrix} 2 - \lambda & \frac{3}{2} \\ \frac{3}{2} & 1 - \lambda \end{vmatrix} = 0,$$

so that one λ is $+ve$ and other $-ve$ and the quadratic form on L.H.S of the equation is indefinite.

We may now put this equation as :

$$\text{i, e,} \quad \left(x' + \frac{3}{4}y'\right)^2 - \left(\frac{y'}{4}\right)^2 = \text{constant } (a, \text{say})$$

We now write $x'' = x' + \frac{3}{4}y'$ and $y'' = \frac{y}{4}$ so that the above equation becomes:

$$x''^2 - y''^2 = a$$

The asymptotes of the above curve are given by $x'' + y'' = 0$ and $x'' - y'' = 0$

$$\text{i, e,} \quad x + y = 0 \quad \text{and} \quad x + \frac{1}{2}(y + 1) = 0$$

This completes the proof.

1.3   Classification of Solutions : General , Particular and Singular Solutions

A solution or integral or primitive of a differential equation is a relation between the variables, by means of which and the derivatives obtained there from, the equation is satisfied). The form of the solution (1.6) is $f(x, y, A, B) = 0$ and that of solution (1.7) is $f(x, y, g, f, c) = 0$. These two examples bring to light one important feature of the ode — number of arbitrary constants appearing in the general solution equals the order of the ode.

Consider the general equation

$$f(x, y, c_1, c_2, \cdots, c_n) = 0, \qquad\qquad (1.12)$$

which involves n arbitrary constants $c_1, c_2, \ldots, c_n$.

(1.12) being in implicit function form, partial derivatives automatically come into play.

Differentiation of (1.12) successively with respect to x gives

$$\frac{\partial f}{\partial x} + \frac{\partial f}{\partial y} \cdot \frac{dy}{dx} = 0$$

$$\frac{\partial^2 f}{\partial x^2} + 2\frac{\partial^2 f}{\partial x \partial y} \cdot \frac{dy}{dx} + \frac{\partial^2 f}{\partial y^2}\left(\frac{dy}{dx}\right)^2 + \frac{\partial f}{\partial y}\frac{d^2 y}{d^2 x} = 0$$

$$\cdots\cdots\cdots\cdots\cdots\cdots\cdots\cdots\cdots\cdots\cdots\cdots$$

$$\cdots\cdots\cdots\cdots\cdots\cdots\cdots\cdots\cdots\cdots\cdots\cdots$$

$$\frac{\partial^n f}{\partial x^n} + \cdots\cdots\cdots\cdots\cdots\cdots + \frac{\partial f}{\partial y} \cdot \frac{d^n y}{dx^n} = 0$$

Between (1.12) and its n accessory equations obtained by the partial differentiations, the constants $c_1, \cdots, c_n$ can be eliminated to give the ode :

$$F\left(x, y, \frac{dy}{dx}, \frac{d^2 y}{dx^2} \cdots\cdots\cdots \frac{d^n y}{dx^n}\right) = 0 \qquad\qquad (1.13)$$

of which (1.12) is the primitive. There being n successive differentiations, the resulting equation must contain a derivative of the nth order
and therefore a relation between x and y involving n arbitrary constants
will give rise to an ode free from those arbitrary constants.

The ode is quite independent of the elimination process adopted. The
solution of an nth order ode involves n arbitrary constants because in
each step of n successive integrations enroute the primitive one constant
of integration is introduced.

Example (6) : Solve the ode $\dfrac{d^2y}{dx^2} + y = 0$.

Denoting $\frac{dy}{dx}$ by v, the above ode reads

$$\frac{v\,dv}{dy} + y = 0$$

$$\text{i, e}\quad \frac{d}{dy}\left(v^2 + y^2\right) = 0$$

Integrating with respect to y we get $v^2 + y^2 = a^2$, a being the constant
of integration.

Choosing the negative square root we have,

$$v \equiv \frac{dy}{dx} = -\sqrt{a^2 - y^2}\ ,$$

which on integration gives : $\cos^{-1}\left(\dfrac{y}{a}\right) = x + b,$

b being another constant of integration.

Thus, the primitive reads : $y = a\cos(x + b)$, involving two arbitrary
constants of integration, viz, a and b. This form on expansion gives the
linear form

$$y = A\cos x + B\sin x.$$

provided we identify $a\cos b$ as A and $-a\sin b$ as B. Had we started from
the solution in the form $y = A\cos x + B\sin x$, the above compact form
could be attained by putting $A = a\cos b$ and $B = -a\sin b$. However to
the reader it might seem to be a sleight of hand.

Remark : The introduction of v translates the given second order linear
ode to the equation $v\frac{dv}{dy} + y = 0$ which is of first order, first degree

but non-linear. This example also indicates that as per our needs, we may incorporate linearity into an ode and enhance its order or we may introduce non-linearity into the ode and by the way reduce its order. All these salient features are summed up in 'Riccati substitution' that will be addressed to while we discuss the special features of a non-linear ode in the next chapter.

The solution which contains a number of arbitrary constants equal to the order of the differential equation considered, is called its **general solution** or **primitive** or **complete integral**. Solutions obtained from the general solution by assigning particular values to the arbitrary constants will be referred to as **particular solutions**. (Do not confuse it with particular integrals appearing in the general solutions of inhomogeneous odes discussed later on). For some differential equations there exist solutions that involve neither any arbitrary constant nor can they be identified as particular solutions. These solutions are called **singular solutions**. Briefly speaking, they owe their origin to the non-uniqueness of solutions to the ode.

Example (7) : Show that $y = kx + k^3$, k being arbitrary parameter represents a family of curves that are characterised by the ode

$$\left(\frac{dy}{dx}\right)^3 + x\frac{dy}{dx} - y = 0$$

Also show that $x^2 + 4y = 0$ is a solution to this ode.

Differentiating $y = kx + k^3$ w.r. to x we get, $\frac{dy}{dx} = k$, so that k-eliminant ode is obtained.

Similarly direct differentiation can show that $x^2 + 4y = 0$ is also a solution to the same ode. This solution being not obtainable from the general solution is of the singular type. This first order ode is an example of Clairaut's equation, for which existence of singular solution is guaranted.

Example (8) : Find the value(s) of n for which the ode

$$x^3y''' + 2x^2y'' - 10xy' - 8y = 0 \left(\text{where } y' = \frac{dy}{dx}; \ y'' = \frac{d^2y}{dx^2} \text{ etc.} \right)$$

has solutions of the form $y = x^n$.

Differentiating $y = x^n$ successively we have :

$$y' = nx^{n-1} \; ; \; y'' = n(n-1)x^{n-2} \; ; \; y''' = n(n-1)(n-2)x^{n-3},$$

which on substitution into the original ode yields the equation

$$(n+1)(n+2)(n-4) = 0.$$

Hence the admissible values of n are -1, -2, 4.

Remark : This result is apparently very straight forward. However, in our later discussions in Chapter 6 we will observe that any solution of the homogeneous differential equations (the chosen equation is a third order homogeneous differential equation) is of the form x^n, where n is a constant, not necessarily an integer.

1.4   More about Solutions of an ODE

Recall that solutions (primitives) of an ode were defined as relations. Relations might or might not be elevated to the status of functions. Depending on whether the solution of an ode can be cast in the form of a function or not we call it an 'implicit solution' or a 'formal solution'. For implicit solutions at least one function form is available which is referred to as an 'explicit' solution. Often we quote explicit solutions as only 'solutions'. The explicit solutions of an ode is a function $y = y(x)$ which if substituted along with its derivatives into the ode satisfies it for all x in some specified interval called the 'domain'.

For illustration let's consider the unit circle $x^2 + y^2 = 1$.

On differentiation w.r.to x, we have, $x + y\frac{dy}{dx} = 0$, which is a first order ode. $x^2 + y^2 = 1$ is the primitive of this ode. Moreover we can see that both $y = \pm\sqrt{1-x^2}$ satisfy the same ode provided $\mid x \mid \leq 1$. Hence $x^2 + y^2 = 1$ is an implicit solution to the ode while two explicit solutions exist to the ode. Observe that these two explicit solutions give semicircles which unite to form the unit circle.

Example (9) : Show that $(5x^2 - 2x^3)y^2 = 1$ is an implicit solution of the differential equation $x\frac{dy}{dx} + y = x^3y^3$ for $x < \frac{5}{2}$.

We observe that $x \neq 0$ and $x \neq \frac{5}{2}$ is a must. We differentiate the relation given w.r. to x to have

$$
\begin{aligned}
x\frac{dy}{dx} &= xy^3(3x^2 - 5x) \quad (\because (5x^2 - 2x^3)y^2 = 1) \\
&= x^3y^3 - y^3(5x^2 - 2x^3) \\
&= x^3y^3 - y,
\end{aligned}
$$

showing that the given relation is a primitive to this ode. Since for $x < \frac{5}{2}$ we can write down the relation given as $y = \pm\frac{1}{x\sqrt{5-2x}}$, we conclude that extraction of two explicit functions is possible from the above primitive. Hence the primitive given is an implicit solution to the ode.

Consider a first order ode in the normal form : $\dfrac{dy}{dx} = f(x,y)$

Let $g(x, y, c) = 0$ be its implicit solution, c being the family parameter. Hence there exists an explicit solution $y = \phi(x, c)$ to the above ode in some interval. The graphs of these explicit solutions give us the family of integral curves which is in one-to-one correspondence with the range of parameter c. But how to single out a desired integral curve within the swarm? The answer may be found in geometric approach. We choose a point x_0 in the domain common to several solutions. Draw the line $x = x_0$ and mark the point P where it cuts the desired solution curve. The ordinate y_0 of P is deterministic. This artifice of picking a point (x_0, y_0) on the given integral curve for its identification is often called 'prescription' of initial condition $y(x_0) = y_0$' to an ode, or more precisely, initial value problem.

Example (10) : The general solution of the first order ode $(x^2 - 1)y' = xy$ represents a family of ellipses $x^2 + c^2y^2 = 1$. Since this equation can be cast as $y = \pm\frac{1}{c}\sqrt{1 - x^2}$ for $|x| \leq 1$, the relation $x^2 + c^2y^2 = 1$ represents an implicit-solution family while $y = \frac{1}{c}\sqrt{1 - x^2}$ for $|x| \leq 1$ is representing a family of explicit solution. We choose a particular $x_0 \in [-1, 1]$ and get a corresponding value $y_0 = \frac{1}{c}\sqrt{1 - x_0^2}$ of y if c is preassigned. Hence for this preassigned c, the desired integral curve is identified by the point (x_0, y_0).

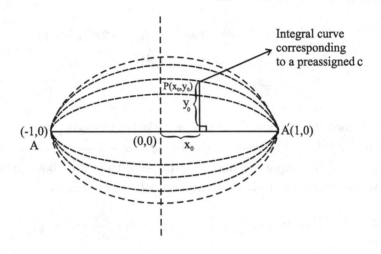

Fig 1.4: Family of ellipses $x^2 + c^2 y^2 = 1$, c being family parameter

From the above discussion it transpires that the explicit solution $y(x)$ which we intend to single out has two liabilities, viz,

(a) satisfying the ode given.

(b) satisfying the supplementary initial condition.

Example (11) : Consider the problem in Example (10) annexed with the supplementary condition $y\left(\dfrac{3}{5}\right) = \dfrac{8}{5}$.

In the given IVP, $c^2 = \frac{1}{4}$ and so $x^2 + \frac{1}{4}y^2 = 1$ is a formal solution. Although $y = \pm 2\sqrt{1 - x^2}$ for $\mid x \mid \leq 1$, we cannot treat the formal solution as an implicit solution until one of the two explicit functions $y = \pm 2\sqrt{1 - x^2}$ satisfies the supplementary condition. Observe that the explicit solution $y = -2\sqrt{1 - x^2}$ is inadmissible as in no circumstances it can satisfy the given condition. So $y = 2\sqrt{1 - x^2}$ can be the only admissible explicit solution.

Note : Had the initial condition been chosen $y\left(\dfrac{3}{5}\right) = -\dfrac{8}{5}$,

$y = -2\sqrt{1 - x^2}$ would be the explicit solution to the resulting IVP.

If now we pass on to the second order ode, obviously the general solution will involve two arbitrary constants. If we try to isolate a particular integral curve from this family we are to fix up values of these constants. However the identification of a particular integral can be done in three possible ways :

(i) Picking up one point (x_0, y_0) on the desired curve and prescribing the slope to it at that point.

(ii) Demanding that the desired curve should pass through the given points (x_0, y_0) and (x_1, y_1).

(iii) Demanding that the desired curve should have preassigned slopes at two given points (x_0, y_0) and (x_1, y_1).

In case (i) the problem is called is called **Initial Value Problem (IVP)** or more appropriately 'one-point boundary value problem'.

In case (ii) and (iii) the problem is called 'two-point boundary-value-problem' or more succinctly **Boundary Value Problem (BVP)** As there are two supplementary conditions to be used for the latter, their consistency with respect to the given ode is a prime factor. Through the following illustrations we shall prove this point.

Example (12) : Find solution to the b.v.p. :

$$\frac{d^2y}{dx^2} + y = 0 \quad \text{where} \quad y(0) = 1 \text{ and } y\left(\frac{\pi}{2}\right) = 5.$$

General solution to the ode is $y(x) = A \cos x + B \sin x$

Using the supplementary conditions we get $A = 1$ and $B = 5$.

So the desired particular solution is $y(x) = \cos x + 5 \sin x$ and the supplementary conditions are consistent.

Example (13) : Consider the b.v.p :

$$\frac{d^2y}{dx^2} + y = 0 \quad \text{subject to } y(0) = 1 \text{ and } y(\pi) = 3.$$

The supplementary conditions are inconsistent as it gives two different values of the arbitrary constant A while keeps B unrestricted. Hence the b.v.p. has no solutions.

Example (14) : Test whether $y = e^x + 2e^{-2x}$ is a solution to the ode.

$$\frac{d^2y}{dx^2} + \frac{dy}{dx} - 2y = 0$$

subject to the initial conditions $y(0) = 3$ and $y'(0) = -3$. What is your inference regarding $y = 2e^x + e^{-2x}$?

By direct differentiation we can show that both $y = e^x + 2e^{-2x}$ and $y = 2e^x + e^{-2x}$ satisfy the ode but the former satisfies also both the initial conditions while the latter satisfies one of the two supplementary conditions. Hence $y = e^x + e^{-2x}$ is a solution to the i.v.p. but $y = 2e^x + e^{-2x}$ is not.

Remark :

(a) Near the earth's surface, the gravitational force experienced by a particle is nearly constant and given by the equation of motion $\frac{d^2x}{dt^2} = -g$, where x is the vertical height attained at subsequent time t and the g is a constant known as acceleration due to gravity. To fix up the trajectory of a particle one needs specify the initial position and initial velocity of the particle, i.e., x and $\frac{dx}{dt}$ at $t = 0$. If you are very fastidious you will say that without prescription of the supplementary initial conditions the law of motion is not deterministic as quantifying the force mass of the particle is not the whole of particle dynamics.

(b) The different types of two-point boundary value problems dealt with so far are only particular cases of a more general problem referred up as **Sturm-Liouville** problems, the details of which will be taken up in Chapter-6.

1.5   Existence-Uniqueness Theorem for Cauchy Problem.

Till now we never addressed the question of existence and uniqueness of solutions of an ordinary differential equation. An overambitious reader might think that the study of ordinary differential equations is primarily confined to the hunt of the solutions followed by their qualitative analysis. However this idea might land him into trouble as there are many

IVP's or BVP's for which either no solution or a non-unique solution exists. It is therefore of immediate concern for us to go into circumstances under which the solution to a Cauchy problem exists and find the criteria so as to ensure its uniqueness. Essentially we pose two questions that seem to be quite relevant :

(a) Is there a solution to the problem ?

(b) If at all a solution exists, is it the only one?

In the following discussions we focus on the need for a theory of existence and uniqueness of solutions to an ordinary differential equation. In order to be useful, a mathematical model must admit solution because otherwise it can never represent any problem of practical interest. As the differential equations model the dynamical systems that we often come across in real life, consistency of the model together with the existence of a unique solution is a prime need. In mathematical modelling 'hypothesis of determinism' is a basic idea which states that a particular set of initial conditions must result in exactly one solution. Applied to an ode, this means that there should exist exactly one solution for a preassigned set of initial conditions.

Thus a mathematical model representing a physical process should have three main features.

(a) A solution satisfying the given initial conditions must exist.

(b) Each set of initial conditions leads to a unique solution. Hence two solutions which obey the same initial conditions are identical.

(c) The solutions depend continuously on the initial conditions.

Remark :

The existence of solutions ensure practical utility of the model. The property of uniqueness is very much desirable because it corresponds to the 'hypothesis of determinism' in absence of which we shall obtain different outcomes even if we iterate the expriment under identical initial conditions. This would allow randomness into the experimental set up and make the model probabilistic. The continuous dependence of solutions on the supplementary initial conditions speaks out the sensitivity of the deterministic model. It has been shown by many a mathematician that a wide class of odes obey these requirements, even for which there exists no possible analytic methods for finding explicit solutions.

The above three prerequisites of the mathematical models are embodied in the following **Existence-Uniqueness Theorem** of the odes:

Let $\frac{dy}{dx} = f(x,y)$ be an ode subject to the initial condition $y(x_0) = y_0$.
If (a) $f(x,y)$ be continuous for every $(x,y) \in D$, a rectangle bounded by the straight lines $x = x_0 \pm a$; $y = y_0 \pm b$ in $\mathbb{R}$

(b) $f(x,y)$ satisfies the Lipschitz condition (of order 1), namely,

$$\mid f(x,y_1) - f(x,y_2) \mid < K \mid y_1 - y_2 \mid,$$

where K is a constant (dependent on D), then there exists a unique solution $y = \overline{y}(x)$ to the ode that satisfies

$$y_0 = \overline{y}(x_0) \text{ for all } x \in [x_0 - \delta,\ x_0 + \delta]$$

where $\qquad \delta < \min\left\{a, \dfrac{b}{M}, \dfrac{1}{K}\right\} \qquad$ and $\qquad M = \max_{(x,\ y)\in\ D} f(x,y).$

Remark :

(a) The Lipschitz condition (b) appearing in the statement of the above theorem might be replaced by the relatively simple condition of bounded partial derivative $\frac{\partial f}{\partial y}$ in the rectangle D as the two are interrelated by the Lagrangean Mean Value Theorem. This is why many authors substitute condition (b) by the condition of continuity of $\frac{\partial f}{\partial y}$ in D. This equivalent statement of existence uniqueness theorem is mostly referred to while working out problems and is known as **Picard's Existence Theorem**. However, there are functions $f(x,y)$ for which Lipschitz condition of order one holds good but the partial derivative $\frac{\partial f}{\partial y}$ is non-existent at some points (e.g, $f(x,y) = \mid y \mid$ does not have $\frac{\partial f}{\partial y}$ for points on the x-axis). Thus Lipschitz condition will be applicable to a broader class of functions.

(b) It is quite reasonable to ask if the above existence-uniqueness theorem applies to domains other than rectangles. Hopefully the answer is in the affirmative. In case of elementary figures, that is, the subsets of $\mathbb{R}^2$ that are representable as set theoretic unions of a finite number of rectangles $D \subseteq \mathbb{R}^2$ any two of which either do not intersect or intersect only along parts of their boundaries, the above theorem is equally valid. More generally, if D is an open connected set in $\mathbb{R}^2$ we have the guarantee of its applicability. We shall not dwell with these anymore as it would require more ideas about the topology of $\mathbb{R}^2$.

(c) The two conditions of the existence-uniqueness theorem are sufficient but not necessary. This is really a boon to us as often we exploit the sufficiency of conditions to use alternative equivalent conditions that serve our purpose. For instance, the following existence-uniqueness criterion turns out be very handy and easily verifiable :

"Every first order linear differential equation with a given initial condition possesses a unique solution if the co-efficients of y and $\frac{dy}{dx}$ and also the terms free of y are continuous functions of x" (proof deferred to next chapter). It is really surprising that this criterion together with its parallel results for higher order differential equations save us from being plagued by the existence-uniqueness problem in a special class of odes.

The following examples clarify the importance of the above theorem.

Example (15) : Apply the existence-uniqueness theorem to establish that the IVP given by

$$\left.\begin{array}{c} \frac{dy}{dx} = 2y^2 - xy \\ y(0) = 1 \end{array}\right\}$$

possesses a unique solution.

Here $f(x,y) \equiv 2y^2 - xy$ being a polynomial in x and y is continuous everywhere in $\mathbb{R}^2$. $\frac{\partial f}{\partial y} = 4y - x$ is a linear function in x and y and so continuous. Clearly the two sufficient conditions of Picard's theorem are satisfied and so unique solution exists to the given I.V.P.

Example (16) : Does the IVP given by $\frac{dy}{dx} = \sqrt{y}$; $y(0) = 0$

admit of unique solution?

Here $f(x,y) \equiv \sqrt{y}$ is continuous everywhere but $\frac{\partial f}{\partial y}$ is not defined at $(0,0)$. Clearly centering $(0,0)$ we cannot have any domain D, however small, that excludes $y = 0$ straight line. Thus one of the hypotheses in Picard's theorem fails. So drawing any definite conclusive remark to the nature of solution is impossible.

Example (17) :   Has the IVP given by

$$\left.\begin{array}{c} \frac{d^2y}{dx^2} + P(x)\frac{dy}{dx} + Q(x)y = 0 \\ y(-1) = 0 \end{array}\right\}$$

a unique solution $(P(x)$ and $Q(x)$ being polynomials in $x)$?

The answer is in the affirmative because the parent ode is linear.

Example (18) :   Apply the existence-uniqueness theorem to the I.V.P.

$$\left.\begin{array}{c} \frac{dy}{dx} = \frac{x-y}{x+y} \\ y(0) = 0 \end{array}\right\}$$

Here $f(x, y) \equiv \dfrac{x-y}{x+y}$. Since $\lim\limits_{x \to 0}\lim\limits_{y \to 0} \dfrac{x-y}{x+y} \neq \lim\limits_{y \to 0}\lim\limits_{x \to 0} \dfrac{x-y}{x+y}$, the

simultaneous limit $\lim\limits_{(x,y) \to (0,0)} f(x, y)$ does not exist.

Hence no question of continuity of $f(x, y)$ at $(0,\ 0)$. Existence and uniqueness theorem (in any version) fails to draw a definite conclusion for this I.V.P. Simple workout shows that $y = (\pm\sqrt{2} - 1)x$ are solutions to the I.V.P showing that indeed there are two solutions to the IVP.

Example (19) :   Apply Picard's theorem to check whether the I.V.P.

$$\left.\begin{array}{c} \frac{dy}{dx} = \frac{x+y}{x-y} \\ y(0) = 0 \end{array}\right\}$$

possess a unique solution or not?

The function $f(x, y) = \frac{x+y}{x-y}$ is not continuous as the simultaneous limit

$\lim\limits_{(x,y) \to (0,\ 0)} f(x, y)$ is non-existent. So one of the basic postulates of the

existence uniqueness theorem of Picard fails. No conclusion can be drawn hereform. However by analytic solution method one may establish that there is no solution to satisfy the initial condition $y(0) = 0$.

Example 20 : Apply existence-uniqueness theorem to the I.V.P :

$$\frac{dy}{dx} = \frac{1}{\{2-(x-1)^2\}\{5-(y-5)^2\}} ; \quad y(1) = 5$$

Observe that $f(x,y) \equiv \dfrac{1}{\{2-(x-1)^2\}\{5-(y-5)^2\}}$ is well-defined

everywhere in $\mathbb{R}^2$ with the exception of points lying on the lines $x = 1 \pm \sqrt{2}$ and $y = 5 \pm \sqrt{5}$. Hence we shall apply the existence uniqueness theorem in a region lying well within the rectangle formed by these lines. For sake of definiteness let's choose a rectangular region

$$D \equiv \{(x,y) : \mid x-1 \mid \le 1 \text{ and } \mid y-5 \mid \le 2\}$$

Obviously for $(x,y) \in D$, $f(x,y)$ is continuous and $\mid f(x,y) \mid \le 1$. Hence without loss of generality we may define

$$M = \max_{(x,y) \in D} f(x,y) = 1$$

Again for any $(x, y_1), (x, y_2) \in D$ we have :

$$\mid f(x,y_1) - f(x,y_2) \mid$$
$$= \frac{1}{\{2-(x-1)^2\}} \left| \frac{1}{5-(y_1-5)^2} - \frac{1}{5-(y_2-5)^2} \right|$$
$$= \frac{1}{\{2-(x-1)^2\}} \cdot \frac{\mid y_1 - y_2 \mid \mid y_1 + y_2 - 10 \mid}{\{5-(y_1-5)^2\}\{5-(y_2-5)^2\}}$$
$$\le 4 \mid y_1 - y_2 \mid \text{ (using triangle inequality)}$$

Thus $f(x,y)$ satisfies Lipschitz condition (of order 1) in D with Lipschitz constant $K = 4$.

Therefore, one can apply the existence-uniqueness theorem to the given initial value problem and conclude that $\exists$ a unique solution $y = \bar{y}(x)$ to it that satisfies the condition $\bar{y}(1) = 5$ for all $x \in [1-\delta, 1+\delta]$ with $\delta < \min\{1, 2, \frac{1}{4}\} \equiv \frac{1}{4}$

Example (21) : Apply Picard's theorem to show that IVP given by

$$\left. \begin{array}{l} \frac{dy}{dx} = \frac{y^2}{x-2} \\ y(1) = 0 \end{array} \right\}$$

possesses a unique solution defined on a sufficiently small interval $\mid x - 1 \mid \leq h$ about the point $x = 1$. $f(x,y) = \frac{y^2}{x-2}$ is continuous everywhere except at $x = 2$ while $\frac{\partial f}{\partial y} = \frac{2y}{x-2}$ is continuous everywhere except at $x = 2$.

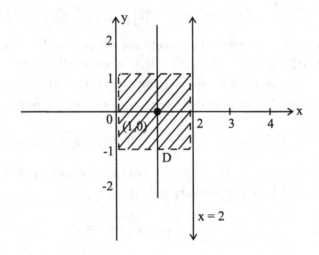

Fig 1.5 : Shaded region D bounded by $x = 0$ and $x = 2$

Accordingly the problem satisfies both hypotheses of Picard Theorem in a squarish domain D (shaded region in the figure) centred at $(1,0)$ if width of D is less than 2, so that $x = 2$ lies outside the region D. Thus within this shaded region D, $(1,0)$ lies and moreover, unique solution of the given first order ode exists.

1.6   Importance of Lipschitz's condition involved in existence uniqueness theorem in the light of a comparative study between Radiactive decay and Leaky bucket problems :

Radioactive decay is governed by the ode $\frac{dN}{dt} = -\lambda N, \lambda > 0$ where N stands for the total number of radiactive particles present at time t. If it is known that N_0 radioactive particles were present at time t_0, then solution to the IVP will be

$$N(t) = N_0 e^{-\lambda(t-t_0)} \qquad (1.14)$$

(1.14) gives the unique solution of the first order ode and so there exists no problem, at least mathematically, to go temporally backwards from any preassigned initial condition. This feature of having a unique solution makes radioactive decay very useful in carbon-dating employed in palaeontology.

We now discuss briefly the **leaky-bucket problem**. Consider a cylindrical bucket having cross-section A. At its bottom, there is a circular hole of radius r as shown in the figure. The problem may be formulated in the following way.

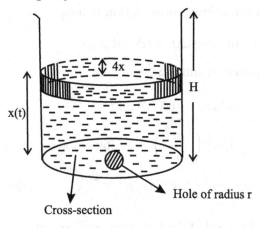

Fig 1.6 : Leaky-bucket

Suppose, at some instant, one discovers the bucket empty. Will he be able to figure out whenever it was full? (compare with the situation of radioactive decay). To give a satisfactory answer to this question, we need develop the ordinary differential equation that governs the physics of the problem.

Through the hole, the water flows continually out of the bucket. However, to study the rate at which the water level drops, we require a physical assumption regarding the velocity with which the water is ejected. Although seems to be rather over simplified, it will not sbe out of place to assume that the velocity $v(t)$ of the water coming out of the bucket depends on the height of water remaining in the porous bucket at the time instant t. Intitively, fuller the bucket is, faster is the flow as greater depth of water exerts more hydrostatic pressure on the bottom. Due to the principle of continuity, volume of water receding in the bucket over any time period equals the volume of water leaving through the hole of radius r over the same span of time.

$$\therefore \qquad rv(t) = A.\frac{dx}{dt} \qquad (1.15)$$

The potential energy lost due to gutting of a small amount of water in time t equals the kinetic energy of an equal amount of water leaving the bucket through the hole at the bottom. Water being incompressible a fluid, its density ρ is time-invariant.

Due to the energy-conservation principle,

P.E. lost in time $\Delta t = (\Delta x.A.\rho)gx$

K.E. gained in time $\Delta t' = \frac{1}{2}(\Delta x.A.\rho)v^2$

$\therefore \ (\Delta x.A.\rho)gx = \frac{1}{2}(\Delta x.A.\rho).v^2$

i,e, $\left(\dfrac{dx}{dt}\right)^2 = \left(\dfrac{2gr^2}{A^2}\right)x$ (using equation (1.15))

or, $\dfrac{dx}{dt} = -\mu x^{1/2}$, where $\mu^2 \equiv \dfrac{2gr^2}{A^2}$. $(\because x$ deceases with $t)$

The water-level of the leaky bucket obeys approximately the equation $\frac{dx}{dt} = -\mu\sqrt{x}$ and is known as Toricelli's law in Physics. Let's now solve the ode $\frac{dx}{dt} = -\mu\sqrt{x}$ along with the full-bucket initial condition $x(0) = H$, say. The solution reads :

$$x(t) = \frac{\mu^2}{4}\left(t - \frac{2}{\mu}\sqrt{H}\right)^2$$

If $t = t_e$, the time required for the full bucket to be emptied, then $t_e = \frac{2}{\mu}\sqrt{H}$. Substituting it back in the above equation, the particular solution turns out to be

$$x(t) = \begin{cases} \frac{\mu^2}{4}(t - t_e)^2 & \text{if } 0 \leq t \leq t_e \\ 0 & \text{if } t = t_e \end{cases} \qquad (1.16)$$

This solution (1.16) for a full-bucket initial condition $x(0) = H$ uniquely determines the height of water at any time $t \geq 0$. The problem of non-uniqueness arises when we look backwards from any empty-bucket initial condition, i.e., if we are assigned the initial condition $x(t_e) = 0$. In this case, one of the solutions is the solution curve (1.16) obtained for full-bucket initial condition $x(0) = H$; the remaining infinite number of solutions being the horizontal translates (towards the left) of that full-bucket solution that reaches the t-axis at $t = t_e$ (see fig. 1.7 below).

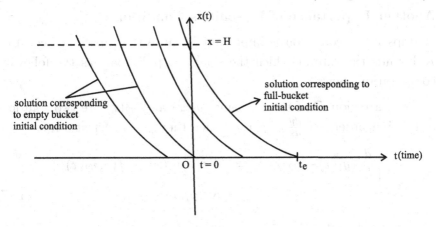

Fig 1.7 : Graph of the solution curves for Leaky Bucket

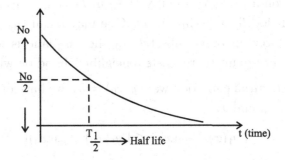

Fig 1.8 : Radioactive Decay

The equation (1.16) governing the physics of the leaky bucket cannot foretell when the bucket was full because of the non-uniqueness to its solution when associated with empty-bucket initial condition. Obviously one may ask what is the basic mathematical criterion that is responsible for the contrasting behaviour of the two IVPs

$$\frac{dN}{dt} = \lambda N, \ \lambda > 0, \ N = N_0 \ at \ t = t_0$$

and $\quad \dfrac{dx}{dt} = -\mu\sqrt{x}, \ \mu > 0, \ x = 0 \ at \ t = 0.$

One observes that the basic difference between the radioactive decay and the leaky bucket examples is that the former satisfies Lipschitz's condition while the latter does not over a region of the t-x plane.

Another Importance of Lipschitz's Condition :

Lipschitz's condition is important because the Lipschitz's constant K bounds the rate at which the solutions pull apart as the following computation shows :

If in a region D of the xy-plane, $y_1(x)$ and $y_2(x)$ are two solutions of the first order ode $\frac{dy}{dx} = f(x, y)$, then they pull apart at a rate

$$\left| \frac{d}{dx} (y_1(x) - y_2(x)) \right| = | f(x, y_1(x)) - f(x, y_2(x)) |$$
$$\leq K | y_1(x) - y_2(x) |, \tag{1.17}$$

so that we should make K as small as possible.

Note : (i) In the remark following the existence uniqueness theorem we stated that the Lipschitz's condition can be replaced by the relatively simple condition of continuity of $\frac{\partial f}{\partial y}$ in D. This is equivalent to the fact that f is locally Lipschitz in D. One may recall that a function $f(x, y)$ defined over an open subset U of the xy-plane is locally Lipschitz if about every point there exists a neighbourhood on which f is Lipschitz.

(ii) If the initial condition were $y(x_0) = y_0$, we have on integrating (1.17) between x_0 and x,

$$| y_1(x) - y_2(x) | \leq | y_1(x_0) - y_2(x_0)|.e^{K(x-x_0)} \tag{1.18}$$

Inference : Any two solutions of the ode diverge from each other atmost exponentially fast. Observe that in case the two solutions $y_1(x)$ and $y_2(x)$ satisfy the same initial conditions, i.e., if $y_1(x_0) = y_2(x_0)$, we would have from (1.18) : $y_1(x) = y_2(x)$, thereby asserting uniqueness of solutions.

1.7   First Order Ode and some of its Qualitative Aspects

(a) Consider the ordinary differential equation $\frac{dy}{dx} = f(x, y)$ where $f(x, y)$, commonly referred to as 'rate function', is a real-valued function on a region D in $\mathbb{R}^2$. The above equation has a simple geometrical interpretation. If (x_0, y_0) is a point in D through which passes an integral curve $y(x)$ of the given equation, i,e, if $y(x_0) = y_0$, then the differential equation specifies the slope $y'(x_0) = f(x_0, y_0)$ of the curve at that point. This eventually provides us the idea of 'line element' that is a triplet (x, y, y') associated to every point of D. The line element (x, y, y') consists of the point (x, y) and the slope y' of a line through (x, y). The set of all line elements is known as 'slope field' or 'direction-field' that is mathematically embodied as $\{(x, y, y')/(x, y) \in D$ and $y' = f(x, y)\}$. To construct a slope-field associated with an ode, the method of isoclines is the most useful one. An isocline is just a locus of equal inclination— in other words, it is the locus of those points in D where the line elements are parallel and hence have same preassigned slopes. At each point of an isocline, the tangents of the desired integral curves preserve a constant direction. However, in general, the isoclines are not solutions of the differential equations. In following we cite two examples of ode in one of which the isoclines are also integral curves and for the other, they are not integral curves.

(i) Consider $\frac{dy}{dx} = \frac{x}{y}$. The integral curves are $y^2 = x^2 + c$, c being an arbitrary constant. In case $c = 0$, we have the particular integral curve $y^2 = x^2$ which represents a pair of mutually orthogonal lines through the origin. The isoclines of this ode form a one-parameter family of straight lines $y = kx$. Clearly, $k = \pm 1$ gives us the solution curves $y = \pm x$.

(ii) Consider the integral curves of the ode $\frac{dy}{dx} = xy$. These are the exponential curves $y = Ce^{\frac{x^2}{2}}$, C being arbitrary constant. However, the isoclines are the hyperbolas $xy = k$ which decomposes to a pair of straight lines $x = 0$ and $y = 0$ if $k = 0$. In this case no isocline is an integral curve.

We now prove the following result involving Lipschitz condition.

(b) Let D be either $\{(x, y) \in \mathbb{R}^2/ \mid x - x_0 \mid \leq a$ and $\mid y - y_0 \mid \leq b\}$ or $\{(x, y) \in \mathbb{R}^2/ \mid x - x_0 \mid \leq a, \mid y \mid < \infty\}$, the former being a closed rectangle while the latter being an infinite rectangular strip.

If f is a real-valued function defined on D such that $\frac{\partial f}{\partial y}$ exists, continuous and satisfies the condition $\left| \frac{\partial f}{\partial y} \right| \leq K \ \forall \ (x, y) \in D$, then f satisfies the Lipschitz condition on D with K as the Lipschitz constant.

We prove this from the result

$$f(x, y_2) - f(x, y_1) = \int_{y_1}^{y_2} \frac{\partial f}{\partial y}(x, v)dv.$$

Taking modulus on both sides, we get

$$\mid f(x, y_2) - f(x, y_1) \mid \leq \int_{y_1}^{y_2} \left| \frac{\partial f}{\partial y}(x, v) \right| dv \leq K \mid y_2 - y_1 \mid$$

for all (x, y_1) and (x, y_2) in D. This asserts the claim.

As an example of a continuous function not satisfying a Lipschitz condition on a rectangle $D = \{(x, y) \in \mathbb{R}^2 : \mid x \mid \leq 1 \wedge \mid y \mid \leq 1\}$, consider $f(x, y) = \mid y \mid^{\frac{p}{q}}$, where $p < q$ and p, q are positive integers relatively prime to each other. Indeed if $y_2 > 0$,

$$\frac{\mid f(x, y_2) - f(x, 0) \mid}{\mid y_2 \mid} = \frac{\mid y_2 \mid^{\frac{p}{q}}}{\mid y_2 \mid} = \frac{1}{\mid y_2 \mid^{(1-\frac{p}{q})}},$$

which is unbounded as $y_2 \to 0$. Hence in this case there exists no Lipschitz constant K. If on the otherhand $p > q$, then of course Lipschitz's condition would be satisfied. This is why the initial-value problem of solving $\frac{dy}{dx} = \mid y \mid^{\frac{p}{q}}$ subject to $y(0) = 0$ has unique solution or infinitely many solutions according as $p > q$ or $p < q$, p and q being positive integers relatively prime to each other. This IVP has no unique solution as both $y(x) = 0$ and $y(x) = \left(\frac{2}{3}x\right)^{\frac{3}{2}}$ are its solutions. When uniqueness fails, the geometrical interpretations of the ode collapses as the so called 'phase point' is not sure how to move or which way to move − along the semicubical parabola $y^2 = \left(\frac{2}{3}x\right)^3$ or just get stuck at $(0,0)$?

(c) Often people become crazy in finding the asymptotic behaviour of the solutions. In this regard one poses the statutory question : 'What will be the solution as x tends to infinity?' However, this might be out of context for some problems where either the solutions are not defined for large x, or they blow up for some finite value of x− hinting the existence of vertical asymptotes. For instance, the general solution of $\frac{dy}{dx} = y^2$ is $y = \frac{1}{A-x}$, which represents two functions− one for $(-\infty, A)$ and other over (A, ∞). The first function has a vertical asymptote at $x = A$.

(d) If a differential equation of first order be invariant under some transformation like translation, rotation or dilation, then the family of its integral curves is also invariant under the same transformation. This result owes its origin to the theory of Lie groups that is beyond our scope. We just try to give a glimpse of that invariance property of odes and their solution family through simple illustrations.

Consider the ode
$$\frac{dy}{dx} = f(x) \tag{1.19}$$

Equation (1.19) is invariant under the translation $y \to y + k$ along the y-axis. The family of integral curves of (1.19) is :

$$y = \int f(x)dx + c \tag{1.20}$$

c being an arbitrary constant that distinguishes between individual members of the family.

Equation 1.20 is clearly invariant under the transformation $y \to y+k$, implying that every member of the family is shifted parallel to itself along the y-axis. (see figure 1.9 below):

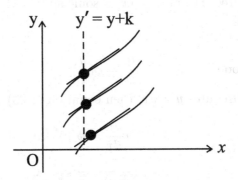

Fig 1.9 : Invariance of the family of integral curves of $\frac{dy}{dx} = f(x)$ under the transformation $y \to y + k$

Consider the ode
$$\frac{dy}{dx} = f(y) \tag{1.21}$$

Equation (1.21) is invariant under the translation $x \to x + h$ along the x-axis and consequently the family of integral curves of (1.21) is :

$$\ln(f(y)) = x + c', (c' \text{ being family parameter}) \tag{1.22}$$

Observe that (1.22) is invariant under the same transformation given in normal form (see figure 1.10).

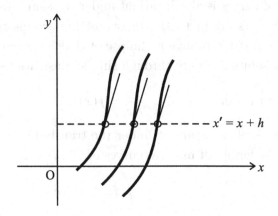

Fig 1.10 : Invariance of the family of integral curves of $\frac{dy}{dx} = f(y)$ under the
transformation $x \to x + h$

Lastly, we turn our attention to the ode $\frac{dy}{dx} = f(x, y)$ given in normal form where $f(x, y)$ is invariant under dilation, i,e, $f(tx, ty) = f(x, y)$ whenever, $t > 0$. This in turn ensures that $f(x, y)$ is a homogeneous function of degree zero in x, y, i,e, $\exists$ some suitable function ϕ such that $f(x, y) = \phi(\frac{y}{x})$.

Consider the ode,
$$\frac{dy}{dx} = \phi\left(\frac{y}{x}\right) \tag{1.23}$$

Now if one substitutes $y = vx$, then equation (1.23) reduces to :

$$x\frac{dv}{dx} = \phi(v) - v$$

$$\text{i, e,} \quad \frac{dv}{\phi(v) - v} = \frac{dx}{x}.$$

The above equation gives on integration :

$$\ln(cx) = F(v) = F\left(\frac{y}{x}\right), \ (c \text{ being family parameters}) \tag{1.24}$$

Observe that this equation (1.24) is invariant under dilation.

All these results are bare illustrations of the celebrated theorem : **Any symmetry of a given differential equation transforms any integral curve of the differential equation into some other integral curve of the same differential equation.**

[For details, the reader is referred to the book 'Geometrical Methods in the Theory of Ordinary Differential Equations' by V.I. Arnold.]

Exercise 1

1. Classify each of the following differential equations as ordinary or partial; state its order and determine whether the equation under consideration is linear or non-linear. Justify with reasons.

(a) $x^2\dfrac{dy}{dx} + y^2 = 0$

(b) $\dfrac{d^3y}{dx^3} - 2\dfrac{d^2y}{dx^2} - 4\dfrac{dy}{dx} + 8y = 0$

(c) $\dfrac{d^2y}{dx^2} + y\sin x = e^y$

(d) $\dfrac{\partial^2 u}{\partial x^2} = c^2\dfrac{\partial^2 u}{\partial t^2}$

(e) $\dfrac{\partial^2 u}{\partial x^2} + \dfrac{\partial^2 u}{\partial y^2} = 0$

(f) $\left(\dfrac{\partial^2 \phi}{\partial x^2} - \dfrac{\partial^2 \phi}{\partial y^2}\right)^2 + \left(\dfrac{\partial^2 \phi}{\partial x \partial y}\right)^2 = 4k^2$

(g) $\left(2y^{\frac{3}{2}} + 1\right)dx + x^{\frac{1}{2}}(3x^{\frac{1}{2}}y^{\frac{1}{2}} - 1)dy = 0$

(h) $x^2 dx + y^2 dy = \ln(xy)$

(i) $x^3\dfrac{d^2y}{dx^2} + \cos x\dfrac{dy}{dx} + y\sin x = 0$

(j) $\dfrac{d^2y}{dx^2} + x^2\dfrac{dy}{dx} + x\sin y = 0$

2. In (1.1) you found an example of first order ode that is linear or non-linear according as we select the independent variable. Does this feature hold good for first order pde? Give an illustration to clarify your viewpoint.

3. Obtain the differential equation associated with the primitive $y = c_1e^x + c_2e^{2x} + c_3e^{4x}$.

4. Eliminate the arbitrary constants c_1 and c_2 from the relation $y = e^x(c_1 \ln x + x + c_2)$

5. Form the differential equation of all parabolas each of which has latus rectum $4a$ and whose axes are parallel to the line $y = x$.

6. Show that $x^3 + 3xy^2 = 1$ is an implicit solution to the ode $2xy\dfrac{dy}{dx} + x^2 + y^2 = 0$ on the interval $0 < x < 1$.

7. Enlist all the admissible values of m so that $g(x) = e^{mx}$ is a solution to the ordinary differential equation

$$\frac{d^3y}{dx^3} - 3\frac{d^2y}{dx^2} + 4y = 0$$

8. Does the initial-value problem $\frac{dy}{dx} = x^3 + y^4$; $y(1) = 2$ possess a unique solution in the region D containing $(1, 2)$?

9. Apply the existence-uniqueness theorem to show that the IVP

$$\frac{dy}{dx} = y^2 \cos x; \ y(2) = 1$$

possesses a unique solution defined on a sufficiently small interval $|x - 2| \le h$ about $x = 2$.

10. Show that $f(x, y) = |y|$ satisfies Lipschitz condition in the neighbourhood of $(0, 0)$ although $\frac{\partial f}{\partial y}$ is not defined at $(0, 0)$.

11. (a) Does the initial-value problem

$$\left.\begin{array}{c} \frac{dy}{dx} = \frac{3x+2y}{2x-3y} \\ y(0) = 0 \end{array}\right\}$$

admit of a unique solution? Give suitable reasons.

(b) Had the suplementary condition been altered to $y(0) = 1$, will the same conclusion hold?

12. Is the existence-theorem of Picard applicable to the following IVP's:

(a) $\dfrac{dy}{dx} = xy^{\frac{1}{5}}$; $y(3) = 0$ (b) $\dfrac{dy}{dx} = xy^{\frac{1}{5}}$; $y(1) = 2$

Chapter 2

Equations of First Order and First Degree

2.1 Introduction

Without bothering too much with the persistent problem of existence and uniqueness of solutions of an ordinary differential equation we now concentrate on finding the solutions of differential equations on the basis of assumption that they exist. Right now our main aim will be to make the most use of small repertoire of functions (polynomials, trigonometric functions, exponential and logarithmic functions) to find the analytic solutions of the odes. However, we shall not confine ourselves only to theoretical details of mathematical artifices but try to look into the qualitative aspects of the problems from a practical world viewpoint. By the way we shall touch upon the 'existence-uniqueness' feature so that our mathematical exercises get justification.

First of all, we shall take up a first order ode which is nothing but a simplified mathematical model of a one dimensional dynamical system. This ode might appear in different forms as shown below.

(i)
$$\frac{dy}{dx} = f(x, y),$$

the normal derivative form, f being a real-valued C^1 function of two variables x and y defined on an open set $U \subseteq \mathbb{R}^2$, the xy-plane.

(ii)
$$M(x, y)dx + N(x, y)dy = 0,$$

where M and N are C^1 functions on an open set $U \subseteq \mathbb{R}^2$.

(iii)
$$M(x, y)\frac{dx}{dt} + N(x, y)\frac{dy}{dt} = 0,$$

where M and N are C^1 functions on an open set $U \subseteq \mathbb{R}^2$, with x and y themselves being functions of another variable t.

In all the forms (i) $-$ (iii) C^1 functions are involved. By C^1 **function**
we mean a function that has continuous first order partial derivatives
with respect to its arguments in its domain of definition.

The following example will illustrate how the three forms come to
use under different circumstances.

Consider the canonical equation $x^2 + y^2 = a^2$ of a family of concentric circles, (0,0) being centre and 'a', the family parameter, being the
radius. Taking differentials on both sides, we have $x\,dx + y\,dy = 0$. Hence
$x^2 + y^2 = a^2$ is a solution of the above ode. If we cast the same ode in normal derivative form $\frac{dy}{dx} = -\frac{x}{y}$ ($y \neq 0$ implied), then technically speaking,
$y = \pm\sqrt{a^2 - x^2}$ and not the circle $x^2 + y^2 = a^2$ are solutions. Moreover, the differential equation $\frac{dy}{dx} = -\frac{x}{y}$ is valid only for $\mid x \mid < a$ as
the derivative blows off at $x = \pm a$. To do away with this difficulty we
represent the solution curves in parametric form (such as $x = a\cos t$,
$y = a\sin t$ and the differential equations also in parametric form (such
as $x\frac{dx}{dt} + y\frac{dy}{dt} = 0$).

The statement that a solution to an ode exists is equivalent to the
statement that the ode is solvable or integrable.

A non-constant C^1 function $F : U_1 \to \mathbb{R}$ defined on an open subset
U_1 of U is called a **first integral** or constant of motion for the differential
equation $\frac{dy}{dx} = f(x,y)$ if $F_x(x,y) + F_y(x,y)f(x,y) = 0$ holds good for all
$(x,y) \in U_1 \left[\text{where } F_x(x,y) \equiv \frac{\partial F}{\partial x} \text{ and } F_y(x,y) \equiv \frac{\partial F}{\partial y}\right].$

The differential equation $\frac{dy}{dx} = f(x,y)$, is called **integrable** or **solvable** if it possesses a first integral $F : U_1 \to \mathbb{R}$ with U_1 open and dense
in U and $F(x,y) \neq 0$ for all $(x,y) \in U_1$. A set U_1 is dense in U iff
the closure of U_1 is equal to U. The density criterion is incorporated to
highlight the fact that U_1 is 'essentially' the whole of U. Now the time is
ripe for a zoom into the conditions so far laid down. First of all, trivially
obtainable constant solutions are left out. F is taken to be of C^1-type
as continuity of partial derivatives F_x and F_y in a neighbourhood of
every point (x, y) on the domain U_1 and also the non-vanishing partial
derivative F_y at each point (x, y) of U_1 builds the sufficient conditions
of the Implicit function theorem which ensures that locally (i.e, in a
neighborhood of each point (x, y) of the domain U_1 of F) the relation
$F(x,y) = c$ can be written in the form $y = \phi(x,c)$ where $\phi \in C^1(U_1)$.
Had the neighbourhood of (x, y) been changed, the explicit function

form $y = \phi(x, c)$ would undergo some change accordingly.

The integrable differential equations of first order and first degree are classified as (i) Exact (ii) Separable (iii) Homogeneous (iv) Linear. In course of our detailed discussion on these four species we shall observe their interrelations. Moreover we shall try to furnish the physical insights into what is going on surface in the mathematical mould.

2.2 Exact Differential Equation

Definition 2.1: Suppose M and N are C^1 functions defined on an open set $U \subseteq \mathbb{R}^2$. The differential equation $M(x, y) + N(x, y)\frac{dy}{dx} = 0$ (or its equivalent differential form $M(x, y)dx + N(x, y)dy = 0$) is called 'exact' if there exists a function $F : U \to \mathbb{R}$ so that the conditions $F_x = M(x, y)$ and $F_y = N(x, y)$ on U hold good.

[**Note :** M and N being C^1 functions, the mixed partial derivatives F_{xy} and F_{yx} are not only defined but also continuous. If follows therefrom that $F(x, y)$ is a C^2 function on U and $F_{xy} = F_{yx}$. The assumption of M and N being C^1 functions on $U \subset \mathbb{R}^2$ speaks out something more than even the well-known Schwarz theorem for the equality of mixed partial derivatives.]

$F_{xy} = \frac{\partial N}{\partial x}$ and $F_{yx} = \frac{\partial M}{\partial y} \Rightarrow \frac{\partial M}{\partial y} = \frac{\partial N}{\partial x}$.

So for the exact ode $M(x, y)dx + N(x, y)dy = 0$, $\frac{\partial M}{\partial y} = \frac{\partial N}{\partial x}$.

Contrapositively, we may say that if $\frac{\partial M}{\partial y} \neq \frac{\partial N}{\partial x}$, the ode is not exact. Sometimes the authors claim (without quoting underlying restrictions) that this equality is sufficient for the exactness of ode. However we shall prove the utility of the restrictions through a counter example after the following theorem.

Theorem 2.1: If for the differential equation $M(x, y)dx + N(x, y)dy = 0$ where M and N possess continuous partial derivatives in a rectangular domain D, then the differential equation is exact if and only if $\frac{\partial M}{\partial y} = \frac{\partial N}{\partial x}$ holds good provided the rectangular domain D is a simply-connected region bounded by a simple closed curve, viz, a rectangle Γ.

Proof of Necessity : Let's assume that $Mdx + Ndy = 0$ is an exact differential equation in D. This implies that $M\,dx + N\,dy$ is an exact differential and so there exists a function $F(x,y)$ over D such that

$$dF = M\,dx + N\,dy$$

i.e., $\quad F_x\,dx + F_y\,dy = M\,dx + N\,dy,$

$\therefore \quad F_x = M$ and $F_y = N$, so that $F_{yx} = \dfrac{\partial M}{\partial y}$ and $F_{xy} = \dfrac{\partial N}{\partial x}$;

We observe that by the given condition of theorem, F_{xy} and F_{yx} are both continuous at every point $(\xi, \eta) \in D$ and hence they are equal. So $\frac{\partial M}{\partial y} = \frac{\partial N}{\partial x}$ holds good.

Proof of Sufficiency: We shall furnish two proofs.

First Proof: Since $M \in C^1(D)$, $\int M\partial x$ (partial integral of M w.r. to x) exists. So the function $F(x,y) = \int M\partial x$ is well-defined and F_y exists at every point $(\xi, \eta) \in D$. Moreover, $F_x = M$.

Hence by the conditions given in the theorem,

$$F_{yx} = \frac{\partial M}{\partial y} = \frac{\partial N}{\partial x} \tag{2.1}$$

Since $N(x,y)$ is a C^1 function, F_{yx} given in (2.1) is continuous at every point (ξ, η) of the rectangular domain D. (D being open, every point (ξ, η) of D is an interior point so that there exists a neighborhood of (ξ, η) which lies entirely in D. Moreover F_y exists at every point $(\xi, \eta) \in D$ as stated before (differentiability under sign of integration). In this regard uniform continuity of $F(x,y)$ will come to use. By Schwarz theorem on mixed partial derivatives it follows directly that F_{xy} exists at (ξ, η) and is equal to $F_{yx}(\xi, \eta)$. [Had the openness of D been not there, at least one point in D would fail to be its interior point and hence the applicability of Schwarz theorem would be questionable thereat.]

$$\therefore \quad \frac{\partial N}{\partial x} = F_{yx} = F_{xy} \tag{2.2}$$

Integrating (2.2) partially w.r. to x we have,

$$N(x,y) = F_y + g(y), \tag{2.3}$$

where $g(y)$ is an arbitrary continuous function of y.

$$\therefore \ Mdx + Ndy = F_x dx + (F_y + g(y))dy$$
$$= dF + g(y)dy$$
$$= d\left(F(x,y) + \int g(y)dy\right) \equiv du$$

where $\quad u(x,y) \equiv F(x,y) + \int g(y)dy$

$$= \int M\partial x + \int \left(N - \frac{\partial}{\partial y}\int M\partial x\right)dy \tag{2.4}$$

Making use of the flexibility of choice of independent and dependent variables in a first order ode in the differential form, we could have cast $u(x,y)$ in the alternative form

$$u(x,y) = \int N\partial y + \int \left(M - \frac{\partial}{\partial y}\int N\partial x\right)dx \tag{2.5}$$

[In (2.4), $\left(N - \frac{\partial}{\partial y}\int M\partial x\right)$ being a function of y alone, we have used dy instead of ∂y while in (2.5), $\left(M - \frac{\partial}{\partial x}\int N\partial y\right)$ being a function of x alone we used dx instead of ∂x.]

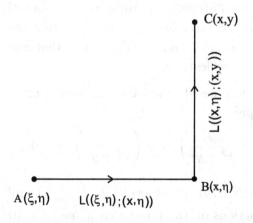

Fig 2.1 : Diagram of $\Gamma = L\left((\xi,\eta);(x,\eta)\right) \cup L\left((x,\eta);(x,y)\right)$

Second proof : Suppose D is simply-connected. Hence all the paths joining any two given points in D will be reconciliable. We consider the points (ξ,η) and (x,y) in D and define the path

$$\Gamma = L\left((\xi,\eta);(x,\eta)\right) \cup L\left((x,\eta);(x,y)\right)$$

where $L\left((\xi,\eta);(x,\eta)\right)$ denotes the directed line segment joining (ξ,η) and (x,η) and $L\left((x,\eta);(x,y)\right)$ denoted the directed line segment joining (x,η) and (x,y).

Denote the line integral $\int_\Gamma (M\,dx + N\,dy)$ by $F(x,y)$

$$\therefore F(x,y) = \int_{L((\xi,\ \eta);\ (x,\ \eta))} M(x,y)\partial x + \int_{L((\xi,\ \eta);\ (x,\ y))} N(x,y)\partial y$$

$$= \int_\xi^x M(x,\eta)dx + \int_\eta^y N(x,y)dy \tag{2.6}$$

$$\therefore F_x = M(x,\eta) + \int_\eta^y \frac{\partial N}{\partial x}dy = M(x,\eta) + \int_\eta^y \frac{\partial M}{\partial y}dy = M(x,y) \tag{2.7}$$

Using these results we have

$$M\,dx + N\,dy = F_x + F_y dy = dF(x,y) \tag{2.8}$$

because $F_y = N(x,y)$ establishing that the given ode is exact. Observe that we have computed the line integral along a suitably chosen Γ joining (ξ,η) to (x,y). We have been able to draw a generalised result (that is curiously path-independent) simply because D is simply-connected.

Remark: (1) This elaborate discussion of the above theorem should not be treated as an outburst of periphrase, but honestly speaking, it is intended to highlight the role played by the conditions stated in the theorem so that we become careful of the pitfalls that may trap us while using it in practical applications.

As promised before we cite the following counter-example which will really be an eye-opener.

Consider the ode $\left(\dfrac{y}{x^2+y^2}\right)dx + \left(\dfrac{-x}{x^2+y^2}\right)dy = 0.$ $\hspace{1cm}$ (∗)

Here $M(x,y)$ and $N(x,y)$ are respectively $\left(\frac{y}{x^2+y^2}\right)$ and $\left(\frac{-x}{x^2+y^2}\right)$. M and N are C^1 functions on the punctured plane $\mathbb{R}^2 - \{0\}$. Hence if D be a region within $\mathbb{R}^2 - \{0\}$ that encompasses $\{0\}$, D is not simply-connected. However at every point $(\xi,\eta) \in \left(\mathbb{R}^2 - \{0\}\right) \cap D$, $\frac{\partial M}{\partial y} = \frac{\xi^2 - \eta^2}{(\xi^2+\eta^2)^2}$. Effectively this lack of simple-connectedness is manifested in the path dependence of line integrals on the choice of Γ_k's in D.

Consider four different paths $\Gamma_1, \Gamma_2, \Gamma_3, \Gamma_4$ in $(\mathbb{R}^2 - \{0\}) \cap D$ joining the points $(1, 0)$ to $(-1, 0)$

(a) Γ_1 : arc of the unit circle : $x^2 + y^2 = 1$; $y \geq 0$

(b) Γ_2 : arc of the unit circle : $x^2 + y^2 = 1$; $y \leq 0$

(c) Γ_3 : arc of the parabola : $y = -1 + x^2$

(d) Γ_4 : polygonal line $L\left((1,0); (0,+1)\right) \cup L\left((0,+1); (-1,0)\right)$

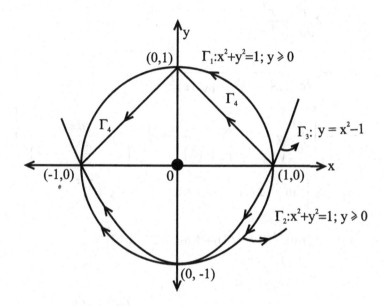

Fig 2.2 : Illustrating four paths $\Gamma_1, \Gamma_2, \Gamma_3, \Gamma_4$ along which the line integrals $\int_{\Gamma_k}(Mdx + Ndy)$ are evaluated, $(k = 1, 2, 3, 4)$

(a)

$$\int_{\Gamma_1} \frac{-ydx + xdy}{x^2 + y^2} = \int_{\theta=0}^{\theta=\pi} d\theta = \pi$$

($\because$ $x = \cos\theta$; $y = \sin\theta \Rightarrow dx = -\sin\theta d\theta$; $dy = \cos\theta d\theta$; $\theta = 0$ corresponds to the point (1,0) while $\theta = \pi$ corresponds to (−1,0))

(b)

$$\int_{\Gamma_2:\ x^2+y^2=1;\ y\leq0} \frac{-ydx + xdy}{x^2 + y^2} = \int_{\theta=0}^{\pi} -d\theta = -\pi$$

($\because$ $x = \cos\theta$; $y = -\sin\theta \Rightarrow dx = -\sin\theta d\theta$; $dy = -\cos\theta d\theta$)

See that $\theta = 0$ corresponds to (1,0) but $\theta = \pi$ corresponds to (−1,0))

(c)

$$\int_{\Gamma_3:\, y\, =\, x^2-1} \frac{-ydx + xdy}{x^2 + y^2}$$

$$= \int_{x=1}^{x=-1} \frac{(x^2 + 1)dx}{x^4 - x^2 + 1} \qquad [\because\ y = x^2 - 1,\ dy = 2xdx]$$

$$= -2\int_0^1 \frac{(x^2 + 1)dx}{(x^4 - x^2 + 1)}$$

$$= -2\int_0^1 \frac{(x^2 + 1)dx}{(x^2 + 1)^2 - \left(\sqrt{3}x\right)^2}$$

$$= -\left[\int_0^1 \frac{dx}{\left(x + \frac{\sqrt{3}}{2}\right)^2 + \left(\frac{1}{2}\right)^2} + \int_0^1 \frac{dx}{\left(x - \frac{\sqrt{3}}{2}\right)^2 + \left(\frac{1}{2}\right)^2}\right]$$

$$= -2\left[\tan^{-1}(2x + \sqrt{3}) + \tan^{-1}(2x - \sqrt{3})\right]\Big|_{x=0}^{x=1},$$

$$= -2\left[\tan^{-1}(2 + \sqrt{3}) + \tan^{-1}(2 - \sqrt{3})\right] = -\pi.$$

(d)

$$\int_{\Gamma_4:L((1,\,0);(0,\,1))\cup L((0,\,1);(-1,\,0))} \frac{-ydx + xdy}{x^2 + y^2}$$

$$= \int_{L((1,0);(0,1))} \frac{-ydx + xdy}{x^2 + y^2} + \int_{L((0,1);(-1,0))} \frac{-ydx + xdy}{x^2 + y^2}$$

$$= \int_{x=0}^{x=1} \frac{dx}{x^2 + (x-1)^2} + \int_{x=0}^{-1} \frac{dx}{x^2 + (x+1)^2}$$

$$= 2\int_{x=0}^{x=1} \frac{dx}{x^2 + (x-1)^2} = 4\tan^{-1}1 = \pi$$

We therefore observe that over paths Γ_1 and Γ_4, the value of the line integral $\int_{\Gamma_k}(Mdx + Ndy)$ are same, viz, $+\pi$ while over the paths Γ_2 and Γ_3, the value of the line integral $\int_{\Gamma_k}(Mdx + Ndy)$ are same, viz, $-\pi$. Thus line integral is path dependent. Hence for exactness of $Mdx + Ndy = 0$, $\frac{\partial M}{\partial y} = \frac{\partial N}{\partial x}$ is a necessary and sufficient condition provided the domains of M and N are simply connected.

All the above discussion clarifies why in the strict sense of the term this first order ode $(*)$ is not exact. However, its solution can be formally determined via the method of partial integration (2.4) as:

$$u(x,y) \equiv \tan^{-1}\left(\frac{x}{y}\right) = \text{constant}$$

It is interesting to see that the ode taken up is 'homogeneous' as it is invariant under the transformation $x \to tx$; $y \to ty$ $(t > 0)$. The family of integral curves $u(x,y)$= constant is also 'homogeneous'; showing an illustration of the claim we made in article 1.7(d).

(2) To avoid confusion regarding interpretation of terminologies we explicitly write out their meanings :

(i) **Region :** An open connected set with its boundary points, taken either all or none or some.

(ii) **Domain :** If no boundary point is included, the region is called 'open region' or more commonly 'domain'. This concept of domain is purely a topolocial idea in contrast to the concept of domain of functions as the latter lacks in algebraic structure.

(iii) **Simply-connected :** An open connected set D is said to be 'simply-connected' if every 'simple-closed curve' Γ in D is 'homotopic' to a single point in D. Roughly speaking, simply connected sets do not contain any holes. A 'simple-closed curve' Γ is the image of a topological map or homeomorphism of a circle. Alternatively, a simple closed curve Γ is one which has no double point other than its end points (fused). In this respect a rectangle is topologically equivalent to a circle. In theories of ode whenever we talk of an open set, this so called domain (topological sense) is usually implied.

(3) If D is merely an open set, then exactness of an ode $Mdx + Ndy = 0$ will imply the vanishing of the curl of the two-component vector field $M\hat{i} + N\hat{j}$, $\hat{i}$ and $\hat{j}$ being unit vectors along positive directions of x and y-axis respectively. But the converse is not true in general i.e, vanishing of curl of a vector field $M\hat{i} + N\hat{j}$ will not ensure exactness of ode $Mdx + Ndy = 0$. Infact, for the converse to hold good, even the assumption of an open-connected D will not suffice as in case D contains holes, the paths joining two preassigned points may or may not be reconcilable so

that the line integral over any Γ joining these points turns out to be path-dependent. This path-dependence goes off once simple-connectedness is imposed on D.

(4) In case the differential equation $Mdx + Ndy = 0$ is exact, we say that the two-component vector field $M\hat{i} + N\hat{j}$ is a **gradient vector field** or **conservative vector field** or **irrotational vector field**. The vector field is 'irrotational' as curl $(M\hat{i} + N\hat{j}) = \overrightarrow{0}$. The vector field is 'conservative' as the line integrals of $(Mdx + Ndy)$ over a Γ joining two preassigned points in domain D are totally independent of the path chosen. Had it been a force-field, we would have said that total energy is conserved for the system or more circumlocutedly, it is a non-dissipative force-field. The vector field is also called 'gradient vector-field' because exactness allows us to express M and N as the components of the gradient of a scalar function $F(x, y)$ which is known as 'potential function' of the vector field $(M\hat{i} + N\hat{j})$. In brief, for a gradient vector-field, the components are derivable from a differentiable single-valued function, unique upto an additive constant. This arbitrariness can be also given a nice explanation from the physical viewpoint.

Example (1) : Check the exactness and find solution of the ode

$$(2x \sin y + y^3 e^x)dx + (x^2 \cos y + 3y^2 e^x)dy = 0$$

$$\text{Here} \quad \left. \begin{array}{rcl} M &=& 2x \sin y + y^3 e^x \\ N &=& x^2 \cos y + 3y^2 e^x \end{array} \right\}$$

$$\therefore \left. \begin{array}{l} \frac{\partial M}{\partial y} = 2x \cos y + 3y^2 e^x \\ \frac{\partial N}{\partial x} = 2x \cos y + 3y^2 e^x \end{array} \right\} \therefore \frac{\partial M}{\partial y} = \frac{\partial N}{\partial x} \text{ for all } (x,y) \in D$$

Thus the given ode is exact. The complete solution will be

$$F(x,y) = \int M\partial x + \int \left(N - \frac{\partial}{\partial y} \int M\partial x \right) dy$$

$$\int M(x,y)\partial x = \int (2x \sin y + y^3 e^x)\partial x = (x^2 \sin y + y^3 e^x) + g(y)$$

$$\therefore \quad \left(N(x,y) - \frac{\partial}{\partial y} \int M(x,y)dx \right)$$

$$= (x^2 \cos y + 3y^2 e^x) - \frac{\partial}{\partial y}(x^2 \sin y + y^3 e^x) - g'(y)$$
$$= x^2 \cos y + 3y^2 e^x - x^2 \cos y - 3y^2 e^x - g'(y) = -g'(y)$$

$$\therefore \quad F(x, y) \quad = \quad x^2 \sin y + y^3 e^x + g(y) - g(y) + c$$
$$= \quad (x^2 \sin y + y^3 e^x) + c, \quad c \text{ being arbitrary constant.}$$

Remark: The function $(x^2 \sin y + y^3 e^x)$ is a particular solution of the p.d.e $\frac{\partial F}{\partial x} = M$. To yield general solution one must use the arbitrary additive differentiable function.

Example (2) : Solve the IVP :

$$(2x \cos y + 3x^2 y)dx + (x^3 - x^2 \sin y - y)dy = 0 \; ; \; y(0) = 2$$

Comparing this ode with the canonical $Mdx + Ndy = 0$ form we get

$$\left. \begin{array}{rcl} M &=& 2x \cos y + 3x^2 y \\ N &=& x^3 - x^2 \sin y - y \end{array} \right\}$$

Since $\frac{\partial M}{\partial y} = -2x \sin y + 3x^2$; $\frac{\partial N}{\partial x} = 3x^2 - 2x \sin y$, exactness is ensured.

The general solution is given by $F(x, y) = c$ (arbitrary constant)

where $\quad F(x, y) = \displaystyle\int M\partial x + \int \left[N - \frac{\partial}{\partial y} \int M\partial x \right] dy$

$$\int M\partial x = \int (2x \cos y + 3x^2 y)\partial x = x^2 \cos y + x^3 y + g(y)$$

$$\therefore \left(N - \frac{\partial}{\partial y} \int M\partial x \right) = (x^3 - x^2 \sin y - y) - \frac{\partial}{\partial y}(x^2 \cos y + x^3 y + g(y))$$
$$= -y - g'(y)$$

and hence $\displaystyle\int \left(N - \frac{\partial}{\partial y} \int M\partial x \right) dy = -\int ydy - g(y) = -\frac{y^2}{2} - g(y) + c$

Again $y(0) = 2$ is to be satisfied, i.e, $c = 2$ is a must.

The particular solution reads : $x^2 \cos y + x^3 y - \frac{y^2}{2} + 2 = 0$.

Example (3) : Solve the ode : $yx^{y-1}dx + x^y \ln x dy = 0$

Comparing the given ode with $M(x,y)dx + N(x,y)dy = 0$, we get

$$\begin{aligned} M(x,y) &= yx^{y-1} &= ye^{(y-1)\ln x} \\ N(x,y) &= x^y \ln x &= e^{y \ln x}(\ln x) \end{aligned}$$

$$\begin{aligned} \therefore \quad \frac{\partial M}{\partial y} &= e^{(y-1)\ln x} + ye^{(y-1)\ln x}.\ln x = e^{(y-1)\ln x}[1 + y \ln x] \\ \frac{\partial N}{\partial x} &= \frac{1}{x}e^{y \ln x} + e^{y \ln x}\left(\frac{\ln x}{y}\right).y \\ &= e^{y \ln x}.e^{-\ln x} + e^{y \ln x}(\ln x)e^{-\ln x}.y \\ &= e^{(y-1)\ln x} + e^{(y-1)\ln x}(\ln x).y \\ &= e^{(y-1)\ln x}[1 + y \ln x] \end{aligned}$$

Thus $\frac{\partial M}{\partial y} = \frac{\partial N}{\partial x}$, ensuring that given ode is exact.

The general solution is :

$$F(x,y) \equiv \int M\partial x + \int \left(N - \frac{\partial}{\partial y}\int M\partial x\right) dy = c, \ c \text{ being a constant.}$$

$$\int M\partial x = \int yx^{y-1}\partial x = x^y + g(y);$$

$g(y)$ being an arbitrary differentiable function of y.

$$\begin{aligned} \int \left(N - \frac{\partial}{\partial y}\int M\partial x\right) dy &= \int \left[x^y \ln x - \frac{\partial}{\partial y}(x^y + g(y))\right] dy \\ &= \int \left[x^y \ln x - \frac{\partial}{\partial y}(x^y)\right] dy - g(y) \\ &= \int (x^y \ln x - x^y \ln x)\, dy - g(y) = -g(y) \end{aligned}$$

So the general solution is : $x^y + g(y) - g(y) = c$

Example (4) : Solve the ode :

$$\left[y\left(1 + \frac{1}{x}\right) + \sin y\right] dx + (x + \ln x + x\cos y)dy = 0$$

Here

$$\left.\begin{aligned} M(x,y) &= y\left(1 + \tfrac{1}{x}\right) + \sin y \\ N(x,y) &= x + \ln x + x\cos y \end{aligned}\right\}$$

$$\therefore \quad \frac{\partial M}{\partial y} = \left(1 + \frac{1}{x}\right) + \cos y \; ; \quad \frac{\partial N}{\partial x} = \left(1 + \frac{1}{x}\right) + \cos y,$$

So the given ode is exact.

The complete solution of this exact ode is $F(x, y) = 0$, where

$$F(x, y) = \int M \partial x + \int \left(N - \frac{\partial}{\partial y} \int M \partial x\right) dy$$

$$\int M \partial x = \int \left[y\left(1 + \frac{1}{x}\right) + \sin y\right] \partial x = y(x + \ln x) + x \sin y + g(y)$$

$$\int \left(N - \frac{\partial}{\partial y} \int M \partial x\right) dy$$

$$= \int \left[(x + \ln x + x \cos y) - \frac{\partial}{\partial y}\{y(x + \ln x) + x \sin y + g(y)\}\right] dy$$

$$= \int \left[(x + \ln x) + x \cos y - (x + \ln x) - x \cos y - g'(y)\right] dy$$

$$= \int -g'(y) dy = -g(y) + C \quad (C \text{ being constant of integration})$$

$$\therefore \quad F(x, y) = y(x + \ln x) + x \sin y + C$$

Sometimes people like to work out exact odes in very tricky but elegant ways — one being the 'grouping of terms' method and the other is the 'mixed term isolation' as shown in examples (5) , (6) and (7) respectively. However, there is no reprieve from the exactness test.

Example (5) : Solve the ode

$$(2x^2 y + 4x^3 - 12xy^2 + 3y^2 - xe^y + e^{2x})dy$$
$$+ (12x^2 y + 2xy^2 + 4x^3 - 4y^3 + 2ye^{2x} - e^y)dx = 0$$

Here comparing with $(M dx + N dy) = 0$ we get

$$\left.\begin{array}{rcl} N(x, y) & = & 2x^2 y + 4x^3 - 12xy^2 + 3y^2 - xe^y + e^{2x} \\ M(x, y) & = & 12x^2 y + 2xy^2 + 4x^3 - 4y^3 + 2ye^{2x} - e^y \end{array}\right\}$$

and obviously $\frac{\partial M}{\partial y} = \frac{\partial N}{\partial x}$ asserted everywhere in $\mathbb{R}^2$.

We regroup the terms of the ode as follows so that within the parentheses perfect differentials appear :

$$3y^2 dy + 4x^3 dx + (2x^2 y dy + 2xy^2 dx) + (e^{2x} dy + 2ye^{2x} dx)$$
$$- (xe^y dy + e^y dx) + (4x^3 dy + 12x^2 y dx) - (4y^3 dx + 12xy^2 dy) = 0$$

$\therefore$ Integrating, we get the solution as :

$$y^3 + x^4 + x^2 y^2 + e^{2x} y - x e^y + 4 x^3 y - 4 x y^3 = c, \text{ some arbitrary constant.}$$

Example (6) : Solve the ode :

$$\left(\frac{1}{y} \sin \frac{x}{y} - \frac{y}{x^2} \cos \frac{y}{x} + 1 \right) dx + \left(\frac{1}{x} \cos \frac{y}{x} - \frac{x}{y^2} \sin \frac{x}{y} + \frac{1}{y^2} \right) dy = 0$$

Comparing with $M dx + N dy = 0$:

$$\left. \begin{array}{rcl} M(x,y) & = & \frac{1}{y} \sin \frac{x}{y} - \frac{y}{x^2} \cos \frac{y}{x} + 1 \\[2mm] N(x,y) & = & \frac{1}{x} \cos \frac{y}{x} - \frac{x}{y^2} \sin \frac{x}{y} + \frac{1}{y^2} \end{array} \right\}$$

It's a routine exercise to check that $\frac{\partial M}{\partial y} = \frac{\partial N}{\partial x}$ to ensure that the given ode is exact. However, we shall solve this ode by regrouping of terms. One can express the ode as:

$$\left(dx + \frac{dy}{y^2} \right) + \sin \frac{x}{y} \cdot \left(\frac{dx}{y} - \frac{x}{y^2} dy \right) + \cos \frac{x}{y} \cdot \left(\frac{-y dx + x dy}{x^2} \right) = 0$$

or, $\left(dx + \dfrac{dy}{y^2} \right) + \sin \dfrac{x}{y} . d \left(\dfrac{x}{y} \right) + \cos \dfrac{y}{x} . d \left(\dfrac{y}{x} \right) = 0$

Integrating, $x - \dfrac{1}{y} - \cos \left(\dfrac{x}{y} \right) + \sin \left(\dfrac{y}{x} \right) = C,$

C being the constant of integration. Hence the integral of the ode is:

$$x - \frac{1}{y} - \cos \left(\frac{x}{y} \right) + \sin \left(\frac{y}{x} \right) = C.$$

Remark: The grouping method is based on trial and error approach and so is not very useful for solving complicated exact first order odes. Method of partial integration used in Examples (1)−(4) is more generic.

Example (7) : Solve the ode

$$(6xy + 2y^2 - 5) dx + (3x^2 + 4xy - 5) dy = 0$$

Here $\left. \begin{array}{rcl} M(x,y) & = & 6xy + 2y^2 - 5 \\ N(x,y) & = & 3x^2 + 4xy - 5 \end{array} \right\}$

Easily verified the exactness. We observe that if we integrate $M(x, y)$ partially with respect to x then automatically the x-involving terms in $N(x, y)$ have been taken into account. Only x-free term (-6) appearing in $N(x, y)$ is to be considered and this integrated w.r.t y.

$\therefore$ Integrating we have the first integral :

$$F(x, y) = 3x^2 y + 2xy^2 - 5x - 6y$$

and the general solution is $F(x, y) = c$

Example (8) : Find the values of the constants A and B so as to make the IVP

$$\frac{1 + Axy^{\frac{2}{3}}}{x^{\frac{2}{3}} y^{\frac{1}{3}}} dx + \frac{2x^{\frac{4}{3}} y^{\frac{2}{3}} - Bx^{\frac{1}{3}}}{y^{\frac{4}{3}}} dy = 0; \quad y(1) = 27 \quad \text{exact. Hence solve it.}$$

Here $\begin{aligned} M(x, y) &= x^{-\frac{2}{3}} y^{-\frac{1}{3}} + Ax^{\frac{1}{3}} y^{\frac{1}{3}} \\ N(x, y) &= 2x^{\frac{4}{3}} y^{-\frac{2}{3}} - Bx^{\frac{1}{3}} y^{-\frac{4}{3}} \end{aligned} \Bigg\}$

$\therefore \quad \begin{aligned} \frac{\partial M}{\partial y} &= -\frac{1}{3} x^{-\frac{2}{3}} y^{-\frac{4}{3}} + \frac{1}{3} Ax^{\frac{1}{3}} y^{-\frac{2}{3}} \\ \frac{\partial N}{\partial x} &= \frac{8}{3} x^{\frac{1}{3}} y^{-\frac{2}{3}} - \frac{1}{3} Bx^{-\frac{2}{3}} y^{-\frac{4}{3}} \end{aligned} \Bigg\}$

For exactness, $A = 8$ and $B = 1$ is a must. The resultant exact ode is :

$$\left(x^{-\frac{2}{3}} y^{-\frac{1}{3}} + 8x^{\frac{1}{3}} y^{\frac{1}{3}} \right) dx + \left(2x^{\frac{4}{3}} y^{-\frac{2}{3}} - x^{\frac{1}{3}} y^{-\frac{4}{3}} \right) dy = 0$$

We now partially integrate w.r.t x the terms of $M(x, y)\big|_{A=8}$ and since $N(x, y)\big|_{B=1}$ involves no x-free term, the first integral is :

$$F(x, y) \equiv 3x^{\frac{1}{3}} y^{-\frac{1}{3}} + 6x^{\frac{4}{3}} y^{\frac{1}{3}} = c$$

Using the initial conditions, $c = 19$ is computable easily. Hence the specific integral curve we are interested in is :

$$3x^{\frac{1}{3}} y^{-\frac{1}{3}} + 6x^{\frac{4}{3}} y^{\frac{1}{3}} = 19$$

Example (9) : Solve $\quad (y \sec^2 x + \sec x \tan x) dx + (2y + \tan x) dy = 0$

Here $\begin{aligned} M(x, y) &= y \sec^2 x + \sec x \tan x \\ N(x, y) &= 2y + \tan x \end{aligned} \Bigg\}$, showing that $\frac{\partial M}{\partial y} = \frac{\partial N}{\partial x} = \sec^2 x$

Write $\quad \left.\begin{array}{ccccc} F_x(x,y) & = & M(x,y) & = & y\sec^2 x + \sec x \tan x \\ F_y(x,y) & = & N(x,y) & = & 2y + \tan x \end{array}\right\}$

Integrating the first relation partially w.r.to x and second relation partially w.r.to y we get

$$\left.\begin{array}{ccl} F(x,y) & = & y\tan x + \sec x + h_1(y) \\ \text{and} \quad F(x,y) & = & y^2 + y\tan x + h_2(x) \end{array}\right\}$$

where $h_1(y)$ and $h_2(x)$ are arbitrary functions. Now reconciliation of these two forms and using method of inspection it follows that $h_1(y) = y^2$ and $h_2(x) = \sec x$

So the first integral reads : $F(x,y) \equiv y\tan x + \sec x + y^2 = c$

Separable Equations : The first order differential equations of the general form $Mdx + Ndy = 0$ is said to be **separable** if both M and N are expressible as product of two functions, one of x and other of y. Explicitly, if $M(x,y) = F_1(x)G_1(y)$ and $N(x,y) = F_2(x)G_2(y)$ then we get separable equations. In case $F_2(x)G_1(y) \neq 0$, we may divide the given equation by it and get the new ode as :

$$\frac{F_1(x)}{F_2(x)}dx + \frac{G_2(y)}{G_1(y)}dy = 0$$

In this new form $\overline{M}(x,y) \equiv \frac{F_1(x)}{F_2(x)}$ is only a function of x so that $\frac{\partial \overline{M}}{\partial y} = 0$ and $\overline{N}(x,y) \equiv \frac{G_2(y)}{G_1(y)}$ is only a function of y so that $\frac{\partial \overline{N}}{\partial x} = 0$.

Thus trivially exactness criterion satisfied. Sometimes we say that all separable equations are exact but we do not pay heed to the division process by $F_2(x)G_1(y)$ that brought the given ode to the separable form. It may happen that through such division some solutions are lost whose track should be kept. In the following examples we shall throw light on this idea.

Once the separable equation $Mdx + Ndy = 0$ has been recast in the form $\overline{M}dx + \overline{N}dy = 0$ as stated before, necessary and sufficient condition for exactness becomes trivially satisfied and hence the general solution is given by $F(x,y) = c$, where

$$F(x,y) = \int \overline{M}\partial x + \int \left(\overline{N} - \frac{\partial}{\partial y} \int \overline{M}\partial x \right) dy$$

$$= \int \frac{F_1(x)}{F_2(x)}dx + \int \left\{ \frac{G_2(y)}{G_1(y)} - \frac{\partial}{\partial y}(\text{a function of } x \text{ only}) \right\} dy$$

$$= \int \frac{F_1(x)}{F_2(x)}dx + \int \frac{G_2(y)}{G_1(y)}dy$$

Example (10) : Solve the ode : $(xy + 2x + y + 2)dx + (x^2 + 2x)dy = 0$

The ode can be written as : $(x+1)(y+2)dx + x(x+2)dy = 0$, that on being divided by $x(x+2)(y+2)$ becomes

$$\frac{(x+1)}{x(x+2)}dx + \frac{dy}{y+2} = 0,$$

which trivially is seen to be exact as it is a first order separable ode. However it works on the tacit assumption that $x(x+1)(y+2) \neq 0$

If now $y = -2$ then the co-efficient of dx would vanish and hence $y = -2$ is a solution.

The first integral $F(x,y)$ of the recast separable equation reads :

$$F(x,y) \equiv \int \frac{(x+1)}{x(x+2)}dx + \int \frac{dy}{y+2} = \frac{1}{2}ln\mid x \mid + \frac{1}{2}ln \mid x+2 \mid + ln \mid y+2 \mid$$

The general solution, after rationalisation becomes :

$\mid x(x+2) \mid (y+2)^2 = c^2$, c being arbitrary constant of integration.

Here $c = 0$ corresponds to $y = -2$ and hence no solution of the original ode gets an overlook due to transition to the separable form.

Example (11) : Solve the initial-value problem :

$$2y\frac{dy}{dx} + (1+y^2)\cos x = 0, \quad \text{subject to initial condition } y(0) = 1.$$

The equation may be rewritten as

$$\frac{2y}{1+y^2}dy + \cos x dx = 0.$$

Integrating this separable form between 0 and x we have

$$\int_1^y \frac{2zdz}{1+z^2} + \int_0^x \cos t \, dt = 0$$

$$\text{or,} \quad ln(1+y^2) - ln2 + \sin x = 0$$

$$\text{i,e} \quad y(x) = [2e^{-\sin x} - 1]^{\frac{1}{2}},$$

where we have chosen positive square root as $y(0) > 0$ is preassigned. Now as is obvious, $y(x)$ is defined only if $e^{-\sin x} \geq \frac{1}{2}$ i.e, if $x \leq \sin^{-1}(\ln 2)$. Infact solution exists in the open interval $\left(-\pi - \sin^{-1}(\ln 2), \sin^{-1}(\ln 2)\right)$ as $y(x) = 0$ for $x = \sin^{-1}(\ln 2)$ or $x = -\pi - \sin^{-1}(\ln 2)$. Existence of such a behavior was suspected because the normal form of given ode, viz, $\frac{dy}{dx} = \frac{-(1+y^2)}{2y} \cos x$ is not defined whenever $y = 0$. Therefore if a solution assumes value zero at some points $x = x_1$ and $x = x_2$, there is no question of spread of the solution outside the interval (x_1, x_2).

Example (12) : Solve the initial value problem :

$$\left(\frac{dy}{dx}\right)^2 + 1 = e^{2y} \quad \text{subject to the initial condition } y(0) = 0.$$

The ode might be recast as :

$$\frac{dy}{\sqrt{e^{2y} - 1}} = dx$$

$$\text{or,} \quad \frac{dz}{z\sqrt{z^2 - 1}} = dx \quad [\text{If we put } z = e^y].$$

Integrating between 0 and x, we have

$$\int_1^z \frac{du}{u\sqrt{u^2 - 1}} = \int_0^x dt$$

$$\therefore \quad y = \ln(\sec x)$$

Here the solution is valid only in the open interval $\left(-\frac{\pi}{2}, \frac{\pi}{2}\right)$ as $x = \pm\frac{\pi}{2}$ are the vertical asymptotes of the solution curve $y = \ln(\sec x.)$

Example (13) : Solve the IVP

$$\frac{dx}{dt} = k(a - x)(b - x), \quad x(0) = 0 \; ; \; a, b > 0$$

The equation can be rewritten in the separable-form :

$$\frac{dx}{(a-x)(b-x)} = kdt$$

or, $$\frac{dx}{\left[x - \frac{1}{2}(a+b)\right]^2 - \left\{\frac{1}{2}(a-b)\right\}^2} = kdt$$

Integrating this equation between 0 and t and noting the underlying initial condition we have

$$\int_0^x \frac{dy}{\left[y - \frac{1}{2}(a+b)\right]^2 - \left\{\frac{1}{2}(a-b)\right\}^2} = k\int_0^t ds = kt \qquad (*)$$

Subcase I : $a = b$

$(*)$ gives : $$\int_0^x \frac{dy}{\left[y - \frac{1}{2}(a+b)\right]^2} = kdt$$

or, $$x(t) = a - \frac{a}{1 + akt} = \frac{a^2 kt}{1 + akt}$$

As $x(0) = 0$ and $\left.\dfrac{dx}{dt}\right|_{t=0} > 0$ when $k > 0$, $x(t)$ should be positive

if $t > 0$. But $x(t) > 0$ when $t > -\dfrac{1}{ak}$. Hence the solution is valid for

$-\dfrac{1}{ak} < t < \infty$.

Subcase II : $a \neq b$

Now $(*)$ gives : $$\int_0^x \frac{dy}{\left[y - \frac{1}{2}(a+b)\right]^2 - \left[\frac{1}{2}(a-b)\right]^2} = kt$$

or, $$\frac{1}{a-b}\left[\ln\left|\frac{y-a}{y+b}\right|\right]_0^x = kt$$

or, $$\frac{1}{a-b}\ln\left|\frac{x-a}{x+b}\right| - \frac{1}{a-b}\ln\left(\frac{a}{b}\right) = kt$$

or, $$\ln\left|\frac{b(x-a)}{a(x+b)}\right| = (a-b)kt$$

$\therefore$ $$x(t) = \frac{ab\left(1 - e^{(a-b)kt}\right)}{\left(b - ae^{(a-b)kt}\right)}$$

As $x(t) \geq 0$, $1 - e^{(a-b)kt}$ and $b - ae^{(a-b)kt}$ will be of same sign— either both positive or both negative. In either case it follows that

$\frac{1}{k(b-a)} \ln \left(\frac{b}{a}\right) < t < \infty$. Hence the interval of existence of the solution to this IVP for $a \neq b$ is $\frac{1}{k(b-a)} \ln \left(\frac{b}{a}\right) < t < \infty$.

Remark : The given IVP reminds us of the 'Law of Mass Action' in Chemistry which states that at a given temperature the rate of chemical change at any instant is directly proportional to the active mass of each of the interacting substances. If A and B react to produce C and D in a simple reversible reaction, the rate r_{AB} of the forward reaction $A + B \to C + D$, according to the law of mass action, varies jointly with molar concentrations $[A]$ and $[B]$ of A and B respectively.

$\therefore$ $r_{AB} = k[A].[B]$, k being the proportionality constant for the forward reaction. Suppose we started at $t = 0$ with 'a' moles of substance A and 'b' moles of substance B. At time t, suppose x moles of A react with x moles of B to form C and D. Thus at time t, molar concentration of A is $(a - x)$, that of B is $(b - x)$ and the forward-reaction rate is $r_{AB} \equiv \frac{dx}{dt}$. Hence we get back

$$\frac{dx}{dt} = k(a - x)(b - x) \text{ with } x(0) = 0.$$

Example (14) : Solve the ode : $4xy\,dx + (x^2 + 1)dy = 0$

Dividing by $y(x^2 + 1)$ we get the newly recast form of ode :

$$\frac{4x}{x^2 + 1}dx + \frac{dy}{y} = 0,$$

which being separable is exact. Integrating we have

$2\ln(x^2 + 1) + ln \mid y \mid = ln \mid c \mid$, some arbitrary integration constant.

Thus the general solution in the separable form reads

$$(x^2 + 1)^2 \mid y \mid = \mid c \mid.$$

By the way, we divided the differential equation by $y(x^2 + 1)$ on tacit assumption that $y \neq 0$. However in the derivative form the ode reads :

$$\frac{dy}{dx} + \frac{4xy}{x^2 + 1},$$ showing that $y = 0$ is also a solution.

Looking back at the general solution, viz, $(x^2 + 1)^2 \mid y \mid = \mid c \mid$, we see that $c = 0$ corresponds to the solution $y = 0$ and hence $y = 0$ is not lost due to the division process. This ensures that

$$(x^2 + 1)^2 \mid y \mid = \mid c \mid$$

stands for the most general solution of the ode given.

Example (15) : Solve the ode :

$$(3x + 8)(y^2 - 1)dx - 4y(x^2 + 4x + 3)dy = 0$$

The equation is written as

$$\frac{3x + 8}{(x + 1)(x + 3)}dx + \frac{4y}{1 - y^2}dy = 0 \quad \text{provided } x \neq -1 \; ; \; x \neq -3 \; ; \; y^2 \neq 1.$$

Under this assumption, the equation is cast into a separable form and so the first integral reads :

$$
\begin{aligned}
F(x, y) &= \int \frac{(3x + 8)}{(x + 1)(x + 3)} \, dx + \int \frac{4y}{1 - y^2} \, dy \\
&= \frac{1}{2} \ln \mid x + 1 \mid + \frac{5}{2} \ln \mid x + 3 \mid - 2 \ln \mid y^2 - 1 \mid \\
&= \ln \mid x + 1 \mid^{\frac{1}{2}} + \ln \mid x + 3 \mid^{\frac{5}{2}} - \ln(y^2 - 1)^2
\end{aligned}
$$

$\therefore$ The general solution reads:

$$\frac{\mid (x + 1)(x + 3)^5 \mid}{(y^2 - 1)^4} = c, \quad c \text{ being arbitrary positive constant of integration.}$$

Observe that no real finite value of c exists that corresponds to the pair of solution $y = \pm 1$. Thus by way of casting the given equation into separable form, we have lost two solutions. Hence the fastidious reader should not call the above solution set as general unless and until it is appended by the two solutions $y = \pm 1$.

Remark: The solutions $y = \pm 1$ that could not be included in the general solution of the above problem are sometimes referred to as **singular solutions** to the ode as they could not be produced by any choice of c.

2.3   Homogeneous Differential Equations

A first order ode $M(x, y)dx + N(x, y)dy = 0$ is said to be **homogeneous** if $M(x, y)$ and $N(x, y)$ are homogeneous functions of the same degree.

Alternatively, we may call the ode to be homogeneous if the function $g(x,y) \equiv \frac{-M(x,\ y)}{N(x,\ y)}$ associated with the derivative form $\frac{dy}{dx} = g(x,y)$ is a homogeneous function of degree zero. A function $g(x,y)$ in x and y is said to be homogeneous of degree k iff $g(tx, ty) = t^k g(x,y)$ or equivalently $g(x,y) = x^k g\left(1, \frac{y}{x}\right)$ holds for all real $t > 0$

For example, $(2xy + 3y^2)dx + (2xy + x^2)dy = 0$ is a homogeneous first order ode because

$$M(x,y) \equiv 2xy + 3y^2 \quad \text{and} \quad N(x,y) \equiv (2xy + x^2)$$

are homogeneous functions of degree 2.

Theorem 2.2: If $M(x,y)dx + N(x,y)dy = 0$ is a homogeneous differential equation, then the change of variables $(x,y) \to (x,v)$ where $v = \frac{y}{x}$ transforms the equation $Mdx + Ndy = 0$ into a separable equation in the variables x and v.

Proof : Since $Mdx + Ndy = 0$ is homogeneous, we put it in the derivative form

$$\frac{dy}{dx} = \frac{-M(x,\ y)}{N(x,\ y)} \equiv g(x,y) \equiv h\left(\frac{y}{x}\right), \text{ say,}$$

Put $y = vx$ so that $\frac{dy}{dx} = v + x\frac{dv}{dx}$ and so the above ode becomes $x\frac{dv}{dx} = h(v) - v$, which is visibly of separable form.

Rewriting this new ode as $\frac{dv}{h(v)-v} = \frac{dx}{x}$ we have on integration,

$$\int \frac{dx}{x} + \int \frac{dv}{v - h(v)} = c, \ c \text{ being an arbitrary constant.}$$

Writing $\int \frac{dv}{v-h(v)} \equiv H(v)$, we get the general solution

$$H\left(\frac{y}{x}\right) + \ln x = c$$

Observe that once separability is ensured, exactness follows and first integral or general solution is found to be existent.

Note : This 'homogeneity' of ode is undoubtedly a kind of symmetry. The transformation $y = vx$ just enables us to highlight this inbuilt symmetry by virtue of separation of variables. From the geometric viewpoint, one may interpret that the homogeneity in x and y has been translated to parallelism in the projective plane $\mathbb{R}P^2$ of x and v. The situation reminds us of the separation of variables technique used more often in potential problems in electrostatics where potential V inherits a spherical/cylindrical symmetry. The same is the story for the derivation of time-independent Schrödinger equation if potential is independent of time and velocity. In two and three dimensional geometry such symmetry exploitation is done off and on whenever changeover from one co-ordinate system to the other is made. For instance, the equation to a circle having origin as centre and b as radius reads $x^2 + y^2 = b^2$ in cartesian system but translates into $r = b$ in polar co-ordinate system. The absence of vectorial angle θ in the equation of circle speaks out the underlying circular/rotational symmetry.

Example (16) : Solve the initial value problem :

$$(y + \sqrt{x^2 + y^2})dx - xdy = 0;\ y(1) = 0$$

Here comparing with the traditional ode $Mdx + Ndy = 0$ we get :

$$\left. \begin{array}{rcl} M(x,y) & = & y + \sqrt{x^2 + y^2} \\ N(x,y) & = & -x \end{array} \right\}$$

$$\left. \begin{array}{rcl} \therefore\ M(tx,ty) & = & ty + t\sqrt{x^2 + y^2} = tM(x,y) \\ N(tx,ty) & = & -tx = tN(x,y) \end{array} \right\}$$

Hence both $M(x,y)$ and $N(x,y)$ are homogeneous functions (of degree unity) in x and y. In the derivative form,

$$\begin{array}{rcl} \dfrac{dy}{dx} & = & \dfrac{y + \sqrt{x^2 + y^2}}{x} = \dfrac{y}{x} + \dfrac{\sqrt{x^2 + y^2}}{x} \\[4mm] & = & \dfrac{y}{x} \pm \sqrt{1 + \left(\dfrac{y}{x}\right)^2} \equiv g\left(\dfrac{y}{x}\right) \end{array}$$

(here $x = \pm\sqrt{x^2}$, depending on sign of x)

In this case the initial condition $y(1) = 0$ might be a clue to the selection of signs : + ve sign is to be accepted.

$$\therefore\ \dfrac{dy}{dx} = \dfrac{y}{x} + \sqrt{1 + \left(\dfrac{y}{x}\right)^2}$$

Put the substitution : $y = vx$ so that

$$v + x\frac{dv}{dx} = v + \sqrt{1 + v^2}$$

Integrating, $ln \mid v + \sqrt{1 + v^2} \mid = ln \mid x \mid + ln \mid c \mid$,

where $ln \mid c \mid$ is the arbitrary constant of integration.

$$\therefore \ v + \sqrt{1 + v^2} = \mid cx \mid$$

Removing modulus sign and substituting back $v = \frac{y}{x}$, we have

$$y + \sqrt{x^2 + y^2} = cx^2$$

Using the assigned initial conditions we have $c = 1$.

Hence the specific solution reads : $y + \sqrt{x^2 + y^2} = x^2$, which after removal of radicals and simplification becomes $y = \frac{1}{2}(x^2 - 1)$.

Example (17) : Solve the ode $(2xy + 3y^2)dx + (2xy + x^2)dy = 0$

It is a homogeneous first order ode that can be cast into the form

$$x\frac{dv}{dx} = -\frac{2v + 3v^2}{2v + 1} - v = \frac{-3v - 5v^2}{2v + 1}$$

or, $\quad \dfrac{(2v + 1)}{v(5v + 3)}dv + \dfrac{dx}{x} = 0$

or, $\quad \dfrac{1}{3}\dfrac{dv}{v} + \dfrac{1}{3}\dfrac{dv}{(5v + 3)} + \dfrac{dx}{x} = 0$

Integrating, $\quad \ln v + \ln \mid x \mid^3 + \dfrac{1}{5}ln \mid 5v + 3 \mid = $ constant

or, $\quad \ln \mid v \mid^5 + \ln \mid x \mid^{15} + \ln \mid 5v + 3 \mid = \ln \mid c \mid; \ c \neq 0$

or, $\quad \mid v \mid^5 \mid x \mid^{15} \mid 5v + 3 \mid = c$

or, $\quad y^5 x^9 (5y + 3x) = c \ (c \neq 0)$

which is the general solution.

Remark: Observe that in solving the above ode we presumed $3x + 5y \neq 0$ as otherwise, the entire process fails to deliver. However, it is easy to verify that $3x + 5y = 0$ satisfies the ode and this solution have been left out in course of the method adopted. Thus one cannot claim the family $y^5 x^9 (3x + 5y) = c$ $(c \neq 0)$ as the general solution of the ode unless we append $3x + 5y = 0$ to it. This truth will again be explored in discussion of integrating factors coming up in the next article. [See also Theorem 2.4].

To solve the differential equations of the form $\frac{dy}{dx} = f\left(\frac{a_1 x + b_1 y + c_1}{a_2 x + b_2 y + c_2}\right)$ we apply the linear transformations, $X = x - x_0$ and $Y = y - y_0$ where (x_0, y_0) is the point of intersection of the straight lines $a_1 x + b_1 y + c_1 = 0$ and $a_2 x + b_2 y + c_2 = 0$. However this intersection does uniquely occur if the lines are not parallel i.e, if $(a_1 b_2 - a_2 b_1) \neq 0$. Under the assumption of non-parallelism, we have

$$x_0 = \frac{b_1 c_2 - b_2 c_1}{a_1 b_2 - a_2 b_1} \text{ and } y_0 = \frac{c_1 a_2 - c_2 a_1}{a_1 b_2 - a_2 b_1} .$$

Obviously the given equation transforms as the homogeneous one :

$$\frac{dY}{dX} = f\left(\frac{a_1 X + b_1 Y}{a_2 X + b_2 Y}\right) \equiv g\left(\frac{Y}{X}\right)$$

In case $a_1 x + b_1 y + c_1 = 0$ and $a_2 x + b_2 y + c_2 = 0$ are parallel, then we write $(a_1 x + b_1 y)$ as v and the given ode turns out to be separable in x and v. The following illustrations will clarify this point.

Remark: In classical mechanics the holonomic constraints are those which are or can be made independent of velocities. The holonomic constraints of the latter type appears in the form of exact odes. For example, if a particle moves in a plane so that its position co-ordinates satisfy

$$(ax + hy + g)x + (hx + by + f)y = 0, \left(x \equiv \frac{dx}{dt}; \ y = \frac{dy}{dt}\right),$$

then it represents a holonomic constraint as the differential equation

$$(ax + hy + g)dx + (hx + by + f)dy = 0$$

is exact. Indeed its complete integral is also a holonomic constraint.

Example (18) : Solve the ode $(x + 2y - 3)dy = (2x - y + 1)dx$

In the derivative form we have

$$\frac{dy}{dx} = \frac{2x - y + 1}{x + 2y - 3}.$$

We observe that $\begin{vmatrix} 2 & -1 \\ 1 & 2 \end{vmatrix} = 5 \neq 0$ and so the two straight lines $2x - y + 1 = 0$ and $x + 2y - 3 = 0$ are intersecting–the point of intersection (x_0, y_0) being $\left(\frac{1}{5}, \frac{7}{5}\right)$.

We carry on the translational transformation $X = x - \frac{1}{5}$ and $Y = y - \frac{7}{5}$ so that the above ode reduces to the following homogeneous form in X and Y :

$$\frac{dY}{dX} = \frac{2X - Y}{X + 2Y}. \qquad (*)$$

The solution method is now a routine exercise–the final solution is:

$$F(x, y) \equiv y^2 + xy - x^2 - 3y - x = c, \ c \text{ being an arbitrary constant.}$$

Example (19) : Solve $(2x + 4y + 3)dy = (x + 2y + 1)dx$

In the derivative form the given ode can be written as :

$$\frac{dy}{dx} = \frac{(x + 2y) + 1}{2(x + 2y) + 3},$$

which shows that lines $x + 2y + 1 = 0$ and $2x + 4y + 3 = 0$ are parallel. Substituting $v = x + 2y$ the above ode reduces as

$$\frac{dv}{dx} = \frac{4v + 5}{2v + 3},$$

which is trivially solved to have the first integral

$$x - 2y \ = \ \ln | 4x + 8y + 5 | + c, \ c \text{ being constant of integration.}$$

Theorem 2.3 If $Mdx + Ndy = 0$ be exact and moreover $M(x, y)$ and $N(x, y)$ are homogeneous functions of the same degree, then $(Mx + Ny) = $ constant is the general solution.

Proof : The necessary and sufficient condition for exactness is known to be $\frac{\partial M}{\partial y} = \frac{\partial N}{\partial x}$. The general solution is given by $F(x, y) = C$, where

$$F(x, y) = \int M \partial x + \int \left[N - \frac{\partial}{\partial y} \int M \partial x \right] dy.$$

Since $M(x, y)$ and $N(x, y)$ are both homogeneous functions of degree k, say, we have by Euler's theorem,

$$x \frac{\partial M}{\partial x} + \frac{\partial M}{\partial y} = kM$$

$$x \frac{\partial N}{\partial x} + y \frac{\partial N}{\partial y} = kN$$

Case I. $k = -1$.

In this case the statement of the Euler's theorem reduces to

$$\left. \begin{array}{rcl} \frac{\partial}{\partial x}(Mx) + y \frac{\partial M}{\partial y} & = & 0 \\ x \frac{\partial N}{\partial x} + \frac{\partial}{\partial y}(Ny) & = & 0 \end{array} \right\}$$

Using condition of exactness, we have its equivalent result

$$\text{i.e.,} \qquad \left. \begin{array}{l} \frac{\partial}{\partial x}(Mx + Ny) = 0 \\ \frac{\partial}{\partial y}(Mx + Ny) = 0 \end{array} \right\}$$

This gives us the general solution $Mx + Ny = $ constant.

Case II. $k \neq -1$

Here solution of the given exact equation reads $F(x, y) = c$, where $F(x, y)$ is given above. However, due to exactness,

$$\int M \partial x = \frac{1}{k} \int kM \partial x = \frac{1}{k} \int \left(x \frac{\partial M}{\partial x} + y \frac{\partial M}{\partial y} \right) \partial x$$

$$= \frac{1}{k} \int \frac{\partial}{\partial x}(Mx + Ny)\partial x - \frac{1}{k} \int M \partial x$$

Hence $$\int M \partial x = \frac{1}{(k+1)}(Mx + Ny)$$

$$\therefore \quad \frac{\partial}{\partial y} \int M \partial x = \frac{1}{(k+1)} \frac{\partial}{\partial y}(Mx + Ny)$$

$$= \frac{1}{(k+1)} \left[\frac{\partial M}{\partial y} x + y \frac{\partial N}{\partial y} + N \right] = N$$

$$\therefore \quad F(x, y) \equiv \frac{1}{(k+1)}(Mx + Ny) = c, \text{ a constant}$$

This ensures that $(Mx + Ny) = $ constant is our general solution.

Remark (a) : In the above example we had, after some workout a first order exact and homogeneous ode in X and Y, viz,

$$(Y - 2X)dX + (X + 2Y)dY = 0 \qquad [c.f(*)]$$

Hence its general solution can be written by using the above result as

$$Y^2 - X^2 + XY = \text{constant},$$

which on being translated back to x-y co-ordinates, reduces to

$$y^2 + xy - x^2 - 3y - x = c, \ (c \text{ being some constant})$$

Remark (b) : Homogeneous differential equations have the general form $\frac{dy}{dx} = g\left(\frac{y}{x}\right)$ where g is a given function of a single variable $\frac{y}{x}$. To make this precise, we need to worry about the domains of definition of the functions involved. This motivates the need for the following :

Definition 2.2 : A subset S of $\mathbb{R}^2$ is called **starshaped** if $(tx, ty) \in S$ for every $(x, y) \in S$ and every $t \in \mathbb{R}$, with $t \neq 0$. Geometrically this means that if a point $(x, y) \in S$, then the entire line through the points $O(0,0)$ and $Q(x,y)$ lies in S (however $(0,0)$ need not be in S).

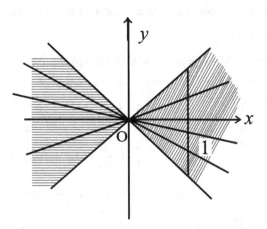

Fig 2.3 : Starshaped region in xy-pane

The differential equation $\frac{dy}{dx} = \frac{-M(x, y)}{N(x, y)}$ is called homogeneous iff the domain S of $\frac{-M(x, y)}{N(x, y)}$ is 'starshaped' and $\frac{-M(tx,ty)}{N(tx,ty)} = \frac{-M(x, y)}{N(x, y)}$ for every $(x,y) \in S$ and every $t \neq 0$ in $\mathbb{R}$.

The homogeneity condition means that the function $\frac{-M(x,\,y)}{N(x,\,y)}$ has constant value along any particular line (ray) in $\mathbb{R}^2$ through the origin (which is in S). Indeed 'homogeneity' implies invariance of the equation under 'same-rescaling' of the dependent and independent variables.

Theorem 2.4: Every homogeneous differential equation $\frac{dy}{dx} = f(x,y)$ has the property that the family of integral curves is invariant under the stretching transformation $x \to tx$, $y \to ty$. More explicitly, it states that under the transformation every integral curve C is transformed to another integral curve C_1, t being non-zero real. Conversely, every differential equation whose family of integral curves has the geometric property of invariance under stretching, is homogeneous.

Proof : Necessary part : Since $\frac{dy}{dx} = f(x,y)$ is a homogeneous ode, we can write $f(x,y) = g\left(\frac{y}{x}\right) = g(z)$, where $z = \frac{y}{x}$.

Hence $\qquad y = xz \quad$ and so $\quad \dfrac{dy}{dx} = z + x\dfrac{dz}{dx} = g(z).$

$$\therefore \quad \frac{dz}{g(z) - z} = \frac{dx}{x}$$

Integrating, $x = A\exp\left[\int \frac{dz}{g(z)-z}\right]$, A being an arbitrary constant. Observe that under the above transformation, $x \to tx$ but z remains unaltered so that the above integral curve (A pressigned) becomes

$$x = \frac{A}{t}\exp\left[\int \frac{dz}{g(z)-z}\right] \equiv A'\exp\left[\int \frac{dz}{g(z)-z}\right]$$

Sufficient part : Without loss of generality, the family of integral curves can be taken as either $y = Bh(z)$ or $x = Ag(z)$, where $g(z)$ and $h(z)$ are functions of $z = y/x$. We prove our proposition for the first case. On differentiation, we have :

$$\frac{dy}{dx} = B\frac{d}{dx}h(z) = Bh'(z)\frac{dz}{dx} = Bh'(z).\left(\frac{1}{x}\frac{dy}{dx} - \frac{y}{x^2}\right)$$

$$\therefore \quad \frac{dy}{dx} = \frac{Bh'(z)y}{x(Bh'(z) - x)} = \frac{z^2 h'(z)}{(zh'(z) - h(z))} \equiv H(z) = H(y/x)$$

Hence the ode is a homogeneous one.

Now for second case, $x = Ag(z)$, so that on differentiation we have:

$$\frac{dx}{dy} = Ag'(z)\frac{dz}{dy} \quad \Rightarrow \quad \frac{dz}{dx} = \frac{1}{Ag'(z)}$$

$$\therefore \quad \left(\frac{1}{x}\cdot\frac{dy}{dx} - \frac{y}{x^2}\right) = \frac{1}{Ag'(z)}$$

$$\text{i, e,} \quad \frac{dy}{dx} = \frac{x}{Ag'(z)} + z = \frac{g(z)}{g'(z)} + z \equiv G(z) = G(y/x)$$

Hence in this case also, the ode is a homogeneous one.

2.4   Integrating Factor

Let's go back to normal derivative form $\frac{dy}{dx} = f(x,y)$ of the first order ode, f being a real-valued C^1 function of two variables (x and y) defined on an open set $U \subseteq \mathbb{R}^2$ or its equivalent differential form $M(x,y)dx + N(x,y)dy = 0$. A real-valued function defined on an open, dense set $U_1 \subseteq U$ is called an integrating factor for the differential equation $\frac{dy}{dx} = f(x,y)$ if $\mu(x,y)$ is never zero on U_1 and the first order ode $-\mu(x,y)f(x,y) + \mu(x,y)\frac{dy}{dx} = 0$ is exact.

Equivalently, the differential equation $M(x,y)dx + N(x,y)dy = 0$ has an integrating factor $\mu(x,y)$ iff

$$\mu(x,y)M(x,y) + \mu(x,y)N(x,y) \text{ is a perfect differential } dF(x,y) .$$

In brief, existence of an integrating factor to a first order ode is a necessary and sufficient condition for its integrability.

To find the condition that $\mu(x,y)$ is an integrating factor of $Mdx + Ndy = 0$ we should have from the condition for exactness,

$$\frac{\partial}{\partial y}(\mu(x,y)M(x,y)) = \frac{\partial}{\partial x}(\mu(x,y)N(x,y))$$

$$\text{or,} \quad \frac{\partial \mu}{\partial y}M + \mu\frac{\partial M}{\partial y} = \frac{\partial \mu}{\partial x}N + \mu\frac{\partial N}{\partial x}$$

$$\text{or,} \quad \mu\left(\frac{\partial M}{\partial y} - \frac{\partial N}{\partial x}\right) = \left(\frac{\partial \mu}{\partial x}N - \frac{\partial \mu}{\partial y}M\right) \tag{2.9}$$

Equation (2.9) being a first order linear PDE for $\mu(x,y)$, its solution can be determined by solving Lagrange's auxiliary equations

$$\frac{dx}{M} = \frac{dy}{-N} = \frac{d\mu}{\left(\frac{\partial M}{\partial y} - \frac{\partial N}{\partial x}\right)}$$

In order to solve the first equality, we have to find a particular solution of the ode $Mdx + Ndy = 0$. However, this is the original first order ode we set out to solve!! This visualises the non solvability of integrating factor in general.

So we can find integrating factors of $Mdx + Ndy = 0$ under special circumstances. There are as many as two possible ways of getting into the act, viz

(i) given a condition satisfied by M and N or given a particular form of the ode, we are to find the integrating factors.

(ii) Demanded the existence of an integrating factor of a particular form, to explore what underlying restrictions need be there on $M(x,y)$ and $N(x,y)$.

Although the first approach is very common in the literature, it has the drawback of very limited range of use as always we will have to be in lookout of the criterion satisfied or not so as to have a particular expression for $\mu(x,y)$. On the otherhand, in the second approach we table our demand of having μ in a desired form and accordingly have our criterion to be satisfied by M and N.

At this stage we table our demand :

Case (I) : μ is a function of x alone.

Hence $\frac{\partial \mu}{\partial y} = 0$ and $\frac{\partial \mu}{\partial x} \equiv \frac{d\mu}{dx}$. (2.9) therefore reduces as

$$\mu = \frac{N \cdot \frac{d\mu}{dx}}{\left(\frac{\partial M}{\partial y} - \frac{\partial N}{\partial x}\right)} \Rightarrow \frac{d}{dx}\left(\ln | \mu |\right) = \frac{\frac{\partial M}{\partial y} - \frac{\partial N}{\partial x}}{N},$$

so that for consistency, the R.H.S should be a function of x alone. Thus if an integrating factor of x alone is demanded, $\frac{\frac{\partial M}{\partial y} - \frac{\partial N}{\partial x}}{N}$ will have to be a function of x alone.

Case (II) : μ is a function of y alone.

Hence $\frac{\partial \mu}{\partial x} = 0$ and $\frac{\partial \mu}{\partial y} \equiv \frac{d\mu}{dy}$. (2.9) now reduces as :

$$\mu = \frac{-M \cdot \frac{d\mu}{dy}}{\left(\frac{\partial M}{\partial y} - \frac{\partial N}{\partial x} \right)} \Rightarrow \frac{d}{dy} \left(\ln \mid \mu \mid \right) = \frac{\frac{\partial N}{\partial x} - \frac{\partial M}{\partial y}}{M},$$

so that for consistency, the R.H.S should be a function of y alone. Thus if an integrating factor of y alone is wanted, $\frac{\frac{\partial N}{\partial x} - \frac{\partial M}{\partial y}}{M}$ will have to be a function of y alone.

Case (III) : μ is function of z alone, where $z = \sqrt{x^2 + y^2}$

Here $\qquad \frac{\partial \mu}{\partial x} = \frac{d\mu}{dz}.2x \quad$ and $\quad \frac{\partial \mu}{\partial y} = \frac{du}{dz}.2y$

$\therefore$ (2.9) reduces as :

$$\mu = 2\frac{d\mu}{dz} \cdot \frac{(Nx - My)}{\left(\frac{\partial M}{\partial y} - \frac{\partial N}{\partial x} \right)} \Rightarrow \frac{d}{dz} \left(\ln \mid \mu \mid \right) = \frac{1}{2} \left(\frac{\frac{\partial M}{\partial y} - \frac{\partial N}{\partial x}}{Nx - My} \right),$$

Again for sake of consistency, R.H.S is to be a function of $\sqrt{x^2 + y^2}$ alone. Thus if μ is demanded to be a function of z alone, $\frac{1}{2} \left(\frac{\frac{\partial M}{\partial y} - \frac{\partial N}{\partial x}}{Nx - My} \right)$ should be a function of z only.

Inference : The approach is seen to be flexible since the conditions needed for having the integrating factor in a special form are derivable just by consistency of relation (2.9).Different demands in the form of μ will incorporate different sufficient criteria to be satisfied by M and N.

Example (20) : Find an integrating factor of the ode $Mdx + Ndy = 0$, where M and N are separable functions of x and y, i.e.,

$$M(x, y) = f_1(x)g_1(y) \text{ and } N(x, y) = f_2(x)g_2(y).$$

In case $\mu(x, y)$ is an integrating factor, the recast ode

$$\mu \left(f_1(x)g_1(y) \right) dx + \mu \left(f_2(x)g_2(y) \right) dy = 0$$

is exact. Because of the separability, our demand will be to make $\mu f_1(x)g_1(y)$ a function of x alone, say $A(x)$, while $\mu f_2(x)g_2(y)$ a function of y alone, say $B(y)$.

$$\therefore \ \mu f_1(x)g_1(y) = A(x), \ \text{implying} \ \mu g_1(y) = \frac{A(x)}{f_1(x)}$$

Further, $\mu f_2(x)g_2(y) = B(y)$, so that $\mu f_2(x) = \dfrac{B(y)}{g_2(y)}$

$$\therefore \ \frac{A(x)}{f_1(x)} \cdot \frac{f_2(x)}{g_1(y)} = \frac{B(y)}{g_2(y)} \Leftrightarrow \frac{A(x)f_2(x)}{f_1(x)} = \frac{B(y)g_1(y)}{g_2(y)} = \text{some constant } k$$

because on one side function only of x and on the other side, function only of y appear in this relation.

Hence $\mu = \frac{k}{f_2(x)g_1(y)}$, which is the desired integrating factor to make the given ode exact. This envisages that thumb rule of division which we used for separation of variables is effectively an introduction of required integrating factor.

Remark : Here $\exists$ a tacit assumption that $f_2(x)g_1(y) \neq 0$. In case there exists some $y_0 \in R$ such that $g_1(y_0) = 0$, then in the derivative form, we have :

$$-\frac{dy}{dx}\Big|_{y=y_0} f_2(x)g_2(y_0) = f_1(x)g_1(y) = 0$$

Therefore $\frac{dy}{dx}\Big|_{y=y_0}$ i.e, $y = y_0$ is a solution to the given ode. This solution $y = y_0$ may or may not be a member of the general solution obtained for the separable resultant ode.

Example (21) : If $Mx - Ny \neq 0$ and the equation $Mdx + Ndy = 0$ can be written as $f_1(xy).ydx + f_2(xy).xdy = 0$, then $\frac{1}{Mx-Ny}$ is an integrating factor of the equation.

Method I : Here $M(x,y) = f_1(xy)y$ and $N(x,y) = f_2(xy)x$. Put $z = xy$. Hence $dz = xdy + ydx$. We now rewrite the given ode as

$$f_1(z).ydx + f_2(z).xdy = 0$$
$$\Leftrightarrow f_1(z).\frac{z}{x}dx + f_2(z).\left(dz - \frac{z}{x}dx\right) = 0$$
$$\Leftrightarrow (f_1(z) - f_2(z)).\frac{z}{x}dx + f_2(z)dz = 0$$

Trivially, by using result of example (20), we see that the above ode (in variables x and z) has an integrating factor $\mu = \frac{1}{(f_1(z)-f_2(z))z}$.

$$\therefore \ \mu(x,y) = \frac{1}{\{xy(f_1(xy) - f_2(xy))\}} = \frac{1}{(Mx - Ny)},$$

Method II : Checking if $\frac{M}{Mx-Ny}dx + \frac{N}{Mx-Ny}dy = 0$ is exact or not.

$$\frac{\partial}{\partial y}\left[\frac{f_1(xy)y}{xy\{f_1(xy)-f_2(xy)\}}\right] = \frac{1}{x}\frac{\partial}{\partial y}\left[\frac{f_1(xy).xy}{xy\{f_1(xy)-f_2(xy)\}}\right]$$

$$= \frac{d}{dz}\left[\frac{f_1(z).z}{z\{f_1(z)-f_2(z)\}}\right] \quad (z=xy)$$

$$= \frac{(f_1'(z)f_2(z)-f_1(z)f_2'(z))}{(f_1(z)-f_2(z))^2}$$

In the same way,

$$\frac{\partial}{\partial x}\left[\frac{f_2(xy).x}{xy\{f_1(xy)-f_2(xy)\}}\right] = \frac{d}{dz}\left[\frac{f_2(z)}{f_1(z)-f_2(z)}\right] \quad (z=xy)$$

$$= \frac{f_1'(z)f_2(z)-f_1(z)f_2'(z)}{(f_1(z)-f_2(z))^2}$$

This ensures exactness as expected.

Remark: Method II adopted above is very weak as it gives no clue to finding an integrating factor. A more powerful and elegant method of establishing that $\frac{1}{Mx-Ny}$ is an integrating factor for the given type of ode follows from the simple observation that the ode is invariant under the transformation $x \to e^t x$; $y \to e^{-t}y$ $\forall t \in \mathbb{R}$. However, the proof of this owes its origin to Lie group theory that is beyond scope.

Example (22) : If $Mx + Ny \neq 0$ and the ode $Mdx + Ndy = 0$ be homogeneous, then $\frac{1}{Mx+Ny}$ is an integrating factor.

Since $Mdx + Ndy = 0$ is homogeneous, M and N are homogeneous functions of the same degree, say k. Thus we have

$$\left.\begin{array}{l} M(x,y) = x^k\overline{M}\left(\frac{y}{x}\right) \\ N(x,y) = x^k\overline{N}\left(\frac{y}{x}\right) \end{array}\right\}, \text{ where } \overline{M} \text{ and } \overline{N} \text{ are some new functions of } \frac{y}{x}.$$

Written $z = \frac{y}{x}$ so that the given equation reduces to

$$x^k\overline{M}(z)dx + x^k\overline{N}(z)(zdx + xdz) = 0$$

or, $\quad x^k(\overline{M}(z) + z\overline{N}(z))dx + x^{k+1}\overline{N}(z)dz = 0,$

To make this ode separable, we have to use the integrating factor

$$\mu = \frac{1}{x^{k+1}(\overline{M}(z) + z\overline{N}(z))}$$

$$= \frac{1}{x(\overline{M}(z)x^k + z\overline{N}(z)x^k)}$$

$$= \frac{1}{x(M + zN)} = \frac{1}{Mx + Ny}$$

Example (23) : Find a sufficient condition for $\frac{1}{Mx+Ny}$ to be an integrating factor of the first order ode $Mdx + Ndy = 0$.

Since $\frac{1}{Mx+Ny}$ is desirably an integrating factor, the criterion

$$\frac{\partial}{\partial y}\left(\frac{M}{Mx + Ny}\right) = \frac{\partial}{\partial x}\left(\frac{N}{Mx + Ny}\right)$$

is to be fulfilled.

This equality holds provided

$$\frac{\frac{\partial M}{\partial y}(Mx + Ny)}{(Mx + Ny)^2} - \frac{M}{(Mx + Ny)^2}\left[\frac{\partial M}{\partial y}x + \frac{\partial N}{\partial y}y + N\right]$$

$$= \frac{\frac{\partial N}{\partial x}(Mx + Ny)}{(Mx + Ny)^2} - \frac{N}{(Mx + Ny)^2}\left[\frac{\partial M}{\partial x}x + \frac{\partial N}{\partial x}y + M\right]$$

or, $\quad M.\dfrac{\partial M}{\partial y}.x + Ny\dfrac{\partial M}{\partial y} - Mx.\dfrac{\partial M}{\partial y} - My\dfrac{\partial N}{\partial y} - MN$

$$= Mx\frac{\partial N}{\partial x} + Ny\frac{\partial N}{\partial x} - MN - N\frac{\partial M}{\partial x}.x - Ny\frac{\partial N}{\partial x}$$

or, $\quad Ny\dfrac{\partial M}{\partial y} - My\dfrac{\partial N}{\partial y} = Mx.\dfrac{\partial N}{\partial x} - Nx\dfrac{\partial M}{\partial x}$

i,e $\quad \dfrac{1}{M}\left(x\dfrac{\partial M}{\partial x} + y\dfrac{\partial M}{\partial y}\right) = \dfrac{1}{N}\left(x\dfrac{\partial N}{\partial x} + y\dfrac{\partial N}{\partial y}\right).$

At this juncture, Euler's theorem for homogeneous functions gives us the clue. If M and N were both homogeneous functions (of x and y) of the same degree, then obviously this criterion is satisfied.

Hence homogeneity of M and N is a sufficient condition for having $\frac{1}{Mx+Ny}$ as an integrating factor.

Example (24) : Find an integrating factor of $x^\alpha y^\beta(mydx + nxdy) = 0$. Here each of $M(x,y) = m.x^\alpha.y^{\beta+1}$ and $N(x,y) = n.x^{\alpha+1}.y^\beta$ is visibly seen to be in the product form of two functions, one of x and other of y. The result of example (20) enables us to conclude that the integrating factor of this separable ode is of the form $\mu_1(x)\mu_2(y)$.

$$\therefore \quad \frac{\partial}{\partial y}\left(m\mu_1(x)\mu_2(y)x^\alpha.y^{\beta+1}\right) = \frac{\partial}{\partial x}\left(n\mu_1(x)\mu_2(y).x^{\alpha+1}.y^\beta\right)$$

or, $$x^\alpha m\mu_1(x)\left[\frac{d\mu_2}{dy}.y^{\beta+1} + (\beta+1)\mu_2(y)y^\beta\right]$$

$$= ny^\beta.\mu_2(y)\left[\frac{d\mu_1}{dx}.x^{\alpha+1} + (\alpha+1)\mu_1(x)x^\alpha\right]$$

or, $$\frac{1}{n}\left[\frac{d}{dy}(\ln \mid \mu_2(y) \mid)y + (\beta+1)\right] = \frac{1}{m}\left[\frac{d}{dx}(\ln \mid \mu_1(x) \mid)x + (\alpha+1)\right]$$

Observe that L.H.S is a function of y alone while R.H.S is a function of x alone. Treating x and y both on the same footing, we consider each equal to same constant, say k.

$$\therefore \quad \frac{d}{dy}(\ln \mid \mu_2(y) \mid).y = kn - \beta - 1 \quad \text{so that} \quad \mu_2(y) = y^{kn-\beta-1}$$

and $$\frac{d}{dx}(ln \mid \mu_1(x) \mid).x = (km - \alpha - 1) \quad \text{so that} \quad \mu_1(x) = x^{km-\alpha-1}$$

$$\therefore \quad \mu(x,y) = \mu_1(x)\mu_2(y) = x^{km-\alpha-1}y^{kn-\beta-1} \quad \text{is the required I.F.}$$

Example (25) : If the first order ode is expressible as

$$x^{\alpha_1}y^{\beta_1}(m_1ydx + n_1xdy) + x^{\alpha_2}y^{\beta_2}(m_2ydx + n_2xdy) = 0,$$

find an integrating factor.

Using the outcome of example (26) we see that the first part, viz, $x^{\alpha_1}y^{\beta_1}(m_1ydx+n_1xdy)$ has an integrating factor $x^{k_1m_1-\alpha_1-1}.y^{k_1n_1-\beta_1-1}$ while that of the 2nd part, viz, $x^{\alpha_2}y^{\beta_2}(m_2ydx + n_2xdy)$ is given by $x^{k_2m_2-\alpha_2-1}.y^{k_2n_2-\beta_2-1}$. In this case k_1 and k_2 are constants not yet determined. To make the two integrating factors consistent and compatible, we must restrict k_1 and k_2 to satisfy the linear equations

$$\left.\begin{array}{l} k_1m_1 - \alpha_1 - 1 = k_2m_2 - \alpha_2 - 1 \\ k_1n_1 - \beta_1 - 1 = k_2n_2 - \beta_2 - 1 \end{array}\right\},$$

Once this system is solved, we freely say that the integrating factor of either part serves for the integrating factor of the original ode.

Finally we try to answer the question of non-uniqueness of the integrating factors of a first order ode. In fact if μ be an integrating factor of $Mdx + Ndy = 0$ and $u(x,y) = $ constant be the implicit form of the general solution or primitive then any differentiable function $f(u)$ of u, when multiplied by μ, stands as an integrating factor. The proof can be furnished as follows :

As $u(x,y) = $ constant is the primitive of $Mdx + Ndy = 0$, $\frac{\partial u}{\partial x}dx + \frac{\partial u}{\partial y}dy = 0$ wherefrom it follows that $M\frac{\partial u}{\partial y} = N\frac{\partial u}{\partial x}$. Since μ is an integrating factor, $(\mu M)dx + (\mu N)dy = 0$ is exact and so $\frac{\partial}{\partial y}(\mu M) = \frac{\partial}{\partial x}(\mu N)$ holds good. We will show that $\mu f(u)$ is also an integrating factor, and hence $\frac{\partial}{\partial y}(\mu f(u).M) = \frac{\partial}{\partial x}(\mu f(u).N)$ is to be established.

$$
\begin{aligned}
\therefore \frac{\partial}{\partial y}(\mu f(u)M) &= \frac{\partial f(u)}{\partial y}.(\mu M) + f(u).\frac{\partial}{\partial y}(\mu M) \\
&= \frac{\partial u}{\partial y}f'(u).\mu M + f(u).\frac{\partial}{\partial x}(\mu N) \\
&= \mu f'(u).N.\frac{\partial u}{\partial x} + f(u)\frac{\partial}{\partial x}(\mu N) \quad \left[\because M\frac{\partial u}{\partial y} = N\frac{\partial u}{\partial x}\right] \\
&= \left(f'(u)\frac{\partial u}{\partial x}\right).(\mu N) + f(u)\frac{\partial}{\partial x}(\mu N) \\
&= \frac{\partial}{\partial x}(f(u)).(\mu N) + f(u)\frac{\partial}{\partial x}(\mu N) \\
&= \frac{\partial}{\partial x}(\mu f(u).N)
\end{aligned}
$$

Remark: This non-uniqueness of the integrating factor is very important from not only mathematical but also from the physical viewpoint.

Example (26) : Solve the first order ode

$$(y^3 - 2x^2y)dx + (2xy^2 - x^3)dy = 0$$

The differential equation can be rewritten by regrouping the terms:

$$y^2(ydx + 2xdy) - x^2(2ydx + xdy) = 0$$

This form resembles the one discussed in the previous example provided we identify

$$
\left.
\begin{array}{l}
\alpha_1 = 0; \quad \beta_1 = 2; \quad m_1 = 1 \quad \text{and} \quad n_1 = 2 \\
\alpha_2 = 2; \quad \beta_2 = 0; \quad m_2 = -2 \quad \text{and} \quad n_2 = -1
\end{array}
\right\},
$$

The general integrating factor for the first part, viz, $y^2(ydx + 2xdy)$ will be $x^{k_1-1}.y^{2k_1-3}$, k_1 being any real number.

For consistency of these two factors, we must restrict k_1 and k_2 to satisfy the equations $k_1-1 = -2k_2-3$ and $2k_1-3 = -k_2-1$, wherefrom it follows that $k_1 = 2$ and $k_2 = -2$

Therefore the integrating factor will be xy for the given ode. Multiplying the given ode by xy we have the resultant exact ode :

$$xy^2(y^2 - 2x^2)dx + x^2y(2y^2 - x^2)dy = 0$$

$\therefore$ The first integral will be $F(x,y) = c$, where

$$F(x,y) = \int xy^2(y^2 - 2x)\partial x = \frac{1}{2}x^2y^4 - \frac{2}{3}x^3y^2$$

(Needless to say that partial integration of $M(x,y)$ of the resultant exact equation suffices for determining the first integral as $N(x,y)$ of that equation involves no x-free term).

Example (27) : Solve the ode : $y(xy+2x^2y^2)dx+x(xy-x^2y^2)dy = 0$

Clearly the form of the ode resembles that in example (21). We identify $f_1(xy) = xy + 2x^2y^2$ and $f_2(xy) = xy - 2x^2y^2$;

Thus vide example (21), $\frac{1}{xy(f_1(xy)-f_2(xy))} = \frac{1}{3x^3y^3}$ is an I.F.

Alternatively, one can rewrite the given ode as

$$xy\,d(xy) + x^2y^2(2ydx - xdy) = 0,$$

where the first term being exact, the integrating factor should be a function of xy only. By inspection we observe that on multiplying by $\frac{1}{x^3y^3}$, the given ode becomes separable and hence exact. So $\frac{1}{x^3y^3}$ is an I.F. Thus both the approaches lead us essentially to one and the same integrating factor.

Multiplying given ode by this I.F, the resultant equation becomes

$$\left(\frac{1}{x^2y} + \frac{2}{x}\right)dx + \left(\frac{1}{xy^2} - \frac{1}{y}\right)dy = 0,$$

which on integration gives : $\ln\left(\frac{x^2}{y}\right) - \frac{1}{xy} = c,$ a constant.

Example (28) : Find two integrating factors of the ode and solve :

$$\left(x\cos\frac{y}{x}+y\sin\frac{y}{x}\right)ydx+\left(x\cos\frac{y}{x}-y\sin\frac{y}{x}\right)xdy=0$$

$$
\begin{aligned}
M(x,y) &\equiv \left(x\cos\tfrac{y}{x}+y\sin\tfrac{y}{x}\right)y\\
N(x,y) &\equiv \left(x\cos\tfrac{y}{x}-y\sin\tfrac{y}{x}\right)x
\end{aligned}
$$

For any $t>0$, we observe that

$$
\begin{aligned}
M(tx,ty) &= t^2\left(x\cos\tfrac{y}{x}+y\sin\tfrac{y}{x}\right)y &= t^2M(x,y)\\
N(tx,ty) &= t^2\left(x\cos\tfrac{y}{x}+y\sin\tfrac{y}{x}\right)x &= t^2N(x,y)
\end{aligned}
$$

ensuring that M and N both are homogeneous functions of degree 2 in x and y. Hence $\frac{1}{Mx+Ny}$ is an integrating factor of the ode.

As
$$
\begin{aligned}
Mx+Ny &= xy\left(x\cos\frac{y}{x}+y\sin\frac{y}{x}\right)-xy\left(x\cos\frac{y}{x}-y\sin\frac{y}{x}\right)\\
&= 2x^2y\cos\frac{y}{x},
\end{aligned}
$$

$$\therefore\quad \frac{1}{Mx+Ny}=\frac{1}{2x^2y\cos\left(\frac{y}{x}\right)}\quad\text{is an I.F.}$$

Alternatively, $\dfrac{\partial M}{\partial y}= x\cos\dfrac{y}{x}+y\sin\dfrac{y}{x}+\dfrac{y^2}{x}\cos\dfrac{y}{x}$

$$\frac{\partial N}{\partial x}=2x\cos\frac{y}{x}+\frac{y^2}{x}\cos\frac{y}{x}$$

$$\frac{\frac{\partial M}{\partial y}-\frac{\partial N}{\partial x}}{N}=\frac{\left(y\sin\frac{y}{x}-x\cos\frac{y}{x}\right)}{-x\left(y\sin\frac{y}{x}-x\cos\frac{y}{x}\right)}=-\frac{1}{x}.$$

So another integrating factor of the ode is $\exp\left[-\displaystyle\int\frac{dx}{x}\right]=\dfrac{1}{x}$

Using the first I.F, viz, $\frac{1}{2x^2y\cos\frac{y}{x}}$, we get the recast exact form as :

$$
\begin{aligned}
0 &= \frac{1}{2x^2\cos\frac{y}{x}}\left(x\cos\frac{y}{x}+y\sin\frac{y}{x}\right)dx+\frac{1}{2xy\cos\frac{y}{x}}\left(x\cos\frac{y}{x}-y\sin\frac{y}{x}\right)dy\\
&= \frac{dx}{2x}+\frac{y}{2x^2}\tan\left(\frac{y}{x}\right)dx+\frac{dy}{2y}-\frac{1}{2x}\tan\left(\frac{y}{x}\right)dy\\
&= \frac{1}{2}\left(\frac{dx}{x}+\frac{dy}{y}\right)+\frac{1}{2x^2}\tan\left(\frac{y}{x}\right)(ydx-xdy)\\
&= \frac{1}{2}\left(\frac{dx}{x}+\frac{dy}{y}\right)+\frac{1}{2}\tan\left(\frac{y}{x}\right)d\left(\frac{y}{x}\right)
\end{aligned}
$$

Integrating and dropping out factor $\frac{1}{2}$ we get :

$$\ln(xy) - \ln\left(\sec \frac{y}{x}\right) = \text{constant}$$

i, e, $xy \cos\left(\frac{y}{x}\right) = \text{constant}$ is the general solution.

If on the otherhand we use the second I.F. viz, $\frac{1}{x}$, we get :

$$\left(y \cos\frac{y}{x} + \frac{y^2}{x}\sin\frac{y}{x}\right) dx + \left(x\cos\frac{y}{x} - y\sin\frac{y}{x}\right) dy = 0$$

or, $(ydx + xdy)\cos\left(\frac{y}{x}\right) + \left(\frac{y^2}{x}dx - ydy\right)\sin\left(\frac{y}{x}\right) = 0$

or, $d(xy)\cos\left(\frac{y}{x}\right) + xy\, d\left(\frac{y}{x}\right)\sin\left(\frac{y}{x}\right) = 0$

$\therefore$ $d\left[xy\cos\left(\frac{y}{x}\right)\right] = 0$

Integrating, the general solution is : $xy \cos\left(\frac{y}{x}\right) = \text{constant}$.

2.5   Linear Equations and Bernoulli Equations

The first order **linear ode** is of the form

$$\frac{dy}{dx} + P(x)y = Q(x),$$

where $P(x)$ and $Q(x)$ are continuous functions over their respective domains. This restriction of demanding $P(x)$ and $Q(x)$ as continuous functions will have two-fold advantages viz, the integrability of $P(x)$ and $Q(x)$ and uniqueness of solutions of the linear ode.

In the differential form, the above equation reads :

$$(P(x)y - Q(x))dx + dy = 0 \qquad\qquad (2.10)$$

Comparing with the canonical form $M(x,y)dx + N(x,y)dy = 0$ we find that $M(x,y) = P(x)y - Q(x)$ and $N(x,y) = 1$ so that $\frac{\partial M}{\partial y} = P(x)$ and $\frac{\partial N}{\partial x} = 0$. Thus unless $P(x) = 0$, we have no question of exactness of the ode. However we shall be in lookout of the integrating factor(s) that renders exactness. The choice of $\mu(x,y)$ can be motivated in the following way :

Observe that $\frac{\left(\frac{\partial M}{\partial y} - \frac{\partial N}{\partial x}\right)}{N} = P(x)$, a function of x alone, and so $\frac{d}{dx}(ln \mid \mu \mid)N = P(x)$; Because $P(x)$ is a continuous function, its Riemann integrability is ensured and hence $\mu(x) = exp\left[\int P(x)dx\right]$. Thus the linear equation is a special illustration of case (I) discussed earlier.

$$\therefore \quad \mu(x)(P(x)y - Q(x))dx + \mu(x)dy = 0$$

is an exact equation. From the necessary and sufficient condition of exactness, we get

$$\frac{\partial}{\partial y}\{\mu(x)(P(x)y - Q(x))\} = \frac{\partial}{\partial x}(\mu(x))$$

$$\text{or,} \quad \mu(x)P(x) = \frac{d}{dx}(\mu(x)) \tag{2.10a}$$

We are now at the doorstep of another vital observation about the ode (2.10). This ode remains invariant under the transformation

$$x \to \bar{x} = x \ ; \ y \to \bar{y} = y + \frac{\lambda}{\mu(x)} \ , \ \lambda \text{ being arbitrary constant} \tag{2.11}$$

as can be verified directly as follows :

$$(P(\bar{x})\bar{y} - Q(\bar{x})) \, d\bar{x} + d\bar{y}$$
$$= \left\{P(x)\left(y + \frac{\lambda}{\mu(x)}\right) - Q(x)\right\}dx + dy - \frac{\lambda \, d(\mu(x))}{(\mu(x))^2}$$
$$= \{(P(x)y - Q(x))dx + dy\} + \frac{\lambda}{(\mu(x))^2}\left\{P(x)\mu(x) - \frac{d(\mu(x))}{dx}\right\} = 0,$$

(using (2.10) and (2.10a))

However, $\frac{1}{\mu(x)}$ satisfies homogeneous ode (2.10a). Hence the invariance of the ode (2.10) under the transformation (2.11) expresses simply the superposition principle for the first order linear ode that states that to any solution of (2.10), one can always add up a solution of the corresponding homogeneous ode (2.10a) multiplied by an arbitrary constant λ.

The general solution can be had by multiplying the originally given ode with the factor $\mu(x)$:

Therefore $\mu(x)\dfrac{dy}{dx} + \mu(x).P(x)y = \mu(x)Q(x)$ is an exact ode.

Hence, $\dfrac{d}{dx}\left[y.exp\left(\int P(x)dx\right)\right] = exp\left(\int P(x)dx\right).Q(x)$

Integrating both sides w.r.t x, we have :

$$y.exp\left(\int P(x)dx\right) = \int exp\left(\int P(x)dx\right)Q(x)dx + C,$$

C being some arbitrary constant of integration.

$$\therefore \ y = \exp\left(\int -P(x)dx\right)\left[\int exp\left(\int P(x)dx\right)Q(x)dx\right] + C,$$

or equivalently, $\quad \psi(x,y) \equiv \int Q(x)\mu(x)dx - y\mu(x) = c \qquad (2.12)$

Observe that the action of (2.11) is to effect a translation $\psi \to \psi - \lambda$ on the family of solutions (2.12) which is the general solution.

The general solution can also be had by the 'Method of Variation of Parameters' where we work with the associated homogeneous linear ode, viz,

$$\frac{dy}{dx} + P(x)y = 0.$$

[The existence of no y-free term will be referred to as the homogeneity of the ode. It is unfortunate that the term 'homogeneity' were used in another context before, but we all have to bear this brunt of ambiguous christening because mathematicians failed to coin a substitute title!].

This separable form of homogeneous ode can be integrated to have $y = C\exp(-\int P(x)dx)$, C being an arbitrary constant. At this point we introduce the term $Q(x)$ into the act and demand that the solution of the initially tabled ode $\frac{dy}{dx} + P(x)y = Q(x)$ should resemble the form of the solution of the associated homogeneous linear ode. This quest is satisfied once we replace C, the constant function by a more general differentiable function, say $C(x)$. The choice of $C(x)$ will be restrained by the fact that $y = C(x)\exp\left(-\int P(x)dx\right)$ is a solution of the original linear ode.

$$\frac{dy}{dx} = \frac{dC}{dx}\exp\left(-\int P(x)dx\right) - C(x)P(x)\exp\left(-\int P(x)dx\right)$$

$$\text{or,} \quad \frac{dC}{dx} = Q(x)\exp\left(-\int P(x)dx\right)$$

$$\therefore \quad C(x) = Q(x)\exp\left(\int P(x)dx\right) + C';$$

C' being some arbitrary constant. Putting back the expression for y we get the general solution of the original linear ode :

$$y = C' \exp\left(-\int P(x)dx\right) + \exp\left(-\int P(x)dx\right) \times$$

$$\int \left(Q(x)\exp\left(\int P(x)dx\right)\right) dx,$$

which indeed is same as that obtained by using integrating factors. This method is very efficient for linear odes even of higher orders as it envisages how the particular integrals (P.I) corresponding to the non-homogeneity term $Q(x)$ arise out of the complementary function (C.F) corresponding to the associated homogeneous linear ode.

Example (29) :   Solve the ode : $x\dfrac{dy}{dx} + \dfrac{2x+1}{x+1}y = (x-1)$

In the derivative form, the above equation can be recast as :

$$\frac{dy}{dx} + \frac{2x+1}{x^2+x}y = \frac{x-1}{x}$$

Since the above ode matches with $\frac{dy}{dx} + P(x)y = Q(x)$ in letter and spirit as far as the form is concerned, the given equation is linear in y having $P(x) \equiv \frac{2x+1}{x^2+x}$ and $Q(x) \equiv \frac{x-1}{x}$.

$$\therefore \quad \text{Integrating factor} \quad \mu(x) = \exp\left[\int \frac{2x+1}{x^2+x}dx\right] = (x^2+x)$$

The general solution reads :

$$y(x^2+x) = \int \frac{x-1}{x}(x^2+x)dx = \int (x^2-1)dx = \frac{1}{3}x^3 - x + c,$$

c being an integration constant.

Example (30) :   Solve the IVP $(x^2+1)\dfrac{dy}{dx} + 4xy = x$; $y(2) = 1$

Rewriting the equation in the form $\frac{dy}{dx} + \frac{4x}{x^2+1}y = \frac{x}{x^2+1}$ and proceeding as in the previous example we can at once have $\mu(x) = (x^2+1)^2$ and consequently the general solution.

$$y(x^2+1)^2 = \frac{1}{4}x^4 + \frac{1}{2}x^2 + c$$

However initial condition $y(2) = 1$ gives $c = 19$. Hence the required solution reads :

$$y(x^2 + 1)^2 = \frac{1}{4}x^4 + \frac{1}{2}x^2 + 19$$

In Example (14) we solved the ode $4xy\,dx + (x^2 + 1)dy = 0$ which is seen to be the homogeneous[1] linear equation associated with ours. We found the general solution to be

$$y(x^2 + 1)^2 = c \text{ [modulus signs in } y \text{ and } c \text{ are omitted here]}.$$

We now apply the method of variation of parameters and replace c by a differentiable function $c(x)$ that is to be determined in a way so that $y = \frac{c(x)}{(x^2+1)^2}$ becomes a solution of the original inhomogeneous ode.

Here $\dfrac{dy}{dx} = \dfrac{-4x}{(x^2 + 1)^3} c(x) + \dfrac{dc}{dx} \cdot \dfrac{1}{(x^2 + 1)^2}$

$\therefore$ $\dfrac{dy}{dx} + \dfrac{4xy}{x^2 + 1} = \dfrac{dc}{dx} \cdot \dfrac{1}{(x^2 + 1)^2}$

$\therefore$ $\dfrac{x}{x^2 + 1} = \dfrac{dc}{dx} \cdot \dfrac{1}{(x^2 + 1)^2}$ (using the given ode)

$\therefore$ $c(x) = \displaystyle\int x(x^2 - 1)dx = \frac{1}{4}x^4 + \frac{1}{2}x^2 + c'$

Putting back $c(x)$ in y we have the general solution :

$$y = \frac{1}{(x^2 + 1)^2} \left(\frac{1}{4}x^4 + \frac{1}{2}x^2 + c' \right),$$

c' being an arbitrary constant, that can be evaluated by using the initial conditions.

Example (31) : Solve the equation $y^2 dx + (3xy - 1)dy = 0$.

In the derivative form the above ode reads :

$$\frac{dy}{dx} + \frac{y^2}{3xy - 1} = 0$$

and so it is a non-linear equation independent variable y. We reverse the role of x and y and hence regard y as independent variable while x as dependent variable. This flexibility of identifying independent and

---

[1]'homogeneous' in the sense R.H.S is zero.

dependent variables enables us to cast the given equation in the form of a linear one :

$$\frac{dx}{dy} + \frac{3}{y}x = \frac{1}{y^2}, \text{ assuming for the time being } y \neq 0.$$

Integrating factor $\mu(y) = \exp\left[\int \frac{3}{y}dy\right] = y^3$ and the differential equation has its (general!) solution in the form :

$$xy^3 = \frac{1}{2}y^2 + c, \quad c \text{ being an arbitrary constant of integration.}$$

$$\therefore x = \frac{1}{2y} + \frac{c}{y^3} \text{ is the (general!) solution.}$$

Remark : Look that one cannot tell the above solution as the general one because on translating the differential form to the normal derivative form we were forced to divide the entire equation by y^2 and thereby lost the solution $y = 0$ of the given ode. Should we not add this feather to the body of general solution?

Example (32) : Solve the ode : $\frac{dy}{dx} = x \sec y - \tan y$

This is a first order non-linear ode in independent variable x and dependent variavle y. We may change the dependent variable y to $z = \sin y$.

Hence the given ode can be written as :

$$\cos y\frac{dy}{dx} + \sin y = x \qquad \text{i, e,} \quad \frac{dz}{dx} + z = x \ ,$$

which being a first order linear ode in independent variable x and dependent variable z has an I.F e^x. This leads to the general solution:

$$ze^x = \int xe^x dx = (x - 1)\, e^x + c \ (c \text{ being arbitrary constant})$$

i, e, $\quad e^x \sin y = (x - 1)\, e^x + c$

Remark (a) : In the last two examples we observed that even if a first order ode $\frac{dy}{dx} = f(x, y)$ be not linear in y, it can be transformed to the linear form either by recasting the ode in the form $\frac{dx}{dy} = \frac{1}{f(x,y)}$ or, by changing the dependent variable to some suitable $z = g(y)$.

Remark (b) : For the linear ode, the inhomogeneity term $Q(x)$ may be sectionally continuous over its domain, say, some interval (a, b). This means that $\exists$ a finite number of points $x_1, x_2, \cdots, x_n$ in (a, b) where $Q(x)$ has jump discontinuity and everywhere else in (a, b) it is continuous. In other words, $\exists$ a partition $\{a \equiv x_0, x_1, \cdots, x_n, x_{n+1} \equiv b\}$ of $[a, b]$ such that in each of the subintervals $[x_{i-1}, x_i]$, $Q(x)$ is continuous, has jump discontinuities at the interior partition points and $\underset{x \to a+0}{\mathrm{Lt}} Q(x)$, $\underset{x \to b-0}{\mathrm{Lt}} Q(x)$ both exist.

Example (33) : Solve the initial value problem :

$$\frac{dy}{dx} + y = f(x), \quad \text{subject to } y(0) = 0$$

where $f(x) = \begin{cases} 2x, & 0 \le x < 2 \\ 4, & x \ge 2 \end{cases}$

The IVP is to be solved for $x \in (0, \infty)$. The integrating factor $\mu(x)$ will be e^x. On being multiplied by this integrating factor, the given ode becomes $\frac{d}{dx}(ye^x) = f(x).e^x$.

We integrate the above relation from 0 to x to have :

$$
\begin{aligned}
ye^x - y(0) &= \int_0^x e^x f(x) dx \\
&= \begin{cases} \int_0^2 2xe^x dx + \int_2^x 4e^x dx &, \quad \text{if } x \ge 2 \\ \\ \int_0^x 2xe^x dx &, \quad \text{if } 0 \le x < 2 \end{cases} \\
\therefore \quad ye^x &= \begin{cases} 4e^x - 2e^2 + 2 & \text{if } x \ge 2 \\ \\ 2(1 + xe^x - e^x) & \text{if } 0 \le x < 2 \end{cases}
\end{aligned}
$$

Example (34) : Solve the linear ode $\frac{dy}{dx} + y = 2\sin x + 5\sin 2x$

Before we go straight way into the solutions, we prove the very simple but useful result :

If $f_1(x)$ be a solution to the linear ode

$$\frac{dy}{dx} + P(x)y = Q_1(x) \tag{2.13a}$$

and $f_2(x)$ be a solution to the linear ode

$$\frac{dy}{dx} + P(x)y = Q_2(x) \tag{2.13b}$$

then $f_1(x) + f_2(x)$ is a solution to the linear ode

$$\frac{dy}{dx} + P(x)y = Q_1(x) + Q_2(x) \tag{2.13c}$$

Indeed this happens for all points lying in the intersection of domains of $Q_1(x)$ and $Q_2(x)$.

$f_1(x)$ being a solution of 2.13(a) and $f_2(x)$ being that 2.13(b),

$$\left.\begin{array}{rcl} \frac{df_1}{dx} + P(x)f_1 & = & Q_1(x) \\ \frac{df_2}{dx} + P(x)f_2 & = & Q_2(x) \end{array}\right\}$$

Adding up, $\frac{d}{dx}(f_1 + f_2) + P(x).(f_1 + f_2) = Q_1(x) + Q_2(x)$ showing that 2.13(c) is satisfied.

Now we come to the problem itself.

Observe the above result ensures us that the solution of the given linear ode is the sum of solutions of the equations

$$\left.\begin{array}{rcl} \frac{dy}{dx} + y & = & 2\sin x \\ \frac{dy}{dx} + y & = & 5\sin 2x \end{array}\right\}$$

These equations have a common homogeneous part and thus they have the same I.F, viz, e^x. Solution of the first linear ode will be

$$y = e^{-x} \int 2\sin x.e^x dx = (\sin x - \cos x) + c_1 e^{-x}$$

In the same vein, solution of the second linear ode will be

$$y = e^{-x} \int 5\sin 2x.e^x dx = (\sin 2x - 2\cos 2x) + c_2 e^{-x}$$

Hence the general solution reads :

$$y = \sin x - \cos x + \sin 2x - 2\cos 2x + ce^{-x}$$

provided we agree to write $(c_1 + c_2)$ as c.

Theorem 2.5 : Existence-Uniqueness Theorem for Linear Odes.

The existence-uniqueness theorem enunciated in § 1.5 can be strengthened for the linear differential equations as follows :

For the IVP comprising of the first order linear ode

$$a_0(x)\frac{dy}{dx} + a_1(x)y = f(x)$$

and the initial condition $y(x_0) = y_0$, if $a_0(x), a_1(x), f(x)$ are all continuous over some common interval I containing the point x_0 and the co-efficient function $a_0(x)$ of $\frac{dy}{dx}$ is non-zero over I, then the IVP possesses a unique solution $y(x)$ defined throughout I.

Since $a_0(x)$ is non-zero over I, we can carry out division by $a_0(x)$ so as to have the normal form of IVP :

$$(DE) \quad \frac{dy}{dx} + P(x)y = Q(x) \quad \text{and} \quad (IC) \quad y(x_0) = y_0$$

Hence if $P(x)$ and $Q(x)$ are continuous over I, then the above IVP admits of a unique solution throughout I.

In determining the unique solution of the IVP, we can begin with an integrating factor that includes the given initial condition as a part of it. So our I.F will be

$$\mu(x) = exp\left[\int_{x_0}^{x} P(t)dt\right]$$

On multiplying the given linear ode by $\mu(x)$ we have :

$$\frac{d}{dx}[y(x)\mu(x)] = \mu(x)\, Q(x)$$

Integrating w.r.to x between x_0 and x we get :

$$y(x)\mu(x) - y_0 = \int_{x_0}^{x} \mu(s)\, Q(s)\, ds$$

$$\therefore \ y(x) = \frac{1}{\mu(x)}\left[y_0 + \int_{x_0}^{x} \mu(s)\, Q(s)\, ds\right]$$

$$= exp\left[-\int_{x_0}^{x} P(t)dt\right]\left(y_0 + \int_{x_0}^{x} Q(s)exp\left[-\int_{x_0}^{s} P(t)dt\right] ds\right)$$

Remark : Observe that the linearity of the ode guarantees the existence of a unique solution over the entire interval I instead of just a small neighborhood of x_0.

The following example illustrates the use of above theorem.

Example (34) : Solve the : $\dfrac{dy}{dx} - 2xy = 3$ subject to $y(0) = 2$

The integrating factor of this ode is $\mu(x)$, where

$$\mu(x) = exp\left[\int_0^x -2x\ dx\right] = e^{-x^2}$$

On multiplying the ode by $\mu(x)$ we get :

$$\frac{d}{dx}\left(ye^{-x^2}\right) = 3e^{-x^2}$$

Integrating between 0 to x we have :

$$ye^{-x^2} - 2 = 3\int_{t=0}^{t=x} e^{-t^2}\ dt = \frac{3}{2}\sqrt{\pi}\ Erf(x)$$

where $\quad Erf(x) \equiv \dfrac{2}{\sqrt{\pi}}\displaystyle\int_0^x e^{-t^2}\ dt\ $ is a non-elementary function.

So the unique solution of the IVP is : $y(x) = 2e^{x^2} + \frac{3}{2}\sqrt{\pi}\ Erf(x)$.

Important Results for Linear Equations :

(A) Consider the two first order odes $\quad y'+ay = q_1(x)$ and $\quad y' + ay = q_2(x)$, where $a = a_r + ia_i$ is a complex constant and $q_1(x), q_2(x)$ are continuous functions over $[0, \infty)$ such that $|\ q_1(x) - q_2(x)\ | \leqslant k\ \forall\ x \in [0, \infty]$ for some constant $k > 0$. If $y_1(x)$ be a solution of $y' + ay = q_1(x)$ and $y_2(x)$ be a solution of $y' + ay = q_2(x)$, and moreover $y_1(0) = y_2(0)$, then

$$|\ y_1(x) - y_2(x)\ | \leq \frac{k}{a_r}[1 - e^{-a_r x}]\quad \forall\ x \in [0, \infty),\quad a_r \neq 0.$$

Proof : From the given condition, it follows that

$$y_1' + ay_1(x) = q_1(x)\ ;\ y_2' + ay_2(x) = q_2(x)$$

$$\therefore\quad \frac{d}{dx}(y_1(x) - y_2(x)) + a(y_1(x) - y_2(x)) = q_1(x) - q_2(x)$$

or $\quad \dfrac{d}{dx}\{e^{ax}(y_1(x) - y_2(x))\} = e^{ax}(q_1(x) - q_2(x))$

Integrating both sides over $[0, x]$ we have :

$$\int_0^x e^{au}(y_1(u) - y_2(u))du = \int_0^x (q_1(u) - q_2(u))e^{au}du$$

$$e^{ax}(y_1(x) - y_2(x)) = \int_0^x (q_1(u) - q_2(u))e^{au}du\ (\because\ y_1(0) = y_2(0))$$

$$\therefore \quad |e^{ax}(y_1(x) - y_2(x))| = \left| \int_0^x e^{au}(q_1(u) - q_2(u))du \right|$$

$$\Rightarrow \quad |e^{a_r x}.e^{ia_i x}(y_1(x) - y_2(x))| = \left| \int_0^x e^{a_r u}.e^{ia_i u}(q_1(u) - q_2(u))du \right|$$

i, e, $\quad |(y_1(x) - y_2(x))| \, e^{a_r x} \le k \int_0^x e^{a_r u} du = \dfrac{k}{a_r}(1 - e^{-a_r x})e^{a_r x}$

Hence the result.

Conclusion on the above result : It speaks out the sensitivity of the system in responding to an impulse. When $q_2(x)$ lies within the range $[q_1(x) - k, q_1(x) + k]$, the solution/response lies within an interval $[y_1(x) - k_1, y_1(x) + k_1]$, where $k_1 = \frac{k}{a_r}(1 - e^{-a_r x})$ provided $a_r \ne 0$. Here in keeping tradition with engineering mathematics we refer to the inhomogeneous terms as 'impulse' and the complete solution of the linear equation as 'response'. In brief, the above result implies that a finite change in the impulse will always yield a finite change in the corresponding response.

(B) Consider the first order ode $y' + ay = q(x)$, where $q(x)$ is a continuous function over $[0, \infty)$ that tends to the constant β as $x \to \infty$ and a is a constant such that $a_r \equiv Re(a) > 0$. To prove that every solution of this equation tends to $\frac{\beta}{a}$ as $x \to \infty$.

Proof : The given linear equation has the integrating factor e^{ax}, so that we have the general solution $\phi(x)$ given in $[0, \infty)$ as

$$\phi(x) = \left(\phi(0) + \int_0^x b(t)e^{at}dt \right) e^{-ax}$$

$$= \dfrac{1}{e^{ax}} \left[\phi(0) + \int_0^x b(t)e^{at}dt \right]$$

$$\therefore \lim_{x\to\infty} \phi(x) = \lim_{x\to\infty} \dfrac{\phi(0) + \int_0^x b(t)e^{at}dt}{e^{ax}} \quad \left(\dfrac{\infty}{\infty} \text{ form} \right)$$

$$= \lim_{x\to\infty} \dfrac{b(x)e^{ax}}{ae^{ax}} = \lim_{x\to\infty} \dfrac{b(x)}{a} = \dfrac{\beta}{a}$$

(C) **Extensibility of solutions :**
Consider the first order linear non-homogeneous differential equation $\frac{dy}{dx} + P(x)y = Q(x)$, where $P(x)$ and $Q(x)$ are real-valued continuous functions in an open interval $I \subseteq domP \cap dom\ Q$. If $y = \phi(x)$ be any solution of this linear ode over a subinterval $I' = \{x \in R \mid x_1 < x < x_2\}$, then $\phi(x)$ can be extended uniquely to the entire interval I as a solution

of the above differential equation.

Note : **(i)** Proof is based on Bellmann-Gronwall's inequality[2] and some deeper global properties of solutions.

(ii) This feature is not limited to first order linear non-homogeneous odes—it is equally applicable to higher order linear odes.

(iii) For non-linear odes, such an extension of solutions cannot be expected as sometimes the solutions may locally blow up.

Bernoulli Equation : It is a kind of generalisation of the linear first order ordinary differential equation, the general form of which reads :

$$\frac{dy}{dx} + P(x)y = Q(x)y^n,$$

where $P(x)$ and $Q(x)$ are continuous functions defined on an interval I and $n \neq 0$ or 1. In case $n = 0$, we get back linear equations, while $n = 1$ gives us a separable first order ode, both cases been discussed earlier. The domain of the associated vector field $f(x,y) \equiv -P(x) + Q(x)y^n$ is $I \times O$, where O should depend on the value of n. The continuity of $\frac{\partial f}{\partial y} = -P(x) + nQ(x)y^{n-1}$ on its domain is necessary for unique solution to exist and this requires choosing O as follows :

(i) If n is a non-negative integer, then $O = \mathbb{R}$, the full real axis.

(ii) If n is a negative integer, then $O = \mathbb{R} - \{0\}$, the real line with origin removed.

(iii) If n is not an integer, then $0 = \mathbb{R}^+$, the positive part of real axis.

These choices are justified by mere inspection.

We now consider a kth degree parabolic transformation so as to reduce the above equation into the linear one.

Put $y^k = u$, $k \neq 0$, $k \neq 1$ so that $y = u^{\frac{1}{k}}$ & $\dfrac{dy}{dx} = \dfrac{1}{k}u^{\frac{1}{k}-1}\dfrac{du}{dk}$

The chosen Bernoulli equation therefore changes to the following :

$$\frac{1}{k}u^{\frac{1}{k}-1}\frac{du}{dk} + P(x)u^{\frac{1}{k}} = Q(x)u^{\frac{n}{k}}$$

$$\text{or,} \quad \frac{du}{dx} + P(x)ku = Q(x)ku^{1-\frac{1}{k}(1-n)}$$

---

[2]See for details Lemma in § 6.6

If this equation is to be a linear one, R.H.S should not involve u and hence one should choose $k = 1 - n$.

The linear differential equation in x and u therefore reads

$$\frac{du}{dx} + (1 - n)P(x)u = (1 - n)Q(x)$$

After the Bernoulli equation is transformed into a linear equation, the next steps towards its general solution are in letter and spirit the repetition of the previous workouts for linear equations.

Example (35) : Solve the IVP :

$$\frac{dy}{dx} + \frac{y}{2x} = \frac{x}{y^3} \ ; \ y(1) = 2.$$

The given differential equation is of the Bernoulli type. Multiplying both sides by $4y^3$ $[\because n = -3$ here$]$ we get at once

$$\frac{d}{dx}(y^4) + \frac{2}{x}y^4 = 4x$$

Putting $u = y^4$, the above equation reduces as

$$\frac{du}{dx} + \frac{2}{x}u = 4x,$$

which is a linear equation with x as independent and u as the dependent variable.

$$\therefore \quad \text{I.F} \quad \mu(x) = \exp\left[\int \frac{2}{x}dx\right] = x^2$$

On multiplying the equation by I.F and integrating, we have

$$x^2 u \ = \ \int 4x^3 dx = x^4 + c,$$

$$\text{or, } u \ = \ x^2 + \frac{c}{x^2}, \ c \text{ being the integration constant.}$$

Since $y(1) = 2$, $c = 15$, so that the particular solution reads

$$y^4 = x^2 + \frac{15}{x^2}$$

Example (36) : Solve the ode

$$\cos x \frac{dy}{dx} - y = y^2(\sin x - 1)\cos x$$

The given ode can be written as

$$\frac{dy}{dx} - y\sec x = y^2(\sin x - 1),$$

which is a first order ode of Bernoulli type. Hence multiplying by $-\frac{1}{y^2}$ we get :

$$\frac{d}{dx}\left(\frac{1}{y}\right) + \left(\frac{1}{y}\right)\sec x = (1 - \sin x)$$

The above ode is first order linear in $\frac{1}{y}$ and so its I.F is

$$\mu(x) = e^{\int \sec x\, dx} = (\sec x + \tan x)$$

$$\therefore (\sec x + \tan x)\left\{\frac{d}{dx}\left(\frac{1}{y}\right) + \frac{1}{y}\sec x\right\} = (1 - \sin x)(\sec x + \tan x)$$

is an exact ode.

i, e,　　$(\sec x + \tan x)\frac{d}{dx}\left(\frac{1}{y}\right) + \frac{d}{dx}(\sec x + \tan x)\cdot\frac{1}{y} = \cos x$

or,　　$\frac{d}{dx}\left[\frac{1}{y}(\sec x + \tan x)\right] = \cos x$

Integrating with respect to x, $\dfrac{\sec x + \tan x}{y} = \sin x + c,$

c being constant of integration.

Hence, $y = \dfrac{\sec x + \tan x}{\sin x + c} = \dfrac{1 + \sin x}{\cos x(\sin x + c)}$ is the general solution.

2.6　Integrating Factors Revisited

(a) While discussing exact differential equations of the first order, we told that the statement '$Mdx + Ndy = 0$ is exact' means M and N are components of an irrotational vector field in two dimensions and the function $\phi(x, y)$ which appears in the general solution of the exact ode is the potential function where from the field components are derivable. If $Mdx + Ndy = 0$ were not exact, we would have interpreted M and N to be components of a vortex-field. Mathematically, $\frac{\partial M}{\partial y} - \frac{\partial N}{\partial x} \neq 0$ and is precisely a measure of the vorticity. To make this given ode exact, we seek integrating factors $\mu(x, y)$ so that $\mu Mdy + \mu Ndy = 0$ becomes exact. This implies that $\mu(x, y)$ gives rise to an accessory vector field

$\overrightarrow{F'}$ having components (M', N') that kills the vorticity of existing vector field and in the process makes it conservative.

In vectorial notation, μ is given by $curl\overrightarrow{F} = \overrightarrow{F} \times \overrightarrow{\nabla}(\ln \mu)$ where $\overrightarrow{F} \equiv M\hat{i} + N\hat{j}$. The accessory vector field of $\overrightarrow{F'} \equiv M\hat{i} + N\hat{j}$ is given by $\overrightarrow{F'} \equiv (\mu - 1)\overrightarrow{F}$ so that we can vouchsafe say that it should also lie in the xy-plane but is not parallel to $\overrightarrow{F}$.

The determination of the accessory vector field $\overrightarrow{F'}$ is easily constructed for the linear differential equations appearing in the differential form

$$(P(x)y - Q(x))\, dx + dy = 0$$

because here the integrating factor is $\exp\left[\int P(x)dx\right]$. Simple calculations establish that $\overrightarrow{F'} = \left(\int P dx\right)\hat{j}$ is the accessory vector field required to make the resultant vector field curl-less/irrotational. Incidentally, this vector field is solenoidal and $\int P(x)dx$ being $\ln(\mu)$, we may look into μ as the logarithmic potential of the accessory vector field.

For example, let's consider the ode $y' + xy = x$. In the differential form, this appears as $(xy - x)dx + dy = 0$. Hence $P(x) = x$ here and $\overrightarrow{F'} = \frac{x^2}{2}\hat{j}$. As is obvious from the expression itself, $div\overrightarrow{F'} = 0$ and $\mid \overrightarrow{F'} \mid = \ln$ (integrating factor $e^{\frac{x^2}{2}}$).

It should be noted that the multiplication of the ode by an integrating factor $\mu(x, y)$ may introduce either new discontinuities in the co-efficients of the equation or may introduce extraneous solution to the given ode, i.e, the curves along which the integrating factor is zero. For example, for the innocent ode, $xdy - ydx = 0$, x^{-2} is an integrating factor which becomes discontinuous at $x = 0$. Thus $xdy - ydx = 0$ is defined over $\mathbb{R}$ but

$$d\left(\frac{y}{x}\right) \equiv \frac{xdy - ydx}{x^2} = 0$$

is defined over $\mathbb{R} - \{0\}$. Now $\mathbb{R} - \{0\}$ is a disconnected set comprised of two disjoint intervals R^+ and R^-. This discontinuity introduced through the integrating factor compels us to treat the resultant differential equation $d\left(\frac{y}{x}\right)$ for $\mathbb{R}^+$ and $\mathbb{R}^-$ separately. The general solution then should appear as :

$$y = \begin{cases} c_1 x, & \text{if } x > 0 \\ c_2 x, & \text{if } x < 0 \end{cases}$$

where c_1 and c_2 are independent constants of integration. Through this little discussion we have an important lesson : the total number of independent arbitrary constants involved in the general solution depends not just on the order of the ode but also on the space of functions to which the general solution should belong. The introduction of integrating factor x^{-2} thus changed the topological behavior of the underlying space of the ode from connected to disconnected.

Example (37) : Find the general solution of the ode

$$(y^2 - 1)dx + (y^3 - y + 2x)dy = 0.$$

This is a linear ode in independent variable y and dependent variable x. Observe that the term $(y^3 - y)dy$ is exact in its own right and will remain exact even if we multiply both sides by some differentiable function $\mu(y)$.

Hence $\quad \mu(y)(y^2 - 1)dx + 2x\mu(y)dy$ is a perfect differential.

$$\therefore \quad \frac{\partial}{\partial y}\{\mu(y)(y^2 - 1)\} = \frac{\partial}{\partial x}\{2x\mu(y)\}$$

After simple workout we have : $\mu(y) = \dfrac{1}{(y+1)^2}$.

Obviously $\mu(y)$ introduces a discontinuity at $y = -1$ in the resultant exact ordinary differential equation

$$\frac{y-1}{y+1}dx + \left[\frac{2x}{(y+1)^2} + \frac{y(y-1)}{(y+1)}\right] dy = 0$$

So we shall write down the general solution for $y \neq -1$ as :

$$2x(y-1) = \begin{cases} (y+1)\{4y - y^2 - \ln(y+1)^4 + c_1\} & \text{if } y > -1 \\ (y+1)\{4y - y^2 - \ln(y+1)^4 + c_2\} & \text{if } y < -1 \end{cases}$$

This solution set should be appended by the isolated solution $y + 1 = 0$ as it satisfies the given ode. Here also c_1 and c_2 are two independent constants of integration corresponding to the disjoint intervals $(-1, \infty)$ and $(-\infty, -1)$ on y-axis.

2.7 Riccati Equation

Riccati equations are first order non-linear equations that contain the Bernoulli equations and linear first order equations as special case. Moreover, it can always be converted to second order homogeneous equations with variable co-efficients by means of suitable transformations. Interestingly, the Riccati equations have got plenty of uses in mathematical physics.

The general form of a Riccati equation is :

$$\frac{dy}{dx} = A(x)y^2 + B(x)y + C(x) \qquad (2.14)$$

where $A(x)$, $B(x)$ and $C(x)$ are assumed to be continuously differentiable functions of independent variable x over some interval I. If $C(x) = 0$, the above equation reduces to Bernoulli equation with $n = 2$. If $A(x) = 0$, the above equation reduces to the familiar linear ode.

This important non-linear equation reduces to a linear equation if we apply the transformation $y = f + \frac{1}{v}$ to it, f being a preassigned solution to the given Riccati equation.

Observe that as f is any solution to the ode (2.14), we must get

$$\frac{df}{dx} = A(x)f^2 + B(x)f + C(x) \qquad (2.15)$$

Again $y = f + \frac{1}{v}$ reduces equation (2.14) to

$$\frac{df}{dx} - \frac{1}{v^2}\frac{dv}{dx} = A(x)\left(f + \frac{1}{v}\right)^2 + B(x)\left(f + \frac{1}{v}\right) + C(x)$$

$$= (A(x)f^2 + B(x)f + C(x)) + \frac{2A(x)f}{v^2} + \frac{B(x)}{v} + \frac{A(x)}{v^2}$$

$$\therefore \quad -\frac{1}{v^2}\frac{dv}{dx} = \frac{2A(x)f}{v^2} + \frac{B(x)}{v} + \frac{A(x)}{v^2} \quad (\text{using}(2.15))$$

$$\text{or,} \quad \frac{dv}{dx} + (2A(x)f + B(x))v = -A(x),$$

a linear equation with x as independent and v as the dependent variable.

Theorem 2.6 : Every Riccati equation can be transformed to a homogeneous second order linear differential equation with variable coefficients.

Proof : Let's consider the Riccati equation

$$\frac{dy}{dx} = A(x)y^2 + B(x)y + C(x),$$

where $A(x) \neq 0$ for any $x \in I$ and moreover, $A(x)$ is continuously differentiable over I. We apply the transformation $y = \frac{-u'(x)}{u(x)A(x)}$ to have the new form of given ode as

$$\frac{-u''(x)}{u(x)A(x)} + \frac{u'^2(x)}{u^2(x)A(x)} + \frac{u'(x)A'(x)}{u(x)A^2(x)} = \frac{A(x)u'^2(x)}{u^2(x)A^2(x)} - \frac{B(x)u'(x)}{u(x)A(x)} + C(x)$$

i, e, $\quad A(x)u'' - (A'(x) + A(x)B(x))u'(x) + C(x)A^2(x)u(x) = 0 \quad (2.16)$

which is the required second order linear homogeneous ode.

Remark : If $u_1(x)$ and $u_2(x)$ be two fundamental solutions of equation (2.16), then the general solution might be taken as $cu_1(x) + u_2(x)$, c being an arbitrary constant. Substituting this into the formula $\frac{-u'(x)}{u(x)A(x)}$, we have the general solution of (2.14) as

$$y(x) = -\frac{cu_1'(x) + u_2'(x)}{cu_1(x)A(x) + u_2(x)A(x)} \quad (2.17)$$

As the relation (2.17) shows, $y(x)$ represents a one parameter family of curve (c being parameter). Thus the general solution of any Riccati equation is given by the one-parameter family of curves of the form

$$y(x) = \frac{cg(x) + h(x)}{cG(x) + H(x)}. \quad (2.18)$$

The converse is also true, i.e, the differential equation of a one-parameter family of the form (2.17) is a Riccati equation.

Theorem 2.7 : If we know any three particular solutions of a Riccati equation, its general integral is determined.

Proof : Let $\frac{dy}{dx} = A(x)y^2 + B(x)y + C(x)$ be a Riccati equation and let y_1, y_2, y_3 be any three particular solutions of this equation.

$$\frac{dy_1}{dx} = A(x)y_1^2 + B(x)y_1 + C(x) \quad (2.18a)$$

$$\frac{dy_2}{dx} = A(x)y_2^2 + B(x)y_2 + C(x) \quad (2.18b)$$

$$\frac{dy_3}{dx} = A(x)y_3^3 + B(x)y_3 + C(x) \quad (2.18c)$$

Subtracting 2.18(b) from the main equation we have

$$\frac{d}{dx}(y - y_2) = A(x)(y + y_2)(y - y_2) + B(x)(y - y_2)$$

Equivalently, $\quad \dfrac{y' - y_2'}{y - y_2} = A(x)(y + y_2) + B(x) \quad (2.19a)$

where prime($'$) denotes differentiation w.r.to x

Further, subtracting 2.18(c) from the main equation we have :

$$\frac{y' - y_3'}{y - y_3} = A(x)(y + y_3) + B(x) \qquad (2.19b)$$

Now subtracting 2.19(b) from 2.19(a) we get

$$\frac{y' - y_2'}{y - y_2} - \frac{y' - y_3'}{y - y_3} = A(x)(y_2 - y_3),$$

On integrating this w.r.to x yields

$$ln\left(\frac{y - y_2}{y - y_3}\right) = \int A(x)(y_2 - y_3)dx + c_1$$

Following the same sequence of steps with 2.18(a) in lieu of the main equation we get

$$ln\left(\frac{y_1 - y_2}{y_1 - y_3}\right) = \int A(x)(y_2 - y_3)dx + c_2$$

From these last two relations one gets

$$\frac{(y - y_2)(y_1 - y_3)}{(y - y_3)(y_1 - y_2)} = \exp(c_1 - c_2) \equiv K, \text{ say}$$

This shows that if we know three particular solutions, we can find the general solution by the above formula. Sometimes we are forced to transform a Riccati equation to the linear equations, as seen before because the most general form in which our Riccati equation appears is not integrable even by method of quadratures. Now we see that once we know these particular solutions to a Riccati equation, even the need of the method of quadrature is not felt! Had there been four known solutions, say, y_1, y_2, y_3, y_4, we can establish from above that **cross-ratio** or **anharmonic ratio** is a constant.

i.e, $$\frac{(y_2 - y_4)(y_1 - y_3)}{(y_3 - y_4)(y_1 - y_2)} = K.$$

Example (38) : Solve the ode $\frac{dy}{dx} = (1 - x)y^2 + (2x - 1)y - x$, given that one solution is $y(x) = 1$.

Since the ode is seen to be a Riccati equation, we apply the substitution $y = 1 + \frac{1}{v}$, v being a differentiable function of x. Applying differentiation, $y = 1 + \frac{1}{v}$ yields $\frac{dy}{dx} = -\frac{1}{v^2} \cdot \frac{dv}{dx}$ which in turn transforms ode as

$$-\frac{1}{v^2} \cdot \frac{dv}{dx} = (1 - x)\left(1 + \frac{1}{v}\right)^2 + (2x - 1)\left(1 + \frac{1}{v}\right) - x$$

$$\therefore \quad \frac{dv}{dx} + v = (x - 1),$$

which being a linear differential equation in dependent variable v, we have the general solution

$$v = 1 + (x - 2) + ce^{-x},$$

c being an arbitrary constant of integration. Substituting v in terms of y we have the general solution of the ode given as;

$$y = \frac{1}{(x - 2) + ce^{-x}}$$

Remark : **(a)** Although pedantic, it will not be out of place to mention here that Riccati equations find many uses in supersymmetric quantum mechanics, an important branch of physics. In that study, the so-called supersymmetric partner potentials $V_\pm(x)$ are related to the superpotential by the Riccati equation:

$$V_\pm(x) = W^2(x) \pm \frac{dW}{dx}$$

As a concrete important example we might quote the case of 3D-isotropic oscillator for which

$$V_-(x) = -w_c\left(l + \frac{3}{2}\right) + \frac{l(l + 1)}{x^2} + \frac{1}{4}w_c^2 x^2$$

and so superpotential $W(x) = \frac{1}{2}w_c x - \frac{(l+1)}{x}$ is the solution of the corresponding Riccati equation. (symbols bear usual meaning).

(b) Every second order linear homogeneous equation can be converted to a Riccati equation by means of so called 'Riccati substitution'.

Let a second order linear homogeneous ode be given as :

$$y'' + p_1(x)y' + p_2(x)y = 0,$$

where $p_1(x)$, $p_2(x)$ are continuous functions defined over some interval. We substitute $v = \frac{y'}{y}$, so that $y' = vy$ and hence $y'' = vy' + v'y$. From this we have the given ode transformed as :

$$vy' + v'y + p_1(x)y' + p_2(x)y = 0$$
$$\Leftrightarrow \quad v' + \frac{y'}{y}v + p_1(x)\frac{y'}{y} + p_2(x) = 0 \quad \text{(dividing both sides by } y\text{)}$$
$$\Leftrightarrow \quad v' + v^2 + p_1(x)v + p_2(x) = 0$$
$$\Leftrightarrow \quad v' = -p_2(x) - p_1(x)v - v^2,$$

last equation being the familiar form of Riccati equation. Thus the Riccati substitution $v = \frac{y'}{y}$ enables us to reduce the order of the homogeneous differential equation by unity but induces non-linearity in the resultant differential equation.

As an illustration, let's pick up the simple harmonic motion governed by the equation $x'' + k^2x = 0$. (prime denoting differentiation w.r. to t) We make the Riccati substitution $v = \frac{x'}{x}$.

Hence $\quad x'' = xv' + x'v = x(v' + v^2) \quad$ where $v' = \frac{dv}{dt}$.

$$\therefore \quad \frac{dv}{v^2 + k^2} = -dt, \quad \text{which on integration}$$

$v = k \tan\left(k(t_0 - t)\right)$, $\quad t_0$ being a constant of integration.

$$\therefore \quad \frac{x'}{x} = k \tan\left(k(t_0 - t)\right)$$

Integrating both sides w.r.to t,

$$\ln | x | \quad = \quad \int k \tan\left(k(t_0 - t)\right) dt = \ln | \cos\left(k(t_0 - t)\right) | + c$$
$$\therefore \quad x \quad = \quad c \cos\left(k(t_0 - t)\right), \quad c \text{ being the integration constant.}$$

This is the general solution and it involves two arbitrary constants (t_0 and c) as expected. Infact the Riccati substitution is one way of exploiting non-linearity to get any higher order linear ode solved.

Some General Features of Non-linear Differential Equations

(a) The principle of superposition of solutions of linear differential equations not applicable to non-linear equations. (c.f (f) of § 2.8).

(b) Unless all the terms of a non-linear differential equation are of the same degree in y, no constant multiple of a solution of any homogeneous non-linear equation is again a solution. For instance, if $y = y_0$ be a solution of the non-linear Vanderpol's equation

$$y'' + \epsilon(y^2 - 1)y' + y = 0,$$

then ky_0 is not a solution.

(c) If y_1 and y_2 be two linearly independent solutions of a non-linear equation, their sum is not a solution in general. Indeed the concept of linear independence of solutions is useless for non-linear differential equations.

If we have a second order linear homogeneous equation of the form

$$y'' + p_1(x)y' + p_2(x)y = 0,$$

then it might be considered to have a general solution, obtainable through two integrations and so involving two arbitrary constants. For the non-linear equations of the second order, the general solution, if found at all, consists of a number of arbitrary constants but they appear not in a simple form analogous to the linear odes. To illustrate this, let us consider the following ode

$$y\frac{d^2y}{dx^2} + \lambda\left(\frac{dy}{dx}\right)^2 = 0, \ \lambda \text{ being a constant.}$$

Writing out $\frac{dy}{dx} \equiv p$, the equation appears in a separable form $yp\frac{dp}{dy} + \lambda p^2 = 0$, which on integrating w.r.to x, yields $y^\lambda p = A$, some constant. Putting back $p = \frac{dy}{dx}$ we have on further integration,

$$y = [(\lambda + 1)(Ax + B)]^{\frac{1}{\lambda + 1}}$$

B being the second constant of integration.

This form is in sharp contrast to the form $y = Ay_1 + By_2$ available as solutions of linear homogeneous differential equations.

2.8   Application of Differential Equations of First Order

(a) Hanging Chain : Imagine a homogeneous, ductile but inextensible cable, like a TV cable, suspended between two supports at equal heights and hanging freely. The linear specific weight, i.e, the weight of cable per unit length is equal to γ. We choose a co-ordinate system in the plane of the cable where the x-axis is horizontal, the positive y-axis vertically upwards, the force of gravity straight down. Let A be the lowest point (trough) of the cable and $P(x,y)$ be an arbitrary point of the cable. C is the midpoint of the arc AP.

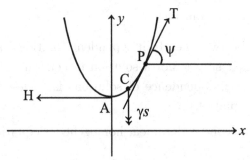

Fig 2.4 : Hanging chain in the form of a catenary.

Part AP of the hanging cable is in equilibrium under the action of three forces, viz, (i) the tension T acting along the tangent at P, subtending an angle ψ with the x-axis. (ii) the horizontal tension H acting at the lowest point A and (iii) the weight of the string γs, acting vertically downwards, s being the arcual distance of P from A. By Lami's theorem,

$$\frac{T}{\sin\frac{\pi}{2}} = \frac{H}{\sin(\frac{\pi}{2}+\psi)} = \frac{\gamma s}{\sin(\pi-\psi)} \quad \text{so that } \tan\psi = \frac{dy}{dx} = \frac{\gamma s}{H}.$$

Differentiating once again with respect to x, we get:

$$\frac{d^2y}{dx^2} = \frac{\gamma}{H}\frac{ds}{dx} = \frac{\gamma}{H}\sqrt{1+\left(\frac{dy}{dx}\right)^2},$$

which is a non-linear second order differential equation. However it reduces to the first order ode (separable form)

$$\frac{dp}{\sqrt{1+p^2}} = \frac{\gamma}{H}dx$$

provided we write p for $\frac{dy}{dx}$. Integrating both sides w.r.to x, we get:

$$p = \frac{dy}{dx}\sinh\left(\frac{\gamma}{H}x\right)$$

Integrating once again with respect to x, we have

$$y = \frac{H}{\gamma} \cosh\left(\frac{\gamma}{H}x\right)$$

provided initial conditions befitting the physical situation are employed. The curve is known as **catenary**.

(b) Inverted Cycloid : Identify the curve for which the differential of the arc is k times the differential of the angle made by its tangent with the x-axis, multiplied by the cosine of this angle and determine the constant of integration so that the curve touches the x-axis at the point from which the arc is measured.

According to the condition of the problem,

$$ds = k\cos\psi d\psi, \quad \text{i.e, } \rho = \frac{ds}{d\psi} = k\cos\psi$$

In cartesian form, this yields

$$\frac{\left\{1 + \left(\frac{dy}{dx}\right)^2\right\}^{\frac{3}{2}}}{\frac{d^2y}{dx^2}} = \frac{k}{\sqrt{1 + \left(\frac{dy}{dx}\right)^2}} \quad \left(\text{Assuming that } \frac{d^2y}{dx^2} \neq 0\right)$$

$$\therefore \quad k\frac{d^2y}{dx^2} = \left(1 + \left(\frac{dy}{dx}\right)^2\right)^2$$

Writing $\frac{dy}{dx} = p$, we get on integration with respect to x,

$$x = k\int \frac{dp}{(1+p^2)^2}.$$

Substituting $p = \tan\left(\frac{\theta}{2}\right)$ one finds that

$$x = \frac{k}{4}(\theta + \sin\theta) \quad \text{and} \quad y = \frac{k}{4}\int(1 + \cos\theta)\tan\frac{\theta}{2}d\theta = \frac{k}{4}(1 - \cos\theta),$$

where we tacitly used the conditions $x = 0$ and $y = 0$ at $\theta = 0$.

In this way we have the parametric equation

$$\left.\begin{array}{rcl} x & = & \frac{k}{4}(\theta + \sin\theta) \\ y & = & \frac{k}{4}(1 - \cos\theta) \end{array}\right\}$$

of the **Inverted cycloid** whose generating circle has radius $\frac{k}{4}$.

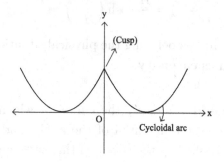

$$\textbf{Fig 2.5} : \text{Inverted cycloid} : \begin{cases} x & = & \theta + \sin\theta \\ y & = & 1 - \cos\theta \end{cases}$$

Remark : An inverted cycloid is both a tautochrone and a brachistochrone in the sense that on one hand the time taken by a particle to move from a point on a smooth inverted cycloid to its bottom under the force of gravity only is independent of the point chosen and on the otherhand to minimise the time taken in going from one point to a lower point, the path followed must be an inverted cycloid.

(c) A pursuit curve : A dog D standing at the point $(d, 0)$ on the x-axis, located a cat C at the point $(c, 0)$ and running parallel to the y-axis with a uniform velocity u cm/sec. The dog began to pursue the cat with a uniform speed v cm/sec, always moving in the direction of the cat so that at any time, CD is along the tangent to the path of the cat. Find the differential equation of the trajectory of pursuit. When and where does the dog catch up the cat?

From the figure in the next page it follows that

$$\tan(\pi - \psi) = \frac{ut - y}{x - c}$$

$$\therefore \quad (c - x)\frac{dy}{dx} + y = ut \qquad \left[\because \frac{dy}{dx} = \tan\psi \right] \qquad \text{(a)}$$

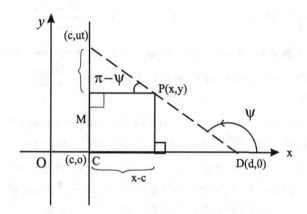

Fig 2.6 : Pursuit curve.

∴ Differentiating w.r.to x,

$$(c - x)\frac{d^2y}{dx^2} = \frac{u}{\frac{dx}{dt}} = -\frac{u}{v}\sqrt{1 + \left(\frac{dy}{dx}\right)^2},$$

which can be rewritten as a first order ode in x and p as

$$\frac{dp}{\sqrt{1 + p^2}} = k\frac{dx}{x - c} \qquad \text{(b)}$$

provided we agree to write $k \equiv \frac{u}{v}$ and $p \equiv \frac{dy}{dx}$. (b) gives us the first order ode of the trajectory of pursuit.

Integrating (b) and using the initial condition $p = 0$ whenever $x = d$, we get that

$$p = \sinh\left(k\ln\left(\frac{x - c}{d - c}\right)\right) \qquad \text{(c)}$$

By use of fundamental theorem of integral calculus we have from (c)

$$y(c) = \int_d^c p\,dx = \int_d^c \sinh\left(k\ln\left(\frac{x - c}{d - c}\right)\right)dx$$

$$= -(d - c)\int_0^1 \sinh(k\ln z)dz \qquad \left(\text{putting } \frac{x - c}{d - c} = z\right)$$

$$= \frac{d-c}{2} \int_0^1 (z^{-k} - z^k) dz$$

$$= \frac{d-c}{2} \left(B(-k+1,1) - B(k+1,1) \right) = \frac{(d-c)k}{(1-k^2)}$$

($0 < k < 1$ is a must for convergence of these improper integrals).

Thus the dog catches up the cat at a distance $\frac{(d-c)uv}{v^2-u^2}$ from the x-axis provided $0 < u < v$.

Since the line $x = c$ is the path of the cat, it follows from equation (a) itself that after $\frac{v(d-c)}{v^2-u^2}$ seconds the dog will intercept.

Note : If $k \geq 1$, i,e, if $u \geq v$, we face the mathematical problem of non-convergence of improper integral $\int_0^1 (z^{-k} - z^k)$ and effectively this corresponds to the infeasibility of the event 'catch up'.

(d) LR circuit : The following diagram shows an electrical circuit having self inductance L (in henries) (represented symbolically as a coil) and resistance R (in ohms) (represented by a wavy line). If i (in amperes) be the intensity of the instantaneous current flowing in the circuit and V (in volts) be the constant potential difference impressed in the direction of i at the terminals of the circuit, then from Ohm's law we get

$$\frac{di}{dt} + \frac{R}{L} i = \frac{V}{L}$$

This linear first order ode can be integrated between 0 and t to have

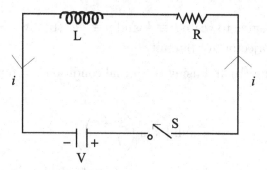

Fig 2.7(a) : LR-circuit with switch S.

$$i(t) e^{\frac{R}{L}t} - i(0) e^{\frac{R}{L} \cdot 0} = \frac{V}{L} \int_0^t e^{\frac{R}{L}t} dt = \frac{V}{R} \left(e^{\frac{R}{L}t} dt - 1 \right)$$

Therefore, $i = \frac{V}{R} \left(1 - e^{\frac{R}{L}t} \right)$ $(\because i = 0$ at $t = 0)$

The graph of solution curve shown below.

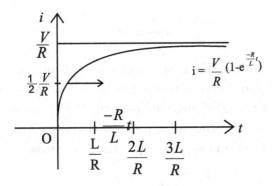

Fig 2.7(b) : Graph of the current (i) growth with time (t).

From this solution curve it follows that the current i is less than $\frac{V}{R}$ at any finite time t but approaches the steady state value when t becomes infinitely large. Infact $\frac{V}{R}$ is the current that would flow in the circuit if either there was a steady current (i.e, $\frac{di}{dt} = 0$) or there were no inductance (i,e, $L = 0$) in the circuit.

As a passing observation we find that the time required by the current in the LR circuit to assume half the steady state value is independent of the voltage V and given by $t_{\frac{1}{2}} = \frac{L}{R}\ln 2$ (Compare the situation with half-life of radioactive elements).

Remark : This LR circuit is also comparable to the problem of resisting medium in classical dynamics when resistance is proportional to velocity of the projectile. In that case 'terminal velocity' plays the analogue of the 'steady-state value' of current.

(e) Logistic Equation in Population Model : Here we start with a plausible assumption that the rate of change of the population at any time is directly proportional to the size of the population at that time. The mathematical analogue of this hypothesis is the equation:

$$\frac{dN}{dt} = kN, \quad k \text{ being constant of proportionality.}$$

We may consider the change in size of the population as a consequence of two different phenomena, viz, birth & death. Hence if we let

b denote the birth-rate (i.e, population-growth rate) and m denote the death-rate (i.e, population-annihilation rate), we can write

$$\frac{dN}{dt} = (b-m)N \tag{a}$$

Equation (a) elucidates the meaning of k only. If now we become more realistic and regard b and m themselves as depending upon the size of the population then (a) changes its pattern. If we assume the resources to be limited, usually m increases with N. Had been it assumed that m is proportional to N and so $m = m_1 N$, then (a) reads:

$$\frac{dN}{dt} = (b - m_1 N)N \tag{b}$$

which is a Bernoulli equation. If we restrict ourselves to the case where b is a constant, then we might give the solution of (b) by barely using separation of variable technique. If b is a function of time t, then there is no respite from inculcating treatment of Bernoulli equation.

In case b is a constant, we get

$$N = \frac{bN_0 e^{bt}}{e^{bt} m_1 N_0 + (b - m_1 N_0)}$$

and so if $t \to \infty$, $N \to \frac{b}{m_1}$, a constant which seems rather unrealistic (N_0 being initial population size).

In case b is a function of time t, say $b(t)$, then

$$\frac{dN}{dt} - b(t)N = -m_1 N^2$$

$$\text{or,} \quad -\frac{1}{N^2}\frac{dN}{dt} + \frac{b(t)}{N} = m_1$$

Put $y = \frac{1}{N}$ to have $\frac{dy}{dt} + b(t)y = m_1$,

which has the integrating factor $\exp\left[\int b(t)dt\right] = a(t)$, say.

$\therefore$ General solution is $y = \frac{1}{a(t)}\left\{\int m_1 a(t)dt + \text{constant}\right\}$,

which in turn gives N, the population size at time t.

(f) Varying Mass : A tennis ball of mass m_0 and downward velocity V_0 is dropped into a reservoir of goo which deposits matter onto the ball at a constant rate λ. If only the vertical motion is taken into account, the goo offers a resistance proportional to cube of the velocity, find how far the ball would dip into the goo?

Initial velocity being V_0 and initial mass being m_0, at subsequent time t, the mass is $(m_0 + \lambda t)$. The vertical distance (measured from the point of projection) is $x(t)$ and according to the given problem,

$$\frac{d}{dt}\left[(m + \lambda t)\frac{dx}{dt}\right] = -(m + \lambda t)g - k\left(\frac{dx}{dt}\right)^3 \qquad \text{(a)}$$

k being the proportionality constant.

On writing $\frac{dx}{dt} = v$ we have :

$$\frac{d}{dt}[(m + \lambda t)v] = -(m + \lambda t)g - kv^3$$

so that the above equation reduces to

$$\frac{dv}{dt} + \frac{\lambda v}{m + \lambda t} = -g - \frac{kv^3}{m + \lambda t} \qquad \text{(b)}$$

From the equation (b), it transpires that it is a non-linear first order differential equation which becomes linear had k been set zero, i,e, resistance offered by the goo is ignored while it becomes a Bernoulli equation had the acceleration due to gravity been ignored. However, it is never possible to piece together these two subcases because non-linear equations do not in general permit superposition of solutions. In this case obviously qualitative analysis is indispensable.

Remark : In physical world we often come across non-conservative forces like friction. This force presumably depends on velocity, i,e, $F = F(v)$. The simplest possible assumption is that $F \propto v$, which corresponds to fluid friction at low velocity. This is indeed the Stoke's law for viscous drag that is applicable also to the slow motion of dust particles in air or to the motion of the electrons in a conductor (c.f, Millikan oil-drop method for determining the charge of an electron). For moderate and higher speeds we sometimes find quadratic drag, i,e, $F \propto v^2$ (c.f. motion of parachute and short range missiles). This brief discussion and the previous examples, viz, (e) and (f), show that Bernoulli equation is more involved in the physical problems and their modelling.

(g) Newton's law of gravitation and Planetary motion

It is well-known a result in classical mechanics that to a first approximation, each planet moves round the sun in a bounded orbit, the underlying force being the attractive gravitational force of the inverse-square type. If M be the mass of the sun, m be the mass of the planet and r be the distance between these heavenly bodies at any instant $t-$ the force of attraction between them is $\frac{GMm}{r^2}$, where G stands for the universal constant of gravitation. The acceleration of the Sun towards the planet is $\frac{Gm}{r^2}$ while the acceleration of the planet towards the Sun is $\frac{GM}{r^2}$. Since both act along the radius vector joining sun and planet in opposite directions, the relative acceleration of the planet towards the sun is $\frac{G(M+m)}{r^2}$. If we write $G(M+m)$ as μ, we get the effective equation of motion of the planet as

$$\vec{F} \equiv F_r \hat{r} + F_\theta \hat{\theta} = -\frac{\mu}{r^2} \hat{r},$$

where $\hat{r}$ and $\hat{\theta}$ stand for unit vectors along radial and cross-radial directions respectively.

The component equations read :

$$F_r \equiv \frac{d^2 r}{dt^2} - r \left(\frac{d\theta}{dt} \right)^2 = \frac{-\mu}{r^2} \tag{i}$$

$$F_\theta \equiv \frac{1}{r} \frac{d}{dt} \left(r^2 \frac{d\theta}{dt} \right) = 0 \tag{ii}$$

Integrating (ii) w. r. to t, we have $r^2 \frac{d\theta}{dt} = h$, a constant of motion.

$h \neq 0$ because otherwise θ will be a constant and the planet will fall onto the sun after executing rectilinear motion. Physically h is very significant as it corresponds to 'conservation of angular momentum' in the plane of the orbit.

Substituting $r = \frac{1}{u}$ in (1), we get after a bit of algebra,

$$\frac{d^2 u}{d\theta^2} + u = \frac{\mu}{h^2} \tag{iii}$$

Rewrite $\dfrac{d^2u}{d\theta^2} = \dfrac{d}{du}\left(\dfrac{du}{d\theta}\right) \cdot \dfrac{du}{d\theta}$ and integrate (iii) to w.r. to u to have

$$\left(\frac{du}{d\theta}\right)^2 + \left(u - \frac{\mu}{h^2}\right)^2 = \frac{A^2}{h^4} \ , \quad \text{a constant.}$$

This peculiar form of integration-constant is chosen to make the final form of the orbit equation neater. This equation can be put in the separable form as

$$\frac{du}{d\theta} = \pm\sqrt{\frac{A^2}{h^4} - \left(u - \frac{\mu}{h^2}\right)^2}$$

Integrating both sides w.r. to θ, one finds

$$u = \frac{\mu}{h^2} + \frac{A}{h^2}\cos(\theta \pm \beta),$$

β being another constant of integration. So equation of the orbit is

$$\frac{h^2/\mu}{r} = 1 + \frac{A}{\mu}\cos(\theta \pm \beta).$$

Absorbing $\pm$ sign in the constants A and β we could have written the above polar equation of the orbit in the form

$$\frac{l}{r} = 1 + e\cos(\theta - \beta) \tag{iv}$$

provided we agree to write $\frac{h^2}{\mu} = l$ and $\frac{A}{\mu} = e$.

Equation (iv) is the general polar equation of a conic section with 'l' as semi latus rectum and 'e' as eccentricity. The origin works as the focus of the conic while the line $\theta = p$ is the axis of symmetry of the conic. The constant A is also physically significant as it stands for the magnitude of 'Runge-Lenz' vector, an important constant of motion. The angle $(\theta - \beta)$ is called 'true anomaly' and is the angle between radius vector and the Runge-Lenz vector. Since the orbit of the planet is bounded, $e = \frac{A}{\mu} < 1$. (As $e = 1$ corresponds to parabolic orbits followed by comets and $e > 1$ corresponds to hyperbolic orbits followed by some high-speed heavenly bodies passing through the solar system). Had e been 0, i,e, $A = 0$, as expected, the orbit would be circular—a possibility that is eliminated from 'specific-energy' consideration (Ref : § 4.7: Kepler's Problem of "Classical Mechanics", N Rana & P Joag).

(h) Parabolic Shaving Mirror : To find the shape of a mirror that reflects, parallel to a given direction, all the rays emanating from a given point source.

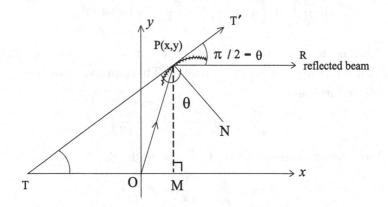

Fig 2.8 : Reflection of beam by the curved mirror

θ = angle of incidence TPT' : tangent at P

P = point of incidence PN : Normal at P

O = Source point $\overrightarrow{OP}$: Incident beam.

Let's choose the origin at the source O and the direction of the reflected beam as the direction of positive x-axis. In the figure, $\overrightarrow{OP}$ the incident ray, $\overline{PR}$ (parallel to the x-axis) is the reflected ray. $T'PT$ is the tangent and PN is the normal to the mirror.

$\therefore$ $\angle OPN$(angle of incidence) $= \angle RPN$(angle of reflection) $= \theta$, say.

From the figure, $\angle TPR = \angle PTO = \frac{\pi}{2} - \theta$ and $\angle OPT = \frac{\pi}{2} - \theta$.

Hence $\triangle POT$ is an isoceles.

If P has co-ordinates (x, y), the slope of $T'PT$ is $\frac{dy}{dx}$, where

$$\frac{dy}{dx} = \tan \angle PTO = \frac{PM}{TM} = \frac{PM}{OM + OP} = \frac{y}{x + \sqrt{x^2 + y^2}}$$

Our job is to solve this first order homogeneous ode in x and y. The ode is written in the standard differential form as

$$ydx - (x + \sqrt{x^2 + y^2})dy = 0$$

Here $M = y$ and $N = -x - \sqrt{x^2 + y^2}$ are homogeneous function of degree one in x and y.

Example (22) at once tells us that $\dfrac{-1}{y\sqrt{x^2+y^2}}$ is an integrating factor. Thus multiplying the above ode by this integrating factor, we have a resultant exact ode :

$$\frac{-dx}{\sqrt{x^2 + y^2}} + \left(\frac{x}{y\sqrt{x^2 + y^2}} + \frac{1}{y} \right) dy = 0,$$

which on integration gives

$$\ln \left(\frac{y}{x + \sqrt{x^2 + y^2}} \right) + \ln y = \ln c, \ \ln c \text{ being the integration constant.}$$

Since logarithm is injective, this is equivalent to

$$\sqrt{x^2 + y^2} = (c + x), \quad \text{i.e,} \quad y^2 = 2cx + c^2,$$

Thus we have the same general solution $y^2 = 2cx + c^2$ as before.

The next article is a very important type of applications of the first order differential equations. Because of its many use in geometry and potential theory, we discuss it rather elaborately and also point out the objectives of studying these examples. Next section is devoted to this.

2.9   Orthogonal Trajectories and Oblique Trajectories

In geometry and potential field theory one finds a family of plane curves each of which intersects the members of a given family at right angles. The two families are called "orthogonal" to each other and the individual members of one family are said to be orthogonal trajectories of the other family. For example, a family of concentric circles and a family of coplanar rays through their centre constitute two families of orthogonal trajectories. If individual members of one family intersect each of the members of the other family at a constant angle $\alpha \neq 90^0$, we call the first family to be oblique trajectories to the second family.

How to find orthogonal trajectories?

Let $F(x, y, c) = 0$ be the implicit equation of a family of curves, c being the family parameters. c identifies individuals of a family. If we differentiate the equation $F(x, y, c) = 0$ w.r. to x, we have

$$F_x + F_y \frac{dy}{dx} = 0 \; ;$$

Infact, without loss of generality one may assume that this equation also contains c, the family parameter. Eliminating c from the above two equations we have $f\left(x, y, \frac{dy}{dx}\right) = 0$, which is supposed to be rewritten in the form $\frac{dy}{dx} = g(x, y)$. This differential equation gives us the local behaviour of the family, as far as geometry is concerned. The differential equation of the family of curves that are orthogonal to members of the given family, reads :

$$\frac{dy}{dx} = -\frac{1}{g(x, y)} \quad \text{(assumed that } g(x, y) \neq 0\text{)}.$$

On integration, we have a family of orthogonal trajectiones.

The above procedure is only applicable for finding the orthogonal trajectories in rectangular cartesian co-ordinates. For polar co-ordinates, the approach will be a bit different in form. Here curves being $r = f(\theta)$ and $r = g(\theta)$,

$$\tan \phi_1 = r \frac{d\theta}{dr} = \frac{r}{\frac{dr}{d\theta}} = \frac{f(\theta)}{f'(\theta)}$$

$$\tan \phi_2 = \frac{g(\theta)}{g'(\theta)} \; .$$

$\therefore$ Angle between two curves $= \phi_1 \sim \phi_2$

$$= \tan^{-1} \left[\tan(\phi_1 \sim \phi_2)\right]$$

$$= \frac{\tan \phi_1 \sim \tan \phi_2}{1 + \tan \phi_1 \tan \phi_2}$$

$$= \frac{\frac{f(\theta)}{f'(\theta)} \sim \frac{g(\theta)}{g'(\theta)}}{1 + \frac{f(\theta)g(\theta)}{f'(\theta)g'(\theta)}}$$

$$= \frac{f(\theta)g'(\theta) \sim f'(\theta)g(\theta)}{f(\theta)g(\theta) + f'(\theta)g'(\theta)}$$

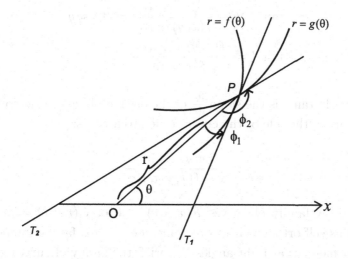

Fig 2.9 : Two curves $r = f(\theta)$ **and** $r = g(\theta)$ **intersecting at** (r, θ).

For orthogonality of the two curves, $f(\theta)g(\theta) + f'(\theta)g'(\theta) = 0$

$$\therefore \qquad \left(r \frac{d\theta}{dr} \right)_1 \cdot \left(r \frac{d\theta}{dr} \right)_2 = -1$$

where $\left(r \frac{d\theta}{dr} \right)_1$ is $r \frac{d\theta}{dr}$ computed for first curve i.e., $r = f(\theta)$ and $\left(r \frac{d\theta}{dr} \right)_2$ is the corresponding one for second curve i,e, $r = g(\theta)$.

Thus ode for the orthogonal trajectories of the curves given by a differential equation in the polar form is obtainable by replacing $r^2 \frac{d\theta}{dr}$ by $-\frac{dr}{d\theta}$. On integrating this new ode we have the family of orthogonal trajectories.

How to find oblique trajectories?

To get into act we have to find out as before the ordinary differential equation of the given family of curves. Without loss of generality we assume it to be expressible in the form $\frac{dy}{dx} = f(x, y)$.

Now the inclination of the tangent line at (x, y) with the +ve x-axis is $\tan^{-1}(f(x, y))$; we assume that the required family of curves makes a constant angle α with the given one. Hence the angle of inclination of the tangent to the desired curve at (x, y) is

$$\alpha + \tan^{-1}(f(x, y)) = \tan^{-1}(\tan \alpha) + \tan^{-1}(f(x, y))$$

$$= \tan^{-1}\left[\frac{f(x,y) + \tan\alpha}{1 - f(x,y)\tan\alpha}\right] = \beta, \quad say$$

$$\therefore \quad \tan\beta = \frac{f(x,y) + \tan\alpha}{1 - f(x,y)\tan\alpha}.$$

Incidentally $\tan\beta$ is the slope $\frac{dy}{dx}$ of the desired oblique trajectory at (x,y) and hence the ode of the curve to be given reads;

$$\frac{dy}{dx} = \frac{f(x,y) + \tan\alpha}{1 - f(x,y)\tan\alpha}.$$

Remark :   A family of curves $F(x,y,c) = 0$ or $G(r,\theta,c) = 0$ will be treated as self-orthogonal iff every member of this family intersects every other member at right angles. In brief, the family of curves given by $F(x,y,c) = 0$ will be self-orthogonal provided the corresponding differential equation of the family is invariant under the transformation

$$\frac{dy}{dx} \longrightarrow -\frac{dx}{dy}$$

For polar co-ordinate system, self-orthogonality is asserted only if the corresponding differential equation of the family is invariant under the transformation

$$r^2\frac{d\theta}{dr} \longrightarrow -\frac{dr}{d\theta}$$

Example (39) :   Find the orthogonal trajectories of the family of circles which are tangent to the y-axis at the origin.

The one-parameter a family of circles that are tangent to the y-axis at the origin reads

$$x^2 + y^2 + 2gx = 0, \text{ g being the parameter.}$$

The ode representing a family of circles that are tangent to the y-axis at the origin reads

$$x^2 + y^2 + 2gx = 0, \text{ g being the parameter.}$$

The ode representing the family of circles will be given by

$$(y^2 - x^2) - 2xy\frac{dy}{dx} = 0 \tag{$*$}$$

The ode of the family of trajectories orthogonal to $(*)$ reads :

$$\frac{dy}{dx} = \frac{2xy}{y^2 - x^2} \qquad\qquad (**)$$

$(**)$ being homogeneous first order ode, $y = vx$ substitutional trick works. On solving $(**)$ we get the cartesian equation of the orthogonal family as :

$$x^2 + y^2 - ky = 0, \qquad k \text{ being family parameter.}$$

Obviously this represents a family of circles that are tangent x-axis at the origin.

Example (40) : Find the value of n such that $x^n + y^n = c_1$ are the orthogonal trajectories of the family $y = \frac{x}{1-c_2 x}$. The differential equation of the family of curves $x^n + y^n = c_1$ will be $\frac{dy}{dx} = -\frac{x^{n-1}}{y^{n-1}}$ while the differential equation of the second family, viz, $\frac{1}{x} - \frac{1}{y} = c_2$ will be $\frac{dy}{dx} = \frac{y^2}{x^2}$.

Due to orthogonality, product of these two slopes will be -1.

$$\therefore \quad \frac{x^{n-1}}{y^{n-1}} \cdot \frac{y^2}{x^2} = 1 \Rightarrow \left(\frac{x}{y}\right)^{n-3} = 1 \Rightarrow n = 3$$

Hence $x^3 + y^3 = c_1$ and $y = \frac{x}{1-c_2 x}$ are mutually orthogonal families.

Example (41) : Show that the system of confocal conics given by

$$\frac{x^2}{a^2 + \lambda} + \frac{y^2}{b^2 + \lambda} = 1, \quad \lambda \text{ being family parameter, contains its own}$$

orthogonal trajectories, i.e., the family is self-orthogonal. Differentiating the given equation w.r. to x gives

$$\lambda = -\frac{(a^2 yy' + b^2 x)}{(x + yy')}$$

Substituting this value of λ in the original equation gives the ode of the given family as :

$$x(x + yy') - y(x + yy') = a^2 - b^2 \qquad\qquad [a]$$

To find the orthogonal family of curves w.r.t the given ones we have to substitute $y' \equiv \frac{dy}{dx}$ by $-\frac{1}{y'} \equiv \frac{dx}{dy}$.

However it is transparent that the equation [a] is invariant under the prescribed substitution, ensuring our claim of self orthogonality.

Example 42 : Show that the family of confocal parabolas $y^2 = 2cx + c^2$ is self-orthogonal.

Differentiating the given equation w.r.t x, we have $c = y\frac{dy}{dx}$ which gives the ode of the given family of confocal parabolas as

$$y = 2x\frac{dy}{dx} + y\left(\frac{dy}{dx}\right)^2 \qquad [\text{i}]$$

∴ The differential equation of the family of orthogonal trajectories w.r.to the family of parabolas reads

$$y = 2x\frac{dy}{dx} + y\left(\frac{dy}{dx}\right)^2 \qquad [\text{ii}]$$

[ii] being identical with [i], self-orthogonality follows.

Note : Combining the results of examples (41) and (42) one concludes that a system of confocal conics is self-orthogonal. In case of the family of conics discussed in (41) orthogonality occurs in the following sense:

$\lambda > 0$ correspond to ellipses only while $\lambda < 0$ corresponds to either ellipses or hyperbolas according as the relative magnitudes of a, b, λ stand. Here every individual member of the subfamily of ellipses cuts very individual member of the subfamily of hyperbolas orthogonally. The same result is easily verifiable in polar co-ordinate system. The equation of any system of confocal conics (pole being the focus of all members and axis of the conic coincident with the polar axis) reads $\frac{l}{r} = 1 - e\cos\theta$, where l stands for semi-latus rectum and e the eccentricity of the conic.

However l depends on e and explicity its expression stands as

$$\begin{aligned} l &= a(1 - e^2) & \text{if } e < 1 \text{ (ellipses)} \\ &= a(e^2 - 1) & \text{if } e > 1 \text{ (hyperbolas)} \\ &= 2a & \text{if } e = 1 \text{ (parabolas)} \end{aligned} \qquad [\text{a}]$$

where a is an arbitrary constant. (for the last case we treat 'a' as our parameter of the family).

Hence the polar equation of

$$\left.\begin{array}{llll} \text{ellipse} & : & r & = \frac{a(1-e^2)}{1-e\cos\theta} \\[3mm] \text{hyperbola} & : & r & = \frac{a(e^2-1)}{(1-e\cos\theta)} \\[3mm] \text{parabola} & : & r & = \frac{2a}{1-\cos\theta} \end{array}\right\} \qquad [\text{b}]$$

The differential equation of the ellipse or hyperbola (obtained through elimination of parameter e) reads

$$r^2\frac{d\theta}{dr} = a\left[\left(r\frac{d\theta}{dr}\right)^2\sin\theta - \sin\theta + 2r\cos\theta\frac{d\theta}{dr}\right] \qquad [\text{c}]$$

while the corresponding differential equation of the parabola (here a is itself a parameter) reads :

$$(1-\cos\theta)\frac{dr}{d\theta} + r\sin\theta = 0 \qquad [\text{d}]$$

Clearly both [c] and [d] are invariant under the transformation $r^2\frac{d\theta}{dr} \rightarrow -\frac{dr}{d\theta}$ so that [c] and [d] represent family of curves which are self-orthogonal. In particular we find that the ode obtained by the prescribed substitution will again represent confocal and coaxial parabolas with the sense of axis inverted.

Example (43) : Find orthogonal trajectories of the system of curves ,

$$r^n\sin n\theta = a^n, \quad a \text{ being family parameter.}$$

Taken logarithm on both sides to have

$$n\,ln\,r + ln\sin n\theta = ln\,a^n$$

Differentiation w.r.t θ gives the governing ode of the family :

$$r^2\frac{d\theta}{dr} = -r\cot n\theta$$

The ode of the family of curves orthogonal to the given one will be obtainable by replacing $r^2\frac{d\theta}{dr}$ with $-\frac{dr}{d\theta}$. On integration we get, after a sleight of hand a nice form, $r^n = b^n\sin n\theta$, b being a positive constant. This new family is called sine spiral and is very important as different values of n yields different familiar curves :

(i) $n = -1$ corresponds straight line.

(ii) $n = +1$ corresponds circle.

(iii) $n = -2$ gives a rectangular hyperbola.

(iv) $n = -\frac{1}{2}$ gives parabola.

Example (44) : Find a family of oblique trajectories that intersect the family of circles $x^2 + y^2 = c^2$ at angle 45°.

The differential equation of the family of circles is : $\frac{dy}{dx} + \frac{x}{y} = 0$. Thus the slope of the tangent to the circle $x^2 + y^2 = c^2$ at (x, y) is equal to $-\frac{x}{y}$ and hence the slope of the tangent to the oblique trajectory that makes an angle 45° with the circles $x^2 + y^2 = c^2$ is $\tan^{-1}\left(\frac{1-\frac{x}{y}}{1+\frac{x}{y}}\right) = \tan^{-1}\left(\frac{y-x}{y+x}\right)$.
Hence $\frac{dy}{dx} = \frac{y-x}{y+x}$, which can be solved by the standard substitution $y = vx$ to have

$$\ln(x^2 + y^2) + 2\tan^1\frac{y}{x} = 2k,$$

that represents the family of oblique trajectories.

Remark :

(a) From a physical viewpoint, the main aim of finding oblique or orthogonal trajectories to a given family of curves is to estimate directional derivative of a function in x and y. For the function $f(x, y)$ whose directional derivative we want to compute, the equation $f(x, y) = c$ represents the contour lines (i.e, treated as a pressigned family of curves) and their orthogonal or oblique trajectories quantify the directional derivatives.

(b) For a conservative force-field, the equipotential surfaces and the lines of force form a pair of mutually orthogonal families, for instance, if a point charge +q is placed at a point, the lines of force emanate from the charge. The family of concentric circles with centre at the charge are orthogonal to the lines of force and form the equipotential curves (the contour lines).

Exercise 2A

1. Check whether the following differential equations are exact or not. If exact, find the solutions in the implicit form.

    (a) $(3x^2 + 3xy)dx + (x + y^2)dy = 0$

(b) $(2x \sin y + y^3 e^x)dx + (x^2 \cos y + 3y^2 e^x)dy = 0$

(c) $(ye^{xy} \cos 2x - 2e^{xy} \sin 2x + 2x)dx + (xe^{xy} \cos 2x - 3)dy = 0$

(d) $\cos x \cos^2 ydx - \sin x \sin 2ydy = 0$

(e) $\left(\dfrac{1 + 8xy^{\frac{2}{3}}}{x^{\frac{2}{3}}y^{\frac{2}{3}}}\right)dx + \left(\dfrac{2x^{\frac{4}{3}}y^{\frac{2}{3}} - x^{\frac{1}{3}}}{y^{\frac{4}{3}}}\right)dy = 0$

(f) $(x \, lny + xy)dx + (y \ln x + xy)dy = 0$, x and y both positive.

(g) $(3x^2 \, ln \mid x \mid + x^2 + y)dx + xdy = 0$

2. Solve the following initial value problems :

(a) $\left(\dfrac{y - 3}{x^2}\right)dx + \left(\dfrac{2x - y^2}{xy^2}\right)dy = 0$, $y(-1) = 2$

(b) $(2ye^{2x} + 2x \cos y)dx + (e^{2x} - x^2 \sin y)dy = 0$, $y(0) = 3$

(c) $(2x \cos y + 3x^2 y)dx + (x^3 - x^2 \sin y - y)dy = 0$, $y(0) = 2$

3. Find the value of the constant λ for which the following equations are exact. Then solve these equations with those values of λ .

(i) $(xy^2 + \lambda x^2 y)dx + (x + y)x^2 dy = 0$

(ii) $\left(\dfrac{\lambda y}{x^3} + \dfrac{y}{x^2}\right)dx + \left(\dfrac{1}{x^2} - \dfrac{1}{x}\right)dy = 0$

(iii) $(2y^3 + 2)dx - 3\lambda xy^2 dy = 0$

4. Why is it necessary to exclude the origin from the domain of definition of the differential equation $\dfrac{xdy - ydx}{x^2 + y^2} = 0$? Give ample reasons.

5. Exact differential equations $M(x, y)dx + N(x, y)dy = 0$ can be solved alternatively as follows :

Compute $\int M(x, y)\partial x$ and $\int N(x, y)\partial y$ separately $- f_1(y)$ and $f_2(x)$ being the respective constants of partial integration. By mere inspection find $f_1(y)$ and $f_2(x)$. The implicit form of the solution is $F(x, y) = C$, where $F(x, y) \equiv \int M\partial x + f_1(y) = \int N\partial y + f_2(x)$.

Solve the following problems by making use of this technique :

(a) $(x^2 - 4xy - 2y^2)dx + (y^2 - 4xy - 2x^2)dy = 0$; $x, y \in \mathbb{R}^2$

(b) $(2y^{2x} + 2x \cos y)dx + (e^{2x} - x^2 \sin y)dy = 0$; $x, y \in \mathbb{R}^2$

(c) $(3x^2 y + 8xy^2)dx + (x^3 + 8x^2 y + 12y^2)dy = 0$; $y(2) = 1$

Exercise 2B

1. Find suitable integrating factors of the given equations and solve. Indicate also the respective regions over which exactness occur.

   (a) $(3x^2y + 2xy + y^3)dx + (x^2 + y^2)dy = 0$

   (b) $(5x^3y^2 + 2y)dx + (3x^4y + 2x)dy = 0$

   (c) $(y^2 + 2xy)dx - x^2dy = 0$

   (d) $(x^3y^3 - 3xy^2)dx + 3x^2ydy = 0$

   (e) $(xy\sin(xy) + \cos(xy))ydx + (xy\sin(xy) - \cos(xy))ddy = 0$

   (f) $y(xy + 2x^2y^2)dx + x(xy - x^2y^2)dy = 0$

   (g) $(3x + 2y^2)ydx + 2x(2x + 3y^2)dy = 0$

   (h) $(x + yf(x^2 + y^2))dy + (xf(x^2 + y^2) - y)dx = 0$,

   where f is a continuously differentiable function of $(x^2 + y^2)$

   (i) $(x - 2)y^3dx + (y^2 - 3)x^2dy = 0$.

2. (a) Consider the differential equation $M(x, y)dx + N(x, y)dy = 0$, where M and N belong to the space $C^1(D)$, D being some rectangular domain in $\mathbb{R}^2$. Prove that a function $\mu(x, y) \in C^1(D)$ is an integrating factor of the above equation provided

$$\mu\left(\frac{\partial M}{\partial y} - \frac{\partial N}{\partial x}\right) = \left(N\frac{\partial \mu}{\partial x} - M\frac{\partial \mu}{\partial y}\right)$$

   Is the condition sufficient also? Justify your answer.

   (b) If z be an analytic function of x and y and we want the integrating factor $\mu(x, y)$ as a function of z, show that

$$\mu = \mu(z) = \exp\left[\int \frac{\left(\frac{\partial M}{\partial y} - \frac{\partial N}{\partial x}\right)}{N\frac{\partial z}{\partial x} - M\frac{\partial z}{\partial y}} dz\right]$$

   provided the integrand itself is a function of z only.

   Now try to write down the criterion to be satisfied between $M(x, y), N(x, y)$ and their partial derivatives if we choose $z = x, y, (x + y), xy$ and $\frac{y}{x}$ in succession.

(c) Show that the following equations has an integrating factor of the form $\mu(x), \mu(y), \mu(x+y), \mu(xy)$ or $\mu\left(\frac{y}{x}\right)$ and then solve.

(i) $dx + \{1 + (x+y)\tan y\}dy = 0$

(ii) $(x^2 - y^2 + 1)dx + (x^2 - y^2 - 1)dy = 0$

(iii) $(3x^5 y^8 - y^3)dx + (5x^6 y^7 + x^3)dy = 0$

(iv) $(x^3 - 2y^3 - 3xy)dx + 3x(x + y^2)dy = 0$

(v) $(7x^3 + 3x^2 y + 4y)dx + (4x^3 + 5y + x)dy = 0$

(vi) $(2y^2 + xy + a^2)xdy + (y + 2x)(y^2 + a^2)dx = 0$

(vii) $(1 + xy)ydx + (1 - xy)xdy = 0.$

3. Consider the differential equation : $y^2 dx + x^2 dy = 0$

Show that both $\frac{1}{x^2 y^2}$ and $\frac{x^2 y^2}{(x+y)^4}$ are integrating factors of this equation. Does this different choice of integrating factors have any effect on the general solution?

4. If the differential equation $(x^2 + 2y)dx + f(x)dy = 0$ is known to have an integrating factor equal to x, find the generalised form of the function $f(x)$.

5. Consider the differential equation

$$\frac{1}{x^2} + \frac{1}{y^2} + \frac{\lambda x + 1}{y^3} \cdot \frac{dy}{dx} = 0$$

Establish that $\lambda = -2$ makes the equation exact and for this case the solution is

$$y(x) = \pm \left[\frac{x(2x-1)}{2(1+cx)}\right]^{\frac{1}{2}}, \quad c \text{ being integration constant.}$$

If $\lambda \neq -2$, suggest an integrating factor of this equation, if possible.

6. Solve the following equations by method of term-regrouping :

(a) $(ye^{xy} - 2y^3)dx + (xe^{xy} - 6xy^2 - 2y)dy = 0$

(b) $(x^{\frac{5}{2}} + x^{-\frac{1}{2}}y^4)dx + 8x^{\frac{1}{2}}y^3 dy = 0$

7. Solve the following differential equations :

    (a) $(x+4)(y^2+1)dx + (x+1)(x+2)ydy = 0$

    (b) $(e^y + 1)\cos x dx + e^y(1+\sin x)dy = 0$

    (c) $\left(x^3 + y^2\sqrt{x^2+y^2}\right)dx - xy\sqrt{x^2+y^2}dy = 0$

    (d) $(y^4 - 2x^3y)dx + (x^4 - 2xy^3)dy = 0$

    (e) $xyln\left(\dfrac{x}{y}\right)dx + \left(x^2ln\left(\dfrac{x}{y}\right) - y^2\right)dy = 0$

8. Show that the first order homogeneous differential equation $M(x,y)dx + N(x,y)dy = 0$ can be put into the separable form $A(r)dr + B(\theta)d\theta = 0$ by changing over to polar co-oprdinates by means of the substitution $x = r\cos\theta$, $y = r\sin\theta$. Apply this technique to express differential the equation

$$x\sin\left(\frac{y}{x}\right)(ydx + xdy) = 2y\cos\left(\frac{y}{x}\right)(xdy - ydx)$$

in polar form. Is the converse true, i.e., given a separable equation $f_1(r)dr + f_2(\theta)d\theta = 0$ in polar co-ordinates, can it be always cost into a homogeneous equation in cartesian co-ordinates? If yes, prove it. If not, cite a counterexample.

9. Solve the initial value problems :

    (a) $(3x+2y-5)dx + (2x+3y-5)dy = 0$; $y(0) = 2$

    (b) $(3x^2 + 9xy + 5y^2)dx - (6x^2 + 4xy)dy = 0$; $y(2) = -6$

    (c) $2(y+1)dy + (3x^2 + 4x + 2)dx = 0$; $y(0) = -3$

    (d) $(3x+8)(y^2+4)dx - 4y(x^2+5x+6)dy = 0$; $y(1) = 2$.

    (e) $(x^2+y^2)(xdx + ydy) + (x^2+y^2-2x+2y)(ydx - xdy) = 0$; $$y(1) = 0$$

Exercise 2C

1. Solve the following differential equations :

    (a) $\dfrac{dy}{dx} - 2y = x^2 + x$

    (b) $\dfrac{dy}{dx} + y = \dfrac{1}{1+e^x}$

(c) $\quad x\dfrac{dy}{dx} + 2y = \sin x$

(d) $\quad \dfrac{dy}{dx} + \left(2 + \dfrac{1}{x}\right)y = e^{-2x}$

(e) $\quad x(1 - x^2)\dfrac{dy}{dx} + (2x^2 - 1)y = ax^3$

(f) $\quad \dfrac{dy}{dx} + (gx)y = \cos x$

(g) $\quad \dfrac{dy}{dx} + y\tan x = x\sin 2x, \quad -\dfrac{\pi}{2} < x < \dfrac{\pi}{2}$

(h) $\quad (x\cos y + 2)\dfrac{dy}{dx} + \sin y = 0$

(i) $\quad y^4 dx + (1 + xy)dy = 0$

(j) $\quad \dfrac{dy}{dx} - 2xy = xy^2$

(k) $\quad x\dfrac{dy}{dx} + y = -2x^6 y^4$

(l) $\quad \dfrac{dy}{dx} + y\cos x = y^n \cos 2x, \quad n \in Z$

(m) $\quad (4xy^4 - 8x)dx + y^3 dy = 0.$

2. Show that general solution of the linear ode $\frac{dy}{dx} + P(x)y = Q(x)$ can be expressed as $y = v + t(u - v)$, where t is a constant and u, v are its two particular solutions.

3. If u and v be two solutions of the linear ode $\frac{dy}{dx} + P(x)y = Q(x)$; where $P(x)$ and $Q(x)$ are continuous functions of x and $v = uz$, then ,

$$z = 1 + A\exp\left(-\int \dfrac{Q(x)dx}{u}dx\right), \quad A \text{ being arbitrary constant.}$$

4. Solve the following initial value problems :

(a) $\quad (x + 2)\dfrac{dy}{dx} + y = g(x), \quad g(x) = \begin{cases} e^{-x}, & 0 \le x \le 2; y(0) = 1 \\ e^{-2}, & x \ge 2 \end{cases}$

(b) $\quad \dfrac{dy}{dx} + 3y = g(x), \quad g(x) = \begin{cases} x, & \text{if } 0 \le x \le 1; y(0) = \frac{2}{3} \\ x^2, & \text{if } x \ge 1 \end{cases}$

(c) $\quad \dfrac{dy}{dx} + 2y = g(x), \quad g(x) = \begin{cases} 2, & \text{if } 0 \le x \le 1; y(0) = 0 \\ 0, & \text{if } x \ge 1 \end{cases}$

(d) $\quad x\dfrac{dy}{dx} + P(x)y = xe^{-x}, \quad P(x) = (\Theta)(x) \; ; \; y(-1) = 1,$

$(\Theta)(x)$ being the Heaviside function

5. (a) In case only one of the functions $p(x)$ and $q(x)$ appearing in the equation $\frac{dy}{dx} + p(x)y = q(x)$ fails to be continuous, can the solution be made continuous? What kind of discontinuity may be perimitted for $p(x)$ and $q(x)$? If ξ be a point of discontinuity of $p(x)$ or $q(x)$ or even both, how the solutions on either side of $x = \xi$ are matched so as to have $y(x)$ continuous at that point? Can $\frac{dy}{dx}$ be continuous at $x = \xi$?

   (b) Find suitable condition that makes the solution $y(x)$ continuous everywhere on the entire domain of definition if

$$\frac{dy}{dx} + p(x)y = q(x), \quad p(x) \;=\; \begin{cases} 2, & 0 \le x \le 1 \\ 1, & x \ge 1 \end{cases}$$

$$\text{and} \quad q(x) \;=\; \begin{cases} e^{-2x}, & if \;\; x \le \frac{1}{2} \\ e^{x}, & if \;\; x > \frac{1}{2} \end{cases}$$

6. Apply the so-called 'superposition principle' to solve the equations

   (a) $\dfrac{dy}{dx} + y = \sin x + 3\cos 2x$

   (b) $\dfrac{dy}{dx} + 3y = e^{ix} + x^2$

   (c) $x^2 \dfrac{dy}{dx} + 2xy = \cos 5x + 2$

7. Consider the equation $L\frac{dy}{dx} + Ry = E_0 \sin \omega t$, where L, R, E_0 and ω are positive constant. Incidentally this equation represents the e.m.f. equation of the A.C. circuit containing a resistor with resistance R, a pure inductor with inductance L and a sinusoidal alternating source of e.m.f. represented by the term $E_0 \sin \omega t$.

Compute the solution $i(t)$ satisfying $i(0) = 0$. Show that this solution may be written in the form

$$i(t) = \frac{E\omega L}{R^2 + \omega^2 L^2} e^{-(R/L)t} + \frac{E}{\sqrt{R^2 + \omega^2 L^2}} \sin(\omega t - \phi),$$

where ϕ is the 'phase-lag' angle given by

$$\sin \phi = \frac{\omega L}{\sqrt{R^2 + \omega^2 L^2}} \quad \text{and} \quad \cos \phi = \frac{R}{\sqrt{R^2 + \omega^2 L^2}}$$

8. (a) Show that if p and λ be positive constants, and a any real number, then every solution of the equation

$$\frac{dy}{dx} + py = ae^{-\lambda x}$$

has the general property that $\lim\limits_{x \to \infty} y(x) = 0$.

(b) Show that if p be a constant and $q(x)$ be a continuous function on $[0, \infty)$, satisfying the condition that $|q(x)| \leq K, K > 0$, then solve the differential equation

$$\frac{dy}{dx} + py = q(x),$$

subject to the initial condition $y(0) = 0$. In case $Re(p) \neq 0$, show that this particular solution satisfies the condition

$$|y| \leq \frac{K}{Re(p)} \left[1 - e^{-(Re(p)x)}\right]$$

9. Consider the periodic differential equation with period 2π.

$$\frac{dy}{dx} = (\sin x + \alpha)y - \beta e^{\cos x} y^2.$$

It is a Bernoulli equation. Translate it into a first order linear equation by the standard 'Leibnitz trick' of substituting $u = \frac{1}{y}$. Solve the resultant equation by method of variation parameters. Use periodicity to evaluate the constants of integration.

10. Solve the following Riccati equations when a particular solution is known. In the problems $y_1(x)$ denotes this known solution.

    (a) $(2\cos^2 x - \sin^2 x + y^2)dx - 2\cos x dy = 0$; $y_1(x) = \sin x$

    (b) $(x^6 - x^4 y^2 - y)dx + x dy = 0$; $y_1(x) = x$

    (c) $\dfrac{dy}{dx} = -8xy^2 + 4x(4x+1)y + (1 - 4x^2 - 8x^3)$; $y_1(x) = x$

    (d) $\dfrac{dy}{dx} = xy^2 + \left(\dfrac{1}{x} - x^3\right)y + x$; $y_1(x) = x^2$

Exercise 2D

1. Find the orthogonal trajectories of

   (i) $r = 2a(\sin\theta + \cos\theta)$;

   (ii) $r^n = a^n \cos n\theta$

   (iii) $r^2 = a\cos\theta$,

   where 'a' stands for family parameter.

2. Show that the family of curves which intersect the lemniscate $r^2 = a^2\cos 2\theta$ at a constant angle α is $r^2 = c^2\cos(2\theta + \alpha)$.

3. Prove that the ellipse $3x^2 + 2y^2 = a^2$ and the semicubical parabola $y^3 = kx^2$ are orthogonal.

4. Find the orthogonal trajectories of the family of circles that are tangent to the line $y = x$ at the origin.

5. Find the orthognal trajectories of a family of straight lines that pass through the point $(-g, -f)$

6. Find the plane curves such that for each point on the curve, the y-axis bisects that part of the tangent line between the point of tangency and the x-axis.

7. Find the value of n so that the curves $x^n + y^n = c_1$ (c_1 being family parameter are orthogonal tyrajectories of the family $y = \frac{x}{1-c_2 x}$, c_2 being arbitrary constant.

8. Show that the equation of orthogonal trajectories of the system of ellipses $\frac{x^2}{a^2} + \frac{y^2}{a^2+\lambda} = 1$, λ being the family parameter, is $x^2 + y^2 - 2a^2 lnx + c = 0$.

9. Find a family of oblique trajectories that intersect the family of parabolas $y^2 = cx$ at an angle $\frac{\pi}{3}$. c being family parameter.

10. Find equation of the curve which cuts at a constant angle $\tan^{-1}\left(\frac{u}{v}\right)$ all the circles touching a given straight line at a given point.

11. Show that the orthogonal trajectories of a family of confocal and coaxial parabolas is again a system of confocal and coaxial parabolas. Does it imply self-orthogonality–give reasons to your answer.

12. Heat transfer : Newton's Law of Cooling :
   It's our daily experience that with the passage of time, a cup of hot beverage cools down to the room temperature. In contrast, a bottle of cold drinks loses half its chill when left in the room. In both the cases, Newton's law of cooling works that states that the rate at which the temperature of a body changes at any given time is approximately proportional to the difference between its own temperature (T) and the temperature (T_s) of the surrounding medium, so that

$$\frac{dT}{dt} = -k(T - T_s), \quad k > 0.$$

Show that the excess of the temperature of the body over that of the surrounding medium decays exponentially. The temperature of a silver beam is $60^\circ C$ above room temperature right now. Ten minutes earlier, it was $70^\circ C$ above room temperature. How far above the room temperature will the beam be (i) 10 minutes from now? (ii) When will the silver be $10^\circ C$ above room temperature?

13. Fick's Law of Diffusion :
   According to this law, the time rate of movement of a solute across a thin membrane is proportional to the area of the membrane as well as the difference in concentrations of the solute on either side of the membrane. Seemingly it is a joint variational problem, but if we assume the membrane to have a constant area and the concentration of the solute on one side to be a constant, viz, c_o, then Fick's law gives

$$\frac{dc}{dt} = -k(c - c_0), \quad k > 0$$

Show that, irrespective of the initial concentration of the solute on the other side, the concentration $c(t)$ tends to c_o after a prolonged period.

The concentration of sodium in a kidney is 0.0024 mg/c.c. The kidney is placed in a vessel containing sodium concentration 0.0035 mg/c.c. In two hours, the sodium concentration in the kidney is boosted to 0.0027 mg/c.c. How long the kidney is to be dipped in the solution so as to boost the sodium concentration to 0.0030 mg/c.c.?

Chapter 3

A Class of First Order Non-Linear Odes

3.1 Introduction

Bernoulli equations discussed in the previous chapter are the first serious non-linear odes taken up. However, the generalised parabolic or hyperbolic transformation enabled us to recast the Bernoulli equation to linear odes. The non-linearity in Bernoulli's equation owed its origin to the terms $x^\alpha y^\beta$ as the derivative-free terms. There are also many non-linear first order differential equations whose non-linearity is due to the occurence of higher powers of $\frac{dy}{dx}$ (p is the conventional abridged notation for $\frac{dy}{dx}$). The general form of the first order n-th degree ode reads

$$p^n + P_1 p^{n-1} + P_2 p^{n-2} + \cdots + P_{n-1} p + P_n = 0 \qquad (3.1)$$

where P_k's are functions of x and y.

In case the equation (3.1) is solvable for either p or x or y, we see that the non-linearity poses no threat as in either case techniques of previous chapter do the trick.

3.2 Non-linear First Order Ode Solvable for p

Let $f(x, y, p) = 0$ be a first order n-th degree equation in p. We regard $f(x, y, p) = 0$ as an n-th degree polynomial equation in p which by means of fundamental theorem of algebra admits of exactly n roots. These n roots of the polynomial equation give rise to n linear factors. Suppose, for definiteness,

$$f(x, y, p) \equiv (p - p_1)(p - p_2) \cdots (p - p_n) = 0,$$

where p_k's, $k = 1, 2, \cdots, n$ are functions of x and y. In this way we get n first order normal equations $\frac{dy}{dx} = p_k(x, y)$. We take it granted that each of these "first order and first degree" differential equations admits of explicit solutions in the form

$$y = g_k(x) + c_k,$$

c_k's being constants of integration. All these c_k's being associated with one differential equation of first order, all are expressible as functions of a single variable/parameter, say c.

$$\therefore \quad c_1 = h_1(c); \ c_2 = h_2(c); \ \cdots\cdots; \ c_n = h_n(c)$$

General solution therefore reads :

$$y = g_k(x) + h_k(c), \ k = 1(1)n.$$

More compactly, all the solutions of the n component odes are included in the relation given by

$$(y - g_1(x) - h_1(c)) \, (y - g_2(x) - h_2(c)) \cdots\cdots (y - g_n(x) - h_n(c)) = 0.$$

Again all $h_k(c)$'s have as their range the 'single infinity.' If we fix up h_k's as being just the identity function, there is no loss of generality. Hence we set $h_k(c) = c$ for all $k = 1, 2, \cdots\cdots, n$ and get

$$\phi(x,y,c) \equiv (y - g_1(x) - c) \, (y - g_2(x) - c) \cdots\cdots (y - g_n(x) - c) = 0.$$

as the general solution or complete primitive.

We therefore arrive at the assertion : If $f(x,y,p) = 0$ be an n-th degree first order equation in p, then its general solution $\phi(x,y,c) = 0$ is an n-th degree equation in c. We get here two salient features, viz,

$$\begin{cases} \text{order of ode} \ = \# \ \text{independent constant of integration.} \\ \text{degree of ode} \ = \ \text{degree of the arbitrary constant of integration} \\ \qquad\qquad\qquad\quad \text{appearing in the complete primitive.} \end{cases}$$

Note : Choosing the solutions in explicit version $y_k = g_k(x) + c_k$ for the component odes $p = p_k(x,y)$ may seem to be rather biased an approach. However, we must keep in mind that from the theoretical standpoint, existence of explicit solutions ensure that the formal solutions are elevated automatically to the status of implicit solutions (cf. § 1.4). Our arguments in the foregoing paragraphs are therefore also tenable if the solutions were taken as $F_k(x, y, c_k) = 0$, $k = 1(1)n$.

Example (1) : Solve the ode

$$p^3 - p(x^2 + xy + y^2) + xy(x + y) = 0$$

The equation is factored as $(p - y)(p - x)(p + x + y) = 0$

Equating each of the three linear factors separately to zero we get the solutions. Therefore

$$\left.\begin{aligned} p - y &= 0 \quad yields \quad y - ce^x &= 0 \\ p - x &= 0 \quad yields \quad y - \tfrac{1}{2}x^2 + c &= 0 \\ p + x + y &= 0 \quad yields \quad x + y - 1 + ce^{-x} &= 0 \end{aligned}\right\}$$

These can be combined to have general solution

$$(y - ce^x)\left(y - \frac{1}{2}x^2 + c\right)(y + x - 1 + ce^{-x}) = 0,$$

c being an arbitrary constant of integration.

Observe that the three component odes being part of a first order and third degree ode, one and only one constant of integration, viz c, is existent and the complete primitive, as evident from the final result, is a cubic polynomial equation in c.

Example (2) : Solve the ode $x^2 p^2 - 2xyp + 2y^2 - x^2 = 0$

The given ode can be written in the form :

$$(xp - y)^2 + y^2 - x^2 = 0$$
$$\text{i, e,} \quad (x dy - y dx) \pm \sqrt{x^2 - y^2}\, dx = 0$$

Assuming $x \neq 0$ we divide both sides by x^2 to have :

$$\frac{x dy - y dx}{x^2} \pm \frac{\sqrt{1 - \frac{y^2}{x^2}}}{x} dx = 0$$

$$\text{or,} \quad d\left(\frac{y}{x}\right) \pm \frac{\sqrt{1 - \left(\frac{y}{x}\right)^2}}{x} dx = 0$$

$$\therefore \quad \frac{d\left(\frac{y}{x}\right)}{\sqrt{1 - \left(\frac{y}{x}\right)^2}} \pm \frac{dx}{x} = 0$$

Integrating both sides we have the two solutions as :

$$\sin^{-1}\left(\frac{y}{x}\right) \pm \ln \mid cx \mid = 0, \ c \text{ being an arbitrary constant.}$$

Thus the general solution (?) will be a combination of the above two :

$$\left(\sin^{-1}\left(\frac{y}{x}\right) + \ln \mid cx \mid\right)\left(\sin^{-1}\left(\frac{y}{x}\right) - \ln \mid cx \mid\right) = 0$$

Note : We worked out the solutions on the basis of assumption that $\mid y \mid < \mid x \mid$. In case $y = \mid x \mid$, the equation is also satisfied. This shows up that all solutions of the ode are not caught in the net of general solution.

Example (3) : If a, b, c be positive constants, find the general solution of the ode :

$$p^4 + ap^3 + bp - c = 0$$

From the theory of equations it follows that this ode represents a biquadratic polynomial equation in p whose constant term is negative and hence it admits of two real roots, one +ve and other -ve.

If $p = p_0$ (constant) be a real root, then $y = p_0 x + k$ is a solution, k being integration constant.

Since p_0 is a root of $p^4 + ap^3 + bp - c = 0$, it follows that

$$p_0^4 + ap_0^3 + bp_0 - c = 0$$

$\therefore$ General solution reads :

$$\left(\frac{y-k}{x}\right)^4 + a\left(\frac{y-k}{x}\right)^3 + b\left(\frac{y-k}{x}\right)^2 - c = 0$$

i, e, $$(y-k)^4 + ax(y-k)^3 + bx^2(y-k)^2 - cx^4 = 0$$

Example (4) : Solve the differential equation

$$p^5 - p^3 + p^2 - p + 2 = 0$$

The given ode is a polynomial equation of degree 5 in p. This polynomial equation admits of at least one negative real root as follows from the Descartes' rule in theory of equations. If $y = p_0$ be a solution (real) of that equation in p then $y = p_0 x + k$ (k being constant of integration) and hence the general solution is :

$$(y-k)^5 - x^2(y-k)^3 + x^3(y-k)^2 - x^4(y-k) + 2x^5 = 0$$

3.3   Non-linear Ode Solvable for y

In case the ode $f(x, y, p) = 0$ be solvable for y, we may write it in the form $y = g(x, p)$, which can be solved by differentiating the equation w.r.t x so that it reduces in general to the form

$$p = G\left(x, p, \frac{dp}{dx}\right).$$

The last equation can be solved to get a solution of the form $\phi(x, p, c) = 0$, c being constant of integration. The p-eliminant of this equation and the given equation $f(x, y, p) = 0$ gives us the required solution. A few illustrations will clarify our motivation of the working rule.

Example (5) :   Solve the ode $y = p \sin x + \cos x$

Here differentiation w.r.t x gives : $p = \sin x \dfrac{dp}{dx} + p \cos x - \sin x$

$$\therefore \qquad \frac{dp}{dx} - p \tan\left(\frac{x}{2}\right) = 1$$

This linear equation (x as independent and p as dependent variable) has an integrating factor $\cos^2\left(\frac{x}{2}\right)$

$$\therefore \qquad p \cos^2 \frac{x}{2} = \int \cos^2 \frac{x}{2} dx = \frac{x + \sin x + c}{2},$$

$$\text{so that,} \quad p = \frac{x + \sin x + c}{1 + \cos x}$$

Now we substitute this expression for p into the original ode to have its general solution :

$$\begin{aligned} y &= \frac{(x + \sin x) + c}{1 + \cos x} \cdot \sin x + \cos x \\ &= (x + c) \tan\left(\frac{x}{2}\right) + 1, \ \ c \text{ being constant of integration.} \end{aligned}$$

When the differential equations solvable for y are found to contain no x, we go into parametric methods. Incidentally we find that p itself works as a parameter here.

Example (6) : Solve : $e^y = p^3 + p$

Here x is absent and thus logarithm being taken, we have $y = \ln(p^3 + p)$. We treat p as parameter 't' and thus $y = \ln(t^3 + t)$. Consequently

$$
\begin{aligned}
x = \int dx &= \int \frac{dx}{dy} dy \\
&= \int \frac{1}{p} dy \\
&= \int \frac{3t+1}{t^2(t^2+1)} dt \\
&= 2\tan^{-1}(t) - \frac{1}{t} + c, \quad \text{say.}
\end{aligned}
$$

Thus the parametric form of solution reads :

$$
\left.
\begin{aligned}
x &= 2\tan^{-1}(t) - \tfrac{1}{t} + c \\
y &= \ln(t^3 + t)
\end{aligned}
\right\}
$$

3.4 Non-linear Ode Solvable for x

In case the differential equation $f(x, y, p) = 0$ be solvable for x, we may write it in the form $x = g(y, p)$ which can be solved by differentiating the equation w.r.to y so as to reduce it to

$$
\frac{1}{p} = G\left(y, p, \frac{dp}{dy}\right)
$$

The last ode can be solved to get a solution of the form $\phi(y, p, c) = 0$, c being constant of integration.

The p-eliminant of this equation and the given equation $f(x, y, p) = 0$ gives us the required solution. We put up a couple of illustrations below to clarify that approach.

Example (7) : Solve the ode $y = p^2 y + 2px$

The equation can be cast as $2x = \frac{y}{p} - py \ (p \neq 0)$

Differentiating both sides w.r.t. y, we have

$$
\frac{1}{p} + p = -y\left(1 + \frac{1}{p^2}\right)\frac{dp}{dy}
$$

$$\text{or,} \quad p\left(1 + \frac{1}{p^2}\right) = -y\left(1 + \frac{1}{p^2}\right)\frac{dp}{dy}$$

$$\therefore \quad \frac{dy}{y} + \frac{dp}{p} = 0 \quad \left[\because 1 + \frac{1}{p^2} \neq 0\right]$$

Integrating we have $yp = c$, c being arbitrary constant.

Substituting we have the general solution : $y^2 = 2cx + c^2$, which represents a family of parabolas as verfied before (c.f : example-31 of Chapter 2).

Example (8) : Solve the ode $p^2 - 2xp + y = 0$

We regard the equation to be solvable for x :

$$x = \frac{p^2 + y}{2p} = \frac{p}{2} + \frac{y}{2p} \quad (p \neq 0)$$

(If p were equal to zero, then $y = 0$ would be a solution). Differentiating both sides w.r.t y, and reshuffling the terms we have :

$$\frac{dy}{dp} + \frac{y}{p} = p,$$

This ode being a linear equation with y as dependent and p as independent variable, its solution is immediately given as

$$y = \frac{p^2}{3} + \frac{c}{p} \ , \quad c \text{ being constant of integration.}$$

Putting this back in the given ode we get $\quad x = \frac{2p}{3} + \frac{c}{2p^2}$

Thus we have furnished a parametric solution to the ode $\left.\begin{array}{l} x = \frac{2p}{3} + \frac{c}{2p^2} \\ y = \frac{p^2}{3} + \frac{c}{p} \end{array}\right\}$

$-p$ symbolically represents the parameter.

Remark : When the differential equations solvable for x are found to contain no y we can also go to parametric methods.

Example (9) : Solve the ode $\quad x^2 = a^2(1 + p^2)$

Put $p = t$, some real parameter.

Therefore, $x^2 = a^2(1 + t^2) \quad$ i,e, $\quad x = \pm a\sqrt{1 + t^2}$

Now $\quad y = \int p\, dx \;=\; \pm\, a \int \dfrac{t.\,2t}{2\sqrt{1+t^2}}\, dt$

$$= \pm\, a \left(\int \sqrt{1+t^2}\, dt - \int \dfrac{dt}{\sqrt{1+t^2}} \right)$$

$$= \pm\, \dfrac{a}{2} \left(t\sqrt{1+t^2} - \ln\left| t + \sqrt{t^2+1}\,\right| \right) + c$$

In fact the solutions (parametric form) will be either

$$\left.\begin{array}{l} x = a\sqrt{1+t^2} \\[2mm] y = \dfrac{a}{2}\left(t\sqrt{1+t^2} - \ln\left| t + \sqrt{t^2+1}\,\right| \right) + c \end{array}\right\}$$

$$\text{or}\quad \left.\begin{array}{l} x = -a\,\sqrt{t^2+1} \\[2mm] y = -\dfrac{a}{2}\left(t\sqrt{1+t^2} - \ln\left| t + \sqrt{t^2+1}\,\right| \right) + c \end{array}\right\}$$

The equations which have been dealt with in this chapter so far are of the general type $f(x,y,p) = 0$. We may look into the given equation as an implicit functional relation connecting three entities x, y and p. Obviously two of them might be independent. This may be reflected once we go into parametric framework and write $x = g_1(u,v)$; $y = g_2(u,v)$ and $p = g_3(u,v)$ — g_1, g_2, g_3 being differentiable w.r.t. each of the variables u and v.

When one variable, say y, is absent, the given differential equation becomes $f(x,p) = 0$ which may be looked into as an implicit functional relation connecting two variables x and p. Obviously one of them may be independent. This is reflected once we go into parametric framework and write $x = g_1(t)$; $y = g_2(t)$. Sometimes for simplification, p is taken equal to t so that $x = g_1(p)$ and $y = \int p g_1'(p)dp$ is the solution. In case the given ode is $f(y,p) = 0$, the line of thought is same as before with solution given parametrically as

$$y = g_2(p) \; ; \; x = \int \dfrac{g_2'(p)}{p}\, dp$$

In all cases, primes denote differentiation w.r.to the argument p.

Example (10) : Solve the differential equation :

$$y = 2px + f(xp^2), \qquad \text{where } f \text{ is a differentiable function.}$$

Looking into the given ode as solvable for y we differentiate the given ode w.r.to x to have

$$\left(p + 2x\frac{dp}{dx}\right)\left(pf'(xp^2) + 1\right) = 0,$$

where f' denotes derivative of f w.r.to the arument xp^2.

Therefore, either $p + 2x\frac{dp}{dx} = 0$ or, $pf'(xp^2) + 1 = 0$

The first component ode gives : $\frac{dx}{x} + 2\frac{dp}{p} = 0$, which on integration gives $xp^2 = $ constant $= c^2$, say.

Putting $xp^2 = c$ back into the given ode we have : $y = 2c\sqrt{x} + f(c^2)$ as the general solution. Note that $\{y - f(c^2)\}^2 = 4c^2x$, which evidently is a quadratic equation in c^2 (mind that c^2 and not c were accepted as the integration constant in this problem).

For the second component equation $pf'(xp^2) + 1 = 0$, we put $xp^2 = t$, say, so that $p = -\frac{1}{f'(t)}$.

$$\therefore \qquad x = t(f'(t))^2$$
$$y = 2px + f(xp^2) = f(t) - 2tf'(t)$$

This parametric solution

$$\left.\begin{array}{rcl} x &=& t(f'(t))^2 \\ y &=& f(t) - 2tf'(t) \end{array}\right\}$$

is seen to contain no arbitrary constant of integration and hence not a part of general solution. Infact, this solution is of the singular type, the origin of which we now intend to explore.

3.5   Existence and Uniqueness Problem

In Chapter 2 we stated that the linear first order ode

$$\frac{dy}{dx} + P(x)y = Q(x)$$

associated with the initial condition $y(x_0) = y_0$ admits of a unique solution if $P(x)$ and $Q(x)$ are C^1 functions on an interval $[a, b]$ containing

x_0. Suppose $f(x, y, p) = 0$ be a first order and n-th degree ode which is solvable for p. In this case at most n integral curves may pass through a given point in $\mathbb{R}^2$, because $f(x, y, p) = 0$ yields n component normal odes of the first order :

$$\frac{dy}{dx} = p_k(x, y), \quad k = 1, 2, \cdots\cdots, n$$

For clarification of our idea we take up Example-1 again : The three first order component odes, viz,

$$p = x \ ; \ p = y \ ; \ p = -(x + y)$$

are all linear each of which admits of unique solution at all points in $\mathbb{R}^2$ (since x, y, $x + y$ are polynomials in x and y).

However, for points on $y = x$, the p's of first and second odes are same, for points on $y = -2x$, the p's of first and third odes are same, for points on $y = \frac{1}{2}x$, the p's of second and third odes are same.

This shows that through each point of $y = x$ straight line there pass two integral curves, viz, $y = ce^x$ for some $c = c_1$ and $y = \frac{1}{2}x^2 - c$ for some $c = c_2$.

Infact, $y = x$ works as a common tangent to these curves. Similar arguments and geometrical interpretation may be furnished for remaining two cases also.

When for a first order non-linear equation solvable for p, we talk of uniqueness of solution, we mean that through every point (x', y') there should pass at most one integral curve in a preassigned direction. In this sense the first order ode of Example 1 violates the criterion of uniqueness along the lines $y = x$; $y + 2x = 0$ and $x + 2y = 0$.

Each of the points where the solution to the ode $f(x, y, p) = 0$ is non-unique in the above sense is called a **singular point** of the ode. The set of all singular points is known as the **singular set**.

$\therefore$ For example 1, the singular set is given by

$$A = \{(x, y) \in \mathbb{R}^2 : y = x \text{ or } y + 2x = 0 \text{ or}, x + 2y = 0\}$$

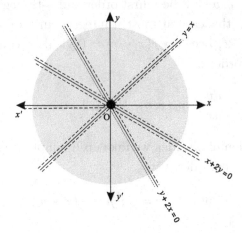

Fig 3.1 : The shaded regions show the zone where solution to ode in example-1 is regular.

As an illustration consider the ode : $xy(p^2 - 1) = (x^2 - y^2)p$,

This ode can be expressed as $(xp + y)(yp - x) = 0$. Observe that both the component odes, viz, $p = -\frac{y}{x}$ and $p = \frac{x}{y}$ have in their right sides functions that are discontinuous at the origin (as (x, y) approaches origin $(0,0)$, both $p = \frac{x}{y}$ and $p = \frac{y}{x}$ blow up). The first ode, viz, $p = -\frac{y}{x}$, gives on integration $xy = c$ (which represents a family of equilateral hyperbolas with co-ordinate axes as their asymptotes) and the straight line $x = 0$ while the second ode, viz, $p = \frac{x}{y}$, gives on integration $x^2 - y^2 = c$(which represents a family of equilateral hyperbolas with $y = \pm x$ as their asymptotes) and the straight line $x = 0$. Clearly $(x = 0, y = 0)$ is a singular point of both the component odes. However, the graph shows that there does not exist any point where more than one integral curve pass in a given direction, ensuring that the singular set of the differential equation under discussion is a singleton $\{(0,0)\}$.

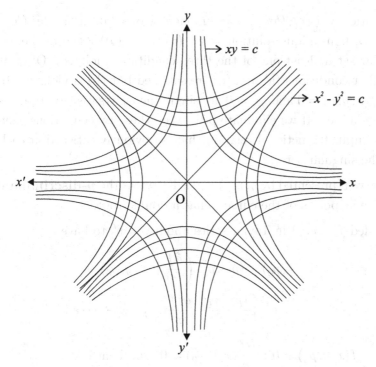

Fig 3.2 : Two families of equilateral hyperbolas, (i) $x^2 - y^2 = c$ (ii) $xy = c$

[centre $(0,0)$ of the hyperbolas is the saddle point]

Theorem 3.1 : If $f(x, y, p) = 0$ be first order ode with $y(x_0) = y_0$ as its associated initial condition and $y'(x_0) = p_0$, where p_0 is one real root of $f(x, y, p) = 0$ (treated as a polynomial equation in p or given (x_0, y_0)), then there exists a unique solution $y = y(x)$ in a sufficiently small neighborhood $N_h(x_0) \equiv (x_0 - h, x_0 + h)$ provided in that small neighborhood of the line element (x_0, y_0, p_0) the function $f(x, y, p)$ satisfies the following three conditions :

(a) $f(x, y, p)$ is continuous w.r.to each of the arguments.

(b) The derivative $\frac{\partial f}{\partial p}$ exists and is non-zero.

(c) The partial derivative $\frac{\partial f}{\partial y}$ is bounded in the sense that there exists a $K > 0$ such that $| \frac{\partial f}{\partial y} | \leq K$ for all points in that neighborhood.

We skip the rigorous proof and for the time being concentrate on the possible outcomes of its violation in specific problems :

Schematically, $(a) \wedge (b) \wedge (c) \Rightarrow$ uniqueness of solution of ode. Contrapositively, non-unique solution $\Rightarrow (\sim (a)) \vee (\sim (b)) \vee (\sim (c))$ i,e, on the singular set at least one of the three conditions violated. Often it happens that conditions (a) and (c) are satisfied but (b) is violated. In this case $f(x, y, p) = 0$ and $\frac{\partial f}{\partial p} = 0$ are simultaneously satisfied at points of the singular set. If we eliminate p between these two equations then we get an implicit functional relation that is obviously satisfied at each point of the singular set.

The curve represented by $g(x, y) = 0$ is known as the **p-discriminant curve** since at points of the curve, p has a double root.

Expanded $f(x, y, p)$ in Taylor series about $p = p_0'$ to have :

$$0 = f(x, y, p) = f(x, y, p_0') + (p - p_0') \left(\frac{\partial f}{\partial p} \right)(x, y, p_0') +$$

$$+ \frac{(p - p_0')^2}{2!} \frac{\partial^2 f}{\partial p^2}(x, y, p_0') + \cdots \text{ad inf.}$$

However, $\quad f(x, y, p_0') = 0 \; ; \; \frac{\partial f}{\partial p}(x, y, p_0') = 0 \quad$ and hence

$(p - p_0')^2 \phi(x, y, p_0') = 0$, i,e, p_0' is a double root of $f(x, y, p) = 0$ assumed solvable for p where $\phi(x, y, p_0') \equiv \frac{1}{2!} \frac{\partial^2 f}{\partial p^2}(x, y, p_0') + \cdots \infty$. This encourages us to pick the terminology 'p-discriminant' curve parallel to the idea of discriminant of roots in the theory of equations.

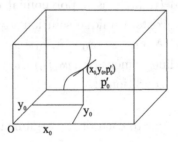

Fig 3.3 : Line element (x_0, y_0, p_0') at (x_0, y_0) in connexion to the ode :
$$f(x, y, p) = 0 \; ; \; y_0 = y(x_0) \; ; \; p_0' = y'(x_0)$$

However, (b) being one of the sufficient conditions of the uniqueness theorem, we cannot conclude that any point on the p-discriminant curve $g(x, y) = 0$ is always a point of the singular set, i,e, we cannot say deterministically that at each point of the p-discriminant curve, the solution is non-unique. If some branch $y = h(x)$ of the p-discriminant curve $g(x, y) = 0$ happens to be an integral curve (i,e, if $y = h(x)$ be an explicit solution of the ode $f(x, y, p) = 0$) and at every point of the curve $y = h(x)$, uniqueness of solution is violated in the sense that two or more integral curves of the ode pass through it, then $y = h(x)$ declared to be a singular solution of the ode. Its graph is called singular integral curve.

In a nutshell, $y = h(x)$ is a **singular integral curve** if all three following conditions simultaneously hold good :

(a) $y = h(x)$ is an integral curve

(b) $y = h(x)$ is a branch of the p-discriminant curve

(c) Every point of $y = h(x)$ is a singular point.

Example (11) : Find the singular solutions, if any, of the ode

$$xyp^2 - (x^2 + y^2)p + xy = 0$$

The ode is expressible as $(xp - y)(yp - x) = 0$, so that the component odes read $p = \frac{y}{x}$ and $p = \frac{x}{y}$ (provided $x \neq 0$ and $y \neq 0$). The ode $p = \frac{y}{x}$ has as the solution a family of coplanar rays : $y = c_1 x$ while $p = \frac{x}{y}$ has as the solution a family of equilateral hyperbolas :

$$x^2 - y^2 = c_2 \ (y = \pm x \text{ serving as their asymptotes})$$

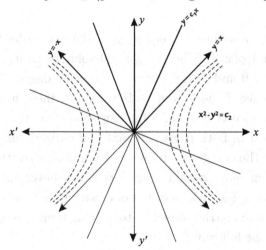

Fig 3.4 : Family of solutions for ode $xyp^2 - (x^2 + y^2)p + xy = 0$.
— trace the first family $y = c_1 x$ while $\cdots\cdots$ trace second family $x^2 - y^2 = c_2 x$

The p-discriminant curve is given by p-eliminant of

$$f(x,y,p) \equiv (xp - y)(yp - x) = 0 \quad \text{and} \quad \frac{\partial f}{\partial p} = 0, \quad \text{i, e,} \quad x^2 - y^2 = 0$$

which represents a pair of straight lines, $y = \pm x$. In fact $c_2 = 0$ corresponds to this pair of straight lines. Moreover, $c_1 = \pm 1$ correspond to the same pair of straight lines. This establishes that the lines $y = \pm x$ represent integral curves to the given ode.

Since $y = \pm x$ represent asymptotes of the hyperbolas, we observe that through every point of $y = x$ or $y = -x$ there passes exactly one solution to the ode, viz, the straight lines themselves and so at no point of the p-discriminanant curve, the uniqueness of solution is violated.

Thus $y = \pm x$ can never be singular integral curves. In fact, no singular solution exists.

3.6   Envelopes and Other Loci

In the remaining part of this chapter our discussion will focus on the singular solutions of the first order odes. However, we must be clear about the type of equations that admit of singular solutions. In general, a first order ode having degree unity cannot have singular solutions. Moreover, an ode $f(x,y,p) = 0$ cannot have singular solution if $f(x,y,p)$ can be resolved into factors that are linear in p and rational in x and y.

If $f(x,y,p) = 0$ be a differential equation of the first order but not of the first degree and $\phi(x,y,c)$ be the general solution to it, then we know that $f(x,y,p) = 0$ and $\phi(x,y,c)$ are of the same degree in p and c respectively and hence if there be a p-discriminant, there must exist a c-discriminant and vice-vice-versa. One can show that the singular solutions are contained in both p-discriminant and c-discriminant relations and also satisfy the ode given. Infact, c-discriminant relations and p-discriminant relation may contain other loci which never satisfy the ode. Because they are not solutions of the ode, we will not be much interested in these loci and restrict ourselves to quoting their types/nature and classification in the following.

(i) **Tac locus :**   It appears in the p-discriminant relation but not in the c-discriminant relation. It does not satisfy the ode.

(ii) Nodal locus : It appears in the c-discriminant relation but not in the p-discriminant relation. It does not satisfy the ode.

(iii) Cuspidal locus : It appears in both the c-discriminant and the p-discriminant relations but fails to satisfy the ode.

Using the symbols E, T, N, C for envelope, tac-locus, nodal locus and cuspidal locus respectively, we represent the c-discriminant relation as $EN^2C^3 = 0$, while p-discriminant relation as $ET^2C = 0$.

Thus **c-discriminant** involves

        (i) envelopes (singular solution)

        (ii) nodal locus squared.

        (iii) cuspidal locus cubed.

while **p-discriminant** involves

        (i) envelopes(singular solution)

        (ii) tac locus squared

        (iii) cuspidal locus.

Finally, we come up with envelopes that geometrically represent the singular solution. A curve E in the xy-plane is said to be an **envelope** of a singly infinite family of curves if it is tangent to an infinite number of members of the family and if at each point of E, at least one member of the family is tangent to E. The envelope satisfies the ode because it is a curve at each point of which the slope and the co-ordinates of the point match with the corresponding quantities of some member of the family of curves representing the general solution.

Importance of envelope is two-fold−

(a) The envelopes (singular solution) of an ode, if exists, separate out regions of no solution from the region where solutions exist. From the geometrical viewpoint we may deem of the envelope as a limiting integral curve to the ode given. This feature has already been discussed in a previous example.

(b) Envelopes enable us to construct new solutions by piecing together parts of general and singular solutions. The following examples illustrate this approach.

Example (12) : Solve the ode : $p^2y + p(x - y) - x = 0$

This ode can be written as $(p - 1)(py + x) = 0$ and so its solutions are $y - x = c$ and $y^2 + x^2 = c$, c being arbitrary constant of integration. The general solution therefore reads :

$$g(x, y, c) \equiv (x - y + c)(x^2 + y^2 - c) = 0$$

We may also rewrite the given ode as :

$$f(x, y, p) \equiv (p - 1)(py + x) = 0$$
$$\therefore \quad \frac{\partial f}{\partial p} = 2py + (x - y) = 0$$

Putting back $p = \frac{y-x}{2y}$ in the given ode we get the p-eliminant as $x + y = 0$ but this does not satisfy the ode. Hence $\nexists$ any singular solution.

Example (13) : Find the general and singular solutions, if any, of the ode : $p^2y^2 \cos^2 \alpha - 2pxy \sin^2 \alpha + (y^2 - x^2 \sin^2 \alpha) = 0$

$$f(x, y, p) \equiv p^2y^2 \cos^2 \alpha - 2pxy \sin^2 \alpha + (y^2 - x^2 \sin^2 \alpha) = 0$$
$$\therefore \quad \frac{\partial f}{\partial p} = 2py^2 \cos^2 \alpha - 2xy \sin^2 \alpha = 0$$

Hence $p = \frac{x}{y} \tan^2 \alpha$ and so the p-eliminant is :

$$\left(\frac{x}{y} \tan^2 \alpha\right)^2 y^2 \cos^2 \alpha - 2\left(\frac{x}{y} \tan^2 \alpha\right) xy \sin^2 \alpha + (y^2 - x^2 \sin^2 \alpha) = 0$$

On simplification this gives : $y^2 = x^2 \tan^2 \alpha$, i,e, $y = \pm x \tan \alpha$. Again it is easy to verify that this p-eliminant satisfies the ode given. Next we solve the ode by looking into it as a quadratic in py :

$$(py)^2 \cos^2 \alpha - 2(py)x \sin^2 \alpha + (y^2 - x^2 \sin^2 \alpha) = 0$$

Therefore, $\quad py = \dfrac{x \sin^2 \alpha \pm \sqrt{x^2 \sin^2 \alpha - y^2 \cos^2 \alpha}}{\cos^2 \alpha}$

i.e, $\quad y \, dy \cdot \cos \alpha = \left(x \sin^2 \alpha \pm \sqrt{x^2 \sin^2 \alpha - y^2 \cos^2 \alpha}\right) dx$

i, e, $\quad \displaystyle\int \frac{d \left(x^2 \sin^2 \alpha - y^2 \cos^2 \alpha\right)}{2\sqrt{x^2 \sin^2 \alpha - y^2 \cos^2 \alpha}} = \mp \int dx$

i, e, $\quad \pm \sqrt{x^2 \sin^2 \alpha - y^2 \cos^2 \alpha} = (x - c \cos \alpha)$ [$c \cos \alpha$ is I.C]

Squaring and simplifying we get the general solution

$$g(x, y, c) \equiv x^2 + y^2 - 2cx + c^2 \cos^2 \alpha = 0$$

$$\therefore \quad \frac{\partial g}{\partial c} \equiv -2x + 2c \cos^2 \alpha = 0$$

This c-eliminant is : $(x^2 + y^2) \cos^2 \alpha - x^2 = 0$ i,e, $y = \pm x \tan \alpha$

We observe that c-eliminant and p-eliminant are identical and satisfy the given ode. So $y = \pm x \tan \alpha$ is the singular solution that is represented geometrically as envelope of the family of general solution representing a family of circles having centres on y-axis.

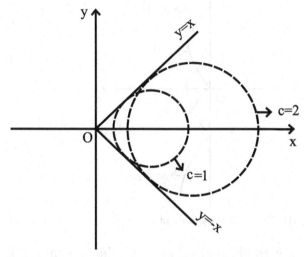

Fig 3.5(a) : For the family of circles $x^2 + y^2 - 2cx + c^2 \cos \alpha = 0$ $\left(\text{with } \alpha = \frac{\pi}{4}\right)$, the straightlines $y = \pm x$ are envelopes. It shows circles for $c = 1$ and $c = 2$.

Example (14) : Solve the ode : $4xp^2 = (3x - \beta)^2$, β being constant.

The given ode can be written as $p = \pm \dfrac{3x - \beta}{2\sqrt{x}}$ so that integrating

$$\int dy = \pm \int \left(\frac{3}{2}\sqrt{x} - \frac{\beta}{2\sqrt{x}} \right) dx$$

i, e, $y + c = \pm \left(x^{\frac{3}{2}} - \beta x^{\frac{1}{2}} \right)$, [c being integration constant]

Squaring both sides, $(y + c)^2 = x(x - \beta)^2$

Let's write $g(x, y, c) \equiv (y + c)^2 - x(x - \beta)^2 = 0$.

As $\frac{\partial g}{\partial c} = 2(y + c)$, the c-discriminant will be : $x(x - \beta)^2 = 0$. Similarly the p-discriminant relation reads : $x(3x - \beta)^2 = 0$.

The p-discriminant and the c-discriminant relations share the common factor $x = 0$ and correspondingly $\frac{1}{p} = 0$ satisfy the given ode. Hence $x = 0$ is the singular solution. Again as c-discriminant relation is representable as $EN^2C^3 = 0$, and the p-discriminant relation is representable as $ET^2C = 0$, it follows that $x - \beta = 0$ is a nodal locus while $3x - \beta = 0$ is a tac locus. There exists no cuspidal locus. If in particular we choose $\beta = 0$, the envelope, tac-locus and nodal locus all become identical, as seen in the figure 3.5(b).

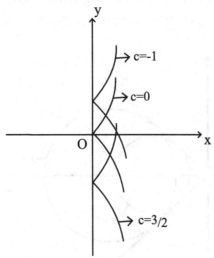

Fig 3.5(b) : Graph of the family of curves $y + c = \pm x^{\frac{3}{2}}$ (as $\beta = 0$)

where c is chosen with different values — yielding a family of semicubical parabolas. Incidentally the y-axis serves as the envelope, tac-locus as well as nodal locus. This diagram is related to example (14).

Example (15) : Reduce the ode $xp^2 - 2yp + x = 0$ to first degree and find all its solutions.

Here assuming $x \neq 0$, we reduce the ode to first degree equations by solving the quadratic in p and getting $p = \frac{y \pm \sqrt{y^2 - x^2}}{x}$.

In this homogeneous form we substitute $y = vx$ to have the separable form $\frac{dx}{x} = \pm \frac{dv}{\sqrt{v^2 - 1}}$, provided $v^2 \neq 1$.

Integrating both sides, we have after simplification,

$$v \pm \sqrt{v^2 - 1} = cx, \ (c \neq 0),$$

c being the constant of integration. If $v^2 = 1$, $y = \pm x$ which satisfy the ode but not obtainable from the general solution $y = vx = \frac{c^2x^2+1}{2c}$ by assigning a particular value to parameter c. The general solution represents a family of parabolas having vertices $\left(0, \frac{1}{2c}\right)$ and foci $\left(0, \frac{1}{c}\right)$. Each member of this family is touched by $y = x$ at $\left(\frac{1}{c}, \frac{1}{c}\right)$ and by $y = -x$ at $\left(-\frac{1}{c}, \frac{1}{c}\right)$. The parabolas for $c > 0$ are shown in figure (a) and those for $c < 0$ are shown in figure (b).

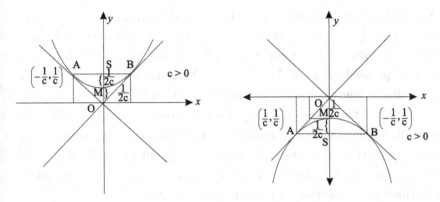

Fig 3.6 (a) : parabola for $c > 0$ **Fig 3.6 (b)** : parabola for $c < 0$

We may now combine the singular solutions/envelope and the general solution to yield further new solutions :

If $c > 0$, as evident from figure (a), the integral curves are defined by

$$y = \begin{cases} -x, & \text{if } -\infty < x \le -\frac{1}{c} \\ \frac{c^2x^2+1}{2c} & \text{if } -\frac{1}{c} \le x \le \frac{1}{c} \\ x & \text{if } \frac{1}{c} \le x \le +\infty \end{cases}$$

If $c < 0$, as clear from figure (b), the integral curves are defined by

$$y = \begin{cases} x, & \text{if } -\infty < x \le \frac{1}{c} \\ \frac{c^2x^2+1}{2c} & \text{if } \frac{1}{c} \le x \le -\frac{1}{c} \\ -x & \text{if } -\frac{1}{c} \le x \le +\infty \end{cases}$$

Each of these curves consists of two singular solutions of the given ode i,e, two mutually perpendicular straight lines and a parabolic arc which are parts of general solution. This curve is a smooth integral curve of the ode as it has continuous derivatives for all $x \in \mathbb{R}$.

Note : We could have obtained the same result by expressing y in terms of x and then applying the procedure of differentiation. However, the

fact that the resulting equation obtained through differentiation involves
a factor which, when equated to zero, leads to the singular solution is
not accidental. The underlying reason will be explained when we go into
discussion of higher order odes in the next chapter.

Example (16) : Find all solutions of the ode : $p^2 + 4y^2 - 1 = 0$

We rewrite the equation as $p^2 = 1 - 4y^2$ and thereby $p = \pm\sqrt{1 - 4y^2}$.
In case $\mid y \mid > \frac{1}{2}$, the slopes become imaginary. Using the separation of
variables, we have the general solution $y = \frac{1}{2}\sin(\pm 2x + c)$, which may
also be expressed as $y = \frac{1}{2}\sin(2x + c)$ because when c runs over a
single infinity, the families $y = \frac{1}{2}\sin(2x + c)$ and $y = \frac{1}{2}\sin(-2x + c)$
are identical. The lines $y = \pm\frac{1}{2}$ are also solutions of the ode. We
observe that these lines are tangent to the sinusoidal curves appearing
in general solutions and hence through every point of the lines $y = \pm\frac{1}{2}$
there pass two integral curves in one and the same direction, violating
the uniqueness of solution to the ode under discussion.

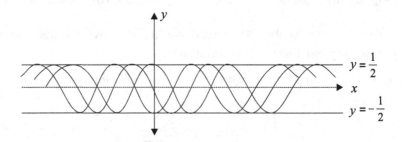

Fig 3.7(a) : Integral curves of the ode $p^2 + 4y^2 = 1$

In the figure 3.7(a) only a few sinusoidal curves $y = \frac{1}{2}\sin(2x + c)$ are
shown, all being enveloped by the lines $y = \pm\frac{1}{2}$. These envelopes might
be tactically used to construct new solutions. For instance, we might
consider $c = \frac{\pi}{4}$ so that a new solution is framed as :

$$y = \begin{cases} -\frac{1}{2} & \text{if } x \leq -\frac{3\pi}{8} \\ \frac{1}{2}\sin\left(2x + \frac{\pi}{4}\right) & \text{if } x \in \left[-\frac{3\pi}{8}, \frac{\pi}{8}\right] \\ x & \text{if } x \geq \frac{\pi}{8} \end{cases}$$

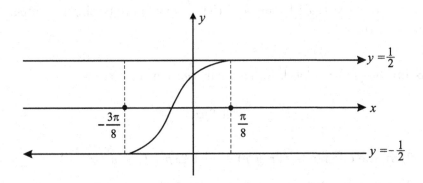

Fig 3.7(b) : Enlarged graph of the smooth solution curve constructed above by piecing together parts of envelopes and a particular integral curve of the general solution family.

Example (17) : Find singular solutions, if any, of the ode

$$x^2 + y^2 + p^2 - 1 = 0$$

Here p-discriminant curve to be determined as before, is the unit circle $x^2 + y^2 = 1$, which being an implicit functional relation has two explicit solution curves, viz, $y = \pm\sqrt{1 - x^2}$. However, direct computations reveal that if $y = \pm\sqrt{1 - x^2}$, $p = \pm\frac{x}{\sqrt{1-x^2}}$ which does not satisfy the given ode. None of the branches of the p-discriminant curve turn out to be integral curves. Hence no singular solution exists.

Example (18) : Find the singular solutions, if any, of the ode

$$p^3 - p(x^2 + xy + y^2) + xy(x + y) = 0.$$

The p-discriminant curve of the above ode, obtained by eliminating p between the given ode and the equation $p^2 = \frac{1}{3}(x^2 + xy + y^2)$.

The p-discriminant curve will have the equation

$$4\left\{(x + y)^2 - xy\right\}^3 = 27x^2y^2(x + y)^2$$

As seen before, the singular set is given by

$$A = \{(x, y) \in \mathbb{R}^2 : y = x \text{ or}, \ y + 2x = 0 \text{ or}, \ x + 2y = 0\}$$

One may check that all the three straight lines comprising the singular set A are branches of the p-discriminant curve. However none of these three curves satisfy the ode. Hence they fail to be singular integral curves of the ode.

Example (19) : Check for singular solutions of the ode

$$y = x^2 + 2px + \frac{p^2}{2}.$$

Here the ode reads $f(x, y, p) \equiv x^2 + 2xp + \frac{p^2}{2} - y = 0$:

Hence p-eliminant of $f(x, y, p) = 0$ and $\frac{\partial f}{\partial p} = 0$ will be $y = -x^2$. Thus the p-discriminant curve is a parabola. Direct substitution ensures that $y = -x^2$ is a solution of the given ode.

Finally we have to check whether there passes through every point of $y = -x^2$ any other integral curve. To verify this literally we have to seek the general solution of the given differential equation.

The general solution will be $\sqrt{2(x^2 + y)} = \pm(x + c)$, which on simplification becomes,

$$y = cx + \frac{c^2 - x^2}{2}, \ \ c \text{ being integration constant.}$$

Parametric equation of $y = -x^2$ being $x = t$, $y = -t^2$, we observe that through the generic point $(t, -t^2)$ of $y = -x^2$ passes the integral curve $y = -tx + \frac{t^2 - x^2}{2}$. Therefore, through every point of the p-discriminant curve two integral curves pass and hence $y = -x^2$ turns out to be a singular solution.

Example (20) : Investigate into the solutions of the ode

$$p^2 - (x + y)p + xy = 0$$

This ode has p-discriminant curve $y = x$ [obtained by eliminating p between the ode given and $2p = x + y$]. However, $y = x$ is not a singular solution because $y = x$ fails to satisfy the given differential equation. Let's now look for the general solution. The given equation is equivalent to two component odes, viz, $p = x$ and $p = y$, the general solution of the former being $y = \frac{x^2}{2} + c$, while that of the latter is $y = ce^x$.

Hence there exist two families of integral curves, one being parabolas while the other being exponential curves. Let's first find the restriction that needs be imposed on c so as to make $y = \frac{x^2}{2} + c$ intersect or touch p-discriminant curve $y = x$. Incidentally $c \leq \frac{1}{2}$ is a must. So for $c > \frac{1}{2}$, there exists no point on the curves of family $y = \frac{x^2}{2} + c$ where abscissa and ordinates are equal. In the sequel, we may conclude that there exists no common tangent to the integral curves at their point of intersection—one member taken from each family. Hence there exists no smooth integral curve that comprises of two parts one parabolic arc and other an exponential arc.

For further illustration, we take up $c = \frac{1}{2}$ case of the first family. Here at point $(1, 1)$ pass the curves $y = e^{x-1}$ and $y = \frac{1}{2}(x^2 + 1)$; $y = x$ being their common tangent. So we have traced out one point $(1,1)$ on $y = x$ where there pass two integral curves, one of each family. We may combine these two parts to have the following smooth integral curve :

$$y = \begin{cases} \frac{1}{2}(x^2 + 1) , & -\infty < x \leq 1 \\ e^{x-1} , & 1 \leq x < \infty \end{cases}$$

If we take up the case $c = 0$, $(0,0)$ and $(2,2)$ are the two points where $y = \frac{1}{2}x^2$ cuts the line $y = x$. Through $(0,0)$ passes the curve $y = 0$ of the second family while through $(2,2)$ passes the curve $y = 2e^{x-2}$ of the second family—common tangent being $y = 0$ itself in the first case while $y = 2(x - 1)$ in the second case. In this way we have discovered more points on where there pass two integral curves, one of each family. We may combine the parts of the integral curves in the following way :

$$y = \begin{cases} \frac{x^2}{2} , & -\infty < x \leq 0 \\ 0 , & 0 \leq x < \infty \end{cases}$$

and

$$y = \begin{cases} \frac{x^2}{2} , & -\infty < x \leq 2 \\ 2e^{x-2} , & 2 \leq x < \infty \end{cases}$$

Thus the points of $y = x$ are points of singular set as more than one (in fact two) integral curves pass through them. The integral curves (pieced together) for cases $c = \frac{1}{2}$ and $c = 0$ are shown in the figures.

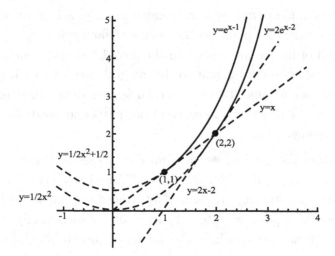

$$\textbf{Fig 3.8}: \text{ For } c = 0,\ y = \begin{cases} \frac{x^2}{2}, -\infty < x \leq 2 \\[2mm] 2e^{x-2}, 2 \leq x < \infty \end{cases}$$

with $y = 0$ being tangent at $(0,0)$ while $y = 2x - 2$ being tangent at $(2,2)$

$$\text{For } c = \frac{1}{2},\ y = \begin{cases} \frac{x^2}{2} + \frac{1}{2}, -\infty < x \leq 1 \\[2mm] e^{x-1}, 1 \leq x < \infty \end{cases} \quad \text{with } y = x \text{ being common tangent at } (1,1)$$

3.7   Clairaut's Equation and Lagrange's Equation

Through the different examples worked out in the previous article we observed that the differential equations of the first order but not of the first degree may or may not possess singular solutions. As a natural curiosity we should be interested to know whether there exists some criterion to examine if an ode of aforesaid type admits of singular solution or not. One clue in this respect has been the Clairaut's equation for which existence of a singular solution is guaranteed. Thus if we manage to transform a given first order ode to Clairaut's form, it should have a singular solution. Contrapositively we may infer that if an ode admits of no singular solution, it can never be transformed to Clairaut's equation. However, we cannot tell that if any first order ode admits of a singular solution, it is always transformable into Clairaut's form. The other unfortunate incident that really troubles us is the absence of definite transformation rules for getting into Clairaut's form where apriori it is known that this can be done for the particular problem. All that we have in hand are empirical transformation rules.

Clairaut's equation has the general form :

$$y = px + f(p) \tag{3.2}$$

where f must be a C^2 function. This is considered as the explicit form of the Clairaut's equation. To solve it we apply the procedure of differentiation to have :

$$p = p + \left(x + f'(p) \right) \frac{dp}{dx}$$

i, e, either, $\qquad x + f'(p) = 0, \qquad$ or, $\quad \dfrac{dp}{dx} = 0$

The second relation, viz, $\frac{dp}{dx} = 0$ can be integrated to give $p = c$, which combined with the Clairaut's equation gives $y = cx + f(c)$, a first degree equation in x and y that represents a family of straight lines.

The first part, viz, $x = -f'(p)$ when combined with the Clairaut's equation, gives the parametric form of a curve :

$$\left. \begin{array}{l} x = -f'(p) \equiv x(p), \\ y = f(p) - pf'(p) \equiv y(p), \end{array} \right\} \tag{3.3}$$

This curve in parametric form is to be identified now. We observe that the general solution $y = cx + f(c)$ and the given ode $y = px + f(p)$ are same in letter and spirit if we look into it as an equation involving a single parameter.

Writing $g(x, y, p) \equiv px - y + f(p)$ we observe that the p-discriminant curve of $g(x, y, p) = 0$ is the p-eliminant of

$$g(x, y, p) = 0 \quad \text{and} \quad \frac{\partial g}{\partial p} \equiv x + f'(p) = 0.$$

Clearly c-discriminant curve is identical with the p-discriminant curve.

Hence the curve

$$\left. \begin{array}{l} x = x(p) \equiv -f'(p) \\ y = y(p) \equiv f(p) - pf'(p) \end{array} \right\} \tag{3.4}$$

is indeed the p as well as c discriminant curve of the differential equation and represents parametrically the singular integral curve, p being the parameter. We will like to have some interpretation of this parameter p as given below. Taking differentials, we have from (3.4)

$$\left.\begin{array}{l} dx = -f''(p)dp \\ dy = -pf''(p)dp \end{array}\right\}$$

so that $\qquad \dfrac{dy}{dx} = \dfrac{pf''(p)}{f''(p)} = p$, only if $f''(p) \neq 0$.

Thus we observe that the parameter p of the singular integral curve (3.4) stands for the derivative $\frac{dy}{dx}$ only if $f''(p) \neq 0$, i,e, where the curve $y = f(x)$ has a non-zero curvature. Note that this result is not at all surprising as curves with zero curvature are just straight lines. Infact, the family of straight lines $y = cx + f(c)$ is tangent to the singular integral curve at every point, ensuring that the latter curve is an envelope to the family of straight lines appearing as general solutions. In brief, the solutions of the Clairaut's equation consists of a curve and the lines tangent to it.

Clairaut's equation in implicit form reads:

$$G(y - xp, p) = 0 \tag{3.5}$$

To solve this equation, we use the substitution $u = y - xp$, so that $\frac{du}{dx} = -x\frac{dp}{dx}$ and the given equation reads : $G(u, p) = 0$.

Differentiating $G(u, p) = 0$ w.r.to x, we have by chain rule:

$$\left(\frac{\partial G}{\partial p} - x\frac{\partial G}{\partial u}\right)\frac{dp}{dx} = 0$$

so that either, $\qquad \dfrac{dp}{dx} = 0$ or, $x = \dfrac{\partial G}{\partial p} \Big/ \dfrac{\partial G}{\partial u}$

For the special explicit form we have dealt with so far,

$$G(u, p) = u + f(p) \text{ and } x = \frac{\partial G}{\partial p} \Big/ \frac{\partial G}{\partial u} = -f'(p).$$

We pick up the following illustrations now:

Example (21) : Find all the possible solutions of the ode :

$$(xp - y)^3 + xp - y + p^2 + 1 = 0$$

Putting $y - xp \equiv u$, and writing

$$G(y - xp, p) \equiv (xp - y)^3 + xp - y + p^2 + 1$$

we have $G(u, p) = -u^3 - u + p^2 + 1$, so that $\frac{\partial G}{\partial p} = 2p$; $\frac{\partial G}{\partial u} = -3u^2 - 1$

Hence applying the procedure of differentiation, we have two equations as above : $\frac{dp}{dx} = 0$ and $x = \frac{-2p}{3u^2 + 1}$

The first equation gives on integration, $p = c$, which in turn yields the general solution : $(xc - y)^3 + xc - y + c^2 + 1 = 0$.

The second equation gives : $3x(y - xp)^2 + 2p + x = 0$, which combined with the given equation $(xp - y)^3 + xp - y + p^2 + 1 = 0$ represents the singular integral curve in parametric form.

Example (22) : Solve the ode $(y - px)^2 = a^2p^2 + b^2$ and obtain its singular solutions, if any.

We put $u = y - xp$ in the equation to have it in the form

$$G(u, p) \equiv u^2 - a^2p^2 - b^2 = 0$$

so that $\frac{\partial G}{\partial p} = -2a^2p$ and $\frac{\partial G}{\partial u} = -2u$. Applying the procedure of differentiation, we have the equations :

$$\frac{dp}{dx} = 0 \quad \text{and} \quad x = \frac{-a^2p}{u} = \frac{-a^2p}{y - xp}$$

Using the given ode we have from first relation, the general solution as

$$(y - cx)^2 = (a^2c^2 + b^2).$$

From the second relation we have, on combining with the given ode,

$$\left.\begin{array}{l} x^2 = \frac{a^4p^2}{a^2p^2 + b^2} \\ y^2 = \frac{b^4}{a^2p^2 + b^2} \end{array}\right\}$$

The p-eliminant being $\frac{x^2}{a^2} + \frac{y^2}{b^2} = 1$, we have the ellipse (canonical form) as the singular integral curve. The general solution represents family of a pair of parallel straight lines as can be seen from elementary geometry. Infact these pair of straight lines represent a pair of parallel tangents to the ellipse.

Geometrical interpretation of Clairaut's equation.

We have not yet established that the general solution $y = cx + f(c)$ and the singular solution (in parametric form) $y = px + f(p)$; $x + f'(p) = 0$ of the Clairaut's equation are distinct.

$$\text{Let} \qquad \left. \begin{array}{l} \phi(x, y, c) \equiv y - cx - f(c) \\ \text{and} \quad f(x, y, p) \equiv y - px - f(p) \end{array} \right\}$$

$\therefore$ By partial differentiation, $\frac{\partial \phi}{\partial x} = -c$; $\frac{\partial \phi}{\partial y} = 1$; $\frac{\partial f}{\partial x} = -p$; $\frac{\partial f}{\partial y} = 1$,

so that $\frac{\partial(\phi, f)}{\partial(x, y)} = \begin{vmatrix} \frac{\partial \phi}{\partial x} & \frac{\partial \phi}{\partial y} \\ \frac{\partial f}{\partial x} & \frac{\partial f}{\partial y} \end{vmatrix} = p - c$, which does not vanish identically

unless $\phi = 0$ and $f = 0$ holds simultaneously. Hence the two equations $\phi = 0$ and $f = 0$ are independent of one another and consequently their respective solutions are distinct.

To have a geometrical interpretation of the first order odes in general, we interpret x, y, p as the co-ordinates in three dimensional space $\mathbb{R}^3$ referred to the rectangular axes with the p-axis vertical. The xy-plane will be then spoken of as the horizontal plane. Consider any point A with co-ordinates $(x_0,\ y_0)$ in this plane, such that the vertical line through A intersects the surface $f(x, y, p) = 0$ in at least one point. Let the height of the point above the horizontal plane be p_0. Thus the point A has a definite direction $p = p_0$ associated with it; if A begins to move forward in this direction, P will begin to move along the surface but the altering value of p will involve a change in the direction of motion of A. If then we suppose that A moves along the horizontal plane (i,e, xy-plane) in such a way that the direction of its motion is measured by AP, the path traced out by A will be an integral curve of the ode, for it will be a continuous curve such that at every point on it, the relation $f(x, y, p) = 0$ is satisfied. This integral curve is the horizontal projection of a certain curve on the surface $f(x, y, p) = 0$. In case of Clairaut's equation in particular, the non-singular integral curves (i,e, general solution curves) are the projections of the intersections of the surface $y = xp + f(p)$ by the family of parallel planes $p = c$.

We now go for a more generalised version of Clairaut's equation, viz, **Lagrange's equation** that appears in the form

$$y = x\phi(p) + \psi(p) \tag{3.5}$$

where ϕ and ψ should be C^2-type functions. To solve the Lagrange's equation we differentiate it w.r.to x :

$$p = \phi(p) + (x\phi'(p) + \psi'(p))\frac{dp}{dx}$$

$$\text{or,} \quad (p - \phi(p)) = (x\phi'(p) + \psi'(p))\frac{dp}{dx} \tag{3.6}$$

$$\text{or,} \quad \frac{dx}{dp} + \frac{\phi'(p)}{\phi(p) - p}x = \frac{\psi'(p)}{p - \phi(p)} \quad \text{provided } \phi(p) \neq p. \tag{3.6a}$$

Under this restriction we get a linear ode of first order with x as dependent and p as independent variables.

An integrating factor $\mu(p)$ of this ode is known to be

$$\mu(p) = \exp\left[\int \frac{\phi'(p)}{\phi(p) - p}dp\right].$$

As expected, without the restriction $\phi(p) \neq p$, the integral for $\mu(p)$ could not be defined. Moreover from the theory of linear odes discussed in Chapter-2, the co-efficient of x, i,e, $\frac{\phi'(p)}{\phi(p)-p}$ and the x-free term, i,e, $\frac{\psi'(p)}{p-\psi(p)}$ are to be continuous everywhere within their respective domains of definition. The restriction $\phi(p) \neq p$ has indeed ruled out the possible points of infinite discontinuity of these co-efficients.

When passing from (3.6) to (3.6(a)) we had to carry out division by $\frac{dp}{dx}$. On division we lose solutions (if they exist) for which p is constant and hence $\frac{dp}{dx} = 0$. If p_k is a real root of the equation, then we have to add up the solutions $y = x\phi(p_k) + \psi(p_k)$ to the solution of ode (3.6(a)).

Note that if $\phi(p) = p$, the Lagrange's equation reduces to the traditional Clairaut's equation. However, Lagrange's equation does not always possess a singular solution as in seen in the following examples :

Example (23) : Consider the Lagrange's equation $y = 3px - p^3$ for checking out existence of singular solutions.

This equation has its p-discriminant curve $y^2 = 4x^3$, a semicubical parabola, but this is not an integral curve as it does not satisfy the original ode. Hence there exists no singular solution.

Following is an example of a Lagrange's equation for which the singular solution exists.

Example (24) : Consider the Lagrange's equation $y - x + \frac{4}{9}p^2 = \frac{8}{27}p^3$ for checking out existence of singular solution.

Differentiation of this given equation w.r.to x gives :

$$(p-1).\left(\frac{8}{9}p\frac{dp}{dx} - 1\right) = 0$$

i, e, either, $p = 1$ or, $8p\frac{dp}{dx} = 9$

Putting $p = 1$, we have from the given ode : $y = x - \frac{4}{27}$, which is free from arbitrary constant of integration. One can easily find that it is a part of the p-discriminant curve. However we cannot say whether this curve $y = x - \frac{4}{27}$ is a singular integral curve unless and until we find the general solution from the equation $\frac{8p}{9}\frac{dp}{dx} = 1$.

From this equation we have : $x = \frac{4p^2}{9} + \text{constant } (c)$

Using it in the given ode we have $y - c = \frac{8}{27}p^3$.

Hence the parametric form of the general solution is obtained. The elimination of p between $x - c = \frac{4p^2}{9}$ and $y - c = \frac{8}{27}p^3$. gives us the semicubical parabola :$(y - c)^2 = (x - c)^2$.

The c-discriminant curve will be obtained by c-eliminant of

$$2(y - c) = 3(x - c)^2 \text{ and } (y - c)^2 = (x - c)^3.$$

Squaring the first relation, we have, after removal of common factors,

$$\left.\begin{array}{l} x - c = \frac{4}{9} \\ y - c = \frac{8}{27} \end{array}\right\} \Rightarrow y = x - \frac{4}{27}$$

Hence c-discriminant curve and p-discriminant curve have the factor $y = x - \frac{4}{27}$ in common and moreover it satisfies the ode given. This shows that $y = x - \frac{4}{27}$ is a singular integral curve and as expected, through each point of this straight line two integral curves pass in one and the same direction—one, the straight line itself and other a semicubical parabola which is tangent to this straight line at the point under consideration.

Note that the p-discriminant curve of this ode also includes $p = 0$ which yields $y = x$ when substituted in the ode. However $y = x$ is not a solution to the ode itself.

Through the above two examples we could establish that unlike Clairaut's equation, Lagrange's equation does not always ensure us of the existence of singular solutions.

In the following we cite example of a Lagrange's equation for which application of the procedure of differentiation for obtaining general solution analogous to that applied to clairaut's equation may cause a loss of the relevant solution of the ode.

Example (25) : Solve the ode : $y = xp^2 + \ln(p^2)$

It is a Lagrange's equation with $\phi(p) = p^2$ and $\psi(p) = \ln p^2$.

Application of the procedure of differentiation yields :

$$p = p^2 + 2xp\frac{dp}{dx} + \frac{2}{p}\frac{dp}{dx}$$

$$\text{or,} \quad (p - p^2) = 2\frac{dp}{dx}\left(xp + \frac{1}{p}\right)$$

$$\text{or,} \quad \frac{dx}{dp} = \frac{2}{p(1-p)}\left(xp + \frac{1}{p}\right) \quad \text{provided } p \neq 1.$$

($p \neq 0$ has been ensured in the ode itself as it involves $\ln(p^2)$ term).

$$\frac{dx}{dp} + \frac{2}{(1-p)}x = \frac{2}{p^2(1-p)},$$

which is a linear ode with x as dependent and p as independent variable.

$\mu(p)$, the integrating factor is computed easily to be $(1 - p^2)$.

$$\therefore \quad x(1-p^2) = \int \frac{2}{p^2}(1-p)dx = 2\int\left(\frac{1}{p^2} - \frac{1}{p}\right)dp = -\frac{2}{p} - \ln p^2 - c$$

$$\text{i, e,} \quad x = \frac{1}{(p^2-1)}\left(\frac{2}{p} + \ln p^2 + c\right)$$

This together with the given ode, viz, $y = xp^2 + \ln p^2$ gives the parametric form of the general solution.

However, still there is a hiccup! if p were equal to 1, we would have $y = x$ as seen from the ode and this indeed would be a solution, Thus the above parametric form of the solution is to be appended by the particular solution $y = x$ to have the general solution.

Note : (a) Clairaut's equation was observed to contain two sets of solutions, viz, the general solution (involving arbitrary constant of integration) and the singular solution (involving no arbitrary constant of integration). If we are interested only in singular solution then we should apply the procedure of differentiation to the equation $y = px + f(p)$, but this time differentiation carried out w.r.to p :

$$\therefore \frac{dy}{dp} = x + p\frac{dx}{dp} + f'(p) \quad \Rightarrow \quad x + f'(p) = 0 \quad \left[\because p \equiv \frac{dy}{dx} \right]$$

Putting this into the ode, we have $y = -pf'(p) + f(p)$.

This approach at once gives the singular solution in parametric form without any knowledge of the general one. p serving as parameter.

(b) So far we have been talking of Clairaut's equation $y = px + f(p)$ with the underlying assumption that x is independent and y is dependent variable. If p were not zero, then $\frac{dy}{dx} = \frac{1}{dy/dx}$ and rewriting $q = \frac{1}{p}$, we have the reformulated equation :

$$x = qy - qf\left(\frac{1}{q} \right)$$

If one agrees to write $-qf\left(\frac{1}{q} \right) \equiv g(q)$, then the recast ode becomes $x = qy + g(q)$ which is again a Clairaut's equation. Hence one may conclude that the Clairaut's form is invariant under the interchange of roles of the two variables x and y involved. From elementary results of calculus one observes that the very assumption of non-vanishing p does guarantee that if y were a function of x, then x could as well be expressed as a function of y i,e, reversal of roles of independent and dependent variables is permissible.

Let's now address the question of possible reduction of an ode of first order but not of first degree to Clairaut's form. As stated earlier, there is no specific general transformation rule to serve our purpose. There are few empirical rules to be applicable to certain odes. For convenience of the reader, let's cite a couple of them in the following.

Rule A : If the ode appears in the form $y^2 = \left(\frac{py}{x}\right)x + f\left(\frac{py}{x}\right)$, f being a differentiable function of its argument, then the substitution $x^2 = u, y^2 = v$ reduces it to Clairaut's form $v = u + f(q) ; q \equiv \frac{dv}{du}$.

Rule B : If the ode appears in the form $e^{by}(bp - a) + f\left(p\frac{e^{by}}{e^{ax}}\right) = 0$, f being a differentiable function of its arguments, then the substitution $e^{ax} = u$; $e^{by} = v$ reduces the above ode to Clairaut's form : $v = qu + \frac{1}{a}f\left(\frac{a}{b}q\right) \equiv qu + g(q)$, say, where again $q = \frac{dv}{du}$.

We leave the verification of these rules to the reader.

Before continuing our illustrative examples, let's observe that the first order odes we begin with are valid in whole of $\mathbb{R}^2$ while the Clairaut's forms to which they are reduced, are valid in $\mathbb{R}^+ \times \mathbb{R}^+ \subset \mathbb{R}^2$.

Following problem [Example 26] admits of the transformation $u = x^2, v = y^2$ because in that case the ode can be cast into a form suited for applying Rule A :

$$y^2 = x\left(\frac{py}{x}\right) - h^2\frac{\left(\frac{py}{x}\right)}{1 + \left(\frac{py}{x}\right)} .$$

In Example (27), the given ode can be put in the form analogous to that given in Example (26). So it is also transformable to the Clairaut's form via the same transformation $u = x^2, v = y^2$.

By applying Rule B we may try the problem 4(e) given in Exercise.

Example (26) : Reduce the following ode to Clairaut's form by using the transformations $u = x^2$ and $v = y^2$:

$$(px - y)(py + x) = h^2p$$

$u = x^2$ implies $du = 2xdx$ while $v = y^2$ implies $dv = 2ydy$.

$$\therefore \ q \equiv \frac{dv}{du} = \frac{yp}{x} \ ;$$

The given ode is rewritten as :

$$(p^2 - 1)xy = (h^2 - x^2 + y^2)p$$

or, $\left(p - \dfrac{1}{p}\right) xy = (h^2 - x^2 + y^2)$ assuming $p \neq 0$

so that $\left(pxy - \dfrac{xy}{p}\right) = (h^2 - x^2 + y^2)$

or, $\left(\dfrac{py}{x}.x^2 - \dfrac{xy^2}{py}\right) = (h^2 - x^2 + y^2)$

or, $qu - \dfrac{v}{q} = (h^2 + v - u)$

or, $v = uq - \dfrac{h^2 q}{(q+1)}$, assuming $q \neq -1$

This is Clairaut's equation in u, v and q. Funnily, if $q = -1$, $x + py = 0$ and so h^2 would become 0, which is an absurdity.

Example (27) : Transform the following ode to Clairaut's form by means of the suitable transformations and hence solve it.

$$xyp^2 - (x^2 + y^2 - 1)p + xy = 0$$

We use the transformation : $u = x^2$ and $v = y^2$.

The given differential equation reads :

$$xy(p^2 + 1) = (x^2 + y^2 - 1)p$$

or, $xy\left(p + \dfrac{1}{p}\right) = x^2 + y^2 - 1$ assuming $p \neq 0$

or, $pxy + \dfrac{xy}{p} = x^2 + y^2 - 1$

or, $\left(\dfrac{px^2y}{x} + \dfrac{xy^2}{py}\right) = x^2 + y^2 - 1$ $\left[q = \dfrac{yp}{x}\right]$

or, $qu + \dfrac{v}{q} = u + v - 1$

or, $u(q - 1) = \dfrac{v}{q}(q - 1) - 1$

or, $v = uq + \dfrac{q}{q-1}$ asssuming $q \neq 1$.

If $q = 1$, then $yp = x$, which gives $x = 0$. Thus the given ode can be converted to Clairaut's form for all points of $\mathbb{R}^2$ with the exception of points along y-axis.

Differentiating resultant Clairaut's equation w.r.to q, we have :

$$\frac{dv}{dq} = u + q\frac{du}{dq} + \frac{1}{q-1} - \frac{q}{(q-1)^2}$$

or, $u + \dfrac{1}{q-1} - \dfrac{q}{(q-1)^2} = 0 \qquad \left[\because q \equiv \dfrac{dv}{du}\right]$

or, $u = \dfrac{1}{(q-1)^2}$

Using this result in the given ode, we have $v = \frac{q^2}{(q-1)^2}$

Thus the parametric form of the solution is :

$$\left. \begin{array}{l} x^2 = \frac{1}{(q-1)^2} \\ y^2 = \frac{q^2}{(q-1)^2} \end{array} \right\}$$

In the non-parametric form, this means $x+y-1=0$ and $x-y+1=0$ are the singular solutions.

The general solution will be $y^2 = cx^2 + \frac{c}{c-1}$, c being an arbitrary constant of integration other than unity.

Example (28) : Transform the equation $y = 2px - p^2y$ to the Clairaut's form by means of the transformation $u = x$ and $v = y^2$.

The given equation might be thought of as a particular type of Lagrange's equation where $\phi(p) = \frac{2p}{1+p^2}$ and $\psi(p) = 0$.

Since $u = x$ and $v = y^2$, $q \equiv \frac{dv}{du} = \frac{dv}{dx} = 2yp$

The recast ode in u, v, q is obtained as : $y = \dfrac{qx}{(1+p^2)y}$

$$\therefore \quad y^2 = \frac{qx}{(1+p^2)} = \left(\frac{q.u}{1+\frac{q^2}{4y^2}}\right) = \left(\frac{qu}{1+\frac{q^2}{4v}}\right)$$

or, $v = qu - \dfrac{q^2}{4}$, $[\because u = x$ and $v = y^2]$

which is a Clairaut's equation. Note that in the above steps we never introduced either squaring or square-rooting to avoid inclusion of extraneous solutions. Infact this caution is necessary to maintain one-one feature of transformation of the odes. Incidentally, the given Lagrange's equation admits of singular solution.

Example (29) : Transform the following ode to Clairaut's form

$$(2x^2 + 1) + p(x^2 + y^2 + 2xy + 2) + 2y^2 + 1 = 0$$

by using transformations $u = x + y$ and $v = xy - 1$ and then solve it.

Applying the procedure of differentiation, we have :

$$\frac{du}{dx} = 1 + p \quad \text{and} \quad \frac{dv}{dx} = y + xp$$

$$\therefore \quad \frac{dv}{du} \equiv q = \frac{dv/dx}{du/dx} = \frac{u + xp}{1 + p}$$

$$\text{i, e,} \quad p = \frac{q - y}{x - q}$$

Substituting this expression into the ode and regrouping the terms, we have after a bit of simplification,

$$q^2(x - y)^2 + q(x + y)(x - y)^2 + (1 - xy)(x - y)^2 = 0$$

Cancelling out the common factor on basis of the assumption $x \neq y$ and putting $x + y = u$ and $xy - 1 = v$, we have :

$$q^2 + qu - v = 0,$$

which is a Clairaut's equation in u, v, q. Its general solution is given by $v = cu + c^2$, c being an arbitrary constant of integration. The singular solution may be verified to be $u^2 = 4v$.

Rewriting these solutions in the old co-ordinates (x, y) we have :

(i) $xy = c(x + y) + c^2 + 1$, representing the general solution. This is seen to be a degenerate conic-family whose each member is a pair of intersecting straight lines.

(ii) $x^2 - 2xy + y^2 + 4 = 0$, representing a non-degenerate conic, viz, a parabola. As expected, members of the family of general solution are tangent to the singular integral curve (parabola).

Example (30) : Solve the ode $y = 2px + y^2p^3$ and also try to convert the equation to Clairaut's form.

Rewrite the given equation as : $x = \frac{y}{2p} - \frac{y^2p^2}{2}$.

Taking derivative of this equation w.r.to y and putting $\frac{dx}{dy} = \frac{1}{p}$, we get on simplification,

$$\frac{1}{2p} + yp^2 = -\frac{y}{p}\left(\frac{1}{2p} + yp^2\right)\frac{dp}{dy} \tag{i}$$

Dividing out $\left(\frac{1}{2p} + yp^2\right)$, assuming it to be non-zero for the time being and integrating, we obtain $y = \frac{c}{p}$, c being the constant of integration. Substituting this expression for y into the given ode, or its equivalent form, we find $x = \frac{c}{2p^2} - \frac{c^2}{2}$. Infact,

$$\left.\begin{array}{rcl} x & = & \frac{c}{2p^2} - \frac{c^2}{2} \\ y & = & \frac{c}{p} \end{array}\right\}$$

parametrically represents the general solution. In non-parametric form, the general solution is a family of confocal parabolas. If now $\frac{1}{2p} + yp^2$ were zero, then also the equation (i) would be satisfied. Elimination of p between given ode and $\frac{1}{2p} + yp^2 = 0$ would yield singular solution

$$27y^4 + 32x^3 = 0$$

The existence of singular solution kindles the hope that the given differential equation might be convertible to Clairaut's form. The following workout shows that our expectations are not belied.

Substitute $u = x$ and $v = y^2$ into the ode given :

Hence $du = dx$ and $dv = 2ydy$

This makes $q \equiv \frac{dv}{du} = 2yp$ and the reformulated ode reads :

$$y^2 = qx + y^3p^3 \qquad \text{i,e,} \quad v = uq + \frac{1}{8}q^3$$

which is a Clairaut's equation in u, v and q.

In the following we cite a nice example that will illustrate a few important points related to singular solutions.

Example (31) : Consider two families of continuous curves viz,

(i) a family of straight lines on which length 'a' is intercepted by the co-ordinate axes,

(ii) a family of ellipses the sum of whose semiaxes is a constant 'a'.

The equation of the first family of curves is given by

$$f(x, y; \alpha) \equiv x \sec \alpha + y \, cosec \, \alpha = a.$$

The envelope of this family will be obtained from $f(x, y; \alpha) = 0$ and

$$\frac{\partial f}{\partial \alpha} \equiv x \sec \alpha \tan \alpha - y \, cosec \, \alpha \, \cot \alpha = 0.$$

Thus the parametric equation of the envelope is obtained to be $x = a \cos^3 \alpha$; and $y = a \sin^3 \alpha$.

Eliminating α we have $x^{\frac{2}{3}} + y^{\frac{2}{3}} = a^{\frac{2}{3}}$, the equation of an astroid.

The equation of the second family is given by

$$g(x, y; \lambda) \equiv \frac{x^2}{(a\lambda)^2} + \frac{y^2}{\{a(1 - \lambda)\}^2} - 1 = 0.$$

The envelope of this family will be obtained through elimination of α from the equations

$$g(x, y; \lambda) = 0 \quad \text{and} \quad \frac{\partial g}{\partial \lambda} \equiv \frac{-2x^2}{a^2 \lambda^3} + \frac{2y^2}{a^2(1 - \lambda)^3} = 0$$

Hence the parametric equation of the envelope is obtained to be

$$x^2 = a^2 \lambda^3; \quad y^2 = a^2(1 - \lambda)^3.$$

Eliminating λ we have $x^{\frac{2}{3}} + y^{\frac{2}{3}} = a^{\frac{2}{3}}$, which is the same equation as for the first family. This workout establishes that the astroid might be thought of as an envelope of two different families of curves.

We now formulate the ordinary differential equations of the two families. Because of the existence of the envelopes it is expected that the governing odes will admit of singular solutions and hence are either Clairaut's equation or transformable into Clairauts equation by means of suitable co-ordinate transformations.

For the first family, $x \sec \alpha + y \, cosec \, \alpha = a.$

On differentiation w.r.to x, $\sec \alpha + p \, cosec \, \alpha = 0.$

$$\therefore \quad \left. \begin{array}{rcl} \sec\alpha & = & -\sqrt{1+p^2} \\[2mm] cosec\,\alpha & = & \dfrac{\sqrt{1+p^2}}{p} \end{array} \right\} \Leftrightarrow p = -\tan\alpha$$

Putting these in the equation $x\sec\alpha + y\,cosec\,\alpha = a$, we get

$$y = xp + \frac{ap}{\sqrt{1+p^2}}$$

which is a Clairaut's equation.

For the second family $\quad \frac{x^2}{(a\lambda)^2} + \frac{y^2}{\{a(1-\lambda)\}^2} = 1,$

we get through differentiation w.r.to x,

$$\lambda^2 = \frac{1}{ap}x(px - y) \text{ and } (1-\lambda)^2 = \frac{1}{a}y(y - px),$$

$$\therefore \quad \lambda = \frac{1}{a}\left[\frac{p + (px - y)(x + yp)}{2p}\right] \quad \text{so that}$$

$$\frac{1}{a}\left[\frac{p + (px - y)(x + yp)}{2ap}\right]^2 = \frac{x(px - y)}{ap}$$

i, e $\qquad [p + (px - y)(x + yp)]^2 - 4apx(px - y) = 0 \qquad\qquad$ [i]

Visibly (i) is the ode of the family of ellipses under discussion and this readily does not appear in Clairaut's form. We now apply the co-ordinate transformation $(x, y) \rightarrow (u, v)$ where $u = x^2$ and $v = y^2$ so that $q \equiv \frac{dv}{du} = \frac{yp}{x}$

In this set up, the ode (i) will change as :

$$\left[\frac{qx}{y} + \left(\frac{x^2 q}{y} - y\right)(x + qx)\right]^2 - \frac{4ax^2 q}{y}\left(\frac{x^2 q}{y} - y\right) = 0$$

or, $\quad [q + (x^2 q - y^2)(1 + q)]^2 - 4aq(x^2 q - y^2) = 0$

$\therefore \quad [q + (uq - v)(1 + q)]^2 - 4aq(uq - v) = 0, \quad (\because u = x^2 \text{ and } v = y^2)$

which is a Clairaut's equation in implicit form involving u, v and q.

Thus the astroid $x^{\frac{2}{3}} + y^{\frac{2}{3}} = a^{\frac{2}{3}}$ stands for the singular solution of two different first order differential equations whose general solutions are quite different in nature. This example is interesting in the sense it envisages that there always does not exist a one-one correspondence between a given ode and its singular solution (when it exist).

[We may quote at this point that the study of Clairaut's equation in details was done with a view to discovering whether or not it admits of any integral (solution) other than the primitive wherefrom it was derived. Incidentally, existence of singular solutions for Clairaut's equation being assured, we observe that there is no one-to-one correspondence between Clairaut's equations and their full set of solutions. Moreover the singular solution of a differential equation, if exists, does the job of partititioning $\mathbb{R}^2$, the plane where the integrals lie into two subsets — one where the solution is available and other where no solution exists.]

In example 22, the singular integral curve is the ellipse $\frac{x^2}{a^2} + \frac{y^2}{b^2} = 1$, in the interior of which no solution exists.

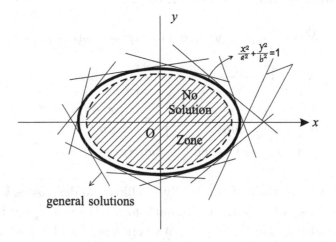

Fig 3.9 : Partitioning of $\mathbb{R}^2$ by singular integral curve $\frac{x^2}{a^2} + \frac{y^2}{b^2} = 1$ of the ode $(y - px)^2 = a^2 p^2 + b^2$ in Example 22.

Similar conclusion will hold true even if the curve representing the singular solution be not closed. As an illustration one may consider the Clairaut's equation $y = px + 2p^2$ whose general solution is the family of straight lines $y = cx + 2c^2$ while the singular solution is the parabola $x^2 + 8y = 0$. In the following diagram we observe that the parabola $x^2 + 8y = 0$ serves as the envelope of the family of straight lines $y = cx + 2c^2$, c being a family parameter and $\mathbb{R}^2$ is getting partitioned into three zones :

$$G_1 \equiv \{(x, y) \in \mathbb{R}^2 : x^2 + 8y > 0\};$$
$$G_2 \equiv \{(x, y) \in \mathbb{R}^2 : x^2 + 8y = 0\}$$
$$\text{and} \quad G_3 \equiv \{(x, y) \in \mathbb{R}^2 : x^2 + 8y < 0\},$$

G_3 indeed emerges as the no solution zone as the singular curve $x^2+8y=0$ cordons off the family of general solutions.

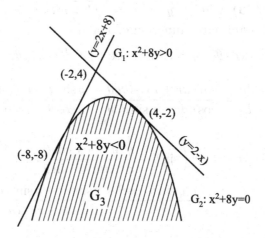

G_1: $x^2+8y>0$

(-2,4)

(4,-2)

(-8,-8)

$x^2+8y<0$

G_3

G_2: $x^2+8y=0$

$(y=-2x+8)$

$(y=-2x)$

Fig 3.10 : Singular integral curve and family of straight lines representing general solution of $y = px + 2p^2$

Example (32) : Solve the equation $x^2p^2 + yp(2x + y) + y^2 = 0$ by means of the substitution $u = y$; $v = xy$

Since $\quad y = u, p = \dfrac{dy}{dx} = \dfrac{du}{dx}$ and $xy = v \Rightarrow \dfrac{dv}{dx} = (xp + y) = (xp + u)$

$\therefore \quad \dfrac{dv}{du} = \dfrac{dv/dx}{du/dx} = \dfrac{xp + u}{p} = x + \dfrac{u}{p} = \dfrac{v}{u} + \dfrac{u}{p}$

$\therefore \quad \left(\dfrac{dv}{du} - \dfrac{v}{u}\right) = \dfrac{u}{p}$ so that $\left(u\dfrac{dv}{du} - v\right) = \dfrac{u^2}{p}$

The given ode is : $\quad x^2p^2 + yp(2x + y) + y^2 = 0$

$\quad\quad$ i, e, $\quad (xp + y)^2 + y^2p = 0$

$\quad\quad$ Therefore, $\quad \left(\dfrac{dv}{dx}\right)^2 + \dfrac{u^2}{p}\left(\dfrac{du}{dx}\right)^2 = 0$

$\quad\quad$ i, e, $\quad v = u\dfrac{dv}{du} + \left(\dfrac{dv}{du}\right)^2$,

which is a Clairaut's equation in independent variable u and dependent variable v. We write $\dfrac{dv}{du}$ as q so that the above Clairaut's equation takes the form : $v = uq + q^2$.

The singular solution is parametrically given by the joint equation $u = -2q$ and $v = -q^2$ so that its non-parametric form is : $u^2 + 4v = 0$

i,e, $y(y + 4x) = 0$ is the singular solution.

Example (33) : Solve $y - px = \sqrt{1 + p^2}\phi(x^2 + y^2)$, where ϕ is an arbitrary differentiable function of x and y.

Put $x = r\cos\theta$; $y = r\sin\theta$ so that $\frac{y}{x} = \tan\theta$.

Again $p = \dfrac{dy}{dx} = \dfrac{\sin\theta dr + r\cos\theta d\theta}{\cos\theta dr - r\sin\theta d\theta} = \dfrac{q\sin\theta + r\cos\theta}{q\cos\theta - r\sin\theta}$, where $q \equiv \dfrac{dr}{d\theta}$

Now $y - px = r\left(\sin\theta - \dfrac{q\sin\theta + r\cos\theta}{q\cos\theta - r\sin\theta}\cos\theta\right)$

$\qquad\qquad = \dfrac{r(q\sin\theta\cos\theta - r\sin^2\theta - q\sin\theta\cos\theta - r\cos^2\theta)}{(q\cos\theta - r\sin\theta)}$

$\qquad\qquad = \dfrac{-r^2}{(q\cos\theta - r\sin\theta)}$

Substituting all these in the given ode we get :

$$\frac{-r^2}{(q\cos\theta - r\sin\theta)} = \frac{\sqrt{q^2 + r^2}}{(q\cos\theta - r\sin\theta)} \cdot \phi(r^2)$$

i, e, $(q\cos\theta - r\sin\theta)\left[r^2 + \sqrt{q^2 + r^2}\phi(r^2)\right] = 0$

Hence either $q\cos\theta = r\sin\theta$ or, $r^2 + \sqrt{q^2 + r^2}\phi(r^2) = 0$

From the second factor we have : $r^4 = (q^2 + r^2)\phi^2(r^2)$

Therefore, $\dfrac{r^4}{\phi^2(r^2)} - r^2 = q^2$ i, e, $q = \pm r\sqrt{\dfrac{r^2 - \phi^2(r^2)}{\phi^2(r^2)}}$

On integrating, $\displaystyle\int \frac{\pm\phi(r^2)dr}{r\sqrt{r^2 - \phi^2(r^2)}} = \int d\theta = \theta + c,$

c being the constant of integration.

If in particular, $\phi\left(x^2 + y^2\right) = \left(x^2 + y^2\right)^{3/4}$, then $\phi(r^2) = r^{\frac{3}{2}}$

Therefore, $\pm\displaystyle\int \frac{r^{\frac{3}{2}}dr}{r\sqrt{r^2 - r^3}} = \theta + c$ so that

$\qquad\qquad \pm\displaystyle\int \frac{2\cos t\sin t dt}{\sqrt{\sin^2 t\cos^2 t}} = \theta + c$ i, e, $\pm 2t = \theta + c$

Hence $\pm 2\sin^{-1}\sqrt{x^2 + y^2} = \theta + c$ $[\because \sin t = \sqrt{r}]$

This gives us the general solution. From $q\cos\theta = r\sin\theta$ we get $r\cos\theta = c \Leftrightarrow x = c$ as another solution.

Example (34) : Solve completely the ode : $y = -xp + x^4p^2$.

From the very appearence of the ode itself, it is clear that the equation is neither of Clairaut's nor of Lagrange's type. However, this ode admits of singular solution.

Differentiating both sides w.r.to x, we get :

$$(2p - 4x^3p^2) + x\frac{dp}{dx}(1 - 2px^3) = 0$$

$$\text{or,} \quad \left(2p + x\frac{dp}{dx}\right)(1 - 2px^3) = 0$$

$$\therefore \quad \text{either,} \left(2p + x\frac{dp}{dx}\right) = 0, \text{ or, } 2px^3 = 1$$

But $2p + x\frac{dp}{dx} = 0$ gives on integration : $x^2p = c$, c being a constant. Hence the general solution reads : $y = -\frac{c}{x} + c^2$.

The singular solution will be the p-eliminant of the given ode and the relation $2px^3 = 1$ and hence is given by $4x^2y + 1 = 0$.

Example (35) : Solve the ode $(8p^3 - 27)x = 12p^2y$.

Arrange the given differential equation in the form

$$y = \left(\frac{2}{3}p - \frac{9}{4p^2}\right)x, \quad p \neq 0$$

Differentiating both sides w.r.to x, we get :

$$\left(2x\frac{dp}{dx} - p\right)\left(\frac{1}{3} + \frac{9}{4p^3}\right) = 0$$

$$\therefore \quad \text{either,} \left(2x\frac{dp}{dx} - p\right) = 0 \quad \text{or,} \quad \left(\frac{1}{3} + \frac{9}{4p^3}\right) = 0$$

But the first relation gives on integration $p^2 = cx$, c being arbitrary constant. Substituting into the given ode we have the general solution :

$$(4cx)^3 = (12cy + 27)^2$$

Further, the second relation gives $4p^3 + 27 = 0$, i,e, $p^3 = -\frac{27}{4}$. Substitution of this in the original differential equation yields $4y^3 + 27x^3 = 0$, which is the equation of the singular integral curve.

Exercise 3

1. Solve the following differential equations :

   (a) $xp^2 - 2yp + ax = 0$

   (b) $4y = x^2 + p^2$

   (c) $p - y = \ln(p^2 - 1)$

   (d) $p^3y^2 - 2px + y = 0$

   (e) $y - 2xp + yp^2 = 0$

   (f) $y = 3px + 6p^2y^2$

   (g) $p^3(x + 2y) + 3p^2(x + y) + p(y + 2x) = 0$

   (h) $y\{x(2x + 1)p - yp^2\} = 2x^3$

   (i) $xy(p^2 - 1) = (x^2 - y^2)p$

   (j) $x^2p^3 + y(1 + x^2y)p^2 + y^3p = 0$

2. Solve the following differential equations completely, i,e, find the general solution and singular solution, if any

   (a) $y = px + \cos p$

   (b) $y = px + a\sqrt{1 + p^2}$

3. Reduce the following differential equations to Clairaut's form by the substitution $u = x^2$; $v = y^2$ and hence solve.

   (a) $axyp^2 + (x^2 - ay^2 - b)p - xy = 0$

(b) $y = px + \frac{p}{x}$

4. Reduce the following differential equations to Clairaut's form by using the substitutions prescribed in the bracket. Hence solve.

(a) $(px^2 + y^2)(px + y) = (1 + p)^2 \cdots [u = x + y ; v = xy]$

(b) $(x^2 - a^2)p^2 - 2xyp - x^2 = 0 \cdots [u = x^2 ; v = y]$

(c) $(xp - y)(xp - 2y) + x^3 = 0 \cdots \left[u = x ; v = \frac{y}{x} \right]$

(d) $\left(\frac{2y}{x} - p \right) = f\left(px - \frac{y}{x^2} \right) \cdots [u = x ; v = y]$

(e) $e^{3x}(p - 1) + p^3 e^{2y} = 0 \cdots [u = e^x ; v = e^y]$

(f) $x^2 - \frac{xy}{p} = f(y^2 - xyp) \cdots \left[u = \frac{1}{x^2} ; v = \left(\frac{y}{x} \right)^2 \right]$

5. (a) Reduce the differential equation $y^2(y - px) = x^4 p^2$ to Clairaut's form by the substitutions $u = y ; v = xy$ and hence solve it for singular solution. show also that $y = 0$ is its tac locus.

(b) Reduce the ode $y^2(y - px) = x^4 p^2$ to Clairaut's form by means of substitutions $u = \frac{1}{x}$ and $v = \frac{1}{y}$ and show that it has no nodal locus. Does it have any tac locus?

(c) Find the singular solution as well as extraneous loci for the following odes :

(i) $4x(x - 1)(x - 2)p^2 = (3x^2 - 6x + 2)^2$

(ii) $p^2(3y - 1)^2 = 4y$

(iii) $2y^2p^2 + 2xyp + x^2 + y^2 = 1$

Chapter 4

Linear Algebraic Framework in Differential Equations

4.1 Introduction

Many of the basic problems of applied mathematics share the property of linearity, and linear spaces alongwith linear operators provide a very generalised framework of the analysis of such problems. The subject of differential equations, which prima facie seem to be a rather mechanical workout of variety of problems with the aid hundreds of artifices, is no exception as its basement is also built on linear spaces and linear operations, specially linear operators on normed linear spaces. Even when in absence of analytical solutions to differential equations we restore to numerical solutions of the same by using various difference schemes, their convergence is found to be based on L_2-norm or the likes. In this chapter we therefore dish out a concised account of functional analysis with special reference to differential operators. However, a beginner may skip this chapter on the first attempt if he/she finds it too intricate.

4.2 Linear Spaces

A non-empty set V is said to form a linear space over a non-trivial scalar field K (might it be $\mathbb{R}$ or $\mathbb{C}$ or something else) if it is associated with two algebraic operations, viz, vector addition (+) and scalar multiplication (denoted by juxtaposition) that satisfy the following constraints.

(i) $(V, +)$ is an abelian group, with 0 as additivity identity
(ii) V is closed under scalar multiplication, i.e., for any $\lambda \in K$, and any $v \in V$, $\lambda v \in V$
(iii) $(\lambda\mu)v = \lambda(\mu v)$ if $\lambda, \mu \in K$ and $v \in V$ (associativity of scalar multiplication)

(iv) $\exists\, 1 \in K$ such that $1v = v$ for any $v \in V$.

(v) Scalar multiplication is distributive over vector addition, i.e.,

$$\left.\begin{aligned}
\lambda(v_1 + v_2) &= \lambda v_1 + \lambda v_2 \text{ for any } v_1, v_2 \in V \\
(\lambda + \mu)v &= \lambda v + \mu v \text{ for any } \lambda, \mu \in K \text{ and } v \in V
\end{aligned}\right\}$$

Observe that a linear space is an algebraic structure that reflects or symbolises the properties associated with usual vector addition and scalar multiplication of vector algebra. The vector spaces are characterised as "real linear space" or "complex linear space" according as $K = \mathbb{R}$ or $K = \mathbb{C}$.

Examples of Linear Spaces

(a) $\mathbb{R}$, the set of all real numbers with usual addition and multiplication as the underlying algebraic operations is a linear space. Hence $\mathbb{R}$ plays a dual role — field as well as linear space. $\mathbb{C}$, the set of complex numbers also plays a similar dual role.

(b) $\mathbb{R}^n$, the set of all n-tuples of real numbers and $\mathbb{C}^n$, the set of all n-tuples of complex numbers are linear spaces under component-wise linear operations.

(c) $P_n([0,1])$, the set consisting of the zero polynomial and all the real polynomials on [0,1] of degree less than or equal to n, forms a real linear space under conventional pointwise addition and scalar multiplication of functions :

$$\left.\begin{aligned}
(f+g)(x) &= f(x) + g(x) \\
(\alpha f)(x) &= \alpha f(x)
\end{aligned}\right\}, x \in [0,1], \alpha \in \mathbb{R};\ f, g \in P_n([0,1])$$

(d) $P([0,1])$, the set of all real polynomials defined on [0,1] is a real linear space under pointwise addition and scalar multiplication of polynomials.

(e) $C([0,1])$, set of all continuous functions defined over [0,1] and having continuous derivatives on this interval upto order n inclusive — the algebraic operations being the pointwise addition and scalar multiplication of functions.

(f) $C\left([0,1]\right)$, set of all possible functions defined on [0,1] that are continuous together with its derivatives upto the n-th order inclusive
forms a vector space under the algebraic operations same as in (e).

(g) $C^{\infty}\left([0,1]\right)$ is the vector space of all infinitely differentiable functions having domain [0,1]— the underlying algebraic operations
being same as in (e). Exponential functions, trigonometric functions belong to this class.

(h) The set $\mathfrak{F}(\mathbf{X},\mathbb{R})$ of all real-valued functions defined on a non-empty
set X forms a linear space under usual pointwise addition and
scalar multiplication:

$$\left.\begin{array}{l}(f+g)(x) = f(x) + g(x)\\ (\alpha f) = \alpha f(x)\end{array}\right\} \text{ where } f,g \in \mathfrak{F}\ (X,\mathbb{R}) \text{ and } \alpha \in \mathbb{R}.$$

In particular if $X = \mathbb{R}$, then $\mathfrak{F}\ (X,\mathbb{R})$ reduces to $\mathfrak{F}\ (X,\mathbb{R},\mathbb{R})$, which we
shall denote by $\mathfrak{F}(\mathbb{R})$ for sake of brevity.

The spaces $P_n\left([0,1]\right), C\left([0,1]\right), C^n\left([0,1]\right), C^{\infty}\left([0,1]\right)$ are all 'function spaces' as their elements are all functions. It is important to note
that the intervals [0,1] and $[a,b]$ are equinumerous because of the bijec-
tive map $y = \frac{x-a}{b-a}$ from $[a,b]$ to [0,1]. This is why consideration of [0,1]
as the domain is no loss of generality for the above function spaces.

Linear Subspaces

A non-empty subset U of the vector space V is said to be a linear
subspace or subspace of V if U becomes a linear space in its own right.
In brief, $U \subset V$ is a linear subspace if

$$0 \in U \text{ and } x_1, x_2 \in U \text{ implies } \lambda_1 x_1 + \lambda_2 x_2 \in U.$$

We state in the following two important results related to subspaces
of linear spaces as they show when and how new subspaces can be generated from given subspaces.

(a) Arbitrary intersection of subspaces of a linear space is again a
subspace of the same linear space. In mathematical symbols, if
$\{W_\lambda : \lambda \in \bigwedge\}$ be any arbitrary family of subspaces of a linear space
V, then $W = \bigcap_{\lambda \in \bigwedge} W_\lambda$ is again a subspace of V.

(b) The union of two subspaces of a linear space need not be a subspace of that linear space in general. This union becomes a subspace only if one of the subspaces in contained in the other, So if W_1 and W_2 be two subspaces of linear space V, then $W_1 \bigcup W_2$ is a subspace of V iff either $W_1 \subseteq W_2$ or $W_2 \subseteq W_1$.

Examples of Subspaces

(a) $\mathbb{R}$ is a subspace of $\mathbb{C}$, no matter whether we choose $\mathbb{R}$ or $\mathbb{C}$ as the underlying field.

(b) Consider the set $C(\mathbb{R})$ of all continuous real-valued functions defined on $\mathbb{R}$. Observe that additive identity of $\mathfrak{F}(\mathbb{R})$, the zero function, is a constant function and so is an element of $C(\mathbb{R})$. The fact that $C(\mathbb{R})$ is closed under pointwise addition and scalar multiplication is easily verifiable. This leads to the conclusion that $C(\mathbb{R})$ is a subspace of $\mathfrak{F}(\mathbb{R})$.

(c) $D([0,1])$, the set of all differentiable functions defined over $[0,1]$, forms a vectors space under usual pointwise addition and scalar multiplication. It can be shown to be a proper subspace of $C([0,1])$ since differentiable functions are necessarily continuous but continuous function need not be differentiable. One can unify the above results to conclude that $P([0,1])$ is a proper subspace of $C([0,1])$.

(d) [0,1] being a closed and bounded subset of $\mathbb{R}$, is compact by Heine-Borel theorem. Again the functions having compact domains are bounded. So every real-valued $f \in C([0,1])$ is bounded but converse is not necessarily true. The set of all bounded functions with domain [0,1] forms a vector space (denoted by $B([0,1])$ under algebraic operations of pointwise addition and scalar multiplication of functions. $C([0,1])$ is a subspace of $B([0,1])$.

(e) If we take a subset W of $D([0,1])$ such that the derivatives of any $f \in W$ are themselves continuous, then W forms a subspace of U. This W is conventionally denoted by $C^1([0,1])$. Note that at the endpoints of [0,1] one-sided derivative is meant. Iterating this approach we can easily claim that $C^n([0,1])$ is a subspace of $C([0,1])$ for every $n \geq 1$. In the same vein, $C^\infty([0,1])$ is a subspace of $C^n([0,1])$ for every $n \geq 1$. These results are summarily expressed by the string of incluion relations :

(i) $P_n([0,1]) \subset P([0,1]) \subset D([0,1]) \subset C([0,1])$ $\forall\, n \geqslant 1$

(ii) $C^\infty([0,1]) \subset C^n([0,1]) \subset C([0,1])$ $\forall\, n \geqslant 1$

Affine Subspace or Linear Manifold

A common feature of linear subspaces is that they all contain the zero element. If we translate the origin of linear subspace, we obtain what is called an affine subspace [linear manifold].

Consider a linear space V defined over some field K (K being $\mathbb{R}$ or $\mathbb{C}$ as per our choice) and let, $v_i, v_2, \cdots, v_n$ be any preassigned vectors in V. A linear combination $\sum_{k=1}^{n} \alpha_k v_k$ with $\sum_{k=1}^{n} \alpha_k = 1$ (1 being the multiplicative identity of field K) is said to be an "affine combination" of $v_1, v_2, \cdots v_n$. An affine combination can therefore be looked into as "weighted average" of the vectors involved. A subset S of V is said to be an **affine subspace** of V provided it is closed under affine combination of vectors in S.

We now show that any affine subspace of a linear space V is of the form $S = v + W$, where v is some preassigned element of V and W is some subspace of V.

First we prove that if S is affine subspace of V then S is expressible in the form $v + W$, when W is a subspace of V and v is some fixed element in V.

We fix up some $v \in S$ and define $W \equiv S - v = \{s - v : s \in S\}$. Our mission is to show that W is non-empty and closed under linear combination of vectors. Observe that $0 \in W$ as $0 = v - v$.

Let $w_1, w_2, \cdots, w_n \in W$ and let $\beta_1, \beta_2, \cdots, \beta_n \in K$ be arbitrarily chosen. From the definition of W it follows that $\exists$ some $v_1, v_2, \cdots, v_n$ in S such that

$$w_1 = v_1 - v \; ; \; w_2 = v_2 - v \; ; \; \cdots \; ; \; w_n = v_n - v$$

$$\text{Now } \sum_{k=1}^{n} \beta_k w_k \;=\; \sum_{k=1}^{n} \beta_k (v_k - v)$$

$$=\; \sum_{k=1}^{w} \beta_k v_k - v \left(\sum_{k=1}^{n} \beta_k \right)$$

$$= \left(\sum_{k=1}^{n} \beta_k \right) \left[\frac{\sum_{k=1}^{n} \beta_k v_k}{\sum_{k=1}^{n} \beta_k} - v \right]$$

$$= \beta \left[\sum_{k=1}^{n} \frac{\beta_k}{\beta} v_k - v \right] \quad \left(\because \beta \equiv \sum_{k=1}^{n} \beta_k, \sum_{k=1}^{n} \frac{\beta_k}{\beta} = 1 \right)$$

$$= \beta(s - v) \quad (\because S \text{ is closed under affine combination})$$

where $s = \sum_{k=1}^{n} \frac{\beta_k}{\beta} v_k$ in S. Again $\beta(s - v) = \beta s + (1 - \beta)v - v$ and S being closed under affine combination, it follows that $\beta s + (1 - \beta)v \equiv s' \in S$

$$\therefore \quad \sum_{k=1}^{n} \beta_k w_k = s' - v \in W$$

This completes the proof of first part.

We now prove the converse part : If $S = v + W$ where v is some element of V and W is a subspace of V, then S is an affine subspace of V.

Take any arbitrary vectors $v_1, v_2, \cdots, v_n$ in S and consider any $\alpha_1, \alpha_2, \cdots, \alpha_n \in K$ such that $\sum_{k=1}^{n} \alpha_k = 1$. We are to show that $\sum_{k=1}^{n} \alpha_k v_k \in S$. By the form of S it follows that $\exists \; w_1, w_2, \cdots, w_n$(uniquely determined) in W such that $v_k = v + w_k$ for $k = 1, 2, \cdots, n$. Now

$$\sum_{k=1}^{n} \alpha_k v_k = \sum_{k=1}^{n} \alpha_k (v + w_k) + \left(\sum_{k=1}^{n} \alpha_k \right) v + \sum_{\alpha=1}^{n} \alpha_k w_k = v + \sum_{k=1}^{n} \alpha_k w_k$$

Since W is a subspace of V, W is closed under linear combinations i,e, $\sum_{k=1}^{n} \alpha_k w_k \in W$. This implies that

$$\sum_{k=1}^{n} \alpha_k v_k = v + w' \in S$$

Hence S is an affine subspace of V. This proves the converse part.

This theoretical result provides an embodiment of any affine subspace of a linear space.

Examples of Affine Subspaces

(a) In Euclidean plane $\mathbb{R}^2$, all lines passing through the origin are subspaces of $\mathbb{R}^2$ while all the lines passing not through the origin are affine subspaces or linear manifolds of $\mathbb{R}^2$.

(b) In Euclidean space $\mathbb{R}^3$, any plane not passing through the origin is an affine space.

(c) In finite dimensional linear spaces, affine subspaces correspond to the solution set of an inhomogeneous linear system. The solution set of an inhomogeneous linear system $A\underset{\sim}{x} = \underset{\sim}{b}$ (A being a matrix, $\underset{\sim}{x}$ and $\underset{\sim}{b}$ being column vectors) is an affine space over the subspace of solutions of the corresponding homogeneous system $A\underset{\sim}{x} = \underset{\sim}{0}$. Similarly the solution of an inhomogeneous linear *ode* $L(D)y = X$ form an affine space over the subspace of solutions of the corresponding homogeneous linear ode $L(D)y = 0$.

If W be a subspace of a linear space V defined over some field K, then the set of all affine subspaces $\{v + W \ : \ v \in V\}$ itself forms a new vector space (commonly known as **quotient space** of W in V) under the algebraic operations

(i) $(v_1 + W) + (v_2 + W) = (v_1 + v_2) + W \quad \forall \ v_1, v_2 \in V$

(ii) $\lambda(v + W) = \lambda v + W \quad \forall \ \lambda \in K \ $ and $ \ \forall \ v \in V$

This quotient space idea is helpful while solving a linear inhomogeneous differential equation with the inhomogeneous term being sum of UC functions (cf : Method of Undetermined co-efficients in Chapter 5).

Dimension of a Linear Space

If V be a real or complex linear space over $\mathbb{R}$ or $\mathbb{C}$ and $S = \{v_1, v_2, \cdots, v_n\}$ be a finite set in V, S is said to be linearly independent iff the relation

$$\lambda_1 v_1 + \lambda_2 v_2 + \cdots\cdots + \lambda_n v_n = 0$$

implies $\lambda_n = 0$ for each $k = 1, 2, \cdots\cdots, n$. Otherwise, they are called linearly dependent. When S is linearly independent, Span $\{v_1, v_2, \cdots\cdots, v_n\}$ is a subspace of V and moreover every element ξ of this spanning set is expressible uniquely as a linear combination of vectors of S. S is declared to be a **basis** of V when Span $\{v_1, v_2, \cdots\cdots, v_n\}$ equals V. The different bases of the same vector space are equipotents i.e., their cardinality are the same. The cardinality of any basis of a vector space is called its "dimension". A linear space is said to have dimension n if the largest number of linearly independent elements accomodated in it is n. When such a natural number n exists, the space is called finite-dimensional. Hence an infinite dimensional vector space is one for which there exists no finite subset whose all possible linear combination span the entire space. In the following we have a few illustrations :

(a) $\mathbb{R}$ is a one-dimensional vector space over $\mathbb{R}$.

(b) $\mathbb{C}$ is a one-dimensional vector space over $\mathbb{C}$ but two dimensional vector space over $\mathbb{R}$. (Note how the dimension is controlled by the underlying scalar field).

(c) $P_n([0,1])$ is a $(n + 1)$-dimensional vector space over $\mathbb{R}$ with $\{1, x, x^2, \cdots\cdots, x^n\}$ serving as its standard or canonical basis. Another basis of $P_n([0,1])$ is given by the set of Legendre Polynomials:

$$\left\{ \frac{1}{k!} \frac{d^k}{dx^k} (x^2 - 1)^k / k = 0, 1 \cdots\cdots, n \right\}$$

$P(\mathbb{R}))$ is an infinite dimensional vector space over $\mathbb{R}$—the infinite sequence of polynomials $\{x^{n-1} : n \in N\}$ forming its basis. Traditionally we write the dimension of $P([0,1])$ as $\aleph_0$, the cardinality of the set of natural numbers.

(d) $C([0,1]), C^n([0,1]), C^\infty([0,1])$ are all infinite dimensional because they contain no finite set of elements that forms a basis. In other words, this implies that for every positive integer n, one can find $(n+1)$ linearly independent elements in each of these linear spaces.

Remark : Zorn's lemma dealing with 'partial ordering relation' on sets ensures that every non-zero linear space has a basis. Non-zero linear space means those linear spaces which involve vectors $v \neq 0$. Incidentally the zero linear space $V = \{0\}$ has no basis and its dimension is hypothetically defined to be **zero**.

4.3   Linear Maps or Transformations

The concept of linear maps is associated with only the linear spaces. A map T from one linear space V to another linear space V' defined over the same non-trivial scalar field K is said to be a **linear transformation** (linear operator) iff

$T(\lambda v_1 + \mu v_2) = \lambda T(v_1) + \mu T(v_2)$ for any $\lambda, \mu \in K$ and $v_1, v_2 \in V$.

From the above definition it follows that if $\alpha_1, \alpha_2, \cdots\cdots \alpha_n \in K$ for $i = 1, 2, \cdots\cdots, n$ then $T\left(\sum_{i=1}^{n} \alpha_i v_i\right) = \sum_{i=1}^{n} \alpha_i T(v_i)$ holds good, no matter what v_i's are chosen in V. The proof follows from the basic definition and principle of mathematical induction.

Examples of Linear Maps

(a) The easiest example of a linear operator is the 'dilation map' characterised by $T : \mathbb{R}^2 \longrightarrow \mathbb{R}^2$ such that $T((x,y)) = (\lambda x, \lambda y)$.

(b) It is well known a result that every real-valued continuous function defined on a compact set is Riemann integrable. The map $T : C([0,1]) \longrightarrow \mathbb{R}$ defined by $Tf = \int_0^1 f(t)dt$ is a linear map as can be verified by the properties of Riemann integral.

(c) The elements of $P_n([0,1])$ are polynomials of degree less than or equal to n and their derivatives are again elements of $P_n([0,1])$. If we define a map $D : P_n([0,1]) \longrightarrow P_n([0,1])$ by $D(x^r) = rx^{r-1}$ for $r = 1, 2, \cdots\cdots, n$ then D is a linear operator.

If we had defined the differentiation operator D from $P([0,1])$ to itself, then also D would be a linear operator.

If T be any linear operator from V to V', it is a simple observation that the null element of V is mapped to the null element of V' i.e, $T(0) = 0$, where the zero(0) on the left side belong to domain linear space V while the zero(0) on the right side belongs to codomain linear space V'. Two important sets related to the linear operator T are :

$$Ker\ T = \{v \in V\ :\ T(v) = 0 \in v'\}$$
$$Im\ T = \{v' \in V'\ :\ v' = T(v) \text{ for some } v \in V\}$$

It is a routine exercise to show that $KerT$ is a subspace of V and ImT is a subspace of V'.

A linear operator $T : V \to V'$ is said to be **injective** if $KerT = \{0\}$ while it is said to be **surjective** iff $Im\ T = V'$.

The differentiation operator D from $C^1([0,1])$ to $C([0,1])$ is a linear map. It is surjective because every continuous function over $[0,1]$ is Riemann integrable and its primitive is differentiable. Hence $Im(D) = C([0,1])$. However, the differentiation operator D is not injective as it maps continuous functions differing from each other by constants to the same function. Clearly $KerD$ consists of set of constant functions. However the operator $\overline{D}\ :\ C^1([0,1]) \to C([0,1]) \times \mathbb{R}$ defined by

$$\overline{D} : v(x) \longrightarrow \begin{pmatrix} v'(x) \\ v(0) \end{pmatrix} ; x \in [0,1]$$

is injective as well as surjective.

4.4 Normed Linear Space

Let V be a linear space over the non-trivial scalar field K. By 'norm' is meant a real-valued function $\| \ . \ \|$ having domain V and satisfying the following criteria :

(i) $\| \ v \ \| \geq 0$ for all $v \in V$ (non-negativity)

(ii) $\| \ v \ \| = 0$ iff $v = 0$ (positive-definiteness or non-degeneracy)

(iii) $\| \ \alpha v \ \| = | \ \alpha \ | \| \ v \ \|$ for every $v \in V$ and any $\alpha \in K$ (homogeneity).

(iv) $\| \ u + v \ \| \leq \| \ u \ \| + \| \ v \ \|$ for any $u, v \in V$ (triangle inequality).

Any map from V to $\mathbb{R}$ that satisfies above four properties is declared to be a 'norm'. A **normed linear space** (or briefly, n.l.s) is a linear space together with a norm defined on it. This definition is after Stephen Banach. More than one norm can be defined on the same linear space to render different topological structures to the same algebraic structure.

Two different norms $\| \ . \ \|_1$ and $\| \ . \ \|_2$ defined on the same vector space V are said to be 'equivalent' iff there exist positive constants C_1, C_2 for which $C_1 \| \ x \ \|_1 \leq \| \ x \ \|_2 \leq C_2 \| \ x \ \|_1$ for all $x \in V$. The main theme behind 'equivalent norms' is the introduction of identical topological properties through either of them into the underlying linear space.

It is worthy to note that the idea of norm originates from our intention to make indirect use of the ordering prevalent in $\mathbb{R}$. Since all norms defined on finite-dimentional vector spaces are equivalent, the main interest about equivalent norms centres around infinite-dimensional vector spaces.

From triangle inequality it immediately follows that for any u and v in linear space V holds the inequality :

$$| \, \| u \| - \| v \| \, | \leqslant \| v - u \|$$

This in turn implies that $\| \cdot \|$ is a continuous map from V to $\mathbb{R}$.

Examples of Normed Linear Spaces

(a) The simplest example of a normed linear space is $\mathbb{R}$ or $\mathbb{C}$ where each element x is endowed with a non-negative real number $| \, x \, |$. Thus $\| x \|$ equals $| \, x \, |$ in this context.

(b) $C([0,1])$ is a normed linear space if we define

$$\| f \|_{C([0,1])} = \sup \, | f(x) |$$

where $f_1 \in C([0,1])$. The verification is based on the standard result of the real number system :

$$\sup \, (A + B) = \sup \, A + \sup \, B \text{ for any } A, B \subset \mathbb{R}$$

Note that $[0,1]$ being a closed and bounded subset of $\mathbb{R}$ is compact and f being continuous (real-valued or complex-valued), the image of $| \, f \, |$ is a bounded subset $\{ | \, f(x) \, | : x \in X \}$ of $\mathbb{R}$. This is why $\sup | \, f(x) \, |$ exists finitely by virtue of order-completeness property of $\mathbb{R}$. This particular norm is called **sup norm** or **Chebyshev norm** or **uniform norm**.

(c) $C([0,1])$ is a n.l.s if we define the norm of $f \in C([0,1])$ as

$$\| f \| = \left\{ \int_0^1 | \, f(x) \, |^2 \, dx \right\}^{\frac{1}{2}}.$$

Observe that

(i) $\| f \| \geq 0$, since the integrand is non-negative.

(ii) $\| f \| = 0$ iff $\int_0^1 | f(x) |^2 dx = 0$ and hence $f = 0$, the zero function almost everywhere on $[0, 1]$, i.e., $f(x) = 0$ for all $x \in [0, 1]$ with the exception of a set of measure zero.

(iii) $\| \alpha f \| = \left\{ \int_0^1 | \alpha f(x) |^2 dx \right\}^{\frac{1}{2}} = | \alpha | \, \| f \|$

(iv) $\| f + g \|^2 \leq \| f \| + \| g \|$

$$\left[\int_0^1 | f(x) + g(x) |^2 dx \right.$$

$$= \int | f(x) |^2 dx + \int_0^1 | g(x) |^2 dx + 2 \int_0^1 | f(x) | \, | g(x) | dx$$

$$\leq \int_0^1 | f(x) |^2 dx + \int_0^1 | g(x) |^2 dx + 2 \left\{ \int_0^1 | f(x) |^2 dx \right\}^{\frac{1}{2}}$$

$$\left. \times \left\{ \int_0^1 | g(x) |^2 dx \right\}^{\frac{1}{2}} \right]$$

(where we made use of Holder inequality for integrals)

Hence four prerequisites are satisfied. This particular norm is known as **L$_2$-norm** in $C([0, 1])$.

(d) Had we defined on $C^1([0, 1])$ the real-valued function $t(.)$ by $t(f) = \sup_{0 \leq x \leq 1} | f'(x) |$, then $t(.)$ is a seminorm but would fail to be a norm because it lacks the property of positive-definiteness. One may find that for any non-zero constant function defined on [0,1], the derivative is zero everywhere.

Since $C^1([0, 1])$ is a subspace of $C([0, 1])$ and $C([0, 1])$ is a normed linear space under the Chebyshev norm, $C^1([0, 1])$ becomes a normed linear space in its own right under the induced norm.

Besides this induced norm, the linear space $C^1([0, 1])$ can be endowed with the norm

$$\| f \| = \sup_{0 \leq x \leq 1} | f(x) | + \sup_{0 \leq x \leq 1} | f'(x) | .$$

It is quite interesting to see a norm and a seminorm add up to yield a new norm on $C^1([0, 1])$.

In an analogous manner, $C^n([0, 1])$ becomes a normed linear space with the norm

$$\| f \| = \sum_{k=0}^{n} \sup_{0 \leq x \leq 1} | f^{(k)}(x) |$$

where $f^{(k)}(x)$ denotes kth derivative of $f(x)$ and $f^{(0)}(x) \equiv f(x)$.

(e) Consider the linear space of all p-summable infinite sequences, i.e, $x = \{x_n\}$ satisfying the condition $\sum\limits_{n=1}^{\alpha} \mid x_n \mid^p < \infty$. If one associates with each x, a non-negative real number defined by

$$\parallel x \parallel = \left(\sum_{n=1}^{\alpha} \mid x_n \mid^p \right)^{\frac{1}{p}},$$ one can make a routine check that $\parallel . \parallel$ satisfies all criteria defining a norm—thanks to Minkowski's inequality.

$$\left(\sum_{n=1}^{\infty} |x_n + y_n|^p \right)^{\frac{1}{p}} \leqslant \left(\sum_{n=1}^{\infty} |x_n|^p \right)^{\frac{1}{p}} + \left(\sum_{n=1}^{\infty} |y_n|^p \right)^{\frac{1}{p}}.$$

This norm is usually called l^p-**norm** and the corresponding normed linear space as l^p-**space**.

Few Important Results :

(a) The space $C(0,1])$ is complete under supnorm.

Proof : To prove this proposition it suffices to establish that every cauchy sequence $\{f_n\}$ in $C([0,1])$ is convergent to some f in $C([0,1])$. This implies that if $\{f_n\}$ be a cauchy sequence, for every preassigned $\varepsilon > 0$, $\exists$ a positive integer $N(\varepsilon)$ such that the inequality

$$\mid f_n(t) - f_m(t) \mid \leq \sup_{0 \leq t \leq 1} \mid f_n(t) - f_m(t) \mid = \parallel f_n - f_m \parallel < \varepsilon$$

holds for each $t \in [0,1]$ and any $m, n \geq N(\varepsilon)$. This shows that $\{f_n(t)\}$ is a cauchy sequence of real numbers and hence due to order-completeness in $\mathbb{R}$, it converges to some real number, $f(t)$ say, for each $t \in [0,1]$. Thus $\{f_n\}$ converges pointwise to f over $[0,1]$. Now keeping n fixed and making m infinitely large it follows that

$$|f_n(t) - f(t)| < \varepsilon \quad \forall \, t \in [0,1] \quad \text{and} \quad \forall \, n \geqslant N(\varepsilon).$$

As a result, $\{f_n\}$ converges uniformly to some function $f \in C([0,1])$. Note that uniform convergence of $\{f_n\}$ to f forces f to be continuous on [0,1]. Therefore,

$$\lim_{n \to \infty} \parallel f_n - f \parallel = \lim_{n \to \infty} \sup_{0 \leq t \leq 1} \mid f_n(t) - f(t) \mid = 0$$

i.e., $\{f_n\}$ converges in norm (sup-norm) to $f \in C([0,1])$. Note that $[0,1]$ being compact, the elements $f \in C([0,1])$ are not only continuous but also uniformly continuous on $[0,1]$.

The function spaces $C^1([0,1])$ and $C^n([0,1])$ are also Banach spaces under sup-norm but their elements are not uniformly continuous functions on their respective domains.

(b) The space $C([0,1])$ of all continuous functions having domain $[0,1]$ is incomplete if the sup-norm is replaced by L_2-norm. To prove the above proposition it suffices to construct a cauchy sequence $\{f_n\}$ in $C([0,1])$ that fails to converge to any element $f \in C([0,1])$ under L_2-norm. Define

$$
\begin{aligned}
f_n(t) &= 0, && \text{if } 0 \leqslant t \leqslant a \\
&= n(t-a), && \text{if } a < t \leqslant a + \frac{1}{n} \\
&= 1, && \text{if } t > a + \frac{1}{n}
\end{aligned}
$$

where $0 < a < 1$ is an arbitrary but fixed number. The graphs of $f_n(t)$ and $f_m(t)(m > n)$ are shown in the following figure. (Observe as m increases, the slope of the middle part also increases)

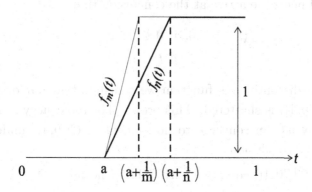

Fig 4.1 : Graph of f_n and f_m in $C[0,1]$

That $\{f_n\}$ is a cauchy sequence in $C[0,1]$ endowed with L_2-norm is verified in the following :

$$
\begin{aligned}
& \| f_m(t) - f_n(t) \| \\
&= \int_a^{a+\frac{1}{m}} (m-n)^2(t-a)^2 dt + \int_{a+\frac{1}{m}}^{a+\frac{1}{n}} \{1 - n(t-a)\}^2 \, dt
\end{aligned}
$$

$$= (m-n)^2 \int_0^{\frac{1}{m}} t^2 dt + \int_{\frac{1}{m}}^{\frac{1}{n}} (1-nt)^2 \, dt$$

$$= \frac{(m-n)^2}{m^2} + \left(\frac{1}{n} - \frac{1}{m}\right)^{\frac{1}{m}} - n\left(\frac{1}{n^2} - \frac{1}{m^2}\right) + \frac{n^2}{3}\left(\frac{1}{n^3} - \frac{1}{m^3}\right)$$

$$= \frac{1}{3n} + \frac{2n^2}{3m^3} - \frac{n}{m^2} \longrightarrow 0 \text{ as } m, n \text{ tend to } \infty.$$

If possible, let there exist a $f \in C([0,1])$ such that

$$\| f_n - f \| = \int_0^1 |f_n(t) - f(t)|^2 \, dt$$

$$= \int_0^1 |f_n(t) - f(t)|^2 \, dt + \int_a^{a+\frac{1}{n}} |f_n(t) - f(t)|^2 \, dt$$

$$+ \int_{a+\frac{1}{n}}^1 |f_n(t) - f(t)|^2 dt$$

$$= \int_0^a |f(t)|^2 \, dt + \int_a^{a+\frac{1}{n}} |n(t-a) - f(t)|^2 \, dt + \int_{a+\frac{1}{n}}^1 |1 - f(t)|^2 dt$$

As per our demand, each of the above three definite integrals should approach zero as $n \to \infty$. The second integral does not eventually contribute and hence we arrive at the conclusion that,

$$f(t) = 0, \text{ if } 0 \leqslant t \leqslant a$$
$$= 1, \text{ if } t > a$$

This f being a discontinuous function we conclude that our dream of having $f \in C([0,1])$ is shattered. This proves that the cauchy sequence $\{f_n\}$ defined by us can converge to no element in $C([0,1])$ under L_2-norm. Hence the conclusion.

(c) The space $C^1([0,1])$ equipped with Sobolev norm

$$\| f \|_{C^1([0,1])} = \sup_{0 \leq x \leq 1} | f(x) | + \sup_{0 \leq x \leq 1} | f'(x) |$$

is a Banach space.

Proof : Let $\{f_n\}$ be a cauchy sequence in $C^1([0,1])$. Hence for any preassigned $\varepsilon > 0$, there corresponds a +ve integer $N(\varepsilon)$ such that $\| f_n - f_m \| < \varepsilon$ whenever $m, n > N(\epsilon)$.

$$\therefore \sup_{0 \leq x \leq 1} | f_n(x) - f_m(x) | + \sup_{0 \leq x \leq 1} | f'_n(x) - f'_m(x) | < \varepsilon$$

Now from the above it follows :

$$| f_n'(x) - f'(x) | \leqslant \| f_n - f_m \| < \varepsilon \qquad \cdots \text{(i)}$$

$$| f_n(x) - f(x) | \leqslant \| f_n - f_m \| < \varepsilon \qquad \cdots \text{(ii)}$$

for any $x \in [0, 1]$ and any $m, n > N(\varepsilon)$. (ii) establishes that the cauchy sequence $\{f_n'\}$ converges to some function g uniformly on $[0, 1]$. As each f_n' is continuous over $[0, 1]$, then their uniform limit, i,e, g is also continuous over $[0, 1]$ and hence integrable thereat.

One can also see from (ii) that $\{f_n\}$ is a cauchy sequence in $C^1([0, 1])$, $\{f_n\}$ converges uniformly to some function $f \in C^1([0, 1])$

Combining the two steps, we have for $0 \leq x \leq 1$:

$$\int_0^x g(t)dt = \lim_{n \to \infty} \int_0^x f_n'(t)dt = \lim_{n \to \infty} (f_n(x) - f_n(0)) = f(x) - f(0) \cdots \text{(iii)}$$

Due to the fact that $g(x)$ is continuous over $[0, 1]$, its primitive is differentiable and

$$\frac{d}{dx} \int_0^x g(t)dt = g(x) \qquad \cdots \text{(iv)}$$

From (iii) and (iv), we get :

$$g(x) = \frac{d}{dx} \int_0^x g(t)dt = \frac{d}{dx}(f(x)) = f'(x), \forall\, x \in [0, 1]$$

$\therefore$ $f' = g \in C^1([0, 1])$, proving the proposition that $C^1([0, 1]$ is a Banach space under the aforesaid norm.

Note : In an analogous manner one can also establish that the space $C^n([0, 1])$ is indeed a Banach space under the Sobolev norm

$$\| f \|_{C^n[0,1]} = \sum_{k=0}^{n} \sup_{0 \leq x \leq 1} | f^{(k)}(x) | \; ; \; f \in C^n([0, 1]).$$

4.5   Bounded Linear Transformation

A linear transformation T from one normed linear space X to another normed linear space Y is said to be **bounded** if there exists a constant $K \geq 0$ so that $\| Tx \|_Y \leq K\| x \|_X$. ($\| x \|_X$ indicates that norm is associated with X while $\| Tx \|_Y$ indicates that norm is associated with Y).K

is called a bound of T. Clearly the set $A = \left\{ \frac{\|Tx\|_Y}{\|x\|_X} / x \in X \text{ and } x \neq 0 \right\}$ is a non-empty set of non-negative real numbers bounded above by K (dependent on T but independent of $x \in X$). By order-completeness of $\mathbb{R}$, sup A exists and is denoted by $\| T \|$. In symbols,

$$\| T \| = \sup_{\substack{x \in X \\ x \neq 0}} \left\{ \frac{\| Tx \|_Y}{\| x \|_X} \right\} = \sup_{\substack{x \in X \\ \|x\|=1}} \| Tx \|_Y$$

For unbounded operators, $\| T \|$ does not exist. This quantity $\| T \|$ is called the **norm** of the linear operator $T : X \to Y$. It also follows from above that $\| Tx \|_Y \leqslant \| T \| \cdot \| x \|_X$. We may verify that the operator norm $\| T \|$ defined above satisfies all properties of usual norm.

(i) **Non-negativity :** Trivially follows from definition itself.

(ii) **Positive definiteness :** Let $T : X \to Y$ be a bounded linear operator such that $\| T \| = 0$. By definition,

$$\| T \| = \sup_{\substack{x \in X \\ x \neq 1}} \left\{ \frac{\| Tx \|_Y}{\| x \|_X} \right\}$$

Therefore, $\| T \| = 0$ implies $\| Tx \|_Y = 0 \ \forall x \in X$ and so $\forall x$, $Tx = 0$ (due to positive-definiteness of $\| \cdot \|_Y$). Hence $T = 0$.

(iii) **Homogeneity :** Let $\alpha \in \mathbb{R}$ and $T : X \to Y$ be a bounded linear operator.

$$\| \alpha T \| = \sup_{\substack{x \in X \\ \|x\|=1}} \frac{\| (\alpha T)(x) \|_Y}{\| x \|_X} = \sup_{\substack{x \in X \\ \|x\|=1}} \frac{| \alpha | \| Tx \|_X}{\| x \|_X} = | \alpha | \cdot \| T \|$$

(iv) **Triangle Inequality :** If T_1 and T_2 be two bounded linear operators from X to Y, then $(T_1 + T_2)$ is again a bounded linear operator and moreover,

$\| T_1 + T_2 \|$
$$= \sup_{\substack{x \in X \\ \|x\|=1}} \frac{\| (T_1 + T_2)(x) \|_Y}{\| x \|_X} \leqslant \sup_{\substack{x \in X \\ \|x\|=1}} \frac{\| T_1 x \|_Y}{\| x \|_X} + \sup_{\substack{x \in X \\ \|x\|=1}} \frac{\| T_2 x \|_Y}{\| x \|_X}$$

Therefore, $\| T_1 + T_2 \| \leqslant \| T_1 \| + \| T_2 \|$

A linear transformation T from one normed linear space X to another normed linear space Y is said to be **continuous** iff for every convergent sequence $\{x_n\}$ in X, the image sequence $\{Tx_n\}$ in Y is also convergent. Moreover, $\{Tx_n\}$ converges to Tx whenever $\{x_n\}$ converges to x if T be continuous. The continuity of the linear operator T over the whole normed linear space X is equivalent to its continuity at any one point, say, the null element of x.

Assume $T : X \rightarrow Y$ to be continuous at $x = 0$. By definition, for any sequence $\{x_n\} \subset X$ convergent to 0, the corresponding image sequence $\{Tx_n\}$ converges to $T(0) = 0$ in $n.l.s$ Y. Choose any arbitrary $x \in X$ and let $\{x_n\} \subset X$ be any sequence convergent to x. Therefore, $\{(x_n - x)\}$ converges to 0 in $n.l.s$ X and because of continuity of T at $0 \in X$, $\{(Tx_n - Tx)\}$ converges to $0 \in Y$, i.e, $Tx_n \longrightarrow Tx$ as $n \rightarrow \infty$. This proves the continuity of T at any $x \in X$.

For linear transformations from one normed linear space to the other, **boundedness** and **continuity** are equivalent as is shown below.

Assume $T : X \rightarrow Y$ to be bounded. Let $\{x_n\}$ be any sequence in $n.l.s$ X that converges to some $x \in X$. As T is bounded, $\exists$ a constant $K > 0$ such that

$$\| Tx_n - Tx \|_Y \leq K \| x_n - x \|_X \longrightarrow 0 \text{ as } n \rightarrow \infty$$

Therefore, $\{Tx_n\}$ converges to Tx in $n.l.s$ Y, ensuring continuity of T.

The proof of the converse part, viz, continuity $\Rightarrow$ boundedness is done contrapositively. So we assume $T : X \rightarrow Y$ to be unbounded. This means $\exists$ a bounded sequence $\{x_n\}$ in X whose image $\{Tx_n\}$ in Y is unbounded. Hence $\| Tx_n \|_Y \longrightarrow \infty$. Without loss of generality one may assume that $Tx_n \neq 0$ $\forall n$. We define a new sequence $\{x'_n\} \subset X$ such that $x'_n = \frac{x_n}{\|Tx_n\|_Y}$. Obviously, $\{x'_n\}$ converges to 0 in X but $\| Tx'_n \|_Y = 1$, implying $T : X \rightarrow Y$ is not continuous. This proves the desired claim.

Important result: If X and Y are normed linear spaces and T is a linear operator from X to Y, then T is continuous if and only if for every convergent series $\sum\limits_{i=1}^{\infty} \alpha_i x_i$ in X,

$$T \left(\sum_{i=1}^{\infty} \alpha_i x_i \right) = \sum_{i=1}^{\infty} \alpha_i T(x_i)$$

Proof : Let $\displaystyle\sum_{i=1}^{\infty} \alpha_i x_i$ be convergent to $x \in X$ and let T be continuous. By

definition of linear operator, for any finite n, $T\left(\displaystyle\sum_{i=1}^{\infty} \alpha_i x_i\right) = T(x)$. If

$\displaystyle\sum_{i=1}^{\infty} \alpha_i x_i = x$ then the partial sum sequence $\left\{\displaystyle\sum_{i=1}^{n} \alpha_i x_i\right\}$ converges to x,

and because T is assumed continuous,

$$T(x) = \lim_{n\to\infty} T\left(\sum_{i=1}^{n} \alpha_i x_i\right) = \lim_{n\to\infty} \sum_{i=1}^{n} \alpha_i\, T(x_i)$$

Therefore, $\displaystyle\sum_{i=1}^{\infty} \alpha_i T(x_i) = T(x) = \left(\sum_{i=1}^{\infty} \alpha_i x_i\right)$

Conversely, let $\displaystyle\sum_{i=1}^{\infty} \alpha_i\, T(x_i) = T(x)$ whenever $\displaystyle\sum_{i=1}^{\infty} \alpha_i x_i = x$; i.e., when

the partial sum sequence $\left\{\displaystyle\sum_{i=1}^{n} \alpha_i x_i\right\}$ converges to x. Conclusion follows

immediately because

$$T(x) = T\left(\sum_{i=1}^{\infty} \alpha_i x_i\right) = \lim_{n\to\infty} \sum_{i=1}^{n} \alpha_i T(x_i) = \lim_{n\to\infty} T\left(\sum_{i=1}^{n} \alpha_i x_i\right)$$

This property of linear transformations is useful in the analysis of ordinary differential equations and is referred to as 'principle of superposition'. However, criteria of continuity and boundedness being equivalent for the linear operators, one can safely conclude that when the linear homogeneous differential equations of the generic form $L(D)y = 0$, $L(D)$ being a polynomial in D with either constant or variable co-efficients employ the 'principle of superposition', the operator should be bounded. In the following examples we shall confine ourselves to the differentiation operator D or its polynomials under choice of various normed linear spaces as its domain and co-domain.

Examples of Bounded and Unbounded Operators :

(a) Let $D : P_n([0,1]) \longrightarrow P_n([0,1])$ be defined by $Df(x) = \frac{df}{dx}$ for all $x \in [0,1]$.

$P_n([0,1])$ is a finite-dimensional normed linear space. D is linear. As every linear operator T defined from one finite dimensional linear space to another is always bounded, D is also bounded.

(b) When $C([0,1])$ is being treated as a normed linear space under the sup-norm and $C^1([0,1])$ is looked as a subspace of this $C([0,1])$ under the induced 'sup-norm', and $D : C^1([0,1]) \to ([0,1])$ is defined by $Df(x) = \frac{df}{dx}$ for all x, then D becomes an unbounded linear operator.

Since $f(x) = x^n$ and $Df(x) = nx^{n-1}$ both are elements of $C([0,1])$,

$$\frac{\|Df\|}{\|f\|} \quad \frac{\sup\limits_{0 \leq x \leq 1} |f'(x)|}{\sup\limits_{0 \leq x \leq 1} |f(x)|} = \frac{n \cdot \sup\limits_{0 \leq x \leq 1} |x^{n-1}|}{\sup\limits_{0 \leq x \leq 1} |x^n|} = n$$

that increases indefinitely when n becomes large.

Hence $\left\{ \dfrac{\|Df\|_{C([0,1])}}{\|f\|_{C([0,1])}} : f \in C([0,1]) ; f \neq 0 \right\}$ is an unbounded

set, proving the claim that D is unbounded operator.

(c) If we define the differentiation operator D from $C^1([0,1])$ endowed with the Sobolev norm, viz,

$$\|f\|_{C^1([0,1])} = \sup\limits_{0 \leq x \leq 1} |f(x)| + \sup\limits_{0 \leq x \leq 1} |f'(x)|$$

to $C([0,1])$ endowed with the standard sup-norm, then D turns out to be a bounded linear operator as shown below :

$$\begin{cases} \|Df\|_{C([0,1])} = \sup\limits_{0 \leq x \leq 1} |f'(x)| \\ \|f\|_{C^1([0,1])} = \sup\limits_{0 \leq x \leq 1} |f(x)| + \sup\limits_{0 \leq x \leq 1} |f'(x)| \end{cases}$$

$$\frac{\|Df\|_{C([0,1])}}{\|f\|_{C^1([0,1])}} = \frac{\sup\limits_{0 \leq x \leq 1} |f'(x)|}{\sup\limits_{0 \leq x \leq 1} |f(x)| + \sup\limits_{0 \leq x \leq 1} |f'(x)|} < 1$$

for any $f \in C^1([0,1])$.

Hence D is bounded. By order completeness of $\mathbb{R}$ it follows that

$$\parallel D \parallel = \sup_{\substack{f \in C^1([0,1]) \\ f \neq 0}} \left\{ \frac{\parallel Df \parallel_{C([0,1])}}{\parallel f \parallel_{C^1([0,1])}} : f \in C^1([0,1]) \right\} \leqslant 1.$$

We shall now prove that $\parallel D \parallel$ is indeed equal to 1. To prove it, let's choose $f(x) = x^n$; $n \in N$ in $C^1([0,1])$ Obviously

$$\parallel D \parallel = \sup_{\substack{f \in C^1([0,1]) \\ f \neq 0}} \left\{ \frac{\parallel Df \parallel_{C([0,1])}}{\parallel f \parallel_{C^1([0,1])}} \right\}$$

$$= \frac{\displaystyle\sup_{0 \leq x \leq 1} n|x^{n-1}|}{\displaystyle\sup_{0 \leq x \leq 1} |x^n| + \sup_{0 \leq x \leq 1} n|x|^{n-1}} = \frac{n}{1+n} \, ,$$

which approaches 1 as n becomes infinitely large. So $\parallel D \parallel = 1$.

(d) The linear operator $L(D)$ defined by

$$L(D) \equiv a_0(x)D^n + a_1(x)D^{n-1} + \cdots\cdots + a_{n-1}(x)D + a_n(x)$$

(where $a_k(x) \in C([0,1]$ and $a_0(x) \neq 0$ for any $x \in [0,1]$) carrying $f \in C^n([0,1])$(endowed with Sobolev norm

$$\parallel f \parallel_{C^n[0,1]} = \sum_{k=0}^{n} \sup_{0 \leq x \leq 1} |f^{(k)}(x)|$$

to some $g \equiv L(D)f \in C([0,1])$(endowed with the sup-norm) is bounded by nature as seen below.

$$\parallel L(D)f \parallel_{C([0,1])} = \sup_{0 \leq x \leq 1} \left| \sum_{k=0}^{n} a_k(x)D^{n-k}f(x) \right|$$

$$\leq \sup_{0 \leq x \leq 1} \sum_{k=0}^{n} |a_k(x)| \, |D^{n-k}f(x)|$$

$$\leq M \sup_{0 \leq x \leq 1} \sum_{k=0}^{n} |D^{n-k}f(x)|$$

$$= M \sup_{0 \leq x \leq 1} \sum_{k=0}^{n} |f^{(n-k)}(x)|$$

$$= M \parallel f \parallel_{C^n([0,1])}$$

where each $a_k(x), k = 0$ (1)n being continuous function over the compact set $[0,1]$ is uniformly continuous and hence bounded by M_k, $k = 0(1)n$ respectively. In our proof $M = \max_{0 \leq k \leq n} M_k$.

As a passing result we observe that D is a bounded operator with norm equal to 1, the proof toeing the line of proof of (c).

An Important result :

If T be a continuous linear operator from a normed linear space X to another normed linear space Y, then $Ker\ T$ is a closed subspace of X.

Proof : Let $\{x_n\}$ be a sequence in $Ker\ T$ and suppose x is the limit to which $\{x_n\}$ converges in X. Hence $T(x_n) = 0\ \forall n \in N$. Now T being continuous, T is bounded and hence

$$\| T(x_n) - T(x) \|_Y \leq \| T \| \| x_n - x \|_X \longrightarrow 0 \text{ as } n \to \infty$$
$$\therefore \quad T(x) = T\left(\lim_{n\to\infty} x_n \right) = \lim_{n\to\infty} T(x_n) = 0,$$

implying that $x \in Ker\ T$. Thus every sequence $\{x_n\}$ in $Ker\ T$ converges to some x in $Ker\ T$. $Ker\ T$ is therefore closed.

Note : The operator $L(D) : C^n([0,1]) \longrightarrow C([0,1])$ being a bounded operator is continuous and hence $Ker L(D)$ is a closed subspace of $C^n([0,1])$. $[Ker\ (L(D))$ being a closed linear subspace of $C^n([0,1])$ (a normed linear space under Sobolev norm), the quantity

$$\| y_p + Ker\ L(D) \| = Inf\ \{\| y_p + y_c \| : y_c \in\ Ker\ (L(D))\}$$

defines a norm and moreover, the quotient space $C^n([0,1])/Ker\ (L(D))$ becomes a normed linear space under this norm].

4.6 Invertible Operators

Let a linear operator T map a linear space X into another linear space Y. If for the operator T, $Ker\ T = \{0\}$ then T is one-to-one (injective) because for every $y \in R(T)$, the operator range, there exists a unique element $x \in X$ so that $Tx = y$. This correspondence can be considered as an operator S defined on $R(T)$ with images filling X. This S is undoubtedly a linear operator. By definition, $(ST)(x) = S(Tx) = x$ for all $x \in X$ and hence S is called "left inverse" of T. In the particular case $R(T) = Y$, S becomes an operator with domain Y and co-domain X, and is referred to as the **inverse operator** of T, denoted by T^{-1}.

Example :   (a) Let $X = C([0,1])$, define an operator T that assigns to every $f \in C([0,1])$, its primitive defined by $F(x) = \int_0^x f(t)dt$. Thus $T(f) = F \in C([0,1])$. We define a second operator S on X by $S(f) = f'$, the derivative of f. S is defined only over a linear manifold of functions which have continuous derivatives, provided we demand $R(S)$, the range of S, to be contained in $C([0,1])$. (This operator S is linear but not continuous because only uniform convergence of a sequence of functions $\{f_n'\}$ in $C([0,1])$ will converge uniformly to f'. Either the sequence $\{f_n'\}$ is not convergent even pointwise, or convergent to some function other than f'. This invites us to stamp the case of 'term-by-term differentiation' in analysis separately.). The operator might be taken as the 'left inverse' of T but not an 'inverse' of T as $R(T)$ is the subspace of $C([0,1])$ consisting of all continuously differentiable functions defined on $[0,1]$ such the their values at 0 are zeros.

(b) Let $X = C^2([0,1])$ and $Y = C([0,1])$. Define a subspace Z of $C^2([0,1])$ which consists of those $f \in C^2([0,1])$ satisfying the conditions $f(0) = f'(0) = 0$. If $p,q \in Y$ and T be defined from Z to Y by $Tf = f'' + pf' + qf$, then T has an 'inverse' operator because

(i) $Tf = 0$ implies $f = 0$,   i.e., $Ker\ T = \{0\}$ and

(ii) p, q being elements of $C([0,1])$, the existence uniqueness theorem for second order differential equations ensures that the inhomogeneous ode of the form $f''(t) + p(t)f'(t) + q(t)f(t) = g(t)$ has a solution such that $f \in Z$ whenever $g \in Y$. Hence $R(T) = Y$, i.e., $T : Z \to Y$ is onto. Had T been defined over X instead of Z, i.e., if the problem were not an IVP, then $Ker\ T \neq \{0\}$ and hence in that case T would not even possess a 'left inverse'. Note that incorporation of suitable initial conditions controls the invertibility of a linear differential operator involving constant or variable co-efficients.

We shall not be content with the so-called 'algebraic inverses' of the linear operators and seek whether these 'inverse' operators preserve the topological properties like boundedness, continuity, compactness etc of the parent operators. For this to investigate, we have to introduce the idea of Banach spaces indispensably. In a nutshell, a **Banach space** is a complete normed linear space. A normed linear space is said to be

complete iff every cauchy sequence $\{x_n\}$ in X (in the sense of prevailing norm in X) is convergent to some $x \in X$ (convergence being accounted for in the sense of the same norm). Explicitly speaking, the sequence $\{x_n\}$ is a cauchy sequence in the normed linear space X iff the sequence $\{\|\ x_n\ \|\}$ is a real cauchy sequence, i.e., iff $\|\ x_n - x_m\ \| \longrightarrow 0$ as m, n tend to infinity. Similarly a sequence $\{x_n\}$ in X is convergent to x in X iff $\|\ x_n - x\ \| \to 0$. The most important feature about convergence here is the 'convergence-in norm'.

When X and Y are ordinary normed linear spaces, the set $\mathbb{B}(X, Y)$ of all bounded linear transformations of X into Y is itself a normed linear space under 'pointwise addition' and 'scalar multiplication', the norm being the usual operator-norm. In particular if X and Y are identical, then the set $\mathbb{B}(X, Y)$ reduces to $\mathbb{B}(X, X)$ or more simply, $\mathbb{B}(X)$. If $T \in \mathbb{B}(X)$, then for any non-negative integer n, $\|\ T^n\ \| \leqslant T\ \|^n$, where the operator $T^n : X \to X$ is defined recursively as :

$$T^0 = I \ ; \ T^n = T(T^{n-1}) \ ; \ T^n(x) = T(T^{n-1}\ x) \quad \forall\ x \in X$$

In theory of differential equations, the differential operator is often required to be approximated by a sequence of operators having simpler form. In such circumstances, it is very useful to consider the limits of convergent sequence of bounded linear operators. This demands to have the normed linear space $\mathbb{B}(X, Y)$ elevated to the status of a complete normed linear space or Banach space.

When X and Y are both Banach spaces, $\mathbb{B}(X, Y)$ is a Banach space. (Infact X need not be Banach space to make $\mathbb{B}(X, Y)$ a Banach space). If in particular, the domain and co-domain are identical Banach space, say X, then $\mathbb{B}(X, Y)$ turns into $\mathbb{B}(X, Y)$ or $\mathbb{B}(X)$, as conventionally denoted in the texts. This $\mathbb{B}(X)$ is not only a Banach space but can be elevated to the status of a Banach algebra provided we introduce operator multiplication and demand the new operation to be associative, distributive and homogeneous.

Fundamental Theorem on Invertibility of Linear Operators.

If T be a linear operator from a normed linear space X to another normed linear space Y with domain X and range $R(T) \subseteq Y$, then the inverse map T^{-1} exists, linear and is continuous on $R(T)$ if and only if there exists a constant $m > 0$ such that $\|\ Tx\ \| \geq m\ \|\ x\ \|\ \forall x \in X$.

Proof : **(Sufficiency)** : Positive-definiteness of norm ensures that

$Tx = 0$ and this together with $\parallel Tx \parallel \geq m \parallel x \parallel$ imply $x = 0$. T is injective (i.e., one to-one) as for any $x', x'' \in X$, $Tx = Tx'$ implies $T(x - x') = 0$ and hence $x = x'$. So T^{-1} exists with $R(T)$ as its domain.

T^{-1} is linear as seen below : Define

$$\begin{aligned} x &= T^{-1}(y_1 + y_2) - T^{-1}(y_1) - T^{-1}(y_2), \ y_1, y_2 \in R(T). \\ \therefore Tx &= T[T^{-1}(y_1 + y_2) - T^{-1}(y_1) - T^{-1}(y_2)] \\ &= (TT^{-1})(y_1 + y_2) - (TT^{-1})(y_1) - (TT^{-1})(y^2) \\ &= y_1 + y_2 - y_1 - y_2 = 0, \end{aligned}$$

and hence (due to injectivity of T) $x = 0$.

$\therefore T^{-1}(y_1 + y_2) = T^{-1}(y_1) + T^{-1}(y_2)$ for any $y_1, y_2 \in R(T)$.

T^{-1} is bounded as seen below :

For any $y \in R(T)$, set $x = T^{-1}y$. By hypothesis this gives

$$\parallel y \parallel = \parallel T(T^{-1}y) \parallel \geq m \parallel T^{-1}y \parallel$$

We therefore conclude that T^{-1} is bounded and hence continuous.

(Necessity) :   Let T^{-1} exist and be continuous on $R(T)$. Because of linearity, T^{-1} is bounded over $R(T)$ so that $\parallel T^{-1}y \parallel \leq K \parallel y \parallel$ for all $y \in R(T)$. For any x, set $x = T^{-1}y$; Hence for some $M > 0$

$Tx = T(T^{-1}y) = y$ and thus $\parallel x \parallel = \parallel T^{-1}y \parallel = \parallel T^{-1}(Tx) \parallel \leq M \parallel Tx \parallel$

This completes our proof.

Note :   In case T and T^{-1} are both bounded and hence continuous, T is 'bicontinuous' i,e, a homeomorphism from X to Y. Had X and Y been both Banach spaces, we would get 'Inverse map' theorem of Banach. According to Banach's Inverse Map theorem a linear operator T is connoted **invertible** iff its 'algebraic inverse T^{-1} exists and is also continuous. The formal statement of the theorem runs as follows :

If X and Y are Banach spaces and $T : X \longrightarrow Y$ is a one-to-one bounded linear operator, then $T^{-1} : Y \longrightarrow X$ exists and is continuous, (i,e, T is a linear homeomorphism from X to Y).

Contraction map and Fixed Point Theorem :

Definition : A linear map T from X to itself is said to be a **contraction map** with Lipschitz constant L if $0 \leq L < 1$ and that for every fixed $\| Tx - Ty \| \leq L \| x - y \|$ holds good.

Theorem 4.1. : 'Contraction mapping theorem' or 'Banach fixed point theorem' states that any contraction map defined on a Banach space X to itself has a unique fixed point i.e., $\exists$ one and only one point $\xi \in X$ so that $T\xi = \xi$.

'Banach's fixed point theorem' or the so-called 'Contraction Principle' is the key for proving the 'Existence and Uniqueness Theorem' for solving ordinary differential equations. We stated the theorem in the first chapter itself, worked a few problems on its basis and illustrated why this theorem is indispensable for handling the real-world problems (c.f. leaky-bucket problem). However, we deferred the details of the proof for so long a period as we lacked enough mathematical sinew to pursue it. The contents so far brought up in this chapter, are, we think, sufficient to mould the proof. Statement of the existence-uniqueness theorem is iterated once again for convenience of the reader.

If $\frac{dy}{dx} = f(x, y)$ together with initial condition $y(x_0) = y_0$ be an IVP, where $f(x, y)$ satisfies the conditions

(i) $f(x, y)$ is continuous for every $(x, y) \in D$, a rectangle bounded by the straight lines $x = x_0 \pm a$ and $y = y_0 \pm b$,

(ii) $f(x, y)$ satisfies the Lipschitz's condition

$$| f(x, y(x)) - f(x, y_2(x)) | \leq K | y_1(x) - y_2(x) |$$

over D, then there exists a unique solution $y = \bar{y}(x)$ to the given ode which satisfies $\bar{y}(x_0) = y_0$ for all $x \in [x_0 - H, \ x_0 + H]$ where $H < \min \left\{ a, \frac{b}{M}, \frac{1}{K} \right\}$ and $M = \underset{(x,y) \in D}{\text{Max}} \ f(x, y)$.

Proof : By hypothesis, $f(x, y)$ is continuous over the two dimensional closed rectangular neighbourhood $D = [x_0 - a, x_0 + a] \times [y_0 - b, y_0 + b]$ of (x_0, y_0) and bounded therein. Thus there exists a constant $M > 0$ such that $| f(x, y) | \leq M \ \forall (x, y) \in D$. We choose $h \leq \min \left\{ a, \frac{b}{M} \right\}$.

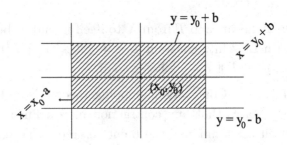

Fig 4.2 : Rectangular region D

Let $\overline{y}(x)$ be a solution of the IVP $\frac{dy}{dx} = f(x,y)$; $y(x_0) = y_0$ over $[x_0, x_0 + h]$. As $f(x,y)$ is continuous over D, $\frac{dy}{dx}$ should also be continuous over $[x_0, x_0 + h]$. Thus integrating $\frac{dy}{dx} = f(x,y)$; $y(x_0) = y_0$ over $[x_0, x] \subseteq [x_0, x_0 + h]$, we have : $y(x) = y_0 + \int_{x_0}^{x} f(s, y(s))ds$ one of whose solution is $\overline{y}(x)$.

Conversely, if $f(x,y)$ be continuous over D and $\overline{y}(x)$ be a continuous solution of the Volterra-type integral equation

$$y(x) = y_0 + \int_{x_0}^{x} f(s, y(s))ds,$$

then   (i) $\overline{y}(x_0) = y_0 + \displaystyle\int_{x_0}^{x} f(s, \overline{y}(s))ds = y_0$

   (ii) $\dfrac{d}{dx}\overline{y}(x) = \dfrac{d}{dx}\left\{ y_0 + \displaystyle\int_{x_0}^{x} f(s, \overline{y}(s))ds \right\}$

$$= \frac{d}{dx}\left[\int_{x_0}^{x} f(s, y(s))ds \right] = f(x, \overline{y}(x)),$$

by fundamental theorem of integral calculus.

Hence the Volterra-type integral equation

$$y(x) = y_0 + \int_{x_0}^{x} f(s, y(s))dx$$

is equivalent to the first order IVP :

$$\frac{dy}{dx} = f(x,y) \; ; \; y(x_0) = y_0$$

Let's consider the normed linear space $C([x_0, x_0 + h])$ endowed with sup-norm. As is well-known, $C([x_0, \; x_0 + h])$ is a Banach space under this norm. We define an operator $T : C([x_0, \ x_0 + h]) \longrightarrow C([x_0, \ x_0 + h])$ as follows :

$y \in C([x_0, \ x_0 + h]) \longrightarrow Ty \in C([x_0, \ x_0 + h])$ where

$$(Ty)(x) = y_0 + \int_{x_0}^{x} f(s, y(s))ds \ \forall \ x \in [x_0, x_0 + h].$$

If $y, z \in C([x_0, x_0 + h])$, then

$$| \ (Ty)(x) - (Tz)(x) \ | \ = \ \left| \int_{x_0}^{x} \{f(s, y(s)) - f(s, z(s))\} \right|$$

$$\leq \ \int_{x_0}^{x} | \ f(s, y(s)) - f(s, z(s)) \ | \ ds$$

$$\therefore \sup_{x_0 \leq x \leq x_0 + h} | \ (Ty)(x) - (Tz)(x) \ | \leq \int_{x_0}^{x_0 + h} | \ f(s, y(s)) - f(s, z(s)) \ | \ ds$$

However, $f(x, y)$ satisfies the Lipschitz's criterion over domain D, i.e.,

$$| \ f(s, y(s)) - f(s, z(s)) \ | \ ds \leq K \parallel y - z \parallel_{\infty}$$

with $\parallel . \parallel_{\infty}$ indicating the supnorm prevalent in $C([x_0, x_0 + h])$.

Combining the last two inequalities and using definition of operator norm, we get,

$$\parallel Ty - Tz \parallel \ \equiv \sup_{x_0 \leq x \leq x_0 + h} | \ (Ty)(x) - (Tz)(x) \ |$$

$$\leq Kh \parallel y - z \parallel_{\infty} = L \parallel y - z \parallel_{\infty}, \text{ where } L \equiv Kh.$$

If T is to be made a contraction map L should be made less than unity, i.e., h should be chosen less than $\frac{1}{K}$. The choice $H < \min\{a, \frac{b}{M}, \frac{1}{K}\}$ as a substitute to h will meet all our demands. So by contraction mapping principle, $\exists$ a unique fixed point of T. The volterra integral equation formulated earlier envisages that $\overline{y}(x)$ itself is that unique fixed point of T. This completes our proof of the fact that over $[x_0, x_0 + H]$, the given IVP admits of a unique solution.

The theorem has been proved for initial data given at the left end point of the interval $[x_0, x_0 + H]$. There is no apriori reason why the initial data cannot be given at the right end point of some interval.

One can furnish the proof for this case if one reminds that over $[x_0 - h, x_0]$ and hence over $[x_0 - H, x_0]$, the Volterra integral equation

$y(x) = \int_{x_0}^{x} f(s, y(s))ds$ is equivalent to the given IVP and moreover that the same T as before is chosen as an endomorphism over the Banach space $C([x_0 - h, x_0])$. The body of the proof is, however, the same in letter and spirit as the previous case.

The conclusion of theorem follows by piecing together the results, of the two subcases.

Note : (a) $(Ty)(x) \in [y_0 - b, y_0 + b]$ since

$$| (Ty)(x) - y_0 | = \left| \int_{x_0}^{x} f(x, y)dx \right| \leq M \mid x - x_0 \mid = Mh \leq b$$

i.e., T is an operator that does not allow its image to go beyond

$$D = [x_0 - a, x_0 + a] \times [y_0 - b, y_0 + b]$$

(b) If the initial data were provided at an interior point x_0 of a closed interval $[\alpha, \beta]$, then the existence and uniqueness of solutions of the IVP under the same conditions can be proved by piecing together one solution on $[\alpha, x_0]$ (with initial data at the right end point x_0) with another on $[x_0, \beta]$(with initial data at the left end point x_0).

(c) The above existence and uniqueness theorem for ordinary differential equations is valid also over on infinite intervals like $[x_0, \infty)$ and $(-\infty, x_0)$. It is known as **Global Picard Existence and Uniqueness theorem for odes**. [For details, see Kripke, 'Introduction to Analysis,' pp. 127.]

(d) One may prove the above theorem of Picard by means of 'Method of Successive Approximation' that basically involves the idea of convergence of the telescoping series over a suitable Banach space.

(e) We observe that in the course of either proof, Banach space holds the center-stage as without the convergence in norm, nothing can survive. This Banach space is also controlling the continuous dependence of solutions on initial data. In brief, the idea of Banach space lies at the heart of all well-posed odes.

Important Theorems and their Consequences :

Theorem 4.2. :   Let T be an element of $\mathbb{B}(X)$ and have a bounded linear inverse T^{-1} in $\mathbb{B}(X)$. Then for every $S \in \mathbb{B}(X)$ satisfying $\| T - S \| < \frac{1}{\|T^{-1}\|}$, S has a bounded linear inverse $S^{-1} \in \mathbb{B}(X)$ and moreover,

$$\| T^{-1} - S^{-1} \| \leq \frac{\| T - S \| \, \| T^{-1} \|^2}{1 - \| T - S \| \, \| T^{-1} \|}.$$

Proof : Denote $T - S$ by A. $A \in \mathbb{B}(X)$. Consider two endomorphisms f and g on $\mathbb{B}(X)$ by

$$\left. \begin{array}{l} f(C) \equiv T^{-1}AT^{-1} + T^{-1}AC \in \mathbb{B}(x) \\ \text{and} \quad g(C) \equiv g(C)T^{-1}AT^{-1} + CAT^{-1} \in \mathbb{B}(x) \end{array} \right\} \text{with } \ C \in \mathbb{B}\ (X)$$

$$\begin{aligned} \therefore \quad \| f(C) - f(C') \| \ &= \ \| T^{-1}A(C - C') \| \\ &\leq \ \| T^{-1}A \| \cdot \| C - C' \| \\ &\leq \ \| T^{-1} \| \cdot \| A \| \cdot \| C - C' \| \\ &= \ L \| C - C' \| \qquad\qquad \cdots \text{(i)} \end{aligned}$$

(In deriving the chain of inequalities we have taken advantage of the truth that $\mathbb{B}(X)$ is a Banach algebra).

In the same vein as before,

$$\| g(C) - g(C') \| \leq L \| C - C' \| \qquad\qquad \cdots \text{(ii)}$$

By hypothesis of the theorem, $L \equiv \| A \| \, \| T^{-1} \| < 1$. Inequalities (i) and (ii) ensure that f and g are contraction maps and hence admit of unique fixed points B and B' , say, respectively due to Banach's fixed point theorem. Hence

$$\left. \begin{array}{l} B \ = f(B) \ = T^{-1}AT^{-1} + T^{-1}AB \\ B' = g(B') \ = T^{-1}AT^{-1}AT^{-1} \end{array} \right\}$$

i.e., $(T - A)(T^{-1} + B) = I$ and $(T^{-1} + B')(T - A) = I$ $\cdots \text{(iii)}$

The relations in (iii) entail that $(T^{-1} + B)$ is the right inverse of $(T - A) \in \mathbb{B}(V, V)$ while $(T^{-1} + B')$ is its left inverse. Moreover,

$$\begin{aligned} (T^{-1} + B') \ &= \ (T^{-1} + B')I \\ &= \ (T^{-1} + B')(T - A)(T^{-1} + B) \\ &= \ ((T^{-1} + B')(T - A))(T^{-1} + B) \\ &= \ (T^{-1} + B) = T^{-1} B, \qquad \text{(using (iii))} \end{aligned}$$

showing that the left and right inverses of $(T - A)$ are one and the same.

$\therefore$ $S = T - A$ is invertible and $S^{-1} = T^{-1} + B \in \mathbb{B}(X)$, where

$$\begin{aligned}
\| B \| &= \| T^{-1}AT^{-1} + T^{-1}AB \| \\
&\leq \| T^{-1}AT^{-1} \| + \| T^{-1}A \| \cdot \| B \| \\
&\leq \| T^{-1}AT^{-1} \| + \| T^{-1} \| \cdot \| A \| \| B \|
\end{aligned}$$

$\therefore$ $(1- \| T^{-1} \| \cdot \| A \|) \| B \| \leq \| T^{-1}AT^{-1} \| \leq \| A \| \cdot \| T^{-1} \|^2$

or, $\| B \| \leq \dfrac{\| A \| \cdot \| T^{-1} \|^2}{1- \| T^{-1} \| \cdot \| A \|} = \dfrac{\| T - S \| \cdot \| T^{-1} \|^2}{1- \| T^{-1} \| \cdot \| T - S \|}$

ensuring that $\| T^{-1} - S^{-1} \| = \| B \| \leq \dfrac{\| T - S \| \cdot \| T^{-1} \|^2}{1- \| T^{-1} \| \cdot \| T - S \|}$

Two important consequences :

(a) The set Q of all invertible elements in $\mathbb{B}(X)$ forms an open subset of $\mathbb{B}(X)$.

Proof : Let $T \in Q \subseteq \mathbb{B}(X)$. Clearly T^{-1} exists and belongs to $\mathbb{B}(X)$ and moreover, $\| T^{-1} \| \neq 0$. The above theorem envisages that the ball $S_{\frac{1}{\|T^{-1}\|}}(T)$ centred about T and having radius $\frac{1}{\|T^{-1}\|}$ wholly lies within Q. Thus T is an interior point of Q. T being our arbitrary choice, the conclusion follows.

(b) The inverse of a linear map, when exists, is a continuous function of the map itself.

Proof : From the theorem 4.2, we have

$$\| T^{-1} - S^{-1} \| \leq \frac{\| T^{-1} \|^2}{1- \| T^{-1} \| \| T - S \|} \cdot \| T - S \|,$$

whenever $\| T - S \| < \frac{1}{\|T^{-1}\|}$. So $\| T^{-1} - S^{-1} \|$ can be made as small as we please by making $\| T - S \|$ small. Moreover, $\frac{\|T^{-1}\|^2.\|T-S\|}{1-\|T^{-1}\|.\|T-S\|}$ being a bound for $\| T^{-1} - S^{-1} \|$, the proposition is verified.

Note : This theorem and its consequences are the pillars for the construction of Green's functions associated with linear inhomogeneous odes. We shall combeack to these details when we discuss 'method of variation of parameters' in the next chapters.

Theorem 4.3. : (Geometric series theorem): When X is a Banach space and $T \in \mathbb{B}(X)$ such that $\| T \| < 1$, then

(i) $(I - T)$ is a linear isomorphism on X.

(ii) $(I - T)^{-1}$ exists, is a bounded linear operator and moreover,

$$(I - T)^{-1} = \sum_{n=0}^{\infty} T^n$$

(iii) $\qquad \| (I - T)^{-1} \| \leqslant \dfrac{1}{1- \| T \|}$

The proof is deliberately dropped and the interested reader may look into any text on 'Functional Analysis' for its details. The generalisation of the above theorem is more interesting in our perspective.

Let $f(x)$ be any function of a real (or complex) variable x such that it admits of a convergent power series expansion about $x = 0$:

$$f(x) = \sum_{n=0}^{\infty} c_n x^n \; ; \; |x| < R.$$

If X is a Banach space and $T \in \mathbb{B}(x)$ such that $\| T \| < R$, the radius of convergence of the power series representing f, then the operator $f(T) = \sum_{n=0}^{\infty} a_n T^n$ is a well-defined element of $\mathbb{B}(X)$.

Eigen-vector of Linear Operators :

Let T be a linear operator from a vector space V to itself. A non-zero vector v of V is said to be an eigenvector if T can only stretch v i.e., iff there exists a scalar $\lambda \in K$, (the field on which V is defined) corresponding to v such that $Tv = \lambda v$. Conventionally, λ is known as the eigenvalue corresponding to the eigenvector v of T.

The following conditions are equivalent :

(i) λ is an eigen value of T.

(ii) $T - \lambda I$ is not an isomorphism.

(iii) $T - \lambda I$ is not invertible.

Proof : λ is an eigenvalue of $T \Leftrightarrow \exists$ a non-zero $v \in V$ such that $Tv = \lambda v \Leftrightarrow$ there exists a non-zero vector $v \in V$ such that $(T - \lambda I)v = 0 \Leftrightarrow (T - \lambda I)$ is not one-to-one $\Leftrightarrow (T - \lambda I)$ is not an isomorphrism is not invertible.

Remark : We shall be making most use of this result when T is chosen as the differentiation operator D, and V is not only a vector space but a suitable Banach space like $C([0, 1])$ etc.

Chapter 5

Differential Equations of Higher Order

5.1 Introduction

Existence and uniqueness of solution, different techniques of finding solutions and various physical application of first order odes were treated in details in the first three chapters. We now like to switch over to higher order differential equations and do the same. However, instead of confining ourselves to a routine development of the subject we shall try to look into it with an analytic mind. The mathematical framework developed in the previous chapter (i,e, Chapter 4) will surely enable us to justify our proceedings as by now we have developed at least a feel that linear algebra is the natural language to express the salient qualitative properties of differential equations. Our mindset will hopefully be reflected once the reader goes through the interpretation of complementary function of any ode as kernel of a differentiation operator, annihilation method and its link with undetermined co-efficients, emergence of the idea of Green's function from the method of variation of parameters and many other important results in the present chapter.

5.2 Theoretical Aspects

In our venture, let's begin with the simplest, viz, linear ordinary differential equations of the nth order that has the generic form

$$a_0(x)\frac{d^n y}{dx^n} + a_1(x)\frac{d^{n-1}y}{dx^{n-1}} + a_2(x)\frac{d^{n-2}y}{dx^{n-2}} + \cdots + a_{n-1}(x)\frac{dy}{dx} + a_n(x)y = b(x),$$

$$(5.1)$$

where the co-efficients $a_k(x), (k = 0, 1, 2, \cdots, n)$ and $b(x)$ are all continuous over some interval I (might it be finite or even infinite) and moreover $a_0(x) \neq 0$ for all $x \in I$. Note that this form of a linear ode appeared in the first chapter itself, but this time the introduction of new assumption of continuity of the co-efficients $a_k(x)$ and $b(x)$ over I ensures that the linear ode admits of a unique solution that is defined over this interval and satisfies the set of initial conditions

$$y(x_0) = y_0; \quad \frac{dy}{dx}\bigg|_{x_0} = y_0'; \quad \cdots\cdots \quad ; \frac{d^{n-1}y}{dx^{n-1}}\bigg|_{x_0} = y_0^{n-1} \text{ where } x_0 \in Int(I)$$

$$(5.2)$$

Remark : (i) This solution has a continuous nth order derivative over I.

(ii) This result is known as **Existence-Uniqueness Theorem** for higher order **linear odes**.

One may recall that these conditions appeared in the discussion of the first order linear odes in the Chapter 2. If the linear ode

$$\frac{dy}{dx} + P(x)y = Q(x)$$

is written in normal form, i.e.,

$$\frac{dy}{dx} = Q(x) - P(x)y \equiv f(x, y), \text{ say,}$$

and $P(x)$, $Q(x)$ are assumed to be continuous over a closed interval I, then $\left|\frac{\partial f}{\partial y}\right| = |P(x)| \leq k$, some constant. This boundedness criterion of the partial derivative $\frac{\partial f}{\partial y}$ over I is needed for the existence and uniqueness of solutions to the first order ode provided initial condition $y(x_0) = y_0$ is prescribed for some $x_0 \in Int(I)$.

We refer to the linear odes appearing in form (5.1) as 'homogeneous' (not to be confused with the homogeneous odes dealt earlier) if $b(x) = 0$ for all x in the interval I. If $b(x) \neq 0 \ \forall \ x \in I$, we call the linear ode 'inhomogeneous'. Often in course of solving a linear ode of higher order we replace $b(x)$ by zero and get the new homogeneous linear ode, commonly called 'associated homogeneous ode' or 'reduced ode'. If the co-efficients $a_k(x)$, $(k = 0, 1, 2, \cdots, n)$ of the equation (5.1) are all chosen constants, we call the corresponding linear ode as 'differential equation with constant co-efficients'. Here the functional dependence of the righthand member $b(x)$ is of no concern. Later on we shall observe that for this class of higher order linear odes with constant co-efficients, the problem of existence and uniqueness simplifies further.

In the following we cite examples of such linear odes :

$$(i) \quad x^2y'' - xy' + e^xy = ln\,|\,x\,|$$

$(ii) \quad y''' - 6y'' + 12y' - 6y = 0$

$(iii) \quad y'' + 2y' + 5y = e^{-x} \cos 2x$

$(iv) \quad y'' + y = \sin x; \quad y(0) = 1; \quad y'(0) = 0$

$(v) \quad y'' + 2y' + 5y = 0$

Here the first ode, i.e., (i) appears in non-normal form while the odes (ii)−(v) are in normal form. The equations (ii) and (v) are homogeneous while the equations (iii) and (iv) are inhomogeneous. The equation (iv) is an initial-value problem as it has two accessory initial conditions. The ode (v), infact, is the 'associated homogeneous ode' corresponding to (iii).

If we agree to write $\frac{a_k(x)}{a_0(x)}$ as $p_k(x)$ for all $k = 1, 2, \cdots\cdots, n;$ and $\frac{b(x)}{a_0(x)} = q(x)$ and $y^{(k)}$ for $\frac{d^k y}{dx^k}$ for all $k = 1, 2, \cdots\cdots, n$ then the equation (5.1) assumes the normal form :

$$y^{(n)} = -p_1(x)y^{(n-1)} - p_2(x)y^{(n-2)} - \cdots\cdots - p_n(x)y + q(x) \qquad (5.3)$$

the initial conditions given in (5.2) appears as

$$y(x_0) = y_0; \quad y'(x_0) = y_0'; \quad \cdots\cdots y^{(n-1)}(x_0) = y_0^{(n-1)} \qquad (5.4)$$

If we like to apply the theory of first order linear odes to higher order odes given in normal form (5.3), then we have to regard the right hand side of (5.3) as a function of $x, y, y', \cdots\cdots, y^{(n-1)}$. Let us introduce again a new notation :

$$y' \equiv y_1; \quad y'' \equiv y_2; \cdots\cdots \quad , y^{(n-1)} \equiv y_{n-1}.$$

This replaces (5.3) by a system of n coupled first order linear odes :

$$y' \ = y_1$$
$$y_1' \ = y_2$$
$$y_2' \ = y_3$$
$$\cdots\cdots \quad \cdots\cdots$$
$$\cdots\cdots \quad \cdots\cdots$$
$$y_{n-2}' \ = y_{n-1}$$
$$y_{n-1}' \ = -p_1(x)y_{n-1} - p_2(x)y_{n-2} - \cdots\cdots - p_n(x)y + q(x) \qquad (5.5)$$

while the newlook initial conditions (5.4) become :

$$y(x_0) = y_0; \ y_1(x_0) = y_0'; \ \cdots\cdots, \ y_{n-1}(x_0) = y_0^{(n-1)}.$$

Observe that the right sides of each of the first order ode appearing in (5.5) are continuous functions and satisfies the condition of existence of bounded derivatives with respect to $y, y_1, y_2, \cdots, \cdots, y_{n-1}$. Hence the criteria of existence of a unique integral (solution) satisfying the preassigned initial conditions are fulfilled. This approach eventually replaces the nth order linear ode by a system of n simultaneous linear first order odes each having its own requisite initial condition.

For the homogeneous ode associated with (5.5), we have the equivalent system of linear odes representable in the matrix form as :

$$\left(\frac{d}{dx}\right)\begin{bmatrix} y \\ y_1 \\ y_2 \\ \cdots \\ \cdots \\ \cdots \\ y_{n-2} \\ y_{n-1} \end{bmatrix} = \begin{bmatrix} 0 & 1 & \cdots & 0 & 0 \\ 0 & 0 & \cdots & 0 & 0 \\ 0 & 0 & \cdots & 0 & 0 \\ \cdots & \cdots & \cdots & \cdots & \cdots \\ \cdots & \cdots & \cdots & \cdots & \cdots \\ 0 & 0 & \cdots & 0 & 1 \\ -p_n & -p_{n-1} & \cdots & -p_2 & -p_1 \end{bmatrix}\begin{bmatrix} y \\ y_1 \\ y_2 \\ \cdots \\ \cdots \\ \cdots \\ y_{n-2} \\ y_{n-1} \end{bmatrix} \qquad (5.6)$$

Loosely, we may say that the matrix appearing on the R.H.S of (5.6) is a realisation of the differentiation operator $\frac{d}{dx} \equiv D$. We now like to explain what we intend to say here.

We assume that $y \in C^n(I)$, the space of all functions that are continuously differentiable upto order n. This implies that $y, y_1, y_2 \cdots, y_{n-1}$ are all elements of $C^1(I)$ and as a consequence, the n-vector $[y, y_1, y_2, \cdots, y_{n-2}, y_{n-1}]^T$ belongs to $C^1(I) \times \cdots\cdots \times C^1(I)$ (n factors), or compactly, $\prod_{1 \leq i \leq n} C^1(I)$, which is the space of continuously differentiable n-vector functions defined on I. Now D being a linear operator on the function space $C^k(I) \ \forall \ k \geq 1$, D is a linear operator on the space $\prod_{1 \leq i \leq n} C^1(I)$.

The $n \times n$ matrix appearing in (5.6) is indeed a realisation of the operator D in the function space $\prod_{1 \leq i \leq n} C^1(I)$. Henceforth this $n \times n$ matrix will be our main concern.

It is useful an observation that $e^{\lambda x}$ is an eigenvector (eigenfunction) of the linear operator $D \equiv \frac{d}{dx}$ corresponding to the eigenvalue λ because $\frac{d}{dx}(e^{\lambda x}) = \lambda e^{\lambda x}$. For any natural number k, $D^k(e^{\lambda x}) = \lambda^k(e^{\lambda x})$. Using these results we seek a solution of the ode in the form $e^{\lambda x}$ for the special case of the equation (5.3) where $p_k(x)$, $k = 1, 2, \cdots n$ are all constants and $q(x)$ is set to be zero. Substitution of y in the associated homogeneous ode ensures that $e^{\lambda x}$ can be a solution of the ode

$$y^{(n)} + p_1 y^{(n-1)} + p_2 y^{(n-2)} + \cdots\cdots + p_{n-1}y' + p_n y = 0 \qquad (5.7)$$

iff λ is chosen a root of the nth degree polynomial equation

$$\lambda^n + p_1\lambda^{n-1} + p_2\lambda^{n-2} + \cdots\cdots + p_{n-1}\lambda + p_n = 0 \qquad (5.8)$$

Equation (5.8) is known as "auxiliary equation" of the ode (5.7) as it owes its origin from the trial solution $y = e^{\lambda x}$ of the ode. The reason why the trial solution to a homogeneous ode with constant co-efficients are always taken in the form $y = e^{\lambda x}$ lies in the truth that $e^{\lambda x}$ is an eigenvector of D for any λ.

As it turns out, (5.8) is the characteristic equation of the matrix

$$A = \begin{bmatrix} 0 & 1 & 0 & \cdots & 0 & 0 \\ 0 & 0 & 1 & \cdots & 0 & 0 \\ 0 & 0 & 0 & \cdots & 0 & 0 \\ \cdots & \cdots & \cdots & \cdots & \cdots & \cdots \\ \cdots & \cdots & \cdots & \cdots & \cdots & \cdots \\ 0 & 0 & 0 & \cdots & 0 & 1 \\ -p_n & -p_{n-1} & -p_{n-2} & \cdots & -p_2 & -p_1 \end{bmatrix}$$

where p_k's are all constants.

The matrix A quoted above is referred to as a 'companion matrix' of the polynomial equation (5.8). At this point we like to make it clear that there are more than one companion matrices associated with a given polynomial equation. Because of this linkage between (5.8) and the above matrix, we often like to call (5.8) as the characteristic equation of the ode (5.7). The equation (5.8) might have its roots all real and distinct. In that case, we label the roots $\lambda_1, \lambda_2, \cdots\cdots, \lambda_n$. Obviously the corresponding eigenvectors $e^{\lambda_1 x}, e^{\lambda_2 x}, \cdots\cdots, e^{\lambda_n x}$ of D are all solutions of ode (5.7) because in the operator form (5.7) appears as :

$$L(D)y \equiv (D^n + p_1 D^{n-1} + p_2 D^{n-2} + \cdots\cdots + p_{n-1}D + p_n)y = 0 \quad (5.7a)$$

$$\therefore \quad (D^n + p_1 D^{n-1} + p_2 D^{n-2} + \cdots\cdots + p_{n-1}D + p_n)e^{\lambda_k x}$$
$$= (\lambda_k^n + p_1\lambda_k^{n-1} + p_2\lambda_k^{n-2} + \cdots\cdots + p_{n-1}\lambda_k + p_n)e^{\lambda_k x} = 0,$$

whenever λ_k is a root of (5.8).

These n solutions $\{e^{\lambda_k x} : k = 1(1)n\}$ are linearly independent for all x as can be established in the following :

Consider an arbitrary linear combination

$$t_1 e^{\lambda_1 x} + t_2 e^{\lambda_2 x} + \cdots\cdots + t_n e^{\lambda_n x} = 0, \quad t_k\text{'s being scalars.}$$

If we can show that t_k's all are zeros, then we will be done. Differentiating the chosen relation successively $(n-1)$ times w.r.to the independent variable x we have altogether n homogeneous equations :

$$\left. \begin{array}{l} t_1 e^{\lambda_1 x} + \cdots + t_{n-1}e^{\lambda_{n-1}x} + t_n e^{\lambda_n x} \quad\quad = 0 \\ \lambda_1 t_1 e^{\lambda_1 x} + \cdots + \lambda_{n-1}t_{n-1}e^{\lambda_{n-1}x} + \lambda_n t_n e^{\lambda_n x} = 0 \\ \cdots\cdots\cdots\cdots\cdots\cdots\cdots\cdots\cdots\cdots\cdots \\ \cdots\cdots\cdots\cdots\cdots\cdots\cdots\cdots\cdots\cdots\cdots \\ \lambda_1^{n-1}t_1 e^{\lambda_1 x} + \cdots\cdots + \lambda_{n-1}^{n-1}t_{n-1}e^{\lambda_{n-1}x} + \\ + \lambda_n^{n-1}t_n e^{\lambda_n x} \quad\quad = 0 \end{array} \right\}$$

In the matrix form these equations read :

$$\begin{bmatrix} 1 & 1 & \cdots & 1 & 1 \\ \lambda_1 & \lambda_2 & \cdots\cdots & \lambda_{n-1} & \lambda_n \\ \cdots & \cdots & \cdots & \cdots & \cdots \\ \cdots & \cdots & \cdots & \cdots & \cdots \\ \lambda_1^{n-1} & \lambda_2^{n-1} & \cdots\cdots & \lambda_{n-1}^{n-1} & \lambda_n^{n-1} \end{bmatrix} \begin{bmatrix} t_1 e^{\lambda_1 x} \\ t_2 e^{\lambda_2 x} \\ \cdots \\ \cdots \\ t_n e^{\lambda_n x} \end{bmatrix} = \begin{bmatrix} 0 \\ 0 \\ \cdots \\ \cdots \\ 0 \end{bmatrix} \quad (5.9)$$

The co-efficient matrix of the equation (5.9) is non-singular as its determinant is the well-known Vandermonde's determinant

$$V(\lambda_1, \lambda_2, \cdots\cdots, \lambda_n) = \prod_{1 \le r < s \le n} (\lambda_r - \lambda_s)$$

that is non-zero as long as λ_r's are all distinct. (5.9) being a homogeneous system of n equations in the n-variables $\{t_r e^{\lambda_r x} : r = 1, 2, \cdots, n\}$ therefore admits of a unique solution, i.e, trivial zero solution. Because $e^{\lambda_r x} > 0$ for all x and real λ_r's, $\{t_r : r = 1, 2, \cdots, n\}$ are all zeros. Hence the conclusion that $e^{\lambda_r x}$'s are all linearly independent. We can therefore conclude that the solution space of the homogeneous ode (5.7) has

a basis
$$\{e^{\lambda_1 x}, e^{\lambda_2 x}, \ldots\ldots, e^{\lambda_n x}\}$$

if λ_r's are all real and distinct. Any solution of the homogeneous ode (5.7) in this case is expressible as a linear combination of these $e^{\lambda_r x}$'s :

$$y = c_1 e^{\lambda_1 x} + c_2 e^{\lambda_2 x} + \cdots\cdots + c_n e^{\lambda_n x}$$

is the general solution of (5.7). As is obvious, if y_1 and y_2 be any two solutions of (5.7), then for arbitrary scalars t_1, t_2 the expression $t_1 y_1 + t_2 y_2$ are also solutions of (5.7). This instigates us to conclude that superposition principle holds for equation (5.7) and the set of all solutions of (5.7) forms an n-dimensional vector space.

If λ_k's are all real but not distinct, then $\{e^{\lambda_k x} : k = 1, 2, \cdots\cdots, n\}$ are not all linearly independent as some of them are identical. So we have to go for hunt of linearly independent solutions of (5.7). When roots of the auxiliary equation are real but not all distinct, without loss of generality, we may presume that first r of them are equal, say, $\lambda = \lambda_r$ while the rest $(n - r)$ are all distinct.

The solutions $e^{\lambda_{r+1} x}, e^{\lambda_{r+2} x}, \ldots\ldots, e^{\lambda_n x}$ of (5.7) are now linearly independent and forms a basis of a $(n - r)$ dimensional vector space. Any element y_r of this vector space is therefore expressible as

$$y_r \equiv t_{r+1} e^{\lambda_{r+1} x} + t_{r+2} e^{\lambda_{r+2} x} + \cdots\cdots + t_n e^{\lambda_n x}.$$

The equation (5.7(a)) therefore reads :

$$(D - \lambda_1)^r (D - \lambda_{r+1})(D - \lambda_{r+2}) \cdots\cdots (D - \lambda_n) y = 0$$
$$\Leftrightarrow \quad (D - \lambda_{r+1})(D - \lambda_{r+2}) \cdots\cdots (D - \lambda_n)(D - \lambda_1)^r y = 0$$
$$\Leftrightarrow \quad (D - \lambda_{r+1})(D - \lambda_{r+2}) \cdots\cdots (D - \lambda_n) v = 0 \qquad (5.10)$$

provided we denote $(D - \lambda_1)^r y$ by v. and tacitly assume the commutativity of the differentiation operator D to bring $(D - \lambda_1)^r$ to the right end. In this new form (5.10), (5.7(a)) gets replaced by a linear ode of order $(n - r)$, the dependent variable just changed from y to v (which effectively is a function of y). Clearly the general solution of (5.10) is of the form y_r given before. Substituting y_r for v in $(D - \lambda_1)^r y = v$ we have a linear ode (of order r):

$$(D - \lambda_1)^r y = t_{r+1} e^{\lambda_{r+1} x} + t_{r+2} e^{\lambda_{r+2} x} + \cdots\cdots + t_n e^{\lambda_n x} \qquad (5.11)$$

If we write $(D - \lambda_1)^{r-1} \equiv \omega$, then (5.11) becomes a first order linear ode in independent variable x and dependent variable ω :

$$\frac{d\omega}{dx} - \lambda_1\omega = t_{r+1}e^{\lambda_{r+1}x} + t_{r+2}e^{\lambda_{r+2}x} + \cdots\cdots + t_n e^{\lambda_n x} \qquad (5.12)$$

(5.12) has an I.F $e^{-\lambda_1 x}$ making the resultant exact ode :

$$\frac{d}{dx}(\omega e^{-\lambda_1 x}) = t_{r+1}e^{(\lambda_{r+1}-\lambda_1)x} + t_{r+2}e^{(\lambda_{r+2}-\lambda_1)x} + \cdots\cdots\cdots + t_n e^{(\lambda_n-\lambda_1)x}$$

When integrated w.r. to x, the above equation gives:

$$\omega = \frac{t_{r+1}}{\lambda_{r+1} - \lambda_1}\, e^{\lambda_{r+1}x} + \frac{t_{r+2}}{\lambda_{r+2} - \lambda_1}\, e^{\lambda_{r+2}x} + \cdots\cdots + \frac{t_n e^{\lambda_n x}}{\lambda_n - \lambda_1} + t_1' e^{\lambda_1 x}$$

(Note t_1' is an integrating constant and $\frac{t_{r+k}}{\lambda_{r+k}-\lambda_1}$ are all well-defined because $\lambda_{r+k} \neq \lambda_1$ when $k = 1, 2, \cdots, \overline{n-r}$.) Thus

$$(D - \lambda_1)^{r-1}y = \omega = t_{r+1}'e^{\lambda_{r+1}x} + t_{r+2}'e^{\lambda_{r+2}x} + \cdots\cdots + t_n'e^{\lambda_n x} + t_1'e^{\lambda_n x},$$
$$(5.13)$$

where $t_{r+k}' \equiv \frac{t_{r+k}}{\lambda_{r+k}-\lambda_1}$.

If we now write $(D - \lambda_1)^{r-2}y = u$, then (5.13) becomes a first order linear ode in independent variable x and dependent variable u :

$$\frac{du}{dx} - \lambda_1 u = t_{r+1}'e^{\lambda_{r+1}x} + t_{r+2}'e^{\lambda_{r+2}x} + \cdots\cdots + t_n'e^{\lambda_n x} + t_1'e^{\lambda_1 x} \quad (5.14)$$

Equation (5.14) is now subjected to the same treatment that was given to (5.12) and the final solution becomes :

$$(D-\lambda_1)^{r-2}y = u = t_{r+1}''e^{\lambda_{r+1}x}+t_{r+2}''e^{\lambda_{r+2}x}+\cdots\cdots+t_n''e^{\lambda_n x}+(t_1'x+t_2')e^{\lambda_1 x},$$

where $t_{k+r}'' = \frac{t_{k+r}'}{\lambda_{k+r}-\lambda_1}$ $\forall\ k = 1, 2, \cdots\cdots, n-r$ and t_2' is a new constant of integration.

Following these same steps for $(r - 1)$ times successively, one can see that the general solution y to (5.7(a)) will be :

$$y = (c_1 + c_2 x + \cdots + c_r x^{r-1})e^{\lambda_1 x} + c_{r+1}e^{\lambda_{r+1}x} + \cdots + c_n e^{\lambda_n x}$$

where $c_1, c_2, \cdots\cdots, c_n$ are n arbitrary constants of integration. We now prove that $e^{\lambda_1 x}, xe^{\lambda_1 x}, \cdots\cdots, x^{r-1}e^{\lambda_1 x}, e^{\lambda_{r+1}x}, \cdots\cdots, e^{\lambda_n x}$ are all linearly independent, and moreover $x^{k-1}e^{\lambda_1 x}$, $k = 1, 2, \cdots\cdots, r$ are all solutions of (5.7(a)) so long $r \leqslant n$.

Consider the linear relation

$$(t_1 + t_2 x + \cdots\cdots + t_r x^{r-1})e^{\lambda_1 x} + t_{r+1}e^{\lambda_{r+1} x} + \cdots + t_n e^{\lambda_n x} = 0$$

where $x \in I$. Multiplication of this relation by $e^{-\lambda_1 x}$ yields :

$$t_1 + t_2 x + \cdots + t_r x^{r-1} + t_{r+1}e^{(\lambda_{r+1} - \lambda_1)x}$$
$$+t_{r+2}e^{(\lambda_{r+2} - \lambda_1)x} + \cdots\cdots + t_n e^{(\lambda_n - \lambda_1)x} = 0$$

We now differentiate this relation r-times successively to have :

$$t_{r+1}(\lambda_{r+1} - \lambda_1)^r e^{(\lambda_{r+1} - \lambda_1)x} + t_{r+2}(\lambda_{r+2} - \lambda_1)^r e^{(\lambda_{r+1} - \lambda_1)x} +$$
$$\cdots\cdots + t_n(\lambda_n - \lambda_1)^r . e^{(\lambda_n - \lambda_1)x} = 0$$

However $\lambda_1, \lambda_{r+1}, \cdots\cdots, \lambda_n$ being all distinct, $e^{(\lambda_{r+1} - \lambda_1)x}, e^{(\lambda_{r+2} - \lambda_1)x}$, $\cdots\cdots, e^{(\lambda_n - \lambda_1)x}$ are all linearly independent and so the above relation can hold good only if $t_{r+1} = t_{r+2} = \cdots\cdots = t_n = 0$. Using this result, the original linear relation is reduced to

$$(t_1 + t_2 x + t_3 x^2 + \cdots\cdots + t_r x^{r-1})e^{\lambda_1 x} = 0$$
and so $$t_1 + t_2 x + t_3 x^2 + \cdots\cdots + t_r x^{r-1} = 0$$

because for real λ_1, $e^{\lambda_1 x} > 0$. Again $\{1, x, x^2, \cdots\cdots, x^{r-1}\}$ being a basis of $P_{(r-1)}(x)$, the vector space of all polynomials of degree less than or equal to $(r - 1)$, the above relation holds true if and only if $t_1 = t_2 = \cdots\cdots = t_r = 0$. This completes the proof of linear independence of the set

$$\{e^{\lambda_1 x}, xe^{\lambda_1 x}, x^2 e^{\lambda_1 x}, \cdots\cdots, x^{n-1}e^{\lambda_1 x}, e^{\lambda_{r+1} x}, \cdots\cdots, e^{\lambda_n x}\}$$

for every $x \in I$.

To prove the second part, we proceed from (5.7(a)) as follows:

Equation (5.7(a)) reads $L(D)y = 0$, where $L(D)$ is a polynomial in D. Using Leibnitz's rule for successive differentiation we can show that the following shift formula

$$L(D)\{e^{\lambda_1 x}v(x)\} = e^{\lambda_1 x}L(D + \lambda_1)(v(x))$$

holds for any differentiable function $v(x)$. In particular, if $v(x)$ is chosen x^{k-1} ($k = 1, 2, \cdots\cdots, r$), then using the factorisation of $L(D)$ we have:

$$L(D)\{x^{k-1}e^{\lambda_1 x}\}$$
$$= e^{\lambda_1 x}L(D + \lambda_1)(x^{k-1})$$
$$= e^{\lambda_1 x}(D - \lambda_1 + \lambda_1)^r(D - \lambda_{r+1} + \lambda_1)\cdots\cdots(D - \lambda_n + \lambda_1)(x^{k-1})$$
$$= e^{\lambda_1 x}(D - \lambda_{r+1} + \lambda_1)\cdots\cdots(D - \lambda_n + \lambda_1)D^r(x^{k-1}) = 0.$$

Hence each of $x^{k-1}e^{\lambda_1 x}$, $k = 1, 2, \cdots\cdots, r$ solution of (5.7(a)). The fact that the remaining ones, viz, $e^{\lambda_{r+1}x}, e^{\lambda_{r+2}x}, \cdots\cdots, e^{\lambda_n x}$ are all solutions follows trivially when we recall equation (5.8).

At this point a fastidious reader may comment that there is no justification of assuming only one root, viz, λ_1 to be a multiple root and the others all simple roots. Can we not take it granted that there are only k distinct real roots $\lambda_1, \lambda_2, \cdots\cdots, \lambda_k$ to (5.8), each having their multiplicity $r_1, r_2, \cdots\cdots, r_k$ respectively with $r_1 + r_2 + \cdots\cdots + r_k = n$? We have then the same foregone conclusion that the entities

$$e^{\lambda_1 x}, xe^{\lambda_1 x}, \cdots\cdots, x^{r_1-1}e^{\lambda_1 x}$$
$$e^{\lambda_2 x}, xe^{\lambda_2 x}; \cdots\cdots, x^{r_2-1}e^{\lambda_2 x}$$
$$\cdots\cdots\cdots\cdots\cdots\cdots\cdots\cdots\cdots$$
$$\cdots\cdots\cdots\cdots\cdots\cdots\cdots\cdots\cdots$$
$$e^{\lambda_k x}, xe^{\lambda_k x}, \cdots\cdots, x^{r_k-1}e^{\lambda_k x}$$

are all linearly independent and moreover are solutions to (5.7(a)).

To prove the 'linear independence' we harbour on method of contradiction and suppose that they are linearly dependent.

$$\therefore \qquad P_1(x)e^{\lambda_1 x} + P_2(x)e^{\lambda_2 x} + \cdots\cdots + P_k(x)e^{\lambda_k x} \equiv 0, \qquad (5.15)$$

where $P_j(x)$ ($j = 1, 2, \cdots\cdots, k$) is a polynomial of degree less than r_j, and at least one of these polynomials, for instance, $P_k(x)$ is not identically zero. Dividing the identity (5.15) by $e^{\lambda_1 x}$ and differentiating the resultant relation successively r_1 times, we observe that the first summand in (5.15) vanishes and we get a linear relation of the same type but with a smaller number of functions:

$$Q_2(x)e^{(\lambda_2 - \lambda_1)x} + \cdots\cdots + Q_k(x)e^{(\lambda_k - \lambda_1)x} = 0 \qquad (5.16)$$

The degrees of the polynomials Q_j and P_j $(j = 2, 3, \cdots\cdots, k)$ are same because $\frac{d}{dx}[P_j(x)e^{mx}] = [P_j(x)m + \frac{d}{dx}(P_j(x))]e^{mx}$ for every m.

In particular, the degrees of $Q_k(x)$ and $P_k(x)$ being same, $Q_k(x)$ is not identically zero. Dividing (5.16) by $e^{(\lambda_2 - \lambda_1)x}$ and differentiating the resultant relation successively r_2 times, we get back a linear relation involving smaller number of functions:

$$R_3(x)e^{(\lambda_3 - \lambda_2)x} + \cdots\cdots + R_k(x)e^{(\lambda_k - \lambda_2)x} = 0 \qquad (5.17)$$

where again $R_3, \cdots\cdots, R_k$ have the same degrees as $Q_3, \cdots\cdots, Q_k$ respectively. Iterating this process $(k-1)$ times successively, we obtain a relation $S_k'(x)e^{(\lambda_k - \lambda_{k-1})x} \equiv 0$, which is impossible because $S_k'(x)$ being a polynomial of the same degree as $P_k(x)$ is never identically zero. Thus $P_k(x)$ must be identically zero over I, and so the above functions must be linearly independent for all $x \in I$.

Note : (a) The basic line of proof and argument will not change had the roots been complex (some/all), no matter whether repeated or distinct.

(b) The operator $L(D)$ appearing in ((5.7(a))) may be thought of as a linear operator from $C^\infty(\mathbb{R})$ to itself. Thus finding all possible solutions of the nth order ode(5.7) or (5.7(a)) is tantamount to finding the kernel of the operator $L(D) : C^\infty(\mathbb{R}) \to C^\infty(\mathbb{R})$. If one agrees to write $Ker(L(D))$ as M, then clearly M is an n-dimensional subspace of the infinite dimensional space $C^\infty(\mathbb{R})$. The general element of subspace M is known as **complementary function** of the ode $L(D)y = q(x)$ and $L(D)$ is known as **linear differential operator of order n**.

5.3   Wronskian

Theorem 5.1 : The n functions $y_1, y_2, \cdots\cdots, y_n$ are linearly independent functions over some interval I iff their Wronskian $W(x)$ defined by the determinant

$$W(x) = \begin{vmatrix} y_1 & y_2 & \cdots\cdots & y_n \\ y_1' & y_2' & \cdots\cdots & y_n' \\ \cdots & \cdots & \cdots\cdots & \cdots \\ \cdots & \cdots & \cdots\cdots & \cdots \\ y_1^{(n-1)} & y_2^{(n-1)} & \cdots\cdots & y_n^{(n-1)} \end{vmatrix}$$

does not vanish for any $x \in I$.

Proof : (**Necessary part**) : Let $y_1, y_2, \cdots, y_n$ be n linearly independent functions defined over I. Hence if we consider a linear relation $c_1 y_1 + c_2 y_2 + \cdots + c_n y_n = 0$, then the only possible set of scalars c_i that satisfies this relation consists of only zeros. Assuming the functions $y_k \in C^n(I)$, we now differentiate the relation $(n-1)$ times successively to have altogether n homogeneous equations in n scalars $c_1, c_2, \cdots, c_n$:

$$\left.\begin{array}{l} c_1 y_1 + c_2 y_2 + \cdots + c_n y_n = 0 \\ c_1 y_1' + c_2 y_2' + \cdots + c_n y_n' = 0 \\ \cdots\cdots\cdots\cdots\cdots\cdots\cdots\cdots \\ \cdots\cdots\cdots\cdots\cdots\cdots\cdots\cdots \\ c_n y_1^{(n-1)} + c_2 y_2^{(n-1)} + \cdots + c_n y_n^{(n-1)} = 0 \end{array}\right\}$$

As stated before, this homogeneous system of n equations in $c_1, c_2, \cdots, c_n$ admits of only the trivial solution. This uniqueness of solution of the system of linear equations therefore can be had only if the co-efficient matrix

$$\begin{bmatrix} y_1 & y_2 & \cdots\cdots & y_n \\ y_1' & y_2' & \cdots\cdots & y_n' \\ \cdots & \cdots & \cdots\cdots & \cdots \\ \cdots & \cdots & \cdots\cdots & \cdots \\ y_1^{(n-1)} & y_2^{(n-1)} & \cdots\cdots & y_n^{(n-1)} \end{bmatrix}$$

is non-singular. In other words, the Wronskian $W(x)$ of these solutions never vanish on I.

Converse: Left as an exercise.

Theorem 5.2 : If $\{f_1, f_2, \cdots, f_n\}$ be linearly independent functions of x that belong to class $C^{(n-1)}(I)$, I being some interval in $\mathbb{R}$ and if v be any function belonging to the same class $C^{(n-1)}(I)$, then $\{v f_1, v f_2, \cdots, v f_n\}$ is also linearly independent.

Proof : To establish this result, it suffices to establish that if

$W(f_1, f_2, \cdots, f_n) \neq 0$, then $W(vf_1, vf_2, \cdots, vf_n) \neq 0$

$$W(vf_1, vf_2, \cdots, vf_n) = \begin{vmatrix} vf_1 & vf_2 & \cdots & vf_n \\ (vf_1)' & (vf_2)' & \cdots & (vf_n)' \\ (vf_1)'' & (vf_2)'' & \cdots & (vf_n)'' \\ \cdots & \cdots & \cdots & \cdots \\ \cdots & \cdots & \cdots & \cdots \\ (vf_1)^{(n-1)} & (vf_2)^{(n-1)} & \cdots & (vf_n)^{(n-1)} \end{vmatrix}$$

$$= v^n \cdot \begin{vmatrix} f_1 & f_2 & \cdots & f_n \\ f_1' & f_2' & \cdots & f_n' \\ f_1'' & f_2'' & \cdots & f_n'' \\ \vdots & \vdots & \vdots & \vdots \\ \vdots & \vdots & \vdots & \vdots \\ f_1^{(n-1)} & f_2^{(n-1)} & \cdots & f_n^{(n-1)} \end{vmatrix}$$

(using simple rules of differentiation and property of determinants)

$$= v^n \cdot W(f_1, f_2, \cdots, f_n) \neq 0$$

Hence the proof.

Note: When in our earlier discussion the functions $e^{\lambda_1 x}, \cdots\cdots, e^{\lambda_n x}$ were shown independent under the assumption that $\lambda_1, \lambda_2, \cdots\cdots, \lambda_n$ and all real and distinct, then we came across Vandermonde's determinant $V(\lambda_1, \lambda_2, \cdots\cdots, \lambda_n)$ which basically is the Wronskian of the solutions $\{e^{\lambda_r x} : r = 1, 2, \cdots\cdots, n\}$ of (5.7) or (5.7(a)).

For linear homogeneous odes, Wronskian plays also a very important role as can be seen in the light of the following theorem.

Theorem 5.3 : If $y_1, y_2, \cdots\cdots, y_n$ be a set of n linearly independent solutions to the nth order homogeneous linear ode

$$y^{(n)} + p_1(x)y^{(n-1)} + p_2(x)y^{(n-2)} + \cdots\cdots + p_n(x)y = 0 \qquad (5.18)$$

where $\{p_k(x) : k = 1, 2, \cdots\cdots, n\}$ are continuous functions over a closed interval I, then Wronskian $W(x)$ of these solutions will be given by

$$W(x) = k \exp\left[-\int p_1(x)dx\right], \ k \text{ being some constant.}$$

If moreover, $x_0 \in I$ and $x > x_0$ but $x \in I$, then we may express $W(x)$ by the relation

$$W(x) = W(x_0) \exp\left[-\int_{x_0}^{x} p_1(x)dx \right].$$

This result is familiar as **Ostrogradsky-Liouville** formula.

Proof: By definition,

$$W(x) = \begin{vmatrix} y_1 & y_2 & \cdots\cdots & y_n \\ y_1' & y_2' & \cdots\cdots & y_n' \\ \cdots & \cdots & \cdots\cdots & \cdots \\ \cdots & \cdots & \cdots\cdots & \cdots \\ y_1^{(n-2)} & y_2^{(n-2)} & \cdots\cdots & y_n^{(n-2)} \\ y_1^{(n-1)} & y_2^{(n-1)} & \cdots\cdots & y_n^{(n-1)} \end{vmatrix}$$

Differentiating this expression with respect to x, we have:

$$\frac{dW}{dx} = \begin{vmatrix} y_1 & y_2 & \cdots & y_n \\ y_1' & y_2' & \cdots & y_n' \\ \cdots & \cdots & \cdots & \cdots \\ \cdots & \cdots & \cdots & \cdots \\ y_1^{(n-2)} & y_2^{(n-2)} & \cdots & y_n^{(n-2)} \\ y_1^{(n)} & y_2^{(n)} & \cdots & y_n^{(n)} \end{vmatrix}$$

$$= \begin{vmatrix} y_1 & y_2 & \cdots & y_n \\ y_1' & y_2' & \cdots & y_n' \\ \cdots & \cdots & \cdots & \cdots \\ \cdots & \cdots & \cdots & \cdots \\ y_1^{(n-2)} & y_2^{(n-2)} & \cdots & y_n^{(n-2)} \\ -\sum_{k=1}^{n} p_k y_1^{(n-k)} & -\sum_{k=1}^{n} p_k y_2^{(n-k)} & \cdots & -\sum_{k=1}^{n} p_k y_n^{(n-k)} \end{vmatrix}$$

(the last step follows from the original ode). If now we carry out the elementary row operation

$$R_n \longrightarrow R_n' = R_n + p_n(x)R_1 + p_{n-1}(x)R_2 + \cdots\cdots + p_2(x)R_{n-1}$$

then we would have

$$\frac{dW}{dx} = -p_1(x)W(x) \qquad (5.19)$$

Now because of the linear independence of solutions $y_1, y_2, \cdots\cdots, y_n$, $W(x)$ is never zero for any $x \in I$ and we can put (5.19) in the separable form and integrate both sides w.r.to x over I to get :

$$W(x) = K \exp\left[-\int p_1(x)dx\right]$$

where K is an integration constant. Since I is a closed interval, and $p_1(x)$ is continuous over I, $\int p_1(x)dx$ exists in the Riemannian sense.

If moreover $x_0 \in I$ and $[x_0, x] \subseteq I$, then integrating (5.19) over this subinterval we have :

$$W(x) = W(x_0)\exp\left[\int_{x_0}^{x} p_1(x)dx\right] \tag{5.20}$$

In the last case however, the restriction of I being a closed interval is redundant.

Now comes the all important result — the choice of the initial conditions can be arbitrary with the only restriction that $y_0, y_0', \cdots\cdots, y_0^{(n-1)}$ should not be all zeros if a non-trivial solution is desirable. Note that the behavior of Wronskian is very worthy — it just requires to check of linear independence of solutions $y_1, y_2, \cdots\cdots, y_n$ at one point $x_0 \in I$ as Wronskian propagates this feature to the whole of I. In this viewpoint, Wronskian works as a propagator. While we discuss Sturm-Liouville's problem for second order ode in next chapter, we shall find enough support to this view.

Note: When we are given n solutions to a linear homogeneous ode of order n we shall straightway use definition of Wronskian of the solutions to check if their linear span forms the solution space of that ode or not.

Example (1) : Show that the functions e^x, xe^x, $\sinh x$ are linearly independent solutions of linear homogeneous ode $(D^2+1)(D-1)^2 y = 0$ where $D \equiv \frac{d}{dx}$ has its usual meaning. Observe that

$$\left.\begin{array}{rcl}
(D^2+1)(D-1)^2(e^x) &=& (D^2+1)(D-1)(e^x) = 0 \\
(D^2+1)(D-1)^2(xe^x) &=& (D^2+1)(D-1)(e^x) = 0 \\
(D^2+1)(D-1)^2(\sinh x) &=& (D-1)^2(D^2+1)(\sinh x) = 0
\end{array}\right\}$$

This shows that $e^x, xe^x, \sinh x$ are solutions. We now check that they are linearly independent as their Wronskian $W(x)$ is given by

$$W(x) = \begin{vmatrix} e^x & xe^x & \sinh x \\ e^x & (x+1)e^x & \cosh x \\ e^x & (x+2)e^x & \sinh x \end{vmatrix}$$

$$= e^{2x} \begin{vmatrix} 1 & x & \sinh x \\ 1 & (x+1) & \cosh x \\ 1 & (x+2) & \sinh x \end{vmatrix}$$

$$= e^{2x} \begin{vmatrix} 1 & x & \sinh x \\ 0 & 1 & \cosh x - \sinh x \\ 0 & 2 & 0 \end{vmatrix}$$

$$= 2e^{2x}(\sinh x - \cosh x) = 2e^x \neq 0 \qquad \text{for any } x.$$

This linear independence ensures that the linear span of $\{e^x, xe^x, \sinh x\}$ forms a three dimensional subspace of the four-dimensional solution space of the ode $(D^2 + 1)(D - 1)^2 y = 0$. One may verify that $\cosh x$ is the fourth linearly independent solution to this ode.

Example (2) : Show that the sets of functions given below are linearly dependent.

(i) $\{\sin 3x, \ \sin x, \ \sin^3 x\}$ \qquad (ii) $\{x^2 - x + 1, \ x^2 - 1, \ 3x^2 - x - 1\}$

The linear dependence of functions of the first set follows directly from the trigonometric formula $\sin 3x = 3\sin x - 4\sin^3 x$ while by a little algebra one can write $3x^2 - x - 1 = 2(x^2 - 1) + x^2 - x + 1$. Wronskian check can be also exercised.

Note : If the Wronskian of a set of n functions defined on a closed interval I is non-zero for at least one point of this interval, then the set of functions is linearly independent there. If $W = 0$ identically on I and if each of the functions is a solution of the same linear ode, then that set of functions is linearly dependent. For example, the ode $(D^2 - 5D + 6) = 0$ has its fundamental set $\{e^{2x}, e^{3x}\}$. Consider the linear combination $(ae^{2x} + be^{3x})$ and $(ce^{2x} + de^{3x})$. These are linearly independent solutions over any I if $(ad - bc) \neq 0$ and dependent if otherwise.

5.4   Working Rules for Homogeneous Linear Ode

For a homogeneous linear ode, one may find a set of linearly independent solutions. Each of these linearly independent solutions forms a vector space of dimension unity. Since any homogeneous linear ode of order n possesses as many as n linearly independent solutions, we may deem of the n-dimensional solution space of that ode as the direct sum of n one-dimensional spaces.

In this chapter so far we have discussed a lot over the theoretical aspects of the homogeneous linear odes with constant co-efficients. However, we shall not be happy to be mesmerised by the flora and fauna of the theoretical results. For practical utility, let us brief up the following working rules and explain them through illustrations.

Recall that the homogeneous ode of order n and having constant co-efficients is given by

$$L(D)y \equiv (D^n + p_1 D^{n-1} + p_2 D^{n-2} + \cdots\cdots + p_{n-1}D + p_n)y = 0$$

The characteristic equation of this ode is :

$$\lambda^n + p_1\lambda^{n-1} + p_2\lambda^{n-2} + \cdots\cdots + p_{n-1}\lambda + p_n = 0$$

Case I: All the roots of this characteristic equation are real and distinct. If $\lambda_1, \lambda_2, \cdots\cdots, \lambda_n$ denote these roots, the general solution will be in the form $y = c_1 e^{\lambda_1 x} + c_2 e^{\lambda_2 x} + \cdots\cdots + c_n e^{\lambda_n x}$, where c_k's are all reals or complex depending on whether we want a real-valued or a complex-valued solution.

Example (3) :   Consider the ode : $(D^4 - 5D^2 + 4)y = 0$

Here auxiliary (characteristic) equation is : $\lambda^4 - 5\lambda^2 + 4 = 0$, so that its roots are $\pm 1, \pm 2$. All roots being real and distinct, the general solution is : $y = c_1 e^x + c_2 e^{-x} + c_3 e^{2x} + c_4 e^{-2x}$.

It is trivial to check that $\{e^x, e^{-x}, e^{2x}, e^{-2x}\}$ is linearly independent. If we agree to write $(D^4 - 5D^2 + 4)$ as $L(D)$ and the solution space of $L(D)y = 0$ as M, $\{e^x, e^{-x}, e^{2x}, e^{-2x}\}$ forms a basis of the four dimensional solution space M. This M, as we pointed out earlier, is nothing but the kernel of linear operator $L(D)$ and is a subspace of C^∞.

Case II: All the roots are real but some are repeated. This case also covers the case in which $L(D)$ is a binomial expansion of $(D - \lambda)^n$ for some real λ. Without loss of generality, if we assume $\lambda_1, \lambda_2, \cdots\cdots, \lambda_k$ as the only distinct roots having $r_1, r_2, \cdots\cdots, r_k$ as their respective multiplicities $(r_1 + r_2 + \cdots\cdots + r_k = n$ of course), then the general solution of the above ode will appear in the form

$$y = P_1(x)e^{\lambda_1 x} + P_2(x)e^{\lambda_2 x} + \cdots\cdots + P_k(x)e^{\lambda_k x},$$

where $P_j(x)$ is any polynomial of degree less than r_j, $j = 1, 2, \cdots\cdots, k$,

i.e, $\quad P_j(x) \equiv (c_{j1} + c_{j2}x + c_{j3}x^2 + \cdots\cdots + c_{jr}x^{r_j-1})e^{\lambda_j x}$,

$c_{j1}, c_{j2}, \cdots\cdots c_{jr}$,'s being arbitrary constants.

Example (4) : Consider the ode $(D^4 - D^3 - 9D^2 - 11D - 4)y = 0$

The characteristic equation is : $\lambda^4 - \lambda^3 - 9\lambda^2 - 11\lambda - 4 = 0$ which has its roots 4 and -1 (multiplicity 3). Here the roots are all real but one of the roots is a triple root; Hence the general solution will be $y = P_1(x)e^{4x} + P_2(x)e^{-x}$, where $P_1(x)$ is a constant and $P_2(x)$ is any generic polynomial of degree less than 3. Hence we can write

$$y = c_1e^{4x} + (c_2x^2 + c_3x + c_4)e^{-x} \ ; \ c_1, c_2, c_3, c_4 \text{ are arbitrary constants.}$$

In the following let's see how knowledge of first order linear ode helps us frame the above solution. In the factor form the given ode reads:

$$(D - 4)(D + 1)^3 y = 0$$

Put $(D + 1)^3 y = u$ so that the given ode turns out to be $(D - 4)u = 0$, the general solution of which is $u = c_1' e^{4x}$

$$\therefore (D + 1)^3 \ y = c_1' e^{4x}$$

If we put $(D+1)^2 y = v$, the the above ode reduces to a first order linear ode having the form
$$(D + 1) \ v = c_1' e^{4x}$$

The general solution of this ode is : $\quad v = \dfrac{c_1'}{3}e^{4x} + c_2'e^{-x}$

Therefore, $\qquad\qquad (D + 1)^2 y = \dfrac{c_1'}{3}e^{4x} + c_2'e^{-x}$

Put $(D+1)y = w$ so that the above linear ode reads :

$$(D+1)w = \frac{c_1'}{3}e^{4x} + c_2'e^{-x}$$

The general solution of this reads : $w = \frac{c_1'}{9}e^{4x} + (c_2'x + c_3')e^{-x}$

Finally, $\frac{dy}{dx} + y = w = \frac{c_1'}{9}e^{4x} + (c_2' + c_3')e^{-x}$,

the general solution of which is :

$$y = \frac{c_1'}{27}e^{4x} + \left(\frac{c_2'}{2}x^2 + c_3'x + c_4'\right)e^{-x}$$

Rewriting $\frac{c_1'}{27} \equiv c_1$, $\frac{c_2'}{2} \equiv c_2$; $c_3' \equiv c_3$; $c_4' \equiv c_4$, we get the solution :

$$y = c_1e^{4x} + (c_2x^2 + c_3x + c_4)e^{-x}$$

Hence in the language of linear algebra one can conclude that $\{e^{4x}, e^{-x}, xe^{-x}, x^2e^{-x}\}$ is a basis of the solution space of the given linear homogeneous ode. As another observation we see that as $D : C^\infty \to C^\infty$ is the linear transformation defined by $D(y) = \frac{dy}{dx}$, then $Ker\left\{(D+1)^3\right\}$ has basis $\left\{e^{-x}, xe^{-x}, x^2e^{-x}\right\}$. This result is a particular case of fact :

If $D : C^\infty \to C^\infty$ is the linear transformation defined by $D(y) = \frac{dy}{dx}$ and μ is a scalar, then $Ker\left\{(D-\mu)^r\right\}$ has basis $\left\{e^{\mu x}, xe^{\mu x}, \cdots, x^{r-1}e^{\mu x}\right\}$.

Case III: The roots are not all real— some of them are complex. If we restrict ourselves to linear homogeneous ordinary differential equations with real constant co-efficients, then the complex roots of the auxiliary equation occur in pairs and so any such equation of odd order has at least one real root. In this case, the general solution might be written in the same pattern as the two previous cases, but this time one often prefers to put the solutions in terms of sinusoidal functions together with exponential factors.

If $a \pm ib$ be the complex roots of the ode $(5.7(a))$, then the corresponding terms of the general solution will be

$$y = Ae^{(a+ib)x} + Be^{(a-ib)x} = e^{ax}(Ae^{ibx} + Be^{-ibx}),$$

a, b being real and A, B are arbitrary constants (complex quantities). We demand y to be real-valued solution of $(5.7(a))$. This compels A

and B to satisfy the conditions $\bar{A} = B$ and $\bar{B} = A$ $\bar{A}, \bar{B}$ denotes complex conjugate of A and B respectively. If one uses Euler's formula $e^{ibx} = \cos bx + i \sin bx$, then we have

$$y = e^{ax} \left[(A + B) \cos bx + i(A - B) \sin bx \right]$$

Now writing $(A + B) = c_1$ and $i(A - B) = c_2$, we observe that both c_1 and c_2 are real. Thus $y = e^{ax}(c_1 \cos bx + c_2 \sin bx)$ where c_1 and c_2 are real arbitrary constants. If we take $c = \sqrt{c_1^2 + c_2^2}$ and $\tan \alpha = \frac{c_1}{c_2}$, then $\sin \alpha = \frac{c_1}{\sqrt{c_1^2 + c_2^2}}$; $\cos \alpha = \frac{c_2}{\sqrt{c_1^2 + c_2^2}}$ and consequently, $y = ce^{ax} \sin(bx + \alpha)$, α being also real. This α might be physically thought of as the epoch angle of a damped harmonic oscillator provided a, the real part of the complex auxiliary root is negative. In that case, physically one tells that $e^{ax} \equiv e^{-|a|x}$ is the damping factor.

Even if this complex pair $a \pm ib$ be p-fold, i.e., has multiplicity p, then situation will be a blend of above discussion and case II discussion.

Example (5) :   Solve the ode : $(D^3 - 3D^2 + 9D + 13)y = 0$

The auxiliary equation will be : $\lambda^3 - 3\lambda^2 + 9\lambda + 13 = 0$ for which one root is real and equal to -1. The complex roots will be given by the quadratic equation $\lambda^2 - 4\lambda + 13 = 0$. The general solution of the above ode is : $y = Ae^{-x} + Be^{2x} \sin(3x + \alpha)$, A, B, α being arbitrary constants.

Remark : Since $\lambda = 2 \pm 3i$, the solution $y = e^{(2 \pm 3i)x}$ are two linearly independent real solutions of this ode in $C^\infty(\mathbb{R})$ and they together with e^{-x} forms a basis of the solution space.

Example (6) :   For the equation $m \frac{d^2 x}{dt^2} + \frac{mg}{c}(x - l) = 0$; l, g and c being constants, find x and $\frac{dx}{dt}$ if $x = x_0$ and $\frac{dx}{dt} = 0$ at $t = 0$.

Cancelling out the common factor m and putting $(x - l)$ as y we have the ode transformed as :

$$\frac{d^2 y}{dt^2} + \frac{g}{c} y = 0,$$

whose general solution is : $y = A \cos \left(\sqrt{\frac{g}{c}} t + \alpha \right)$, where A and α are to be determined from the initial conditions.

$$\therefore \quad x = A\cos\left(\sqrt{\tfrac{g}{c}}\,t + \alpha\right) + l, \quad \text{where } \alpha = 0 \text{ and } A = x_0 - l$$

$$\text{Also } \tfrac{dx}{dt} = (l - x_0)\sqrt{\tfrac{g}{c}}\sin\left(\sqrt{\tfrac{g}{c}}\,t\right)$$

Couple of Remarks :

(a) If the constants p_k $(k = 1, 2, \cdots\cdots, n)$ appearing in the equation (5.7(a)), viz,

$$L(D)y \equiv (D^n + p_1 D^{n-1} + p_2 D^{n-2} + \cdots\cdots + p_{n-1}D + p_n)y = 0$$

be all real, then there can be had a set of n linearly independent real-valued solutions which span an n-dimensional space over $\mathbb{R}$. We skip the formal proof that makes use of

(i) Fundamental theorem of Algebra and

(ii) Linearity of operator $L(D)$,

However, we cite the following example to highlight the result.

Example (7) : Find all the real-valued solutions of the differential equation $(D^3 - iD^2 + D - i)y = 0$

The auxiliary equation is $(\lambda + i)(\lambda - i)^2 = 0$ so that the general solution is given by

$$y = (c_1 + c_2 x)e^{ix} + c_3 e^{-ix},$$

c_1, c_2, c_3 being complex constants. Hence the solution space is a three dimensional complex vector space. If y is to be real-valued, $y = \bar{y}$ and so $c_1 = \bar{c}_3$ and $c_2 = 0$ is a must. Thus the real-valued solution becomes

$$\begin{aligned}
y_r &= c_1 e^{ix} + \bar{c}_1 e^{-ix} \\
&= c_1(\cos x + i\sin x) + \bar{c}_1(\cos x - i\sin x) \\
&= (c_1 + \bar{c}_1)\cos x + i(c_1 - \bar{c}_1)\sin x,
\end{aligned}$$

$\therefore$ y_r belongs to the linear span of $\{\sin x, \cos x\}$ over $\mathbb{R}$.

This example shows how non-reality of the co-efficients of the given differential equation works as the underlying reason behind the dimensional reduction of the solution space after imposition of the real-valuedness demand for solutions.

(b) If a particular solution of $(5.7(a))$ satisfy real initial conditions, it is real-valued.

We outline the proof in the following:

Let $L(D)y = 0$ be appended by the initial conditions

$$y(x_0) = \beta_0;\ y'(x_0) = \beta_1;\ \cdots\cdots;\ y^{(n-1)}(x_0) = \beta_{n-1}\ (\beta_k\text{'s are real}).$$

From the result $0 = \overline{L(D)y} = L(D)\bar{y}$, it follows that $\bar{y}$ is a solution of the equation whenever y is a solution.

$\therefore L(D)(Im\ \bar{y}) = 0$ and moreover,

$$Im\ \bar{y}(x_0) = Im\ \bar{y}'(x_0) = \cdots\cdots = Im\ \bar{y}^{(n-1)}(x_0) = 0,$$

ensuring that $\bar{y}(x) = 0$ (due to uniqueness theorem for solutions).

Example (8) : Find the solution of the initial-value problem

$$(D^3 - 5D^2 + 9D - 5)y = 0,\ y(0) = 0,\ y'(0) = 1 \text{ and } y''(0) = 6$$

The auxiliary equation has its roots $1, 2 \pm i$ so that the fundamental solution set is $\{e^x,\ e^{2x}\cos x,\ e^{2x}\sin x\}$. The general solution is given by

$$y(x) = c_1 e^x + e^{2x}(c_2\cos x + c_3\sin x)$$

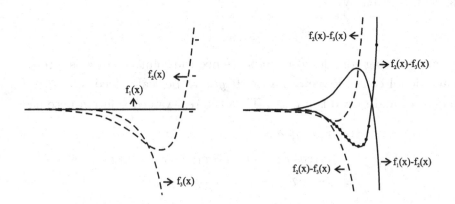

$$\textbf{Fig 5.1 (a)} \qquad\qquad\qquad \textbf{Fig 5.1 (b)}$$

Fig 5.1(a) : shows the graphs of $f_1(x) = e^x; f_2(x) = e^{2x}\cos x; f_3(x) = e^{2x}\sin x$

Fig 5.1(b) : shows the graphs of $f_1 \pm f_2$ and $f_2 \pm f_3$.

Both (a) and (b) show that all these solutions of ode in example(8) tend to zero as

$$x \to -\infty.$$

On successive differentiation and substitution of the initial conditions, we get the set of simultaneous equations as

$$c_1 + c_2 = 0; \quad c_1 + 2c_2 + c_3 = 1; \quad c_1 + 3c_2 + 4c_3 = 6,$$

which give solutions $c_1 = 1$; $c_2 = -1$ and $c_3 = 2$ by Cramer's rule.

∴ The particular solution is

$$y(x) = e^x + e^{2x}(2\sin x - \cos x)$$

(c) If the co-efficients of the differential equation be not all real, there is no assurance that the real or imaginary parts of the solutions will satisfy the differential equation. For example, let us consider

$$\left(D^2 - 4D + (7 + 4i)\right)y = 0.$$

This equation has its fundamental solution set as

$$\{e^x(\cos 2x + i\sin 2x), \ e^{3x}(\cos 2x - i\sin 2x)\}$$

However, none of $e^x\cos 2x$, $e^x\sin 2x$, $e^{3x}\cos 2x$, $e^{3x}\sin 2x$ is a solution to the differential equation. This result points out that the tactical use of complex number system to compute the particular integrals in many cases is delimited to the realm of real-co-efficient differential equations.

5.5   Few Theoretical Results from Linear Algebra.

Once the working rules of solving homogeneous linear odes are mastered, we shall aim to strengthen the theoretical aspects of these rules. This is why the following theorems alongwith their applicability in the theory of linear odes are tabled.

Theorem 5.4 : The kernel of the linear differential operator $(D - \lambda I)$: $C^\infty \to C^\infty$ has $\{e^{\lambda x}\}$ as a basis for each $\lambda \in \mathbb{C}$ and $x \in \mathbb{R}$.

Proof : Consider the linear ode : $(D - \lambda I)y = 0$. Obviously one of its solutions is $e^{\lambda x}$. If $y(x)$ be any other solution of this ode, then $\frac{dy}{dx} = \lambda y$.

$$\text{Define } z(x) = y(x)e^{-\lambda x}$$

$$\therefore \quad \frac{dz}{dx} = \frac{d}{dx}\left(y(x)e^{-\lambda x}\right) = \left(\frac{dy}{dx} - \lambda y\right)e^{-\lambda x} = 0 \quad \forall x$$

Since $\frac{dz}{dx}$ is indentically zero, we conclude that $z(x)$ is a constant only if we assume that domain of the function z is an open connected set in $\mathbb{C}$.

$\therefore$ $z(x) = y(x)e^{-\lambda x}$ is a complex constant , A, say.

$\therefore$ $y(x) = Ae^{\lambda x}$. Hence $\left\{ e^{\lambda x} \right\}$ serves as a basis of $Ker((D - \lambda I))$.

Obviously $\dim (Ker(D - \lambda I)) = 1$.

Theorem 5.5 : For any $\lambda \in \mathbb{C}$, the differential operator $(D - \lambda I)$ from C^∞ to itself is onto.

Proof : To prove onto character of the operator $(D - \lambda I) : C^\infty \to C^\infty$ it suffices to show that any $f \in c^\infty$ has at least one preimage $F \in C^\infty$.

Define $g(x) = f(x)e^{-\lambda x} \ \forall x \in \mathbb{R}$. It is obvious that $g \in C^\infty$ and so continuous. (Without loss of generality one may presume g to be complex-valued). So $\exists$ a function $G : \mathbb{R} \to \mathbb{C}$ that serves as an anti-derivative of g, i,e. $G'(x) = g(x) \ \forall x \in \mathbb{R}$. This $G \in C^\infty$. The function $F : \mathbb{R} \to \mathbb{C}$ defined by $F(x) = G(x)e^{\lambda x} \ \forall x \in \mathbb{R}$ also belongs to C^∞ and moreover,

$$
\begin{aligned}
(D - \lambda I)((F(x))) &= \frac{d}{dx}(F(x)) - \lambda F(x) \\
&= \frac{d}{dx}((G(x)e^{\lambda x})) - \lambda G(x)e^{\lambda x} \\
&= \lambda G(x)e^{\lambda x} + G'(x)e^{\lambda x} - \lambda G(x)e^{\lambda x} \\
&= g(x)e^{\lambda x} = f(x),
\end{aligned}
$$

ensuring that F is a preimage of f under $(D - \lambda I)$. Hence the claim.

Remark : Since composition of any two onto maps is again onto, we can inductively conclude that for any positive integer p , $(D - \lambda I)^p$ is an onto operator from C^∞ to itself.

Theorem 5.6 : If T_1 and T_2 be linear operators on V such that T_2 is onto and $Ker T_1$, $Ker T_2$ are both finite-dimensional and moreover,

$$
\dim(Ker(T_1 T_2)) = \dim(Ker T_1) + \dim(Ker T_2)
$$

Proof : Available in any standard linear algebra text. However, we shall use theorem 5.5 as a guideline to inductively conclude the following:

(i) $\dim(Ker(D - \lambda I)^p) = p$; $\dim(Ker(D - \mu I)^q) = q$ $(p, q \in \mathbb{N})$

(ii) The kernel of the operator $(D - \lambda I)^p (D - \mu I)^q : C^\infty \to C^\infty$ is a finite-dimensional subspace of C^∞ and moreover,

$$\dim(Ker\,((D - \lambda I)^p.(D - \mu I)^q) = p + q$$

One may therefore look into $Ker((D-\lambda I)^p(D-\mu I)^q)$ as the direct sum of $Ker((D-\lambda I)^p)$ and $Ker((D-\mu I)^q)$. If $\lambda_1, \lambda_2, \cdots, \lambda_k$ be only distinct roots of the auxiliary equation $L(\lambda) = 0$ with $r_1, r_2, \cdots, r_k$ being their respective multiplicities, $(i, e.\ r_1 + r_2 + \cdots + r_k = n)$, then the solution space of the ode $L(D)y = 0$ can be deemed as the direct sum of the subspaces $Ker((D-\lambda_1 I)^{r_1}), Ker((D-\lambda_2 I)^{r_2}) \cdots, Ker((D-\lambda_k I)^{r_k})$. If we recall that a basis of $Ker((D-\lambda_1 I)^{r_1})$ is $\{e^{\lambda_1 x}, xe^{\lambda_1 x}, \ldots \ldots, x^{r_1-1}e^{\lambda_1 x}\}$, a basis of $Ker((D - \lambda_2 I)^{r_2})$ is $\{e^{\lambda_2 x}, xe^{\lambda_2 x}, \ldots \ldots, x^{r_2-1}e^{\lambda_2 x}\}$ etc, then a basis of the solution space $Ker(L(D))$ of the ode $L(D)y = 0$ is their union.

Remark : In the subsequent discussions we shall continue to write $(D - \lambda)$ in lieu of $(D - \lambda I)$ etc.

5.6 Symbolic Operator $\frac{1}{L(D)}$ and Particular Integral

We now return to the non-homogeneous ode (represented in the normalised form)

$$y^{(n)} = -p_1(x)y^{(n-1)} - p_2(x)y^{(n-2)} - \cdots\cdots - p_{n-1}(x)y' - p_n(x)y + q(x)$$

and deal with the following theorem that lies at the heart of non-homogeneous differential equations.

Theorem 5.7 : If v be any solution of the non-homogeneous ode

$$y^{(n)} = -p_1(x)y^{(n-1)} - p_2(x)y^{(n-2)} - \cdots\cdots - p_{n-1}(x)y' - p_n(x)y + q(x)$$

and u any solution of the associated homogeneous ode

$$y^{(n)} = -p_1(x)y^{(n-1)} - p_2(x)y^{(n-2)} - \cdots\cdots - p_{n-1}(x)y' - p_n(x)y,$$

then $u + v$ is also a solution of the non-homogeneous ode.

The proof being trivial we skip it and go for its important corollary which states that if y_p be a particular solution of the above non-homogeneous ode involving no arbitrary constant and y_c be a general solution of the associated homogeneous ode (in the sense $y_c = \sum_{k=1}^{n} c_k y_k$, where c_k's are arbitrary constants and y_k's are linearly independent solutions of the homogeneous ode) then the quantity $y_c + y_p$ is by the above theorem, a solution of the non-homogeneous ode. Had we restricted ourselves to the non-homogeneous odes with constant co-efficients, then we would represent it in the operator form as :

$$L(D)y = q(x) \cdots\cdots [c.f. \ (5.7(a))]$$

In this light, y_c satisfies $L(D)y = 0$ while y_p is a solution of $L(D)y = q(x)$. Often we represent $y_p = \frac{1}{L(D)} q(x)$, but should bear in mind the fact that this symbolic operator $\frac{1}{L(D)}$ is not to be confused as an inverse of operator $L(D)$. Recall that non-singularity (i.e. one-to-one and onto character) is a must for invertibility of linear transformations. $L(D)$ has no non-singularity and moreover, $\frac{1}{L(D)}$ is n-fold integration operator as reciprocal of any polynomial is never a polynomial. Despite all these, few authors like to call $\frac{1}{L(D)}$ as inverse of operator $L(D)$ simply because they look into integration as inverse process of differentiation.

The operators $L(D)$ and $\frac{1}{L(D)}$ however commute (!) and

$$L(D) \left(\frac{1}{L(D)} \ q(x) \right) = q(x)$$

for all functions $q(x)$ that are infinitely differentiable.

The following theorem is quite simple but interesting :-

Theorem 5.8 : If $y_1, y_2, \cdots\cdots, y_n$ are linearly independent solutions of the homogeneous ode associated with the non-homogeneous ode $L(D)y = q(x)$ and y_p be a particular solution of the non-homogeneous ode, then $\{y_1, y_2, \cdots\cdots, y_n, y_p\}$ forms a linearly independent set. However, the set of all possible solutions of the non-homogeneous ode does not form a linear space.

Proof : To test the linear independence of the above set we consider the Wronskian of $y_1, y_2, \cdots\cdots, y_n$ and y_p and observe that y_k's are solutions of the homogeneous differential equation $L(D)y = 0$. So for $k = 1(1)n$,

$$y_k^{(n)} = -p_1(x)y_k^{(n-1)} - p_2(x)y_k^{(n-2)} - \cdots\cdots - p_{n-1}(x)y_k' - p_n(x)y_k$$

Wronskian of $y_1, y_2, \cdots\cdots, y_n$ and y_p will be given by

$$W(x) = \begin{vmatrix} y_1 & y_2 & \cdots\cdots & y_n & \frac{1}{L(D)}q(x) \\ y_1' & y_2' & \cdots\cdots & y_n' & \frac{D}{L(D)}q(x) \\ y_1'' & y_2'' & \cdots\cdots & y_n'' & \frac{D^2}{L(D)}q(x) \\ \cdots & \cdots & \cdots\cdots & \cdots & \cdots\cdots \\ \cdots & \cdots & \cdots\cdots & \cdots & \cdots\cdots \\ y_1^{(n-1)} & y_2^{(n-1)} & \cdots\cdots & y_n^{(n-1)} & \frac{D^{n-1}}{L(D)}q(x) \\ y_1^{(n)} & y_2^{(n)} & \cdots\cdots & y_n^{(n)} & \frac{D^n}{L(D)}q(x) \end{vmatrix}$$

$$= \begin{vmatrix} y_1 & y_2 & \cdots & y_n & \frac{1}{L(D)}q(x) \\ y_1' & y_2' & \cdots & y_n' & \frac{D}{L(D)}q(x) \\ y_1'' & y_2'' & \cdots & y_n'' & \frac{D^2}{L(D)}q(x) \\ \cdots & \cdots & \cdots & \cdots & \cdots\cdots \\ \cdots & \cdots & \cdots & \cdots & \cdots\cdots \\ y_1^{(n-1)} & y_2^{(n-1)} & \cdots & y_n^{(n-1)} & \frac{D^{n-1}}{L(D)}q(x) \\ -\sum_{r=1}^{n}p_r y^{(n-r)} & -\sum_{r=1}^{n}p_r y_2^{(n-r)} & \cdots & -\sum_{r=1}^{n}p_r y_n^{(n-r)} & \frac{D^n}{L(D)}q(x) \end{vmatrix}$$

$$= \begin{vmatrix} y_1 & y_2 & \cdots\cdots & y_n & \frac{1}{L(D)}q(x) \\ y_1' & y_2' & \cdots\cdots & y_n' & \frac{D}{L(D)}q(x) \\ y_1'' & y_2'' & \cdots\cdots & y_n'' & \frac{D^2}{L(D)}q(x) \\ \cdots & \cdots & \cdots\cdots & \cdots & \cdots\cdots \\ \cdots & \cdots & \cdots\cdots & \cdots & \cdots\cdots \\ y_1^{(n-1)} & y_2^{(n-1)} & \cdots\cdots & y_n^{(n-1)} & \frac{D^{n-1}}{L(D)}q(x) \\ 0 & 0 & \cdots\cdots & 0 & L(D)\left(\frac{1}{L(D)}q(x)\right) \end{vmatrix}$$

$$= q(x) \begin{vmatrix} y_1 & y_2 & \cdots\cdots & y_n \\ y_1' & y_2' & \cdots\cdots & y_n' \\ \cdots & \cdots & \cdots\cdots & \cdots \\ \cdots & \cdots & \cdots\cdots & \cdots \\ y_1^{(n-1)} & y_2^{(n-1)} & \cdots\cdots & y_n^{(n-1)} \end{vmatrix} \neq 0$$

$y_1, y_2, \cdots\cdots, y_n$ being linearly independent, their Wronskian is non-zero.

Note that in the penultimate step of our computation we applied the elementary row operation

$$R_{n+1} \longrightarrow R_{n+1}' = R_{n+1} + p_1 R_1 + p_2 R_2 + \cdots + p_n R_n$$

while the result $L(D) \left(\frac{1}{L(D)} q(x) \right) = q(x)$ enabled us to get the final result. This completes proof of the theorem's first and main part. The second part follows immediately from the fact that $y_p \notin \text{Span}\{y_1, y_2, \cdots, y_n\}$ and the complete solutions being of the form $y_p + c_1 y_1 + c_2 y_2 + \cdots + c_n y_n$ the vector space structure of the solution set for the associated homogeneous ode gets lost when every element of that space gets a shift through y_p as now there exists no additive identity.

Another important theoretical result− the 'superposition principle', is a special feature of linear differential equations, both homogeneous and non-homogeneous. We brief it up as follows :

Let the non-homogeneous linear differential equation of order n be given by $L(D)y = q(x)$, where co-efficients of D and its higher powers are constants. If $\bar{y}_1(x)$ and $\bar{y}_2(x)$ be the solutions of the equation $L(D)y = q(x)$ when $q(x) = q_1(x)$ and $q(x) = q_2(x)$ respectively, then $L(D)\bar{y}_1 = q(x)$ and $L(D)\bar{y}_2 = q_2(x)$. Obviously

$$L(D) \left(c_1 \bar{y}_1(x) + c_2 \bar{y}_2(x) \right) = c_1 q_1(x) + c_2 q_2(x).$$

This shows that whenever $q(x)$ gets replaced by $c_1 q_1(x) + c_2 q_2(x)$, the linear combination $c_1 \bar{y}_1(x) + c_2 \bar{y}_2(x)$ is a solution of $L(D)y = q(x)$. This result is known as **superposition principle** : "If the righthand member of a non-homogeneous linear ode is a linear combination of two functions $q_1(x)$ and $q_2(x)$, then a particular solution is obtainable as the same linear combination of two solutions of the equation with $q_1(x)$ and $q_2(x)$ respectively as the righthand members. The principle can be extended to linear combinations of any finite number of functions $q_1(x), q_2(x), \cdots\cdots, q_n(x)$." Succinctly, 'Superposition principle' enables us to compute the particular integral of a non-homogeneous linear ode (with constant co-efficients) in small pieces.

If we deem $L(D)y$ as a force producing the displacement y, then the superposition principle states that the several forces $L(D)y_1, L(D)y_2, \cdots$, $L(D)y_n$ producing the respective displacements $y_1, y_2, \cdots\cdots, y_n$, when work in tandem, yield total displacement $y_1 + y_2 + \cdots\cdots + y_n$. This viewpoint is intuitively very appealing as it brings us close to the law of superposition of forces prevalent in electrostatics and Newtonian mechanics (but not in Einstein's general theory of gravitation) where the principle of physical independence of forces is valid. To clarify the analogy one may cite the following example.

Consider a metallic beam projecting horizontally from a wall acted upon by its own weight W_1 and a load W_2 suspended from its free end. The deflection of the beam due to its own weight W_1 alone is y_1 and the deflection due to only the suspended weight W_2 is y_2, so that the net deflection of the beam from the horizontal is $y_1 + y_2$.

A new zoom into the superposition principle :

Any nth order linear differential equation with constant co-efficients involves a nth degree polynomial

$$L(D) = D^n + p_1 D^{n-1} + \cdots\cdots + p_{n-1}D + p_n$$

that maps C^∞ to C^∞. We denote by M the kernel of $L(D)$. M is an n-dimensional subspace of C^∞. We now define a addition modulo relation on C^∞ by $f = g \ mod \ M$ iff $f - g \in M$ for any $f, g \in C^\infty$. This relation is trivially seen to be an equivalence relation. Hence the above modulo relation inflicts a partition of C^∞ into a number of equivalence classes. All the elements of a particular equivalence class are equivalent in the sense they are all solutions to the same inhomogeneous linear differential equation. If we formally define the coset of an element $f \in C^\infty$ by $f + M = \{f + m/m \in M\}$, then the aforesaid equivalence classes are just distinct cosets.

Now following the routine procedure, let's define the addition and scalar multiplication of two equivalence classes, i.e, cosets by

$$
\begin{aligned}
(f + M) + (g + M) &= (f + g) + M \\
\alpha(f + M) &= \alpha f + M
\end{aligned}
$$

These algebraic operations are well-defined and these cosets (known as 'affine space' of M)constitute a linear space, denoted by C^∞/M and is known as the quotient space of C^∞ with respect to M, i,e, with respect to the operator $L(D)$. The addition, or more generally, the linear combination prevailing in the quotient space C^∞/M shows how the general solution of the inhomogeneous linear differential equation $L(D)y = c_1 q_1(x) + c_2 q_2(x)$ is obtained by piecing together the general solution of $L(D)y = q_1(x)$ and that of $L(D)y = q_2(x)$. The discussion of the new outlook towards 'superposition principle' is stopped here for the time being. Now we aim to discuss 'superposition principle' in its general most form in the perspective of 'method of variation of parameters' and Green's functions associated with it.

5.7   Method of Variation of Parameters

The 'method of variation of parameters' is used to integrate an inhomogeneous linear differential equation when the solution of the corresponding homogeneous differential equation, that is the complementary function is known. If the inhomogeneous linear differential equation (5.3) given by

$$y^{(n)}(x) + p_1(x)y^{(n-1)}(x) + \cdots\cdots + p_{n-1}(x)y'(x) + p_n(x)y(x) = q(x)$$

has its complementary function $y = \sum_{k=1}^n c_k y_k(x)$ (where c_k's are arbitrary constants and $\{y_1, y_2, \cdots\cdots, y_n\}$ is the fundamental set of solutions to the related homogeneous equation) then we seek the complete solution in the form $y = \sum_{k=1}^n c_k(x)y_k(x)$ where $c_k(x)$ are arbitrary differentiable functions yet to be determined. These n functions $\{c_k(x) : k = 1, 2, \cdots\cdots, n\}$ should satisfy n constraint equations one of which must be the given inhomogeneous equation itself. The rest of the $(n-1)$ auxiliary constraints are tailored to suit our needs. Infact, these are chosen in such a way that the successive derivatives of y mimic the form they would assume in case of constant c_k's.

For definiteness, if

$$y = \sum_{k=1}^n c_k(x)y_k(x),$$

we should choose c_k's so that the second term on the rightside of

$$y'(x) = \sum_{k=1}^n c_k(x)y_k'(x) + \sum_{k=1}^n c_k'(x)y_k(x)$$

be zero, i,e,

$$\sum_{k=1}^n c_k'(x)y_k(x) = 0$$

and as a result, $y' = \sum_{k=1}^n c_k(x)y_k'(x),$

which is of the same form as in the case of constant c_k's. In the same vein, we demand that the second term in the expression of the second order derivative

$$y'' = \sum_{k=1}^n c_k(x)y_k''(x) + \sum_{k=1}^n c_k'(x)y_k'(x)$$

vanish and thus subject the $c_k(x)$'s to second constraint

$$\sum_{k=1}^{n} c_k'(x)y_k'(x) = 0$$

Iterating the process, we evaluate the derivatives of the function

$$y = \sum_{k=1}^{n} c_k(x)y_k(x)$$

upto order $(n-1)$ inclusive and demanding

$$\sum_{k=1}^{n} c_k'(x)y_k^{(r)}(x) = 0 \quad \text{for} \quad r = 0,1,2,\cdots,\overline{n-2}.$$

$$\therefore \quad y^{(r)} = \sum_{k=1}^{n} c_k(x)y_k^{(r)}(x), \quad r = 0,1,2,\cdots\cdots,\overline{n-1}.$$

Finally

$$y^{(n)} = \sum_{k=1}^{n} c_k(x)y_k^{(n)}(x) + \sum_{k=1}^{n} c_k'(x)y_k^{(n-1)}(x)$$

and consequently

$$\sum_{k=1}^{n} c_k(x)y_k^{(n)}(x) + \sum_{k=1}^{n} c_k'(x)y_k^{(n-1)}(x) + p_1(x)\sum_{k=1}^{n} c_k(x)y_k^{(n-1)} +$$

$$p_2(x)\sum_{k=1}^{n} c_k(x)y_k^{(n-2)} + \cdots\cdots + p_n(x)\sum_{k=1}^{n} c_k(x)y_k(x) = q(x)$$

or, $$\sum_{k=1}^{n} c_k'(x)y_k^{(n-1)}(x) + \sum_{k=1}^{n} c_k(x)[y_k^{(n)} + p_1(x)y_k^{(n-1)}$$

$$+\cdots\cdots + p_n(x)y_k] = q(x)$$

$$\therefore \quad \sum_{k=1}^{n} c_k'(x)y_k^{(n-1)}(x) = q(x)$$

(since by hypothesis y_k's are solutions of reduced ode).

Summarising, we have a system of n inhomogeneous linear equations written in the matrix form as

$$\begin{bmatrix} y_1 & y_2 & \cdots\cdots & y_n \\ y_1' & y_2' & \cdots\cdots & y_n' \\ \cdots & \cdots & \cdots\cdots & \cdots \\ \cdots & \cdots & \cdots\cdots & \cdots \\ y_1^{(n-1)} & y_2^{(n-1)} & \cdots\cdots & y_n^{(n-1)} \end{bmatrix} \begin{bmatrix} c_1'(x) \\ c_2'(x) \\ \cdots\cdots \\ \cdots\cdots \\ c_n'(x) \end{bmatrix} = \begin{bmatrix} 0 \\ 0 \\ \cdots \\ \cdots \\ q(x) \end{bmatrix}$$

This system admits of a unique solution since the Wronskian of the fundamental solutions $\{y_1, y_2, \cdots\cdots, y_n\}$ is non-zero. The solution is obtained by Cramer's rule. The co-efficient c_k's are obtained finally by the method of quadratures. The advantage of the method of variation of parameters lies in its omnipotency of solving inhomogeneous linear odes as it is valid for odes with variable co-efficients and any arbitrary inhomogeneous term (impulse) that is continuous on the interval of interest. However, this method is repugnant owing to its length and that is why one often gets inclined to develop shortcut routes of finding particular integrals for a handful of selective cases. Following illustrations will be useful to the reader from the standpoint of practice and deeper insights into the process.

Example (9) :   Solve by the method of variation of parameters the differential equation $(D^3 + D)y = \sec x$

The auxiliary equation of the corresponding homogeneous equation has its roots $0, \pm i$ and as a result, the complementary function is

$$y_c = c_1 \sin x + c_2 \cos x + c_3;$$

c_1, c_2, c_3 being the arbitrary constants. We seek the solution of the given equation in the form

$$y = c_1(x) \sin x + c_2(x) \cos x + c_3(x),$$

where $c_k(x)$'s are arbitrary differentiable functions of x.

$$\therefore \quad y' = (c_1 \cos x - c_2 \sin x) + (c_1' \sin x + c_2' \cos x + c_3')$$

We impose the restriction

$$c_1' \sin x + c_2' \cos x + c_3' = 0$$

on c_1, c_2, c_3 and as a consequence have

$$y' = c_1(x) \cos x - c_2(x) \sin x$$

Differentiating w.r.to x, we have

$$y'' = -(c_1 \sin x + c_2 \cos x) + (c_1' \cos x - c_2' \sin x)$$

Further impose restriction $c'_1 \cos x - c'_2 \sin x = 0$ on c_1 and c_2 to have

$$y'' = -(c_1 \sin x + c_2 \cos x)$$

Differentiating w.r.to x, we have

$$y''' = -c_1 \cos x + c_2 \sin x - c'_1 \sin x - c'_2 \cos x$$

Finally, putting the expressions of y and y''' in the given equation,

$$\sec x = -(c'_1 \sin x + c'_2 \cos x)$$

However, the other two restrictions, viz,

$$\left.\begin{array}{rcl} c'_1 \sin x + c'_2 \cos x + c'_3 &=& 0 \\ c'_1 \cos x - c'_2 \sin x &=& 0 \end{array}\right\}$$

yield $c'_1 = -\tan x$; $c'_2 = -1$; $c'_3 = \sec x$

Integrating both these relations with respect x we have :

$$\left.\begin{array}{l} c_1 = ln \mid \cos x \mid + A \\ c_2 = -x + B \\ c_3 = ln \mid \sec x + \tan x \mid + C \end{array}\right\}$$

where A, B, C are arbitrary integration constants. Thus the general solution to the inhomogeneous equation is :

$$y = (ln \mid \cos x \mid + A)\sin x + (-x + B)\cos x + ln \mid \sec x + \tan x \mid + C$$

observe that the particular solution can be obtained by setting all the arbitrary constants A, B, C zero simultaneously.

Example (10) : Solve $(x^2 - 1)y'' - 2xy' + 2y = (x^2 - 1)^2$, given that x and $(x^2 + 1)$ are solutions of the associated homogeneous ode

We observe that the Wronskian W of these two solutions, viz, x and $x^2 + 1$ will be $(x^2 - 1)$, that vanishes only at $x = \pm 1$. So x and $(x^2 + 1)$ may be regarded linearly independent everywhere on $\mathbb{R}$ with exception of ± 1. Under this circumstance, we can take the general solution of the associated homogeneous ode as $y_c = c_1(x) + c_2(x^2 + 1)$.

The particular integral is taken as $y_p = c_1(x)x + c_2(x)(x^2 + 1)$, where $c_1(x)$ and $c_2(x)$ are continuously differentiable functions that are to be determined so as to satisfy the given ode.

Differentiating $y_p = c_1(x)x + c_2(x)(x^2+1)$:

$$y_p' = (c_1(x) + 2xc_2(x)) + (xc_1'(x) + (x^2+1)c_2'(x))$$

we impose the constraint $c_1'(x)x + (x^2+1)c_2'(x) = 0$, so that

$$y_p' = c_1(x) + 2xc_2(x)$$

Differentiating the above relation once again,

$$y_p'' = c_1'(x) + 2c_2'(x)x + 2c_2(x)$$

so that from the given ode we have :

$$(x^2-1)(c_1'(x) + 2c_2'(x)x + 2c_2(x)) - 2x(c_1(x) + 2xc_2(x)) +$$
$$2(c_1(x)x + c_2(x)(x^2+1)) = (x^2-1)^2$$

i,e, $c_1'(x) + 2xc_2'(x) = (x^2-1)$.

$\therefore$ $c_1'(x) = -(x^2+1)$; $c_2'(x) = x$, which on integration yields :

$$c_1(x) = -\frac{x^3}{3} - x; \; c_2(x) = \frac{x^2}{2}$$

(Integration constants switched off as we are concerned with P.I.).

$$\therefore \; y_p = x\left(-\frac{x^3}{3} - x\right) + (x^2+1)\frac{x^2}{2} = \frac{x^4}{6} - \frac{x^2}{2}$$

The general solution is therefore

$$y = y_c + y_p = c_1x + c_2.(x^2+1) + \frac{x^4}{6} - \frac{x^2}{2}$$

Example (11) : Solve the ode $(D^2 - 2D + 2)y = e^x(\tan x + \cot x)$

Here the auxiliary equation of the associated homogeneous ode is $m^2 - 2m + 2 = 0$ whose roots are $1 \pm i$. The c.f is

$$y_c = e^x(c_1 \cos x + c_2 \sin x),$$

where c_1 and c_2 are arbitrary constants. To find the general solution one could have made use of the method of variation of parameters directly but we want to work out the particular integral y_p by a special trick that is a blend of shortcut method (coming up in the very next article!) and the method of variation of parameters.

$$y_p = \frac{1}{D^2 - 2D + 2} e^x (\tan x + \cot x)$$

$$= e^x \cdot \frac{1}{(D+1)^2 - 2(D+1) + 2} (\tan x + \cot x)$$

[using the exponential shift]

$$= e^x \cdot \frac{1}{D^2 + 1} (\tan x + \cot x).$$

We consider a second order ode $(D^2 + 1)z = \tan x + \cot x$ and try to find its P.I. with the help of method of variation of parameters. The complementary function of the new equation being $z = A \sin x + B \cos x$, we seek the solution of the new inhomogeneous differential equation in the form

$$z = A(x) \sin x + B(x) \cos x$$

Differentiating both sides w.r.to x and imposing the restriction

$$A' \sin x + B' \cos x = 0$$

we have

$$z' = A \cos x - B \sin x.$$

Differentiating once again, we have

$$z'' = -(A \sin x + B \cos x) + (A' \cos x - B' \sin x),$$

so that $z'' + z = A' \cos x - B' \sin x$

$$\therefore \quad A' \sin x \cos^2 x - B' \sin^2 x \cos x = 1 \quad [\because (D^2 + 1)z = \tan x + \cot x]$$

This condition together with the relation $A' \sin x + B' \cos x = 0$ gives $A' = cosec\ x$ and $B' = -\sec x$, which yield, on integration w.r.to x, $A(x) = ln \mid cosec\ x - \cot x \mid$ and $B(x) = ln \mid \sec x - \tan x \mid$. (Integration constants are switched off as we are working for P.I.).

$$\therefore z_p = \sin x\ ln \mid cosec\ x - \cot x \mid + \cos x\ ln \mid \sec x - \tan x \mid$$

and so $y_p = e^x (\sin x\ ln \mid cosec\ x - \cot x \mid + \cos x\ ln \mid \sec x - \tan x \mid)$.

The complete solution to the given original ode is $y = y_c + y_p$.

We are now in a position to have a close look at the 'method of variation of parameters' applied to the most general second order linear inhomogeneous ode, viz,

$$y'' + p_1(x)y' + p_2(x)y = q(x),$$

where $p_1(x), p_2(x), q(x)$ are continuous functions of x in $[a, b]$. In this method we choose the solution of the given ode in the form :

$$y = c_1(x)y_1(x) + c_2(x)y_2(x), \qquad (5.21)$$

where $y_1(x)$ and $y_2(x)$ are linearly independent solutions of the associated homogeneous ode and $c_1(x), c_2(x)$ are C^1 functions on $[a, b]$.

In course of the method we found that $c_1(x)$ and $c_2(x)$ are subject to the equational constraints

$$\left.\begin{array}{rcl} c_1'(x)y_1(x) + c_2'(x)y_2(x) & = & 0 \\ c_1'(x)y_1'(x) + c_2'(x)y_2'(x) & = & q(x) \end{array}\right\}$$

which in the matrix form may be cast as :

$$\begin{pmatrix} y_1 & y_2 \\ y_1' & y_2' \end{pmatrix} \begin{pmatrix} c_1'(x) \\ c_2'(x) \end{pmatrix} = \begin{pmatrix} 0 \\ q(x) \end{pmatrix}.$$

Due to linear independence of $y_1(x)$ and $y_2(x)$ in $[a, b]$ Wronskian $W(x)$ of $y_1(x)$ and $y_2(x)$ is non-zero and hence by Cramer's rule,

$$\begin{pmatrix} c_1'(x) \\ c_2'(x) \end{pmatrix} = \begin{pmatrix} -y_2(x)q(x)/W(x) \\ y_1(x)q(x)/W(x) \end{pmatrix}.$$

Taking any point $x \in [a, b]$ and integrating the column matrices on both sides between a and x, we have :

$$c_1(x) = -\int_a^x \frac{y_2(x')q(x')}{W(x')} \, dx' + \bar{c}_1$$

$$\hspace{10cm} (5.22)$$

$$c_2(x) = \int_a^x \frac{y_1(x')q(x')}{W(x')} \, dx' + \bar{c}_2$$

$$\therefore \quad y \;\; = \;\; -y_1(x)\int_a^x \frac{y_2(x')q(x')}{W(x')} \, dx' + y_2(x)\int_a^x \frac{y_1(x')q(x')}{W(x')} \, dx' +$$

$$+\bar{c}_1 y_1(x) + \bar{c}_2 y_2(x)$$

$$= \int_a^x \frac{y_2(x)y_1(x') - y_1(x)y_2(x')}{W(x')}\, dx' + \bar{c}_1 y_1(x) + \bar{c}_2 y_2(x)$$

$$= \int_a^x G(x, x')q(x')dx' + \bar{c}_1 y_1(x) + \bar{c}_2 y_2(x),$$

where we write

$$G(x, x') \equiv \frac{y_2(x)y_1(x') - y_1(x)y_2(x')}{W(x')}$$

and call it the **Green's function** for the problem. In fact setting $\bar{c}_1 = \bar{c}_2 = 0$ we have a particular integral

$$y_p = \int_a^x G(x, x')q(x')dx,$$

which is equivalent integral form of P.I. The quantity $G(x, x')$ is known as 'kernel' also. It turns out that $y_p(a) = 0$. Moreover, from differentiation under sign of integration, we get that

$$\frac{dy_p}{dx}\bigg|_{x=a} \equiv y_p'(a) = \int_a^a \frac{\partial G}{\partial x}(x, x')dx' + G(x, x)\bigg|_{x=a} = 0$$

Hence y_p is a particular solution of the original equation that satisfies the initial conditions $y_p(a) = y_p'(a) = 0$.

In particular, had it been a second order linear ode with constant co-efficients, we would have a pleasant surprise— the corresponding Green's function or so-called kernel, is of 'difference type'. This is what we set out to prove for ourselves in the following.

Let $y'' + p_1 y' + p_2 y = q(x)$ be a second order linear ode with constant real co-efficients. The auxiliary equation of the associated homogeneous ode is : $m^2 + p_1 m + p_2 = 0$, whose roots are m_1 and m_2.

Let's now study different cases of the problem :

Case I: If m_1 and m_2 are real and distinct, $e^{m_1 x}$ and $e^{m_2 x}$ forms a fundamental solution set of the homogeneous equation. Here Wronskian of these solutions is

$$W(x) = \begin{vmatrix} e^{m_1 x} & e^{m_2 x} \\ m_1 e^{m_1 x} & m_2 e^{m_2 x} \end{vmatrix} = (m_2 - m_1)e^{(m_1+m_2)x} \neq 0$$

$$\therefore \quad G(x, x') = \frac{e^{m_1 x'}.e^{m_2 x} - e^{m_1 x}.e^{m_2 x'}}{e^{(m_1+m_2)x}.(m_2 - m_1)} = \frac{e^{m_1(x'-x)} - e^{m_2(x'-x)}}{(m_2 - m_1)},$$

which is barely a function of $(x' - x)$.

Case II: If m_1 is a double root of the auxiliary equation, then $e^{m_1 x}$ and $x e^{m_1 x}$ are fundamental solutions of the homogeneous equation and their Wronskian is given by

$$W(x) = \begin{vmatrix} e^{m_1 x} & x e^{m_1 x} \\ m_1 e^{m_1 x} & (1 + m_1 x) e^{m_1 x} \end{vmatrix} = e^{2 m_1 x} \neq 0$$

$$\therefore \ G(x, x') = \frac{e^{m_1 x'} . x e^{m_1 x} - e^{m_1 x} . x' e^{m_1 x'}}{e^{2 m_1 x}} = (x - x') e^{m(x' - x)},$$

which is again a function of only $(x - x')$.

Case III: The roots of the auxiliary equation are complex conjugates, viz. $\alpha \pm i\beta$ so that the fundamental set of solutions read $e^{\alpha x} \cos \beta x$ and $e^{\alpha x} \sin \beta x$. The Wronskian of these two solutions is given by

$$W(x) = \begin{vmatrix} e^{\alpha x} \cos \beta x & e^{\alpha x} \sin \beta x \\ e^{\alpha x}(\alpha \cos \beta x - \beta \sin \beta x) & e^{\alpha x}(\alpha \sin \beta x + \beta \cos \beta x) \end{vmatrix} = \beta e^{2\alpha x}$$

$$\begin{aligned} \therefore \ G(x, x') &= \frac{e^{\alpha x} \sin \beta x . e^{\alpha x'} \cos \beta x' - e^{\alpha x} \cos \beta x . e^{\alpha x'} . \sin \beta x'}{\beta e^{2\alpha x}} \\ &= \frac{-e^{\alpha(x' - x)} \sin \beta(x' - x)}{\beta}, \end{aligned}$$

which is a function of $(x' - x)$ only.

It will not be out of place to mention that when an ordinary linear differential equation with constant co-efficients is reduced to an integral equation by some method (e,g, Fubini's method based on Lagrange's method of variation of parameters), it is found to be a Volterra integral equation with 'difference-type' kernel. These integral equations are commonly called **convolution-type** or 'closed-cycle type' as it carries any periodic function with a preassigned period T into some other periodic function having the same period T. This underlying truth is reflected in the particular integrals of the linear differential equations with constant co-efficients. Stated clearly, a periodic inhomogeneous term/impulse (like $\sin kx, \cos kx$) will produce the periodic functions with same periods in the particular integral/response.

Illustration: The following example of combined spring-mass system shows the use of Green's function of above type.

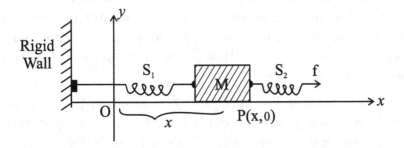

Fig 5.2 : Combined spring-mass system

In the above diagram is shown a heavy metallic block of mass M connected with two springs S_1 and S_2. One spring S_1 is clamped to a rigid wall while a force $f(t)$ is applied to the free end of S_2. In absence of the applied force f, the metal block is at the rest position which we choose to be the origin $x = 0$. Assume that the viscous friction between the block and the surface on which it slides is modelled by $-k\frac{dx}{dt}$ and the combined restoring force of the springs is modelled by $-\mu x$ (otherwise we would have non-linear equation). So the equation of motion reads :

$$\frac{d^2x}{dt^2} + \frac{k}{M}\frac{dx}{dt} + \frac{\mu}{M}x = \frac{f(t)}{M}, \qquad \text{(a)}$$

which is to be solved together with initial conditions $x(0) = 0$ and $\frac{dx}{dt}\big|_{t=0} = 0$. We view the applied force $f(t)$ as an element of $C\left([0,\infty)\right)$.

The problem has its Green's function given by

$$G(t,\tau) = \frac{e^{m_1(\tau-t)} - e^{m_2(\tau-t)}}{m_2 - m_1} \equiv H(t-\tau), (say)$$

where m_1 and m_2 are two roots of the auxiliary equation of the homogeneous ode associated with (a) if we assume m_1 and m_2 real and distinct (otherwise $H(t-\tau)$ would have different expression). Therefore the displacement $x(t)$ of the metal block from the rest position $x(0) = 0$ is given by

$$x(t) = \int_0^t H(t-\tau)f(\tau)d\tau \qquad \text{(b)}$$

Equation (b) defines a linear map $L : C\left([0,\infty)\right) \longrightarrow C\left([0,\infty)\right)$.

Extension: From the above discussion you should not be illuded that the existence of 'difference-type kernel' or Green's function is a peculiarity of only the second order linear differential equations having constant co-efficients. The following workout establishes that this typical form of kernel is an intrinsic feature of any finite order linear differential equations with constant co-efficients. However, we start our discussion anew for convenience of the reader.

Prima facie, passage from $L(D)y = f(x)$ to $y = \frac{1}{L(D)}f(x)$ might seem to be a routine procedure, but it must be emphatically realized that the procedure is valid only if $y(x)$ is a solution of $L(D)y = f(x)$ for which $y(x_0) = y'(x_0) = \cdots\cdots = y^{(n-1)}(x_0) = 0$, $x_0 \in I$ (see (5.4)).

We start our campaign with the first order linear differential equation

$$(D - \lambda_1)y = f(x), \ f(x) \in C(I).$$

Multiplying both sides by $e^{-\lambda_1 x}$ and integrating the resultant equation between x_0 and x, we have

$$y(x)e^{-\lambda_1 x} = \int_{x_0}^{x} e^{-\lambda_1 t}f(t)dt,$$

provided we use the initial conditions $y(x_0) = 0$.

$$\therefore \ y(x) = e^{\lambda_1 x}\int_{x_0}^{x} e^{-\lambda_1 t}f(t)dt = \int_{x_0}^{x} e^{\lambda_1(x-t)}f(t)dt \qquad (5.23)$$

Symbolically, we denote this particular integral (5.23) by $\frac{1}{D-\lambda_1}f(x)$.

Now we extend our approach to the second order linear differential equation given in the normal form

$$(D - \lambda_1)(D - \lambda_2)y = f(x)$$

Therefore $\quad (D - \lambda_2)y = \dfrac{1}{D - \lambda_1}f(x) = \int_{x_0}^{x} e^{\lambda_1(x-t)}f(t)dt \qquad (5.24)$

Multiplying both sides of (5.24) by $e^{-\lambda_2 x}$ and integrating both sides of the new equation between x_0 and x, we have

$$y(x)e^{-\lambda_2 x} = \int_{x_0}^{x} ds. \ e^{(\lambda_1-\lambda_2)s}\int_{x_0}^{s} e^{-\lambda_1 t}f(t)dt \qquad (5.25)$$

provided we use $y(x_0) = 0$. If moreover $\lambda_1 = \lambda_2$, then (5.24) becomes

$$(D - \lambda_1)y = \int_{x_0}^{x} e^{\lambda_1(x-t)}f(t)dt$$

and hence (5.25) gives

$$y(x) = e^{\lambda_1 x} \int_{x_0}^{x} ds \int_{x_0}^{s} dt \, e^{-\lambda_1 t} \, f(t) dt \qquad (5.26)$$

The integrand of the double integration is continuous over the shaded triangular region shown in the figure and so change of order of integration of these iterated integrals is admissible. Once we change the order of integration, (5.26) becomes

$$
\begin{aligned}
y(x) &= e^{\lambda_1 x} \int_{x_0}^{x} dt \int_{t}^{x} ds \, e^{-\lambda_1 t} f(t) dt \\
&= e^{\lambda_1 x} \int_{x_0}^{x} (x-t) f(t) e^{-\lambda_1 t} dt \\
&= \int_{x_0}^{x} (x-t) f(t) e^{\lambda_1 (x-t)} dt \qquad (5.27)
\end{aligned}
$$

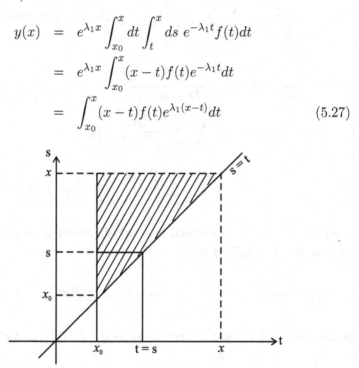

Fig 5.3 : Region of integration after order of integration changed.

Differentiating under the sign of integration we have from (5.27),

$$y'(x) = \lambda_1 \int_{x_0}^{x} (x-t) f(t) e^{\lambda_1 (x-t)} dt + \int_{x_0}^{x} f(t) e^{\lambda_1 (x-t)} dt \qquad (5.28)$$

Equation (5.27) and (5.28) show that $y(x_0) = y'(x_0) = 0$.

We write the solution (5.27) symbolically as $\frac{1}{(D-\lambda_1)^2} f(x)$. Henceforth we shall apply the method of induction to deal with the higher order linear differential equations. For definiteness, we assume the operator $L(D)$ to be of order n and $L(D) = (D-\lambda_1)^{r_1}(D-\lambda_2)^{r_2} \cdots \cdots (D-\lambda_k)^{r_k}$

where $\lambda_1, \lambda_2, \cdots\cdots \lambda_k$ are the distinct roots of the auxiliary equation with $r_1, r_2, \cdots\cdots, r_k$ being their respective multiplicities. Making use of the partial fraction decomposition of

$$\frac{1}{L(D)} \equiv \frac{1}{(D - \lambda_1)^{r_1}(D - \lambda_2)^{r_2} \cdots\cdots (D - \lambda_k)^{r_k}}$$

we have

$$\frac{1}{L(D)} = \sum_{i=1}^{k} \sum_{j=1}^{r_i} \frac{A_{ij}}{(D - \lambda_i)^j}, \quad A_{ij}\text{'s being constants.}$$

$$\begin{aligned}
\therefore \; \frac{1}{L(D)} f(x) &= \frac{1}{(D - \lambda_1)^{r_1}(D - \lambda_2)^{r_2} \cdots\cdots (D - \lambda_k)^{r_k}} f(x) \\
&= \sum_{i=1}^{k} \sum_{j=1}^{r_i} \frac{A_{ij}}{(D - \lambda_i)^j} f(x) \\
&= \sum_{i=1}^{k} \sum_{j=1}^{r_i} A_{ij} \int_{x_0}^{x} \frac{(x - t)^{j-1}}{(j - 1)!} f(t) e^{\lambda_i(x-t)} dt \quad (5.29)
\end{aligned}$$

where in the last step we used method of induction on (5.23) and (5.27) to have the generalised result

$$\frac{1}{(D - \lambda)^j} f(x) = \int_{x_0}^{x} \frac{(x - t)^{j-1}}{(j - 1)!} f(t) e^{\lambda(x-t)} dt, \; j = 1, 2, 3, \cdots\cdots$$

Right side of (5.29) can be rewritten as $\int_{x_0}^{x} f(t) H(x - t) dt$, where

$$H(x) = \sum_{i=1}^{k} \sum_{j=1}^{r_i} \frac{A_{ij} x^{j-1} e^{\lambda_i x}}{(j - 1)!}.$$

The final form shows why we claimed that any finite-order linear differential equation having constant co-efficients posseses 'difference-type' kernel. Some authors like to call the formula

$$\frac{1}{L(D)} f(x) = \int_{x_0}^{x} f(t) H(x - t) dt$$

as 'convolution' or 'Faltung' formula. Obviously any advanced reader might have the reminiscences of 'convolution integral' often encountered in Laplace transform. This will be addressed to in Chapter 7.

5.8   Special Methods for finding Particular Integrals

The particular integral (P.I) of any nth order inhomogeneous linear differential equation with constant co-efficients is determined by tricky short methods only if the inhomogeneous term $q(x)$ in $L(D)y = q(x)$ is of very special type, viz,

  (i) e^{ax}, where 'a' is any constant.

  (ii) x^k, where k is any positive integer.

  (iii) $\sin ax, \cos ax$ where 'a' is any constant.

  (iv) $e^{ax}.V$, where V is any infinitely differentiable function of x.

  (v) xV, where V is any infinitely differentiable function of x.

When we workout the P.I. $\frac{1}{L(D)} q(x)$, we break up the symbolic operator $\frac{1}{L(D)}$ into factors (that commute with each other) or into partial fractions. Had the co-efficients of the different powers of D been functions of x, the commutativity of the factors of $L(D)$ and hence of $\frac{1}{L(D)}$ would not be feasible. However, for inhomogeneous equations with constant co-efficients, no factor $(D - \alpha)$ of $L(D)$ commutes with factor $\frac{1}{D-\beta}$ of $\frac{1}{L(D)}$ if $\alpha \neq \beta$. If $q(x) \in C([0,1])$, $L(D)\frac{1}{L(D)}q(x)$ belongs to $C([0,1])$– i,e, $L(D)\frac{1}{L(D)}$ is an endomorphism over $C([0,1])$. We further observe that $\frac{1}{L(D)}L(D)$ carries an element of $C^{\infty}([0,1])$ to some element of $C^{\infty}([0,1])$ and so is an endomorphism over $C^{\infty}([0,1])$. To make $\left(\frac{1}{L(D)}L(D)\right)q(x)$ well-defined, we need $q(x)$ at least n times continuously differentiable over $[0,1]$. Thus the restriction of $q(x)$ depends on the particular problem at hand. Assuming $q(x) \in C^{\infty}([0,1])$, we shall be free of this brunt. Even if $q(x)$ be infinitely differentiable, does it follow that

$$q(x) = \left(L(D)\frac{1}{L(D)}\right)q(x) = \left(\frac{1}{L(D)}L(D)\right)q(x) \ ?$$

As it turns out, this equality holds under special circumstances. If

$$\left(\frac{1}{L(D)}L(D)\right)q(x) = q(x) + r(x),$$

then operating $L(D)$ on bothsides, we have $L(D)r(x) = 0$, which is possible only if all the constants of integration are made zeros.

Hence in finding $L(D)y = q(x)$, we may use

$$y_p = \frac{1}{L(D)}q(x) = \frac{M(D)}{M(D)L(D)}\,(q(x))$$

and carry out the operation in any order, provided all the constants of integration are taken to be zero. (Hence $M(D)$ is an arbitrary polynomial in D).

Using the symbolic operator $\frac{1}{L(D)}$, there exist two conventional ways to find P.I. of $L(D)y = q(x)$, $L(D)$ being a polynomial of degree n in D with constant co-efficients.

(a) $\dfrac{1}{L(D)}$ is factored as $\dfrac{1}{(D - m_1)(D - m_2)\cdots\cdots(D - m_n)}$ and so

$$y_p = \frac{1}{L(D)}q(x) = \frac{1}{D - m_1}\cdot\frac{1}{D - m_2}\cdots\cdots\frac{1}{D - m_n}\,q(x)$$

$$= e^{m_1 x}\int e^{(m_2 - m_1)x}\int\cdots\cdots\int e^{-m_n x}q(x)(dx)^n \qquad (5.30)$$

(b) In the second method, $\frac{1}{L(D)}$ is broken into partial fractions :

$$\frac{1}{L(D)} = \frac{t_1}{D - m_1} + \frac{t_2}{D - m_2} + \cdots\cdots + \frac{t_n}{D - m_n} \quad \text{and so}$$

$$y_p = \frac{1}{L(D)}q(x) = \sum_{k=1}^{n}\frac{t_k}{D - m_k}q(x)$$

$$= \sum_{k=1}^{n}t_k e^{m_k x}\int e^{-m_k x}q(x)dx \qquad (5.31)$$

In the second method, $\frac{1}{L(D)}q(x)$ belongs to $C^1([0,1])$ while in the first method, $\frac{1}{L(D)}q(x)$ belongs to $C^n([0,1])$ if $q(x)$ is assumed to be an element of $C([0,1])$. This paradox can be resolved only if we assume $q(x) \in C^\infty([0,1])$. In our discussion of the short special methods for determining particular integrals, we shall try to hint out the limitations of these methods and prescribe the alternative(s).

Case I: $q(x) = e^{ax}$, a is any constant.

Traditional workout: Since $D^k(e^{ax}) = a^k e^{ax}$, $L(D)e^{ax} = L(a)e^{ax}$. If 'a' is not a root of the auxiliary equation $L(m) = 0$,

$$y_p = \frac{1}{L(D)}e^{ax} = \frac{1}{L(a)}e^{ax}.$$

If 'a' were a simple root of the auxiliary equation and $L(D) = (D-a)g(D)$, then

$$y_p = \frac{1}{(D-a)g(D)}(e^{ax}) = \frac{1}{g(D)}\left[\frac{1}{D-a}e^{ax}\right] = \frac{1}{g(D)}(xe^{ax}) = \frac{1}{g(a)}xe^{ax},$$

where $g(a) \neq 0$.

If 'a' were a root of the auxiliary equation $L(m) = 0$ with multiplicity r, that is, $L(D) = (D-a)^r f(D)$, with $f(a) \neq 0$, then

$$y_p = \frac{1}{(D-a)^r f(D)}e^{ax} = \frac{1}{f(D)}\cdot\frac{1}{(D-a)^r}(e^{ax}) = \frac{x^r e^{ax}}{r! f(a)}$$

So the working formula

$$y_p \equiv \frac{1}{L(D)}e^{ax} = \begin{cases} \frac{1}{L(a)}e^{ax} & \text{if } L(a) \neq 0 \\ \frac{1}{f(a)}\cdot\frac{x^r e^{ax}}{r!} & \text{if } L(D) = (D-a)^r f(D) \text{ with } f(a) \neq 0 \end{cases}$$

means that under the given conditions, both the sides yield the same result when operated upon by $L(D)$.

Alternatively, we could have taken help of annihilator method in which we construct a new homogeneous ode of order higher than that of $L(D)$ from the given inhomogeneous equation $L(D) = q(x)$. In case $q(x) = e^{ax}$, $(D-a)(e^{ax}) = 0$ and hence $(D-a)L(D)y = 0$. The general solution of this new homogeneous ode will be framed for two subcases :

(i) When $L(a) \neq 0$.

Here $y = y_c + Ae^{ax}$, y_c being the complementary function of the original ode given.

Hence $e^{ax} = L(D)y = L(D)y_c + A.L(D)e^{ax} = AL(a)e^{ax}$,

$$\therefore \quad y = y_c + \frac{1}{L(a)}e^{ax} \quad \left(\because A = \frac{1}{L(a)}\right)$$

(ii) When $L(D) = (D-a)^r f(D)$ with $f(a) \neq 0$, i,e, 'a' is a root of multiplicity r of the auxiliary equation, we have

$$y = y_f + (c_0 + c_1 x + c_2 x^2 + \cdots\cdots + c_r x^r)e^{ax},$$

where y_f is the part of the complementary function that corresponds to the zeros of $f(m)$.

Operating $L(D)$ on both sides, and using its linearity,

$$
\begin{aligned}
e^{ax} = L(D)y &= L(D)y_f + L(D)\{(c_0 + c_1 x + \cdots\cdots + c_r x^r)e^{ax}\} \\
&= f(D)(D-a)^r\{(c_0 + c_1 x + \cdots\cdots + c_r x^r)e^{ax}\} \\
&= c_r . r! f(D)(e^{ax}) \\
&= c_r . r! f(a) . e^{ax}
\end{aligned}
$$

$$
\therefore \quad c_r = \frac{1}{r! f(a)}, \quad \text{and hence}
$$

$$
y = y_f + (c_0 + c_1 x + c_2 x^2 + \cdots\cdots + c_{r-1} x^{r-1})e^{ax} + \frac{x^r}{r! f(a)} e^{ax}
$$

Set $y_f = 0$ and $c_0 = c_1 = \cdots = c_{r-1} = 0$ to get $y_p = \dfrac{x^r}{r! f(a)} e^{ax}$.

The annihilator method thus gives back the same P.I when $q(x) = e^{ax}$ but manages to bypass difficulties encountered earlier.

Example (12) : $(D^2 - a^2)y = e^{ax} + e^{bx}, \ | \, a \, | \neq | \, b \, |$

Here the roots of the auxiliary equation of the corresponding homogeneous ode $(D^2 - a^2)y = 0$ are $\pm a$. The complementary function is:

$$
y_c = c_1 e^{ax} + c_2 e^{-ax}
$$

while
$$
\begin{aligned}
y_p &= \frac{1}{D^2 - a^2}(e^{ax} + e^{bx}) \\[2mm]
&= \frac{1}{D^2 - a^2} e^{ax} + \frac{1}{D^2 - a^2} e^{bx} \quad \text{(using linearity)} \\[2mm]
&= \frac{1}{(D-a)(D+a)} e^{ax} + \frac{1}{D^2 - a^2} e^{bx} \\[2mm]
&= \frac{1}{2a(D-a)} e^{ax} + \frac{1}{b^2 - a^2} e^{bx} \\[2mm]
&= \frac{xe^{ax}}{2a} + \frac{1}{b^2 - a^2} e^{ax} \quad (\because \ | \, a \, | \neq | \, b \, |)
\end{aligned}
$$

Hence the complete solution is

$$
y = y_c + y_p = c_1 e^{ax} + c_2 e^{-ax} + \frac{xe^{ax}}{2a} + \frac{1}{b^2 - a^2} e^{ax}
$$

Alternatively, by the annihilator method we try our solution as follows :

The ode $(D + a)(D - b)(D - a)^2 y = 0$ has e^{ax}, e^{bx} & e^{-ax} as its fundamental solutions. The general solution of this new ode reads :

$$y = c_1 e^{-ax} + c_2 e^{bx} + (c_3 + c_4 x)e^{ax}$$

Applying $(D^2 - a^2)$ on both sides, we have :

$$e^{ax} + e^{bx} = (D^2 - a^2)\{c_1 e^{-ax} + c_2 e^{bx} + (c_3 + c_4 x)e^{ax}\}$$
$$= c_2(b^2 - a^2)e^{bx} + 2c_4 a e^{ax}$$

$$\therefore \quad c_2 = \frac{1}{b^2 - a^2} \ ; \ c_4 = \frac{1}{2a} \quad (\because e^{ax} \text{ and } e^{bx} \text{ are linearly independent}).$$

$$\therefore \quad y = c_1 e^{-ax} + \frac{1}{b^2 - a^2}e^{bx} + c_3 e^{ax} + \frac{1}{2a}x e^{ax},$$

which is the general solution. Setting $c_1 = c_3 = 0$, we get back P.I as

$$y_p = \frac{1}{2a}x e^{ax} + \frac{1}{b^2 - a^2}e^{bx}.$$

Case II: q(x) = x^k, k is a positive integer.

When we are to solve $L(D)y = x^k$, we find as usual the particular integral P.I. $= \frac{1}{L(D)}q(x) = \frac{1}{L(D)}(x^k)$. For practical field computation we raise $L(D)$ to the (-1)th power, arranging the terms in ascending powers of D with several terms thus obtained, operate on x^k; the result will be P.I. corresponding to x^k. It is trivially seen that any term involving D^{k+1} or higher powers of D will yield zero. For exemplification, let us compute $\frac{1}{D+a}x^k$.

$$\frac{1}{D+a}x^k = \frac{1}{a}\left(1 + \frac{D}{a}\right)^{-1} x^k$$
$$= \frac{1}{a}\left(1 - \frac{D}{a} + \frac{D^2}{a^2} - \cdots\cdots + \frac{(-1)^k D^k}{a^k}\right)(x^k),$$

(other terms omitted as they are zeros).

$$\therefore \frac{1}{D+a}x^k = \frac{1}{a}\left(x^k - \frac{k}{a}x^{k-1} + \frac{k(k-1)}{a^2}x^{k-2} - \cdots\cdots + \frac{(-1)^k k!}{a^k}\right)$$

This simple method can make all but a fastidious reader happy. Note that the final result heavily leans on the choice of the polynomial x^k and pays no heed whether the power series in D is convergent or not.

It is a well-known result that the power series expansion of $\left(1 + \frac{D}{a}\right)^{-1}$ is valid only if $\| D \| < | a |$. However, D is unbounded a linear operator so long D is map from $C^1([0,1])$ to $C([0,1])$— both the spaces being endowed with the supnorm. Similarly, as shown earlier, D is bounded with $\| D \|= 1$, if $C^1([0,1])$ is assigned some other norm. (See Chapter 4, page.) Hence under all circumstances, we cannot be too silly to make a power series expansion of $\left(1 + \frac{D}{a}\right)^{-1}$ valid and then ask the operand x^k to do the rest. This simple case has been picked up to warn the reader against the indiscriminate using the power-expansion of $\frac{1}{L(D)}$ as a cue to find P.I.'s. Indeed, despite all these technical drawbacks left untouched, this method fails to deliver when $q(x)$ is of other four types. The fact that D is a nilpotent operator on the space P of all polynomials is the basic theme that clicks in the present case. This nilpotency of D ensures that over this subspace of $C([0,1])$, every polynomial operator in D is invertible.

Example (13) : Solve the ode $(D^3 + 2D^2 + D)y = x^2 + x$

The auxiliary equation of reduced ode $(D^3 + 2D^2 + D)y = 0$ has its roots $0, -1, -1$. Thus the complementary function is

$$y_c = c_1 + (c_2 + c_3 x)e^{-x}, \quad \text{where } c_1, c_2, c_3 \text{ are arbitrary constants.}$$

The particular integral y_p of the given equation are computed successively by three methods in the following :

(i) <u>Power-Series Expansion Method:</u>

$$
\begin{aligned}
y_p &= \frac{1}{D(D+1)^2}(x^2 + x) \\
&= \frac{1}{D}(1+D)^{-2}(x^2 + x) \\
&= \frac{1}{D}(1 - 2D + 3D^2 - \cdots\cdots)(x^2 + x) \\
&= \frac{1}{D}(x^2 - 3x + 4) = \frac{x^3}{3} - \frac{3}{2}x^2 + 4x
\end{aligned}
$$

(ii) <u>Partial Fraction Method:</u>

$$
\begin{aligned}
y_p &= \frac{1}{D(D+1)^2}(x^2 + x) \\
&= \left[\frac{1}{D} - \frac{1}{D+1} - \frac{1}{(D+1)^2}\right](x^2 + x)
\end{aligned}
$$

$$= \left(\frac{x^3}{3} + \frac{x^2}{2}\right) - e^{-x}\int (x^2 + x)e^x dx - e^{-x}\int (x^2 - x + 1)e^x dx$$

$$= \left(\frac{x^3}{3} + \frac{x^2}{2}\right) - (x^2 - x + 1) - (x^2 - 3x + 4)$$

$$= \left(\frac{x^3}{3} - \frac{3}{2}x^2 + 4x\right) - 5$$

(iii) <u>Annihilator Method</u>:

We express $y_p = \frac{1}{D(D+1)^2}(x^2 + x)$ as $(Ax + Bx^2 + Cx^3)$ because 0 is an auxiliary root of the given ode. The choice is justified if we consider the details of the method (see Shepley L. Ross.).

Applying $D(D+1)^2$ on y_p we have

$$\begin{aligned}
x^2 + x &= D(D+1)^2(Ax + Bx^2 + Cx^3) \\
&= (A + 4B + 6C) + (2B + 12C)x + 3Cx^2
\end{aligned}$$

Comparing co-efficients of like powers of x and solving the resultant equations we have $A = 4$; $B = -\frac{3}{2}$ and $C = \frac{1}{3}$.

$$\therefore \quad y_p = \frac{x^3}{3} - \frac{3}{2}x^2 + 4x$$

In light of this example we observe that of the three methods the 'annihilator' method is most accomplished. Partial fraction method is technically correct but rather lengthy. Also note that the resulting particular integral obtained in this partial fraction method differs by a term -5 from the particular integral obtained by other methods. This is however no headache as the extraneous term merges with term(s) in the C.F.

The complete solution can thus be written as :

$$y = c_1 + (c_2 + c_3 x)e^{-x} + \frac{x^3}{3} - \frac{3}{2}x^2 + 4x,$$

where c_1, c_2, c_3 are arbitrary constants.

Case III: $\mathbf{q(x)} = \sin \mathbf{ax}$ or $\mathbf{q(x)} = \cos \mathbf{ax}$

For computation of the P.I of the ode $L(D)y = \sin ax$ we take help of Euler's formula $e^{iax} = \cos ax + i\sin ax$ and use it for Case I. The following illustration shows how this is done.

Example (14) : Solve the ode $(D - 1)(D^2 + 4)y = \cos 2x + \sin x$

The auxiliary equation of the related homogeneous equation has its roots $1, \pm 2i$ and hence the C.F y_c and P.I y_p are given by

$$y_c = c_1 e^x + c_2 \cos 2x + c_3 \sin 2x$$

$$
\begin{aligned}
y_p &= \frac{1}{(D-1)(D^2+4)}(\cos 2x + \sin x) \\
&= Re\left[\frac{1}{(D-1)(D^2+4)}e^{2ix}\right] + Im\left[\frac{1}{(D-1)(D^2+4)}e^{ix}\right] \\
&= Re\left[\frac{1}{(D-1)(D+2i)(D-2i)}e^{2ix}\right] + Im\left[\frac{1}{(D-1)(D^2+4)}e^{ix}\right] \\
&= Re\left[\frac{xe^{2ix}}{(-8-4i)}\right] + Im\left[\frac{-1-i}{2}e^{ix}\right] \\
&= -\frac{x}{20}(\sin 2x + 2\cos 2x) + \frac{1}{6}(\sin x + \cos x)
\end{aligned}
$$

Alternatively, we can workout with the annihilator method :

Since $(D^2+4)(D^2+1)(\cos 2x + \sin x) = 0$, we have a newly derived homogeneous equation $(D-1)(D^2+1)(D^2+4)^2 y = 0$ for which the general solution is:

$$y = c_1 \cos x + c_2 \sin x + c_3 e^x + (c_4 + c_5 x)\cos 2x + (c_6 + c_7 x)\sin 2x$$

As is evident, $c_3 e^x + c_4 \cos 2x + c_6 \sin 2x$ is a solution of the homogeneous equation associated with the given equation. If we are interested only in a particular solution y_p of the given inhomogeneous ode, we can, without loss of generality, assume that

$$y_p = c_1 \cos x + c_2 \sin x + x(c_5 \cos 2x + c_7 \sin 2x)$$

Applying $(D-1)(D^2+4)$ on both sides, we have,

$$
\begin{aligned}
\cos 2x + \sin x &= -3(c_1+c_2)\sin x + 3(c_2-c_1)\cos x \\
&\quad + 4(c_5 - 2c_7)\sin 2x - 4(c_7 + 2c_5)\cos 2x
\end{aligned}
$$

Using the fact that $\sin x$, $\cos x$, $\sin 2x$, $\cos 2x$ are all linearly independent, we have;

$$c_1 = c_2 = -\frac{1}{6}; \qquad c_5 = -\frac{1}{10}; \qquad c_7 = -\frac{1}{20}$$

Therefore $\quad y_p = -\frac{x}{20}(\sin 2x + 2\cos 2x) + \frac{1}{6}(\sin x + \cos x)$

Case-IV: $q(x) = e^{ax} V(x)$, $V(x)$ is an infinitely differentiable function.

For computation of the particular integral of a linear ode $L(D)y = e^{ax}V(x)$, we take help of the following formula, better known as exponential shift, which we quote and prove in the following.

If $L(D)$ be any finite degree polynomial in operator D, then $L(D)(e^{ax}V(x)) = e^{ax}L(D + a)V(x)$, where $V(x)$ is an infinitely differentiable function of x.

By principle of mathematical induction we may show that $D^k(e^{ax}V(x)) = e^{ax}(D + a)^k V(x)$ for any natural number k. Even if $k = 0$, the result holds true. Using this result, we have from definition,

$$
\begin{aligned}
L(D)(e^{ax}V(x)) &= (D^n + p_1 D^{n-1} + \cdots\cdots + p_{n-1}D + p_n)(e^{ax}V(x)) \\
&= e^{ax}\{(D + a)^n + p_1(D + a)^{n-1} + \cdots\cdots + \\
&\qquad\qquad + p_{n-1}(D + a) + p_n\}V(x) \\
&= e^{ax}L(D + a)V(x)
\end{aligned}
$$

In our case $y_p = \frac{1}{L(D)}[e^{ax}V(x)] = e^{ax}\frac{1}{L(D+a)}V(x)$, where the last '=' means that $\frac{1}{L(D)}[e^{ax}V(x)]$ is equivalent to $e^{ax}.\frac{1}{L(D+a)}V(x)$ in the sense that both are solutions of the ode $L(D)y = e^{ax}V(x)$.

$$
\left(\because L(D)\left[e^{ax}.\frac{1}{L(D+a)}V(x)\right] = e^{ax}L(D+a)\frac{1}{L(D+a)}V(x) = e^{ax}V(x)\right)
$$

Example (15) : Find the particular integral of the ode:

$$(D^2 - 6D + 9)y = e^{2x}\sin 3x$$

The particular integral y_p of the above ode is

$$
\begin{aligned}
y_p &= \frac{1}{D^2 - 6D + 9}(e^{2x}\sin 3x) \\
&= e^{2x}.\frac{1}{(D+2)^2 - 6(D+2) + 9}\sin 3x \\
&= e^{2x}.\frac{1}{(D-1)^2}\sin 3x \\
&= e^{2x}.\,Im\left[\frac{1}{(D-1)^2}e^{3ix}\right] \\
&= e^{2x}.\,Im\left[\frac{1}{(-1+3i)^2}e^{3ix}\right]
\end{aligned}
$$

$$= e^{2x}. \, Im\left[\left(\frac{-8+6i}{100}\right)(\cos 3x + i\sin 3x)\right]$$

$$= \frac{e^{2x}}{100}(6\cos 3x - 8\sin 3x)$$

Case-V: q(x) = xV(x), V is infinitely differentiable functions of x.

From product rule of differentiation follows: $D(xV) = xDV + V$. Using the principle of mathematical induction one can show that $D^n(xV) = xD^nV + nD^{n-1}V$ holds good for any finite natural number n. The result is true for $n = 0$ also. Herefrom follows that for any finite degree polynomial $L(D)$ we have :

$$L(D)(xV) = xL(D)V + \frac{d}{dD}\{L(D)\}V = xL(D)V + L'(D)V,$$

where we agree to write $L'(D)$ in lieu of $\frac{d}{dD}\{L(D)\}$. Writing $L(D)V = V_1$ and using the symbolic operator $\frac{1}{L(D)}$ we have $V = \frac{1}{L(D)}V_1$ (where we took advantage of the fact that V and hence V_1 is infinitely differentiable function of x). Using this in the expression for $L(D)(xV)$ we get

$$L(D)\left(x\frac{1}{L(D)}V_1\right) = xL(D)\left(\frac{1}{L(D)}V_1\right) + L'(D)\left(\frac{1}{L(D)}V_1\right)$$

$$= xV_1 + L'(D)\left(\frac{1}{L(D)}V_1\right),$$

so that

$$x\frac{1}{L(D)}V_1 = \frac{1}{L(D)}(xV_1) + \frac{L'(D)}{L(D)}\left(\frac{1}{L(D)}V_1\right)$$

Here '=' means equivalence of the two sides in the sense that they produce the same result when acted upon by $L(D)$.

Using linearity of operators and capitalising on the fact that $V_1(x)$, like $V(x)$, is an arbitrary infinitely differentiable function of x, we have:

$$\frac{1}{L(D)}(xV) = \left(x - \frac{L'(D)}{L(D)}\right)\left(\frac{1}{L(D)}V\right)$$

Taking advantage of the principle of mathematical induction one can further generalise the result as follows :

$$\frac{1}{L(D)}(x^r.V) = \left(x - \frac{L'(D)}{L(D)}\right)^r\left(\frac{1}{L(D)}V\right),$$

r being any natural number.

Example (16) :   Solve $(D^2 - 4D + 4)y = 8x^2 e^{2x} \sin 2x$.

Complementary function is : $y_c = (c_1 + c_2 x)e^{2x}$

Particular integral is : $y_p = \dfrac{1}{(D-2)^2}[8x^2 e^{2x}\sin 2x]$

$$= 8.e^{2x}.\dfrac{1}{(D+2-2)^2}(x^2 \sin 2x)$$

$$= 8.e^{2x}.\dfrac{1}{D^2}(x^2 \sin 2x),$$

(where we used the rule IV in the penultimate step)

$$y_p = 8e^{2x}\left(x - \dfrac{2D}{D^2}\right)^2 \dfrac{1}{D^2}(\sin 2x)$$

$$= -2e^{2x}\left(x - \dfrac{2}{D}\right)^2 (\sin 2x) \quad \text{(using rule V)}$$

$$= -2e^{2x}\left(x - \dfrac{2}{D}\right)\left(x - \dfrac{2}{D}\right)(\sin 2x)$$

$$= -2e^{2x}\left(x - \dfrac{2}{D}\right)(x\sin 2x + \cos 2x)$$

$$= -2e^{2x}\left(x^2 \sin 2x + x\cos 2x - 2\int x \sin 2x dx - 2\int \cos 2x dx\right)$$

$$= -e^{2x}\left((2x^2 - 3)\sin 2x + 4x\cos 2x\right)$$

Alternatively, we may use 'Tabular integration'[1] to compute $\frac{1}{D^2}(x^2 \sin 2x)$ and hence y_p. We write $\frac{1}{D^2}(x^2 \sin 2x) = \frac{1}{D}\left(\frac{1}{D}.\,x^2 \sin 2x\right)$. Now

$$\dfrac{1}{D}(x^2 \sin 2x) = \int x^2 \sin 2x dx = \left(\dfrac{1}{4} - \dfrac{1}{2}x^2\right)\cos 2x + \dfrac{x}{2}\sin 2x$$

as shown below:

x^2 and its derivatives	$\sin 2x$ and its integrals
$x^2 (+)$	$\sin 2x$
$2x (-)$	$-\frac{1}{2}\cos 2x$
$2 (+)$	$-\frac{1}{4}\sin 2x$
0	$\frac{1}{8}\cos 2x$

Tabular integration of x² sin 2x

[1](See Appendix C : For details also referred to 'Calculus' by Thomas Finney, 9th Edn. pp-566)

$\frac{x}{2}$ and its derivatives	$\sin 2x$ and its integrals
$\frac{x}{2}(+)$	$\sin 2x$
$\frac{1}{2}(-)$	$-\frac{1}{2}\cos 2x$
0	$-\frac{1}{4}\sin 2x$

Tabular integration of $\frac{x}{2}\sin 2x$

$\left(\frac{1}{4}-\frac{1}{2}x^2\right)$ and its derivatives	$\cos 2x$ and its integrals
$\left(\frac{1}{4}-\frac{1}{2}x^2\right)(+)$	$\cos 2x$
$-x\,(-)$	$\frac{1}{2}\sin 2x$
$-1\,(+)$	$-\frac{1}{4}\cos 2x$
0	$-\frac{1}{8}\sin 2x$

Tabular integration of $\left(\frac{1}{4}-\frac{1}{2}x^2\right)\cos 2x$

Further

$$\frac{1}{D^2}\left(x^2\sin 2x\right) = \frac{1}{D}\left[\left(\frac{1}{4}-\frac{1}{2}x^2\right)\cos 2x + \frac{x}{2}\sin 2x\right]$$

$$= \int\left(\frac{1}{4}-\frac{1}{2}x^2\right)\cos 2x\,dx + \int\frac{x}{2}\sin 2x\,dx$$

$$= -\frac{1}{8}(2x^2-3)\sin 2x - \frac{x}{2}\cos 2x$$

(Because particular integral is being computed, integration constants are ignored).

$$\therefore\ y_p = -e^{2x}\left((2x^2-3)\sin 2x + 4x\cos 2x\right)$$

5.9   Method of Undetermined Co-efficients

This method is very useful in finding the particular integrals of a non-homogeneous linear differential equation with constant co-efficients. Before we get into the act of the method itself, we point out the restriction which delimits its range of application. If

$$L(D)y = \sum_{k=0}^{n} a_k y^{(n-k)} = q(x)$$

be a nth order differential equation of the above type, then this method works successfully only if $q(x)$ be an element of a finite-dimensional function space $\mathcal{U}$ that is closed under operation of $L(D)$. This implies that whenever $g(x) \in \mathcal{U}$, $L(D)g(x) \in \mathcal{U}$ also. If dimension of $\mathcal{U}$ be r, then hopefully a linear differential operator $M(D)$ of minimal of order r can be availed of so that $(ML)(D)(y) \equiv (M(D)L(D))y = 0$. It means that the given non-homogeneous differential equation $L(D) = q(x)$ of order n gets replaced by a homogeneous differential equation of order $(n + r)$ in course of this method. Most probably this motivates some authors (e,g, Coddington) to name it **Annihilator method.**

We now introduce two preliminary concepts- one of 'UC function' and the other of 'UC set' that are intimately related to the method of undetermined co-efficients. A function is said to be a 'uc function' if it is either a function of the following type:

(i) x^k, k being any positive integer or zero.

(ii) e^{ax}, a being a non-zero constant.

(iii) $\sin(bx + c)$ or $\cos(bx + c)$, b and c being constants.

(iv) a finite product of two or more functions of the above types.

Observe that the successive derivatives of any uc function are again uc functions of the same type. If $L(D)$ operates on x^k, it can atmost give a linear combination of the functions $\{1, x, x^2, \cdots \cdots, x^k\}$ that form a standard basis of $(k + 1)$ dimensional vector space of polynomials of degree less than or equal to k. Incidentally, this basis set $\{1, x, x^2, \cdots \cdots, x^k\}$ is commonly known as a uc set of the uc function x^k. If $L(D)$ operates upon e^{ax}, it yields e^{ax} with atmost a constant multiplicative factor. Hence the one-dimensional vector space generated by e^{ax} is closed under the operation of $L(D)$. Formally, one declares the singleton $\{e^{ax}\}$ as the 'uc set' of the function e^{ax}. In case $L(D)$ operates on either $\sin(bx + c)$ or $\cos(bx + c)$, one gets atmost a linear combination of the functions $\{\sin(bx + c), \cos(bx + c)\}$ that form basis of a two-dimensional function space. Finally when $L(D)$ works on the finite product of two or more functions of the above three categories, we get back an element of a vector space whose basis is the direct product of the bases of component vector spaces involved. As is conventional, this new basis is called a uc set of the given uc function having the product form. For example, we consider the uc function $x^3 \cos x$ and observe that $L(D)(x^3 \cos x)$ is a linear combination of the elements of the vector space $\mathcal{U}$ with basis as

$$\{1, x, x^2, x^3\} \otimes \{\sin x, \ \cos x\}$$
$$= \ \{\sin x, \cos x, x \sin x, x \cos x, x^2 \sin x, x^2 \cos x, x^3 \sin x, x^3 \cos x\}.$$

It shows that $\mathcal{U}$ is an eight-dimensional function space. In general, if h be a uc function defined as the product fg of two basic uc functions f and g, then 'uc set' of the product function h is the set of all the products obtained by multiplying the various members of the uc set of f by the various members of the uc set of g. The following table enlists the useful uc functions and their corresponding uc sets.

If we have a close look at this table we observe that the uc set of a given uc function consists of the function itself and its successive derivatives (ignoring constant multiples)

UC function	UC set
x^k	$\{1, x, x^2, \cdots\cdots, x^k\}$
e^{ax}	$\{e^{ax}\}$
$\sin(bx+c)$, $\cos(bx+c)$	$\{\sin(bx+c), \ \cos(bx+c)\}$
$x^k e^{ax}$	$\{e^{ax}, xe^{ax}, x^2 e^{ax}, \cdots\cdots, x^k e^{ax}\}$
$x^k \sin(bx+c)$, $x^k \cos(bx+c)$	$\{\sin(bx+c), \cos(bx+c),$ $x\sin(bx+c), x\cos(bx+c), \cdots\cdots,$ $x^k \sin(bx+c), x^k \cos(bx+c)\}$
$e^{ax}\sin(bx+c)$, $e^{ax}\cos(bx+c)$	$\{e^{ax}\sin(bx+c), e^{ax}\cos(bx+c)\}$
$x^k e^{ax}\sin(bx+c)$, $x^k e^{ax}\cos(bx+c)$	$\{e^{ax}\sin(bx+c), e^{ax}\cos(bx+c),$ $xe^{ax}.\sin(bx+c), xe^{ax}\cos(bx+c),$ $\cdots\cdots\cdots, x^k e^{ax}\sin(bx+c),$ $x^k e^{ax}\cos(bx+c)\}$

Table of UC functions and UC set

There are n linearly independent solutions of the associate homogeneous equation $L(D)y = 0$. Now two possibilities open up— (i) none of these n linearly independent solutions belong to the uc set of $q(x)$ and (ii) $p(\geq 1)$ of these linearly independent solutions belong to the uc set of $q(x)$. In case (ii) we shall multiply each member of the uc set of $q(x)$ by the lowest (why?) positive integral power of x so that the revised set contains no solution of $L(D)y = 0$. The following illustrations will hopefully clarify these viewpoints.

Example(17) : Solve by UC technique : $(D^3 - 3D + 2)y = xe^x$.

Here the associated homogeneous differential equation is

$$(D^3 - 3D + 2)y = 0,$$

whose fundamental solution set reads $\{e^x, xe^x, e^{-2x}\}$. The right member $q(x) = xe^x$ of the original differential equation has the uc set

$$\{1, x\} \otimes \{e^x\} = \{e^x, xe^x\}.$$

Observe that there are two members of the fundamental solution set that appear in the uc set of $q(x)$. Hence we shall have to modify the uc set by multiplying every member by x^2 so as to have the revised set of linearly independent functions $\{x^2e^x, x^3e^x\}$. Ultimately we shall choose the particular integral as $y_p = (Ax^2 + Bx^3)e^x$, where A and B are constant co-efficients yet to be determined.

By successive differentiation one gets

$$
\begin{aligned}
y'_p &= e^x \left[(A + 3B)x^2 + 2Ax + Bx^3\right] \\
y''_p &= e^x \left[Bx^3 + (A + 6B)x^2 + (4A + 6B)x + 2A\right] \\
y'''_p &= e^x \left[Bx^3 + (A + 9B)x^2 + 6(A + 3B)x + 6(A + B)\right]
\end{aligned}
$$

Substituting these back into the given non-homogeneous ode, we have

$$
\begin{aligned}
&e^x[Bx^3 + (A + 9B)x^2 + 6(A + 3B)x + 6(A + B)] - \\
&3e^x \left[(A + 3B)x^2 + 2Ax + Bx^3\right] + 2e^x[Ax^2 + Bx^3] = xe^x,
\end{aligned}
$$

which on simplification yields $18Bx + 6(A + B) = x$.

Equating co-efficients of like powers of x, we get $A = -\frac{1}{18}$ and $B = \frac{1}{18}$ so that $y_p = \frac{1}{18}e^x(x^3 - x^2)$

Example (18) : Find PI of the ode $(D^2 + 1)y = xe^x \cos 2x$

Observe that the fundamental set of solutions of the associated homogeneous equation $(D^2 + 1)y = 0$ is $\{\sin x, \cos x\}$. The righthand member $q(x) = xe^x \cos 2x$ is a uc function in the product form and so its uc set is given by the direct product

$$
\begin{aligned}
U &= \{e^x\} \otimes \{1, x\} \otimes \{\sin 2x, \cos 2x\} \\
&= \{e^x \sin 2x, e^x \cos 2x, xe^x \sin 2x, xe^x \cos 2x\}.
\end{aligned}
$$

Observe that there is no common element between this uc set and the fundamental solution set and hence the particular integral can be taken in the form

$$
y_p = Ae^x \sin 2x + Be^x \cos 2x + Cxe^x \sin 2x + Exe^x \cos 2x,
$$

where A, B, C and E are four undetermined co-efficients.

Through successive differentiation one gets

$$
\begin{aligned}
y_p' = e^x[(A - 2B + C)\sin 2x + (2A + B + E)\cos 2x + \\
+(C - 2E)x \sin 2x + (2C + E)x \cos 2x]
\end{aligned}
$$
$$
\begin{aligned}
y_p'' = e^x[(-3A - 4B + 2C - 4E)\sin 2x + (4A - 3B + 4C + 2E)\cos 2x \\
-(3C + 4E)x \sin 2x + (4C - 3E)x \cos 2x]
\end{aligned}
$$

which yields through substitution into the ode $(D^2 + 1)y = xe^x \cos 2x$,

$$
\begin{aligned}
xe^x \cos 2x = (-2A - 4B + 2C - 4E)e^x \sin 2x + (4A - 2B + 4C + 2E) \times \\
\times e^x \cos 2x - (2C + 4E)xe^x \sin 2x + (4C - 2E)xe^x \cos 2x
\end{aligned}
$$

Since the uc set consists of only linearly independent functions, we have on equating like terms on both sides the algebraic linear equations:

$$
A + 2B - C + 2E = 0; \quad 2A - B + 2C + E = 0; \quad C + 2E = 0; \quad 4C - 2E = 1.
$$

On solving these, we have $A = -\frac{1}{25}$, $B = \frac{11}{50}$, $C = \frac{1}{5}$ and $E = -\frac{1}{10}$ so that the desired particular integral is

$$
y_p = \frac{1}{50}[(11 - 5x)e^x \cos 2x + (10x - 2)e^x \sin 2x]
$$

One may note that there can be found a second order linear differential operator $M(D)$ which annihilates the righthand member $q(x) = xe^x$ of Example (17) and a fourth order linear differential operator that annihilates the righthand member $q(x) = xe^x \cos 2x$ in Example (18).

The second order linear differential operator whose two linearly independent characteristic functions (or eigenfunctions) are e^x and xe^x (i,e, the elements of the uc set of $q(x) = xe^x$), is $M(D) \equiv (D-1)^2$

Clearly $M(D)q(x) = 0$ and so, the original non-homogeneous differential equation $(D^3 - 3D + 2)y = xe^x$, when pre-multiplied by $M(D)$, reduces to the homogeneous differential equation

$$(D-1)^2(D^3 - 3D + 2)y = 0$$

of order five. This equation has its general solution in the form

$$y = (C_1 + C_2 x + C_3 x^2 + C_4 x^3)e^x + C_5 e^{-2x}.$$

because the auxiliary/characteristic polynomial of this differential equation has its roots -2 and 1 (with multiplicity 4). However we notice immediately that $(C_1 + C_2 x)e^x + C_5 e^{-2x}$ is just a solution of the equation $(D^3 - 3D + 2)y = 0$. As we are interested in only a particular integral y_p of the given equation, we can assume it in the form

$$y_p = (C_3 x^2 + C_4 x^3)e^x.$$

This brings us to the same track of the previous workout of Example (17). One approach is convenient because we made use of the uc set of uc function $q(x)$ of the original differential equation to determine the annihilator $M(D)$. Once $M(D)$ is known, the problem becomes algebraic in nature, no integration etc being needed.

For Example (18), we require a fourth order linear differential operator $M(D)$ as the annihilator. The operator $M(D)$ is determined from the knowledge of uc set of $q(x) = xe^x \cos 2x$ of this problem. As it terns out, the desired operator $M(D)$ is just the one whose four linearly independent eigenfunctions are enlisted in the uc set

$$\{e^x \sin 2x, e^x \cos 2x, xe^x \sin 2x, xe^x \cos 2x\}.$$

So $M(D) = (D^2 - 2D + 5)^2$ which annihilates $q(x) = xe^x \cos 2x$ and consequently the original non-homogeneous differential equation reduces to the homogeneous equation $(D^2 - 2D + 5)^2(D^2 + 1)y = 0$ of order six. This new differential equation has its general solution in the form

$$y = C_1 \cos x + C_2 \sin x + C_3 e^x \sin 2x + C_4 e^x x \sin 2x + C_5 e^x \cos 2x$$
$$+ C_6 e^x x \cos 2x$$

Here the procedure and line of argument towards finding of particular integral is in letter and spirit the same as the example discussed before.

Although determination of $M(D)$ is not very tough an assignment, we put up the following table to enlist some common uc functions and their corresponding annihilators for ready reference.

UC function	Annihilator
x^k	D^{k+1}
e^{ax}	$(D-a)$
$\sin ax$, $\cos ax$	$(D^2 + a^2)$
$x^k e^{ax}$	$(D-a)^{k+1}$
$x^k \sin ax,$ $x^k \cos ax$	$(D^2 + a^2)^{k+1}$
$x^k e^{ax} \sin bx,$ $x^k e^{ax} \cos bx$	$\left((D-a)^2 + b^2\right)^{k+1}$

Extension: If $q(x)$ be the linear combination $\sum_{k=1}^{m} t_k q_k(x)$ of m uc functions $q_k(x)$, $k = 1, 2, \cdots\cdots, m$, then we shall have to proceed in the following sequence of steps.

(a) Find the uc set of each the m uc functions $q_1, q_2, \cdots\cdots, q_m$. Let $\mathcal{S}_k$ denote the uc set of the uc function $q_{k'}$. Any two uc sets $\mathcal{S}_i$ and $\mathcal{S}_j$ are either disjoint or comparable. ($\mathcal{S}_i$ and $\mathcal{S}_j$ are said to be comparable iff either $\mathcal{S}_i \subseteq \mathcal{S}_j$ or $\mathcal{S}_j \subseteq \mathcal{S}_i$ holds). Of two given comparable sets $\mathcal{S}_i$ and $\mathcal{S}_j$, retained the maximal one and omitted the other. In this way we have a class of disjoint uc sets.

(b) Each of these surviving uc sets are treated separately. If one of these uc sets, say $\mathcal{S}_k$, includes one or more members of the fundamental solution set of the associated homogeneous ode $L(D)y = 0$, then we revise $\mathcal{S}_k$ by multiplying each of its members by the lowest positive integral power of x so as to make it free of solutions of the homogeneous equation.

(c) Consider the union of all these uc sets (original or revised as the situation demands). Let's denote it by $\mathcal{U}_0$.

(d) Since we have a finite set of linearly independent functions we choose our particular integral as a linear combination of the members of the set $\mathcal{U}_0$ with unknown co-efficients.

(e) Determine the unknown co-efficients by plugging the linear combination found in the step (d) into the original non-homogeneous differential equation and demanding that it should satisfy the differential equation identically.

Example (19) : Solve $(D^2 + D - 2)y = 6e^{-2x} + 3e^x - 4x^2$ by the method of undetermined co-efficients (for only particular integrals)

Auxiliary equation of the associated homogeneous differential equation $(D^2 + D - 2)y = 0$ has its roots 1 and -2. Hence $\{e^x, e^{-2x}\}$ is its fundamental solution set.

Now $q(x)$ is a linear combination of x^2, e^x and e^{-2x}. These are uc functions having their corresponding uc sets $\{1, x, x^2\}$, $\{e^x\}$ and $\{e^{-2x}\}$ respectively. These uc sets are disjoint. However, since e^x and e^{-2x} are elements of the fundamental solution set $\{e^x, e^{-2x}\}$ we need revision of these uc sets− $\{e^x\}$ replaced by $\{xe^x\}$ while $\{e^{-2x}\}$ replaced by $\{xe^{-2x}\}$. Consider the union of all these uc sets (original or revised). So we have $\mathcal{U}_0 = \{1, x, x^2, xe^x, xe^{-2x}\}$. We choose y_p in the form

$$y = c_0 + c_1 x + c_2 x^2 + c_3 xe^x + c_4 xe^{-2x},$$

where c_0, c_1, c_2, c_3, c_4 are undetermined constants. Successive differentiation, substitution into the given equation, and finally comparing the co-efficients of the linearly independent functions yield the relations

$$-2c_0 + c_1 + 2c_2 = 0; \quad c_1 - c_2 = 0; \quad c_2 = 2; \quad c_3 = 1; c_4 = -2$$

which give the complete solution set

$$c_0 = 3; \quad c_3 = 1; \quad c_1 = c_2 = -c_4 = 2$$

$\therefore$ Particular integral $\quad y_p = 3 + 2x + 2x^2 + xe^x - 2xe^{-2x}$

Example (20) : Solve the equation $(D^4 + D^2)y = 4\sin x - 2\cos x$ by the method of undetermined co-efficients.

Here the fundamental solution set of the associated homogeneous differential equation $(D^4 + D^2)y = 0$ is $\{1, x, \sin x, \cos x\}$. Again the righthand member $q(x)$ is the linear combination of $\sin x$ and $\cos x$. Incidentally both these are uc functions with identical uc sets. Hence we shall consider either of them. However this uc set is included in the fundamental solution set and needs revision. The revised uc set is $\{x \sin x, \ x \cos x\}$ and $y_p = x(A \sin x + B \cos x)$, where A and B are undetermined constants. Imitating the steps in the previous worked out examples one gets $A = 1$ and $B = 2$.

The general solution is $y = y_c + y_p$, where

$$y_c = (c_0 + c_1 x) + (c_2 \sin x + c_3 \cos x) + x(\sin x + 2\cos x)$$

Example (21) : Suggest a suitable form of $y_p(x)$ if the method of undetermined co-efficients is used for solving

$$\frac{d^2 y}{dx^2} + 2\frac{dy}{dx} + 2y = 3e^{-x} + 2e^{-x}\cos x + 4x^2 e^{-x}\sin x$$

Fundamental solution set of the associated homogeneous equation is $\{e^{-x}\cos x, e^{-x}\sin x\}$. uc set of e^{-x} is $\{e^{-x}\}$. uc set of $e^{-x}\cos x$ is $\{e^{-x}\cos x, e^{-x}\sin x\}$ and uc set of $x^2 e^{-x}\sin x$ is

$$\{e^{-x}\sin x, e^{-x}\cos x, xe^{-x}\sin x, xe^{-x}\cos x, x^2 e^{-x}\sin x, x^2 e^{-x}\cos x\}$$

— the former being a subset of the latter, we retain the bigger one. The union $\mathcal{U}_0$ of all these sets (after revision) is

$$\{e^{-x}, xe^{-x}\sin x, x^2 e^{-x}\sin x, x^3 e^{-x}\sin x, xe^{-x}\cos x, x^2 e^{-x}\cos x,$$
$$x^3 e^{-x}\cos x\}.$$

Hence y_p should belong to the linear span of $\mathcal{U}_0$.

$$\therefore \ y_p = Ae^{-x} + x(B_0 + B_1 x + B_2 x^2)e^{-x}\sin x$$
$$+x(c_0 + c_1 x + c_2 x^2)e^{-x}\cos x$$

Useful Discussions

(1) What would be problem had the uc sets $\{e^x\}$ and $\{e^{-2x}\}$ been not revised by multiplying each member by x in example (19)?

In that case we would consider $y_p = c_0 + c_1 x + c_2 x^2 + c_3 e^x + c_4 e^{-2x}$.

On differentiation,

$$y_p' = c_1 + 2c_2 + c_3 e^x - 2c_4 e^{-2x}$$
$$y_p'' = c_3 e^x + 4c_4 e^{-2x},$$

From substitution into the given ode, one has

$$(-2c_0 + c_1 + 2c_2) - 2c_1 x - 2c_2 x^2 = 0.$$

This gives $c_0 = c_1 = c_2 = 0$ and unfortunately c_3 and c_4 remains arbitrary. This renders singularity in the system, indicating that guess was poor. So finding out a seemingly appropriate finite dimensional vector space spanned by the uc set of $q(x)$ is not the end of the journey—restoration of consistency and determinacy of each and every linear algebraic equation involving the undetermined coefficients is also a prime objective. All these demand for inflation of our vector space members by 'suitable' multiplicative factors of the form x^p $(p \geq 1)$.

(2) In Example (20) we found that the uc set $\{\sin x, \cos x\}$ was modified as $\{x \sin x, x \cos x\}$. Had it been modified as $\{x^2 \sin x, x^2 \cos x\}$, would it make any difference?

If y_p of Example (20) were taken as $x^2(A \sin x + B \cos x)$, we get

$$y_p' = (2xA - Bx^2) \sin x + (2Bx + Ax^2) \cos x$$
$$y_p'' = (2A - 4Bx - Ax^2) \sin x + (2B + 4Ax - Bx^2) \cos x$$
$$y_p''' = (-6B - 6Ax + Bx^2) \sin x + (6A - 6Bx - Ax^2) \cos x$$
$$y_p'''' = (-12A + 8Bx + Ax^2) \sin x + (-12B - 8Ax + Bx^2) \cos x$$

Plugging y_p'' and y_p'''' into the original ode has

$$(10A + 4) \sin x + (10B - 2) \cos x + 4Bx \sin x + 4A \cos x = 0,$$

wherefrom follows the set of linear algebraic equations

$$10A + 4 = 0; 10B - 2 = 0; 4A = 0; 4B = 0$$

that are inconsistent. This inconsistency owes its origin from the fact that we revised the uc set $\{\sin x, \cos x\}$ by not using the minimal integral power of x.

(3) The process of undetermined co-efficients fails to deliver when $q(x)$ is $\tan x, \sec x$ or $\ln x$. The reason of this failure is that these functions are not elements of any sufficiently large finite-dimensional vector space closed under a differential operator $L(D)$ of finite order. In other words, the failure is due to the fact that $\tan x, \sec x$, or $\ln x$ are not solutions of any homogeneous differential equation.

It is known to us that the method of undetermined co-efficients is applicable to a linear ode of the form $L(D)y = q(x)$, provided $q(x)$ is a solution of some linear homogeneous ode $M(D)y = 0$. The uc set of $q(x)$ is indeed the fundamental solution set of the linear homogeneous ode $M(D)y = 0$. Now if the fundamental solution set of $L(D)y = 0$ is disjoint with the fundamental solution set of $M(D)y = 0$, then the fundamental solution set of the ode $M(D)L(D) = 0$ is the union of the above two solution sets. On the otherhand, if there is at least one member common to these bases, then the auxiliary equation of the ode $(M(D)L(D))y = 0$ must have at least one root whose multiplicity is greater than or equal to 2. This accounts for why one has to modify the uc set of $q(x)$ by multiplying it with a suitable positive integral power of x, when uc set of $q(x)$ shares at least one term common with the fundamental solution set of $L(D)y = 0$. We hope that the following illustration will clarify this point.

Example (22) : For the ode $(D+1)^3 y = xe^{-x}$ suggest form of y_p

Here $q(x) = xe^{-x}$ and $L(D) = (D+1)^3$

Since the auxiliary equation of the reduced ode $L(D)y = 0$ is $(m+1)^3 = 0$, -1 is its root with multiplicity 3. Hence the c.f. of the given ode is $y_c = e^{-x}(C_1 + C_2 x + C_3 x^2)$, making it clear that the fundamental solution set of $L(D)y \equiv (D+1)^3 y = 0$ is $\{e^{-x}, xe^{-x}, x^2 e^{-x}\}$. By inspection, one observes that the function $q(x) = xe^{-x}$ is a solution of $(D+1)^2 y = 0$. Thus $M(D) \equiv (D+1)^2$. Since the fundamental solution set of $M(D)y \equiv (D+1)^2 y = 0$ is $\{e^{-x}, xe^{-x}\}$, there are two members common between the fundamental solution sets of $L(D)y = 0$ and $M(D)y = 0$. So the ode $M(D)L(D)y = 0$ is $(D+1)^5 y = 0$, where auxiliary equation has root -1 with multiplicity 5. This ensures that the fundamental solution set of $(D+1)^5 y = 0$ is $\{e^{-x}, xe^{-x}, x^2 e^{-x}, x^3 e^{-x}, x^4 e^{-x}\}$.

Since $\{e^{-x}, xe^{-x}, x^2 e^{-x}\}$ is the fundamental solution set of $L(D)y = 0$, the particular integral y_p of $L(D)y = q(x)$ is a linear combination of $x^3 e^{-x}, x^4 e^{-x}$. This justifies why we need modify the uc set of $q(x) = xe^{-x}$ by multiplying each of its terms by x^3 before jotting down the form of y_p in terms of them.

(4) As a by product of our discussion we observe that even after modification, the revised uc set of $q(x)$ remains linearly independent. This result owes its origin to Theorem 5.2 of § 5.3.

(5) From the entire proceeds in the article we may develop a mistaken belief that applicability of this method is confined to the linear differential equations with constant co-efficients. Infact this method applies equally well to some restricted class of linear differential equations with variable co-efficients and even a few non-linear equations (ref. Hubbard & West). The following example is an instance.

Example (23) : By UC method find the particular integral of

$$x \frac{d^2 y}{dx^2} - (1 + x) \frac{dy}{dx} + y = x^2 e^{2x}.$$

vide the 'Method of Inspection' coming up in next chapter one may prove that the fundamental solution set of the associated homogeneous equation is $\{e^x, (x+1)\}$. However $q(x) = x^2 e^{2x}$ is a uc function whose uc set is $\{e^{2x}, xe^{2x}, x^2 e^{2x}\}$. Since none of these belong to the fundamental solution set of the associated homogeneous equation we can try

$$y_p = e^{2x}(a + bx + cx^2), \quad a, b, c \text{ being constants.}$$

By successive differentiation of y_p we have :

$$y_p' = e^{2x}\left((2a + b) + 2(b + c)x + 2cx^2\right)$$
$$y_p'' = e^{2x}\left((4a + 4b + 2c) + (4b + 8c)x + 4cx^2\right)$$

While putting these into the inhomogeneous ode given, we have

$$-(a + b) + 2(a + b)x + (2b + 5c)x^2 + 2cx^3 = x^2,$$

so that $c = 0$; $b = \frac{1}{2}$ and $a = -\frac{1}{2}$.

$$\therefore \ y_p = \frac{1}{2} e^{2x}(x - 1)$$

Example (24) : Can one apply the method of undetermined co-efficients to find a particular integral of $x^2 \frac{d^2y}{dx^2} - 2x \frac{dy}{dx} + 2y = xe^{-x}$?

The answer is negative. But why did it fail to click? The mystery lies in the fact that the given equation is a second order equi-dimensional equation and so transformable into a second order differential equation with constant co-efficients by substituting $z = \ln x$. However the right-member is translated to some non uc function, incorporating trouble. Hence method of variation of parameters is to be implemented. Note that if $q(x)$ in example (24) were just a polynomial in x, we would succeed.

5.10   Fourier Series Method for Particular Integrals

Consider the nth order linear differential equation $L(D)y = f(x)$, where $f(x)$ is bounded and integrable over $[-\pi, \pi]$ and also periodic with period 2π over that interval. In this case, the Fourier Series generated by f is given by

$$f(x) \sim \sum_{k=0}^{\infty} c_k \, q_k(x),$$

where the functions $q_k(x)$ are supposed to form an orthogonal system $\mathcal{S}$ over $[-\pi, \pi]$ and the so-called Euler-Fourier co-efficients are given by

$$c_k = \int_{-\pi}^{\pi} f(x)q_k(x)dx.$$

If the orthogonal system $\mathcal{S}$ is desired to be real-valued, its conventional choice is

$$q_0(x) = \frac{1}{\sqrt{2\pi}}; \quad q_{2k-1}(x) = \frac{1}{\sqrt{\pi}} \cos kx; \quad q_{2k}(x) = \frac{1}{\sqrt{\pi}} \sin kx, k = 1, 2, \cdots.$$

If the orthogonal system $\mathcal{S}$ is desired to be complex-valued, the conventional choice is $q_k(x) = \frac{e^{ikx}}{\sqrt{2\pi}}$, k being any integer positive, negative or zero. In either case, there is no apriori assurance that the Fourier series generated by $f(x)$ would converge to the function $f(x)$ itself for all $x \in [-\pi, \pi]$. However, the above Fourier series converges uniformly to $f(x)$ if $f(x)$ be a continuous periodic function having piecewise continuous derivative over $[-\pi, \pi]$. Infact, the continuity alone of $f(x)$ is not very fruitful a hypothesis as far as convergence of the formal Fourier series of f is concerned. Under the added assumption that f has a piecewise continuous derivative over $[-\pi, \pi]$, we can ensure uniform convergence of the above Fourier series $\sum_{k=0}^{\infty} c_k q_k(x)$ to the sum function $f(x)$. The uniform convergence of the Fourier series of f assures that term-by-term integration of the Fourier series is same as integration of the sum function $f(x)$ over $[-\pi, \pi]$.

Example (25) : Solve the ode $(D^2 + 2D + 2)y = f(x)$ by Fourier series method, where $f(x)$ is a periodic function given by $f(x) =| x |$ if $x \in [-\pi, \pi]$.

Observe that as $f(-\pi) = f(\pi)$, we can define $f(x)$ over the whole real line by demanding $f(x + 2\pi) = f(x)$ to hold for every x. This makes $f(x)$ a continuous periodic function of x with period 2π. Moreover as $f'(x) = Sgn\ x$, $f(x)$ possesses a piecewise continuous derivative over $[-\pi, \pi]$. All these evidences approve implementation of the 'Fourier series method' for determining the particular integral.

The complementary function is $y_c = e^{-x}(A \cos x + B \sin x)$,

A and B being arbitrary constants.

To find the particular integral, we expand $f(x)$ by Fourier series

$$f(x) \sim \frac{\pi}{2} - \frac{4}{\pi}\left(\frac{\cos x}{1^2} + \frac{\cos 3x}{3^2} + \frac{\cos 5x}{5^2} + \cdots ad.inf\right)$$

We next write and solve a set of linear differential equations resembling the one given to us but each having just one term of the above Fourier series on the righthand side: For the first term, a particular integral is $\frac{\pi}{4}$. The remaining part of the above Fourier Series is of the form

$$-\frac{4}{\pi}\frac{\cos(2k - 1)x}{(2k - 1)^2}, \quad k = 1, 2, \cdots\cdots.$$

The particular integral of the component ode

$$(D^2 + 2D + 2)y = -\frac{4}{\pi} \cdot \frac{\cos(2k - 1)x}{(2k - 1)^2}$$

is given as

$$\bar{y}_p^{(k)} = \frac{1}{D^2 + 2D + 2}\left[-\frac{4}{\pi}\frac{\cos((2k - 1)x)}{(2k - 1)^2}\right]$$

$$= -\frac{4}{\pi}\frac{1}{(2k-1)^2}\frac{(D^2-2D+2)}{(D^2+2)^2-4D}\cos((2k-1)x)$$

$$= \frac{4}{\pi}\cdot\frac{1}{(2k-1)^2}\frac{(4k-4k^2+1)\cos(2k-1)x+2(2k-1)\sin((2k-1))}{\{(2k-1)^4+4\}}$$

Particular integral y_p of the given ode will read :

$$-\frac{4}{\pi}\cdot\sum_{k=1}^{\infty}\frac{(4k-4k^2+1)\cos(2k-1)x+2(2k-1)\sin(2k-1)}{\{(2k-1)^2.(2k-1)^4+4\}}$$

Remark :

(1) Had we used the complex exponential form of the Fourier Series, the Fourier Series of f would look different but nevertheless we shall have the same y_p as above.

(2) The 'Fourier series method' is applied when the inhomogeneous term $f(x)$ is a complicated periodic function. This type of situations arise in practice when there is a periodic emf applied to an electrical circuit.

(3) In the literature on Fourier series, often the condition of piecewise smoothness of f over $[-\pi,\pi]$ is replaced by piecewise monotonicity of f over the same interval. The latter is known as Dirichlet's condition for f. The interval $[-\pi,\pi]$ was chosen for the Fourier expansion of f having period 2π. Had the period of f remained unaltered, the interval $[-\pi,\pi]$ could be replaced by $[0,2\pi]$ or any other interval having length 2π. If the period of f were $(b-a)>0$, then we might choose $[a,b]$ as the required interval for Fourier series expansion of f. However, with the help of Schroeder Bernstein theorem one can establish that any interval $[a,b]$ is numerically equivalent $[-\pi,\pi]$; $y = \pi\left(\frac{2x-a-b}{b-a}\right)$ beng the requisite scaling transformation. From this discussion it transpires that length of the interval (over which expansion is done) varies paripassu the periodicity of f but choice of specific interval of a preassigned length is by no means an imposed constraint to the Fourier series expansion of f.

(4) The physicists often use 'Fourier Series method' to determine the particular integrals of the inhomogeneous linear differential equations $L(D)y = f(x)$, $f(x)$ being some complicated periodic function of x. This approach suffers from the drawback that it leans heavily on the mistaken belief that linearity of the differential operator $L(D)$ is omnipotent as far as the convergence etc. of an infinite series is concerned. We should keep in mind that the superposition principle associated with the inhomogeneous linear differential equations works smoothly if their inhomogeneous terms assume the generic form $\sum_{k=0}^{m} c_k q_k(x)$, $q_k(x)$'s being continuous functions. However, if $m \to \infty$, the conclusion of the superposition principle is tentative. This implies that if $\bar{y}_k(x)$'s be the solution of the odes

$$L(D)y = q_k(x), k = 0, 1, 2, \cdots \cdots, m, \cdots \cdots,$$

then for finite m, $\displaystyle\sum_{k=0}^{m} c_k \bar{y}_k(x)$ would be the solution of the ode

$L(D)y = \displaystyle\sum_{k=0}^{m} c_k q_k(x)$ but $\displaystyle\sum_{k=0}^{\infty} c_k \bar{y}_k(x)$ is not necessarily a solution

of $L(D)y = \displaystyle\sum_{k=0}^{m} c_k q_k(x)$. If $\displaystyle\sum_{k=0}^{\infty} c_k q_k(x)$ be the formal Fourier series

of a periodic function $f(x)$ over some interval I, we have to ensure that this Fourier series converges and converges to $f(x)$ over I. As is obvious, the idea of convergence of an infinite series bears no meaning in any linear space unless it is endowed with a topological structure. This need of having a topological structure instigated us to pose the theory of linear differential equations against the backdrop of linear spaces. However, even in presence of an underlying topological structure one cannot vouchsafe state that the expressions $L(D) \left(\displaystyle\sum_{k=0}^{\infty} c_k \bar{y}_k(x) \right)$ and $\displaystyle\sum_{k=0}^{\infty} c_k L(D) \bar{y}_k(x)$ are equivalent as n-fold term-by-term differentiation of an infinite series is not necessarily the same as differentiation of its limit function.

Exercise 5A

1. Find the real-valued solutions of the following equations:

    (a) $\dfrac{d^2 y}{dx^2} + 4y = 0$

(b) $\dfrac{d^2y}{dx^2} - 9\dfrac{dy}{dx} + 9y = 0$

(c) $\dfrac{d^2y}{dx^2} - 7\dfrac{dy}{dx} + 6y = 0$

(d) $\dfrac{d^3y}{dx^3} - 3\dfrac{d^2y}{dx^2} - 6\dfrac{dy}{dx} + 8y = 0$

(e) $4\dfrac{d^3y}{dx^3} + 4\dfrac{d^2y}{dx^2} - 7\dfrac{dy}{dx} + 2y = 0$

(f) $\dfrac{d^4y}{dx^4} - 3\dfrac{d^3y}{dx^3} - 2\dfrac{d^2y}{dx^2} + 2\dfrac{dy}{dx} + 12y = 0$

(g) $\dfrac{d^4y}{dx^4} + 2\dfrac{d^3y}{dx^3} + 6\dfrac{d^2y}{dx^2} + 2\dfrac{dy}{dx} + 5y = 0$

(h) $\dfrac{d^3y}{dx^3} - i\dfrac{d^2y}{dx^2} + \dfrac{dy}{dx} - iy = 0$

(i) $\dfrac{d^3y}{dx^3} - 3i\dfrac{d^2y}{dx^2} - 3\dfrac{dy}{dx} + iy = 0$

2. Solve the following initial-value problems:

(a) $\dfrac{d^2y}{dx^2} - 6\dfrac{dy}{dx} + 8y = 0$, $y(0) = 1$, $y'(0) = 6$

(b) $\dfrac{d^2y}{dx^2} + 2\dfrac{dy}{dx} + 5y = 0$, $y(0) = 2$, $y'(0) = 6$

(c) $\dfrac{d^3y}{dx^3} - 3\dfrac{d^2y}{dx^2} + 4y = 0$, $y(0) = 1$, $y'(0) = -8$, $y''(0) = -4$

(d) $\dfrac{d^3y}{dx^3} + y = 0$, $y(0) = y''(0) = 0$ and $y'(0) = 1$

3. Check whether the following functions are linearly independent or not in the regions indicated. (Give your reasons by computation of Wronskian).

(i) $\phi_1(x) = e^{ix}$, $\phi_2(x) = \sin x$, $\phi_3(x) = 2\cos x$, $-\infty < x < \infty$

(ii) $\phi_1(x) = x$, $\phi_2(x) = e^{2x}$, $\phi_3(x) = x\sin x$, $-\infty < x < \infty$

(iii) $\phi_1(x) = x^3$, $\phi_2(x) = |x|^3$, $-1 \le x \le 1$

(iv) $\phi_1(x) = | \, x \, |, \quad \phi_2(x) = \ln | \, x \, |, | \, x \, | > e$

(v) $\phi_1(x) = 2e^{ix}, \quad \phi_2(x) = \sin x, \quad \phi_3(x) = \cos x \; ; \; x \in \mathbb{C}$

4. (a) Find the interval over which the functions $\sin x$ and $\sin 2x$ forms a linearly independent set.

(b) Is $\{x, x^2, x^3\}$ a linearly independent set over the interval $(-1,2)$? Does the same conclusion hold true over $(1,2)$?

5. If λ be a double root of the auxiliary equation for a differential equation with constant co-efficients, then one of the corresponding fundamental solution is $e^{\lambda x}$. The form of the second fundamental solution can be deduced by use of L Hospital's rule or that of standard limits also from the following argument :

If λ and λ' are two distinct roots of the auxiliary equation,

$$\psi(x; \lambda, \lambda') = \frac{1}{\lambda' - \lambda}(e^{\lambda' x} - e^{\lambda x})$$

is also a solution. When λ is held fixed and $\lambda' = \lambda + \delta\lambda$, we may assume $\psi(x; \lambda, \lambda')$ to be of the form

$$\frac{1}{\delta\lambda}\left(e^{(\delta\lambda)x} - 1\right)e^{\lambda x} \text{ (why?)}.$$

Now compute $\lim_{\delta\lambda \to 0} \psi(x; \lambda, \lambda + \delta\lambda)$ by standard limits to verify that $xe^{\lambda x}$ is the second fundamental solution.

Exercise 5B

1. Find complete solution of the following equations. Use "short" methods for solving the particular integrals.

(a) $\dfrac{d^2y}{dx^2} + \dfrac{dy}{dx} + y = \sin 2x$

(b) $(D^2 + n^2)y = e^x x^4$

(c) $\dfrac{d^3y}{dx^3} - 13\dfrac{dy}{dx} + 12y = x^2$

(d) $\dfrac{d^2y}{dx^2} + 4y = 1 + x + \sin x$

(e) $\dfrac{d^2y}{dx^2} + i\dfrac{dy}{dx} + 2y = 6\sinh 2x - e^{-2x}$

(f) $\dfrac{d^4y}{dx^4} - 4\dfrac{d^3y}{dx^3} + 6\dfrac{d^2y}{dx^2} - 4\dfrac{dy}{dx} + y = e^{2x}\cos x$

(g) $\dfrac{d^3y}{dx^3} + y = \sin^3 x + xe^x$

(h) $\dfrac{d^2y}{dx^2} + 2a\dfrac{dy}{dx} + (a^2 + b^2)y = e^{px}$

(i) $6\dfrac{d^2y}{dx^2} + 5\dfrac{dy}{dx} - 6y = xe^x\cos 3x$

(j) $2\dfrac{d^4y}{dx^4} + 5\dfrac{d^2y}{dx^2} + 2y = x^3e^x + \cos(\sqrt{2}x)$

(k) $\dfrac{d^2y}{dx^2} + 4y = (1-x)\cos 2x$

2. Use the method of partial fractions to yield only the particular integrals of the following equations.

   (a) $(D^2 + a^2)y = \sec ax$

   (b) $(D^2 - 5D + 6)y = \cos 3x$

   (c) $(D^4 + 10D^2 + 9)y = \sin x \sin 2x$

   (d) $(D^2 + D + 1)y = x^3 e^{-x}$

   (e) $(D^3 - 3D^2 + 4)y = e^{3x}$

   (f) $(D^3 - 7D - 6)y = e^{2x}(1 + x)$

3. Solve the initial-value problems, if possible :

   (i) $\dfrac{d^2y}{dx^2} - \dfrac{dy}{dx} - 6y = 8e^{2x} - 5e^{3x}$, $y(0) = 3$ and $y'(0) = 5$

   (ii) $\dfrac{d^2y}{dx^2} + 4y = 8\sin 2x$, $y(0) = 6$ and $y'(0) = 8$

   (iii) $\dfrac{d^2y}{dx^2} - y = x\sin x$, $y(0) = 2$ and $y'(0) = \dfrac{3}{2}$

   (iv) $\dfrac{d^3y}{dx^3} - 3i\dfrac{d^2y}{dx^2} - 3\dfrac{dy}{dx} + iy = 5x^2$,

                          $y(0) = 60i$, $y'(0) = -30$; $y''(0) = 30i$

   (v) $\dfrac{d^2y}{dx^2} - 2\dfrac{dy}{dx} - 3y = 2e^x - 10\sin x$, $y(0) = 2$ and $y'(0) = 4$

4. Find the particular integral of the problem 1(k) in two possible ways, viz.,

(i) The use of rule in Case V for shorter methods and

(ii) The use of Euler formula and rule in Case IV for shorter methods. Are both the results same? If not, how do you justify this difference?

Exercise 5C

1. Find the general solution by the 'method of variation of parameters' of the following differential equations.

(a) $\dfrac{d^2y}{dx^2} + 2\dfrac{dy}{dx} + y = \dfrac{e^{-x}}{x^2},\ x > 0$

(b) $\dfrac{d^2y}{dx^2} - 3\dfrac{dy}{dx} + 2y = \dfrac{e^x}{e^x + 1}$

(c) $\dfrac{d^2y}{dx^2} + y = \dfrac{1}{1 + \sin x}$

(d) $\dfrac{d^2y}{dx^2} + 4\dfrac{dy}{dx} + 5y = e^{-2x}\sec x$

(e) $\dfrac{d^2y}{dx^2} + 4y = 3\,cosec\,2x,\ 0 < x < \dfrac{\pi}{2}$

(f) $\dfrac{d^2y}{dx^2} - 2\dfrac{dy}{dx} + y = x\ln x,\ (x > 0)$

(g) $\dfrac{d^2y}{dx^2} - 2\dfrac{dy}{dx} + 2y = e^x\tan x$

2. Use method of variation of parameters to find the general solution of
$$x^2\dfrac{d^2y}{dx^2} - x(x+2)\dfrac{dy}{dx} + (x+2)y = x^3,$$
given that $y = x$ and $y = xe^x$ are linearly independent solutions of the associated homogeneous equation.

3. Find the general solution of
$$(2x+1)(x+1)\dfrac{d^2y}{dx^2} + 2x\dfrac{dy}{dx} - 2y = (2x+1)^2$$
by the method of variation of parameters, given that $y = x$ and $y = (x+1)^{-1}$ are two linearly independent solutions of the corresponding homogeneous equation.

4. Given that $y = x \sin x$ is a solution of the equation

$$\sin^2 x \frac{d^2 y}{dx^2} - 2 \sin x \cos x \frac{dy}{dx} + (1 + \cos^2 x)y = 0,$$

find by means of method of variation of parameters, the particular integral of the equation

$$\sin^2 x \frac{d^2 y}{dx^2} - 2 \sin x \cos x \frac{dy}{dx} + (1 + \cos^2 x)y = \sin^3 x.$$

5. Show that e^t and $\frac{1}{t}$ are solutions of the homogeneous equation associated with the differential equation

$$(t^2 + t)\frac{d^2 x}{dt^2} - (t^2 - 2)\frac{dx}{dt} - (t + 2)x = t(t + 1)^2$$

and hence find the complete solution by virtue of the method of variation of parameters.

6. Show that the solution of the initial-value-problem

$$\frac{d^2 y}{dx^2} + y = g(x) \text{ subject to } y(0) = y_0 \text{ and } y'(0) = y_1$$

can be put in the form

$$y(x) = \int_0^x g(t) \sin(x - t)dt + y_0 \cos x + y_1 \sin x$$

Exercise 5D

1. Solve (by method of undetermined co-efficients) for particular integrals of the following equations:

(a) $\dfrac{d^2 y}{dx^2} + y = 3 \sin 2x + x \cos 2x$

(b) $\dfrac{d^2 y}{dx^2} + \dfrac{dy}{dx} - 6y = 10e^{2x} - 18e^{3x} - 6x - 11$

(c) $\dfrac{d^2 y}{dx^2} + 4\dfrac{dy}{dx} + 5y = e^{-2x}(1 + 2\cos x)$

(d) $\dfrac{d^2 y}{dx^2} + 2\mu\dfrac{dy}{dx} + \lambda^2 \mu = \cos \omega x, \ \mu^2 < \lambda^2 \text{ and } \lambda^2 = \mu^2 + \omega^2$

(e) $\dfrac{d^2 y}{dx^2} + i\dfrac{dy}{dx} + 2y = 2\cosh x + e^{-2x}$

(f) $\dfrac{d^3y}{dx^3} + 3\dfrac{d^2y}{dx^2} + 3\dfrac{dy}{dx} + y = x^2 e^{-x}$

(g) $\dfrac{d^3y}{dx^3} - 6\dfrac{d^2y}{dx^2} + 11\dfrac{dy}{dx} - 6y = xe^x - 4e^{2x} + 6e^{4x}$

(h) $\dfrac{d^4y}{dx^4} + 10\dfrac{d^2y}{dx^2} + 9y = \sin x \sin 2x$

2. Suggest some suitable form of particular integral y_p in method of undetermined co-efficients for the given odes. You need not compute the unknown constants.

(a) $\dfrac{d^2y}{dx^2} + 4y = x^2 \sin 2x + (6x + 8) \cos x \cos 3x$

(b) $\dfrac{d^2y}{dx^2} + 3\dfrac{dy}{dx} + 2y = (x^2 + 1)e^x \sin 2x + 3e^{-x} \cos x + 4e^x$

(c) $\dfrac{d^4y}{dx^4} + 3\dfrac{d^2y}{dx^2} - 4y = \cos^2 x - \sinh x$

3. Solve the following initial-value problems

(a) $\dfrac{d^2y}{dx^2} + 7\dfrac{dy}{dx} + 10y = 4xe^{-3x}$, $y(0) = 0$ and $y'(0) = -1$

(b) $\dfrac{d^2y}{dx^2} + y = 3x^2 - 5x - 4\sin x$, $y(0) = 0$ and $y'(0) = 1$

(c) $\dfrac{d^3y}{dx^3} - 6\dfrac{d^2y}{dx^2} + 9\dfrac{dy}{dx} - 4y = 8x^2 + 3 - 6e^{2x}$,

$$y(0) = 1, \ y'(0) = 7 \text{ and } y''(0) = 10$$

4. Solve by method of undetermined co-efficients:

(a) $(x^2 - 1)\dfrac{d^2y}{dx^2} - 2x\dfrac{dy}{dx} + 2y = (x^2 - 1)^2$

(b) $x\dfrac{d^2y}{dx^2} + (x - 2)\dfrac{dy}{dx} - 2y = x^3$

(c) $x^3\dfrac{d^3y}{dx^3} - 4x^2\dfrac{d^2y}{dx^2} + 8x\dfrac{dy}{dx} - 8y = 4\ln x$

Chapter 6

Second Order Linear Ode: Solution Techniques & Qualitative Analysis

6.1 Introduction

In this chapter we shall primarily concentrate on the second order linear differential equations, especially their typical solution techniques. Although in most cases we fail to solve differential equations of order higher than the first in finite form, for second order equations we may avail some extra methods of attack.

In course of the discussion on second order linear differential equation one may be very curious to know why people bother so much with the second order differential equation even when first order linear differential equations can be availed. We may not satisfy him fully but hope to meet some his queries. In course of solving a higher order linear differential equation with real constant co-efficients, solving the auxiliary equation is an intermediate step. In this context, we make use of the fact that only non-constant irreducible polynomials over $\mathbb{R}$ are either of degree 1 or of degree 2. Therefore, at least theoretically, under all circumstances, every higher degree polynomial with real co-efficients has quadratic factors. With the intervention of computers, this theoretical possibility is materialised in practice by means of Bairstow's method.[1] Bairstow's method is elegant from the viewpoint that it effectively reduces higher order differential equations to a set of second order differential equations with real constant co-efficients whose solution techniques and qualitative aspects are very rich. Reduction of higher order differential equations to a set of second order differential equations by the above factorization method is also interesting from the viewpoint of physical problems which are mostly governed by second order differential equations.

The present chapter basically consists of three parts. The first part dealing with details of the solution techniques and qualitative aspects

---
[1]see Appendix

of the second order differential equations. The second part focussing on Sturm-Liouville type boundary value problem in course of whose discussions, follows the novel concepts of eigenvalues and eigenfunctions and their characteristic features. The third and final part gives a brief account of Green's function approach to Initial Value Problems (IVP) and Boundary Value Problems (BVP)

6.2   Reduction of Order Method (D'Alembert's Method)

'Reduction of order' is the first useful method that one may apply to solve a second order linear differential equation, no matter whether it is homogeneous or inhomogeneous. A second order linear ode has the general form

$$\frac{d^2y}{dx^2} + p_1(x)\frac{dy}{dx} + p_2(x)y = q(x) \qquad (6.1)$$

where $p_1(x)$, $p_2(x)$, $q(x)$ are continuous functions over some interval I. As is known beforehand, once we determine the complementary function, i.e, the general solution of the associated homogeneous equation

$$\frac{d^2y}{dx^2} + p_1(x)\frac{dy}{dx} + p_2(x)y = 0 \qquad (6.2)$$

we are half way through since there exist several elegant methods to compute the particular integral. To have an idea of how equation (6.2) may be solved, let's try the substitution $y(x) = f(x)v(x) - f(x)$ and $v(x)$ being elements of $C^2(I)$. Making use of Leibnitz's product rule of differentiation, we have from equation (6.2):

$$f(x)\frac{d^2v}{dx^2} + \left(2\frac{df}{dx} + p_1(x)f(x)\right)\frac{dv}{dx} + \left(\frac{d^2f}{dx^2} + p_1(x)\frac{df}{dx} + p_2(x)f\right)v = 0$$
$$(6.3)$$

If $f(x)$ were chosen to be a non-trivial solution of the equation (6.2) itself and $v(x)$ be a non-constant function, then (6.3) would reduce to

$$f(x)\frac{d^2v}{dx^2} + \left(2\frac{df}{dx} + p_1(x)f(x)\right)\frac{dv}{dx} = 0 \qquad (6.3a)$$

Substituting $\frac{dv}{dx}$ by $u(x)$ we can reduce (6.3a) to a first order ode that is easily cast in the following separable form:

$$\frac{du}{u} + \left(2\frac{f'(x)}{f(x)} + p_1(x)\right)dx = 0 \qquad (6.4)$$

Integrating (6.4) we get:

$$\ln u + \int \left(2\frac{f'(x)}{f(x)} + p_1(x) \right) dx = C_1,$$

C_1 being the constant of integration.

$$\therefore \ u(x) = C.\frac{\exp\left[-\int p_1(x)dx\right]}{(f(x))^2}$$

$$\therefore \ g(x) \equiv f(x)\, v(x) = C.\, f(x) \int \frac{\exp\left[\int p_1(x)dx\right] dx}{(f(x))^2}$$

is another solution of (6.2). Because of homogeneity of the equation (6.2), we may leave out the constant C with impunity. However we are to show that $f(x)$ and $g(x)$ are indeed linearly independent. Our surmise is not belied as their Wronskian

$$W\left(f(x), g(x)\right) = \begin{vmatrix} f(x) & g(x) \\ f'(x) & g'(x) \end{vmatrix} = (f(x))^2 v'(x) \neq 0$$

($\because$ $v(x)$ is a non-constant function over I). Hence $f(x)$ and $g(x)$ span the solution space (6.2). Note that $W(f(x), g(x))$ would be zero had $v(x)$ been a constant function over I and our goal would not be satisfied.

Remark (a):   The method transforms the associated homogeneous equation (6.2) to one in which free term of the dependent variable is missing. Effectively this feature is the root of 'order-reduction'.

(b): This method is very fruitful once we have the apriori knowledge of just one solution $f(x)$ of equation (6.2). Otherwise it is of no help. In this perspective, this method is overdependent on the so called 'method of inspection' which we shall describe in the next article.

Example (1) : If $y = x^2$ be a solution of the ode

$$(x^3 - x^2)\frac{d^2y}{dx^2} - (x^3 + 2x^2 - 2x)\frac{dy}{dx} + (2x^2 + 2x - 2)y = 0,$$

find the general solution by reduction of order method.

As a first step to attack the problem we substitute $y = v(x)x^2$ where $v(x)$ is a twice differentiable function over an interval I not containing 0 and 1. By Leibnitz's rule of successive differentiation, we have

$$\frac{dy}{dx} = x^2\frac{dv}{dx} + 2vx$$

$$\frac{d^2y}{dx^2} = x^2\frac{d^2v}{dx^2} + 4x\frac{dv}{dx} + 2v$$

and hence the given ode reduces to

$$x^2(x-1)\left(x^2\frac{d^2v}{dx^2} + 4x\frac{dv}{dx} + 2v\right) - x^2(x^2+2x-2)\left(x\frac{dv}{dx} + 2v\right)$$

$$+ x^2(2x^2 + 2x - 2)v = 0$$

or, $\quad x^2(x-1)\dfrac{d^2v}{dx^2} + (4x^2 - 4x - x^3 - 2x^2 + 2x)\dfrac{dv}{dx}$

$$+ 2(x - 1 - x^2 - 2x + 2 + x^2 + x - 1)v = 0$$

or, $\quad x(x-1)\dfrac{d^2v}{dx^2} + (2x - 2 - x^2)\dfrac{dv}{dx} = 0$

(observe that co-efficient of v is zero)

If we put $\frac{dv}{dx} = u$, the above equation further reduces to

$$x(x-1)\frac{du}{dx} + (2x - 2 - x^2)u = 0,$$

which is a first order linear differential equation in x and u. In the separable form this reads

$$\frac{du}{u} + \left(\frac{2}{x} - 1 - \frac{1}{x-1}\right)dx = 0.$$

Integrating both sides and ignoring the constant of integration we have

$$u = \left(\frac{1}{x} - \frac{1}{x^2}\right)e^x, \text{ so that } v(x) = \int u\,dx = \frac{1}{x}e^x$$

Substituting v back into $y = vx^2$, we get $y = xe^x$ which is another solution of the given ode. We make a check up whether the solutions x^2 and xe^x are linearly independent or not. The Wronskian

$$W(x^2,\ xe^x) = \begin{vmatrix} x^2 & xe^x \\ 2x & (x+1)e^x \end{vmatrix} = x^2(x-1)e^x \neq 0$$

over any interval I not containing 0 and 1. Infact, the restriction we imposed on the choice of I at the very beginning of our workout was just a foresight of this inevitable predicament. Later on we shall see that both these points, viz, $x = 0$ and $x = 1$ are regular singular points of the given ode.

The general solution of the given ode will be the linear combination $Ax^2 + Bxe^x$ of the above two linearly independent solutions, A and B being arbitrary constants.

Remark (a) Do you think I am right in talking of two constants A and B here? If you are nostalgic of topological flavour in your workout, you should argue that choice of A and B must be different across the regular singular points $x = 0$ and $x = 1$.

(b): In this example, the ode does not appear in the normal form

$$\frac{d^2 y}{dx^2} + p_1(x)\frac{dy}{dx} + p_2(x)y = 0$$

but in the more general form

$$a_0(x)\frac{d^2 y}{dx^2} + a_1(x)\frac{dy}{dx} + a_2(x)y = 0.$$

In this case $u(x)$ would assume the form

$$u(x) = \frac{\exp\left[-\int \frac{a_1(x)}{a_0(x)}dx\right]}{(f(x))^2}$$

and accordingly the second linearly independent solution will be

$$g(x) = cf(x)\int \frac{\exp\left[-\int \frac{a_1(x)}{a_0(x)}dx\right]}{(f(x))^2}dx$$

Example (2) : If $y = \sin x$ be a solution of the differential equation

$$\sin^2 x\frac{d^2 y}{dx^2} - \sin x \cos x\frac{dy}{dx} + y = 0,$$

then solve the equation $\quad \sin^2 x\frac{d^2 y}{dx^2} - \sin x \cos x\frac{dy}{dx} + y = -\sin^3 x$ by method of 'reduction of order'.

Our first job is to find the second solution of the given homogeneous equation by reducing the order of equation. Substitute $y = v(x)\sin x$ in the homogeneous equation to have

$$\sin^3 x\frac{d^2 v}{dx^2} + \sin^2 x \cos x\frac{dv}{dx} = 0,$$

which transforms to the first order ode

$$\sin^2 x\left(\frac{du}{dx}\sin x + u\cos x\right) = 0$$

once we substitute u for $\frac{dv}{dx}$. Herefrom we could proceed in the traditional way but we prefer to save time for us and directly take up the main inhomogeneous differential equation. In the present formulation, this inhomogeneous ode reads

$$\sin^2 x \left(\frac{du}{dx} \sin x + u \cos x \right) = -\sin^3 x,$$

so that for points $x \neq n\pi$ (n being integer), we have a first order exact linear ode, viz,

$$\frac{du}{dx} \sin x + u \cos x = -\sin x$$

$$\text{i.e., } \quad \frac{d}{dx}(u \sin x) = -\sin x$$

Integrating both sides w.r. to x we get

$$u(x) = \cot x + c_2 \operatorname{cosec} x,$$

wherefrom follows that

$$v(x) = \int u dx \;\; = \;\; \ln | \sin x | + c_2 \ln | \operatorname{cosec} x - \cot x | + c_1$$

$\therefore$ The general solution of the given ode is:

$$\begin{aligned} y(x) \;\; &= \;\; v(x) \sin x \\ &= \;\; c_1 \sin x + \{ \ln | \sin x (\operatorname{cosec} x - \cot x)^{c_2} | \} \cdot \sin x \end{aligned}$$

where c_1 and c_2 are two arbitrary constants of integration.

An important result: In our earlier discussion (c.f Chapter 2 and Chapter 5) we proved a very important result that every Riccati equation can be transformed to a homogeneous second order linear equation and conversely, every second order homogeneous linear differential equation can be transformed to a Riccati equation. In mathematical terminology, there exists a one-to-one correspondence between them. Consider the following Riccati equation

$$\frac{dy}{dx} = A(x)y^2 + B(x)y + C(x) \tag{6.5}$$

On applying the transformation $y(x) = -\frac{u'(x)}{u(x)A(x)}$, the above equation (6.5) reduces to the following homogeneous linear ode in u:

$$A(x)u''(x) - (A'(x) + A(x)B(x))u'(x) + A^2(x)C(x)u(x) = 0 \tag{6.6}$$

Suppose $u_1(x)$ is known to be a solution of equation (6.6). If we write $-\frac{u_1'}{u_1 A}$ as y_1, then y_1 is a solution of equation (6.5). Following the footsteps of what has been done in the 'reduction of order method' we make the substitution $u(x) = u_1(x)v(x)$ into (6.6) to have

$$v = \int \frac{\exp\left[\int \left(\frac{A'+AB}{A}\right) dx\right]}{u_1^2} dx = \int \frac{A}{u_1^2} \exp\left[\int B dx\right] dx$$

$$\therefore \quad u = u_1 v = u_1 \int \frac{A}{u_1^2} \exp\left[\int B dx\right] dx$$

$$\text{or,} \quad \exp\left[-\int A y dx\right] = u_1 \int \frac{A}{u_1^2} \exp\left[\int B dx\right] dx$$

By virtue of logarithmic differentiation of this relation we have

$$-Ay = \frac{u_1'}{u_1} + \frac{\frac{A}{u_1^2}\exp\left[\int B dx\right]}{\int \frac{A}{u_1^2}\exp\left[\int B dx\right] dx}$$

$$\text{or,} \quad y = -\frac{u_1'}{u_1 A} - \frac{\frac{1}{u_1^2}\exp\left[\int B dx\right]}{\int \frac{A}{u_1^2}\exp\left[\int B dx\right] dx}$$

$$\text{or,} \quad y = y_1 + \frac{1}{\omega}$$

where we agree to define ω by

$$\omega = \frac{-\int \frac{A}{u_1^2}\exp\left[\int B dx\right] dx}{\frac{1}{u_1^2}\exp\left[\int B dx\right]} \tag{6.7}$$

The relation (6.7) can be written as

$$\frac{\omega}{u_1^2}\exp\left[\int B dx\right] + \int \frac{A}{u_1^2}\exp\left[\int B dx\right] dx = 0 \tag{6.7a}$$

Differentiating (6.7a) w.r. to x and simplifying a bit we get

$$\frac{d\omega}{dx} + \left(B - \frac{2u_1'}{u_1}\right)\omega + A = 0$$

$$\text{or,} \quad \frac{d\omega}{dx} + (2Ay_1 + B)\omega + A = 0 \tag{6.8}$$

where in the last step we used definition of y_1. The linear equation (6.8) is nothing new. However, this strenuous workout highlights the link between the transformation applied to the homogeneous ode in the 'reduction of order method' and the transformation applied to the non-linear Riccati equation in the 'reduction of degree method'.

6.3 Method of Inspection for finding one Integral

This method is rather tricky but useful in finding a particular solution of a homogeneous linear second order equation of the form

$$a_0(x)\frac{d^2y}{dx^2} + a_1(x)\frac{dy}{dx} + a_2(x)y = 0 \qquad (6.9)$$

so that the 'reduction of order method' discussed in the previous article gets a nod. The conventional idea is to reduce the given differential equation to a polynomial equation in x effectively. To achieve this goal we substitute simple expressions like $y = e^{\lambda x}$, $y = x^\lambda$ in the given differential equation to be solved and then try to determine the arbitrary constant λ so that the equation becomes satisfied.

If you hurl the question, 'under what conditions can equation (6.9) have a solution of the generic form $e^{\lambda x}$?', I shall pick the cue of my answer from the equation

$$a_0(x)\lambda^2 + a_1(x)\lambda + a_2(x) = 0, \qquad (6.10)$$

obtained by substituting $y = e^{\lambda x}$ into the given differential equation (6.9) itself. Assuming that each of $a_0(x)$, $a_1(x)$, $a_2(x)$ admits of Taylor series expansion in a neighborhood of $x = 0$, we have from (6.10),

$$[a_0(0)\lambda^2 + a_1(0)\lambda + a_2(0)] + x[a_0'(0)\lambda^2 + a_1'(0)\lambda + a_2'(0)]$$
$$+ \frac{x^2}{2!}[a_0''(0)\lambda^2 + a_1''(0)\lambda + a_2''(0)] + O(x^3) = 0$$

Therefore, we have the following system of relations:

$$\left.\begin{array}{l} a_0(0)\lambda^2 + a_1(0)\lambda + a_2(0) = 0 \\ a_0'(0)\lambda^2 + a_1'(0)\lambda + a_2'(0) = 0 \\ a_0''(0)\lambda^2 + a_1''(0)\lambda + a_2''(0) = 0 \\ \cdots\cdots\cdots\cdots\cdots\cdots\cdots\cdots\cdots \\ \cdots\cdots\cdots\cdots\cdots\cdots\cdots\cdots\cdots \end{array}\right\} \qquad (6.11)$$

This system (6.11) should be consistent as they hail from one and the same equation (6.10). One possible way of making (6.11) consistent is to suppose that the first two relations in it are quadratic equations in λ having a common root while the rest are all identities. This ensures that $a_0(x)$, $a_1(x)$, $a_2(x)$ are to be polynomials of degree at most one and moreover they have to satisfy the condition

$$\left(a_2(0)a_0'(0) - a_0(0)a_2'(0)\right)^2$$
$$= \left(a_1(0)a_2'(0) - a_2(0)a_1'(0)\right)\left(a_0(0)a_1'(0) - a_1(0)a_0'(0)\right) \quad (6.12)$$

Here the value of λ for which $y = e^{\lambda x}$ is a solution of (6.9) is given by the following:

$$\lambda = \frac{a_2(0)a_0'(0) - a_0(0)a_2'(0)}{a_0(0)a_1'(0) - a_1(0)a_0'(0)} \quad (6.13)$$

Example (3) : Show by the above method that $y = e^{2x}$ is a solution of the second order ode

$$(2x+1)\frac{d^2y}{dx^2} - 4(x+1)\frac{dy}{dx} + 4y = 0$$

Comparing our given equation with the equation (6.9), we have $a_0(x) = 2x + 1$, $a_1(x) = -4x - 4$ and $a_2(x) = 4$. Little effort shows that we have the condition (6.12) satisfied and hence from (6.13) we have the value of λ:

$$\lambda = \frac{4.2 - 1.0}{1.(-4) - (-4)(2)} = 2$$

Hence the conclusion.

Example (4) : Check whether the equation

$$(x+1)\frac{d^2y}{dx^2} + (4x+5)\frac{dy}{dx} + (8x+6)y = 0$$

admits of a solution of the form $y = e^{\lambda x}$.

Comparing our given equation with the equation (6.9), we have $a_0(x) = (x + 1)$; $a_1(x) = 4x + 5$ and $a_2(x) = 8x + 6$. However, in this case the condition (6.12) is not satisfied- indicating that there can not exist a solution of the equation in the form $e^{\lambda x}$.

In case $a_0(x)$, $a_1(x)$, $a_2(x)$ are all polynomials at least one of which is of degree greater than unity, we seek the solution of differential equation

(6.9) in the form $y = x^\lambda$. Direct substitution of $y = x^\lambda$ into the equation (6.9) yields the polynomial equation

$$\lambda(\lambda - 1)a_0(x) + \lambda x a_1(x) + x^2 a_2(x) = 0 \qquad (6.14)$$

which can be arranged in the form

$$A_0(\lambda) + A_1(\lambda)x + A_2(\lambda)x^2 + A_3(\lambda)x^3 + \cdots\cdots = 0, \qquad (6.14a)$$

where $A_k(\lambda)$'s are themselves polynomials (of degree atmost 2) in λ. In case all these polynomials have common zero(s), $y = x^\lambda$ is a solution of (6.9). Otherwise, a solution of the generic form $y = x^\lambda$ does not exist. As is evident from (6.14a), this process is less laborious if the polynomials $a_0(x)$, $a_1(x)$, $a_2(x)$ are of low-degree.

Example (5) : Show that $y = x$ is a solution of the second order ode

$$(x^2 - x + 1)\frac{d^2y}{dx^2} - (x^2 + x)\frac{dy}{dx} + (x + 1)y = 0.$$

Here $a_0(x) = x^2 - x + 1$; $a_1(x) = -(x^2 + x)$; $a_2(x) = x + 1$. Since two of these three polynomials are quadratic, we seek solution of the given ode in the form $y = x^\lambda$. Direct substitution of $y = x^\lambda$ into the given equation yields the polynomial equation

$$\lambda(\lambda - 1)(x^2 - x + 1) - \lambda x(x^2 + x) + x^2(x + 1) = 0,$$

which can be rearranged as

$$\lambda(\lambda - 1) - \lambda(\lambda - 1)x + (\lambda - 1)^2 x^2 - (\lambda - 1)x^3 = 0$$

It is now clear that $\lambda = 1$ is the only zero common to all the polynomials in λ appearing as λ appearing as co-efficients of different integral powers of x in the above-equation. Hence $y = x$ is solution of the given differential equation.

Example (6) : Check that $y = x^2$, claimed to be a solution of the ode given in Example (3) follows from the above method.

Here $a_0(x) = x^3 - x^2$; $a_1(x) = -(x^3 + 2x^2 - 2x)$; $a_2(x) = 2x^2 + 2x - 2$

From (6.14) it follows that

$$\lambda(\lambda - 1)(x^3 - x^2) - \lambda(x^4 + 2x^3 - 2x^2) + (2x^4 + 2x^3 - 2x^2) = 0,$$

which on being arranged in the ascending powers of x gives

$$x^2(\lambda - 1)(2 - \lambda) + x^3(\lambda - 2)(\lambda - 1) + x^4(2 - \lambda) = 0.$$

It is clear from this relation that the only zero common to all the co-efficient polynomials in λ is $\lambda = 2$. Hence the verification of the claim.

Example (7) : Show that the equation

$$3x^2\frac{d^2y}{dx^2} + (2 + 6x - 6x^2)\frac{dy}{dx} - 4y = 0$$

can have no solution of the form $y = x^\lambda$.

We prove the result by method of contradiction. Had $y = x^\lambda$ been a solution to this equation we would have from (6.14), the relation

$$2\lambda x + (3\lambda^2 + 3\lambda - 4)x^2 - \lambda x^3 = 0$$

However this implies that $\lambda = 0$ should satisfy $3\lambda^2 + 3\lambda - 4 = 0$. This being an absurdity, we assert our conclusion that $\exists$ no solution of the generic form $y = x^\lambda$ to the given equation.

6.4 Transformation of Second Order Ode by changing the Independent Variable

In this article we are interested mainly in transformation of a second order ode with variable co-efficients to one having constant co-efficients by a suitable change of the independent variable. Without any loss of generality we assume the second order linear homogeneous ode to have the form (6.2)

$$\frac{d^2y}{dx^2} + p_1(x)\frac{dy}{dx} + p_2(x)y = 0,$$

where $p_1(x)$ and $p_2(x)$ are continuous functions over some interval I. Let us apply the transformation $z = u(x)$, where for the time being there is no information available regarding the relationship between z and x except the fact that u is differentiable at least twice w.r. to x. It's a routine exercise to show that

$$\left.\begin{array}{rcl}\frac{dy}{dx} &=& u'(x)\frac{dy}{dz} \\ \frac{d^2y}{dx^2} &=& u''(x)\frac{dy}{dz} + (u'(x))^2\frac{d^2y}{dz^2}\end{array}\right\} \tag{6.15}$$

We now demand that $u(x)$ will be such that the equation obtained from (6.15) by substitution of (6.16) becomes a second order homogeneous ode with constant co-efficients in independent variable z and dependent variable y. The transformed differential equation

$$\frac{d^2 y}{dz^2} + \frac{\frac{d^2 z}{dx^2} + p_1(x)\frac{dz}{dx}}{\left(\frac{dz}{dx}\right)^2}\cdot\frac{dy}{dz} + \frac{p_2(x)}{\left(\frac{dz}{dx}\right)^2}\, y = 0 \qquad (6.16)$$

therefore must have constant co-efficients.

Obviously, $\dfrac{dz}{dx} \neq 0$, and the expressions $\dfrac{p_2(x)}{\left(\frac{dz}{dx}\right)^2}$ and $\dfrac{\frac{d^2 z}{dx^2} + p_1(x)\frac{dz}{dx}}{\left(\frac{dz}{dx}\right)^2}$

are to be constants.

In case $\dfrac{p_2(x)}{\left(\frac{dz}{dx}\right)^2}$ be a constant, then $\left(\frac{dz}{dx}\right) \propto -\sqrt{|\,p_2(x)\,|}$ is a must.

If one likes the transformation $z = u(x)$ to be real-valued, it is necessary that $p_2(x)$ is either non-negative everywhere or negative everywhere over I. Thus $\frac{dz}{dx} = A\sqrt{p_2(x)}$, assuming that $p_2(x)\forall x \in I$ is never negative or zero, we have the requisite transformation :

$$z = u(x) = A \int^x \sqrt{p_2(t)}dt \qquad (6.17)$$

Hence $\dfrac{\frac{d^2 z}{dx^2} + p_1(x)\frac{dz}{dx}}{\left(\frac{dz}{dx}\right)^2} = \dfrac{1}{2A}\cdot\dfrac{p_2'(x) + 2p_1(x)p_2(x)}{(p_2(x))^{\frac{3}{2}}} = k,$ say

where k should be a constant—showing that if $p_2(x) \neq 0$ and is differentiable over I, it is feasible to convert a second order ode of the form (6.15) to one having constant co-efficients. In the following we workout a few examples on the above result.

Example (8) : Use the above result to show that the second order ode

$$y'' + xy' + e^{-x^2}y = 0, \quad -\infty < x < +\infty$$

can be transformed to one having constant co-efficients.

Here $p_1(x) = x$ and $p_2(x) = e^{-x^2}$. Moreover we observe that

$$\frac{p_2'(x) + 2p_1(x)p_2(x)}{(p_2(x))^{\frac{3}{2}}} = \frac{-2xe^{-x^2} + 2xe^{-x^2}}{(e^{-x^2})^{\frac{3}{2}}} = 0,$$

establishing that the transformation of the given ode to one with constant co-efficients is feasible, the corresponding transformation is given by

$$z = \int_{-\infty}^{x} e^{-\frac{t^2}{2}} dt$$

Direct differentiation and substitution into the original ode fulfils our mission of having the following ode:

$$\frac{d^2y}{dx^2} + y = 0.$$

Example (9) : Solve the ode $(1 - x^2)y'' - xy' - a^2y = 0$ by making use of the result of this article.

If we cast the given equation in the form (1), we have

$$y'' - \frac{x}{(1 - x^2)}y' - \frac{a^2}{(1 - x^2)}y = 0,$$

where $x = \pm 1$ are singular points. Comparing with (6.1), one gets

$$p_1(x) = \frac{-x}{1 - x^2} \quad \text{and} \quad p_2(x) = \frac{-a^2}{(1 - x^2)}.$$

Observe that if $|x| < 1$, then $p_2(x)$ is strictly negative. Hence we choose $I = (-1, 1)$. Moreover,

$$\frac{p_2'(x) + 2p_1(x)p_2(x)}{(p_2(x))^{\frac{3}{2}}} = \frac{\frac{-2xa^2}{(1-x^2)^2} + \frac{2xa^2}{(1-x^2)^2}}{\frac{-a^3}{(1-x^2)^{\frac{3}{2}}}} = 0,$$

ensuring that the given ode can be transformed to an ode having constant co-efficients, the relevant transformation being given by

$$z = \int^{x} \left| \frac{-a^2}{1 - x^2} \right|^{\frac{1}{2}} dx = 2a \int_{0}^{x} \frac{dx}{\sqrt{1 - x^2}} = 2a \sin^{-1} x$$

The transformed ode becomes

$$\frac{d^2y}{dz^2} + \frac{1}{4}y = 0,$$

whose general solution is

$$y = Ae^{\frac{1}{2}z} + Be^{-\frac{1}{2}z} = Ae^{a\sin^{-1} x} + Be^{-a\sin^{-1} x},$$

A and B being arbitrary constants.

Remark: $\dfrac{p_2'(x) + 2p_1(x)p_2(x)}{(p_2(x))^{\frac{3}{2}}} = $ constant   is a necessary as well as

sufficient condition for (6.15) to be subject to the desired transformation.

Example (10) :   Solve $2x\dfrac{d^2y}{dx^2} + (5x^2 - 2)\dfrac{dy}{dx} + 2x^3y = 0$

We may rewrite the above ode in the form:

$$\frac{d^2y}{dx^2} + \left(\frac{5x^2 - 2}{2x}\right)\frac{dy}{dx} + x^2y = 0,$$

so that on comparison with (6.15) we have:

$$p_1(x) = \frac{5x^2 - 2}{2x} \quad \text{and} \quad p_2(x) = x^2$$

Here we use the transformation $z = u(x)$ so that the transformed ode becomes a linear homogeneous second order ode with constant coefficients having z as new independent variable.

We demand $\dfrac{p_2'(x) + 2p_1(x)p_2(x)}{(p_2(x))^{3/2}}$ and $\dfrac{p_2(x)}{\left(\frac{dz}{dx}\right)^2}$ to be constants.

We observe that the value of the former is 5.

The requisite transformation will be determined from the fact

$$\frac{p_2(x)}{\left(\frac{dz}{dx}\right)^2} = \frac{x^2}{\left(\frac{dz}{dx}\right)^2} = \text{constant} = \frac{1}{k^2}, \text{ say.}$$

$$\therefore \quad \left(\frac{dz}{dx}\right)^2 = k^2x^2$$

$$\text{i, e,} \quad \frac{dz}{dx} = \pm \, kx$$

$$\text{i, e,} \quad z = \pm \, k\frac{x^2}{2} + \text{constant.}$$

We may choose this additive integration constant to be zero and $k = 2$ so that $z = x^2$ is the requisite transformation.

Now $z = x^2 \Rightarrow \frac{dz}{dx} = 2x$ and $\frac{d^2z}{dx^2} = 2$ so that transformed ode becomes:

$$\frac{d^2y}{dz^2} + \frac{5}{4}\frac{dy}{dz} + \frac{1}{4}y = 0,$$

whose auxiliary equation has roots -1 and $-\frac{1}{4}$.

Thus the general solution of the transformed ode is:

$$y = C_1 e^{-z} + C_2.e^{-z/4}$$

C_1 and C_2 being arbitrary constants.

Reverting to xy-form, we have the general solution as:

$$y = C_1 e^{-x^2} + C_2 e^{-x^2/4}$$

Note: (a): Because the ode is homogeneous (i,e, R.H.S $= 0$), dropping $-$ve option in $z = \pm\frac{kx^2}{2}+$ constant does not make a difference.

(b): A statutory example of an ode where this condition is satisfied is the Euler's second order equidimensional equation having the form

$$a_0 x^2 \frac{d^2y}{dx^2} + a_1 x \frac{dy}{dx} + a_2 y = 0 \quad (a_0,\ a_1 \text{ and } a_2 \text{ are constants}). \quad (6.18)$$

Here the Cauchy-Euler type second order ode can be written as:

$$\frac{d^2y}{dx^2} + \left(\frac{a_1}{a_0}\right).\frac{1}{x}\frac{dy}{dx} + \left(\frac{a_2}{a_0}\right).\frac{1}{x^2}y = 0,$$

$$\therefore \quad p_1(x) = \left(\frac{a_1}{a_0}\right).\frac{1}{x}; \quad p_2(x) = \left(\frac{a_2}{a_0}\right).\frac{1}{x^2}$$

We observe that as a_0, a_2 cannot be zero,

$$\frac{\frac{dp_2}{dx} + 2p_1(x).p_2(x)}{(p_2(x))^{3/2}} = \frac{\frac{-2a_2}{a_0}.\frac{1}{x^3} + 2.\left(\frac{a_1}{a_0}\right)\left(\frac{a_2}{a_0}\right).\frac{1}{x^3}}{\left(\frac{a_2}{a_0}\right)^{3/2}.\frac{1}{x^3}} = \frac{2(a_1 - a_0)}{\sqrt{a_2 a_0}}$$

Thus we are ensured that the transformation of the given ode to a new ode having constant co-efficients is feasible — the underlying transformation being given by the demand that $\frac{p_2(x)}{\left(\frac{dz}{dx}\right)^2}$ is a constant.

Hence $\quad \dfrac{\left(\frac{a_2}{a_0}\right).\frac{1}{x^2}}{\left(\frac{dz}{dx}\right)^2} = \left(\frac{a_2}{a_0}\right).a^2$, so that $\left(\frac{dz}{dx}\right)^2 = \dfrac{1}{a^2 x^2}$

$$\therefore \quad \frac{dz}{dx} = \frac{1}{ax} \text{ [without loss of generality we have taken +ve sign]}$$

Integrating w.r to x, we get: $x = Ae^{az}$; A being the constant of integration. To avoid needless complexity we choose $A = 1$ so that the working transformation becomes $x = e^{az}$.

We may now check for ourselves how $x = e^{az}$ transforms the Cauchy Euler type ode (6.18)

$$\text{Since} \quad x = e^{az}, \quad \frac{dx}{dz} = ae^{az} = ax$$

$$\therefore \quad \frac{dy}{dx} = \frac{dy}{dz}.\frac{dz}{dx} = \frac{1}{ax}.\frac{dy}{dz} \quad \text{so that} \quad ax\frac{dy}{dx} = \frac{dy}{dz}$$

Now vide chain rule of differentiation we have:

$$\frac{d}{dx}\left(ax\frac{dy}{dx}\right) = \frac{d}{dx}\left(\frac{dy}{dz}\right) = \frac{1}{ax}.\frac{d^2y}{dz^2}$$

$$\text{i, e,} \quad x^2\frac{d^2y}{dx^2} = \frac{1}{a^2}\left(\frac{d^2y}{dz^2} - a\frac{dy}{dz}\right) = \frac{1}{a^2}\frac{d^2y}{dz^2} - \frac{1}{a}.\frac{dy}{dz}$$

The given ode (6.18) becomes:

$$\left(\frac{a_0}{a^2}\right)\frac{d^2y}{dz^2} + \left(\frac{a_1 - a_0}{a}\right)\frac{dy}{dz} + a_2 y = 0 \tag{6.19}$$

(6.19) is clearly a second order linear homogeneous ode with constant co-efficients. From all the above we conclude:

The transformation $x = e^{az}$ $(a \neq 0)$ reduces a second order Cauchy-Euler type ode to a linear ode with constant co-efficients, and conversely, if we want to transform a Cauchy-Euler type ode to a linear differential equation with constant co-efficients by change of independent variable, then $x = e^{az}(a \neq 0)$ should serve as the requisite transformation. In most of the texts available on ODE, it is customary to use the transformation $x = e^z$ only to the Cauchy-Euler odes for the same purpose — however, we have established above that a more generalised form of transformation could have been applied.

On using the trial solution $y = e^{mz}$, we find the auxiliary equation of the ode (6.19) as:

$$\left(\frac{a_0}{a^2}\right)m^2 + \left(\frac{a_1 - a_0}{a}\right)m + a_2 = 0,$$

the roots being $\dfrac{(a_0 - a_1) \pm \sqrt{(a_0 - a_1)^2 - 4a_0 a_2}}{2 \cdot \left(\frac{a_0}{a_1}\right)}$.

Let's illustrate the above theory through a couple of examples.

Example (11) : Solve the ode $x^2 \dfrac{d^2 y}{dx^2} - x \dfrac{dy}{dx} + y = 0$

If we apply the transformation $x = e^{az}$, we get the new ode as:

$$\frac{1}{a^2} \cdot \frac{d^2 y}{dz^2} - \frac{2}{a} \frac{dy}{dz} + y = 0 \ (\because a_0 = 1; \ a_1 = -1; \ a_2 = 1)$$

i, e, $\quad \dfrac{d^2 y}{dz^2} - 2a \dfrac{dy}{dz} + a^2 y = 0,$

the auxiliary equation of which being $m^2 - 2am + a^2 = 0$, i,e, $(m-a)^2 = 0$, obtained by employing the trial solution $y = e^{mz}$. Thus $m = a, a$ and so the general solution of the ode given is:

$$y = (C_1' + C_2' z) e^{az} = (C_1' + C_2' \cdot \frac{1}{a} \ln x) x = C_1 x + C_2 \cdot x \ln x,$$

where we agree to write $C_1 = C_1'$ and $C_2 = \dfrac{C_2'}{a}$.

Example (12) : Solve the ode $x^2 \dfrac{d^2 y}{dx^2} - 2x \dfrac{dy}{dx} + 2y = x^3$

Here also we apply the transformation $x = e^{az}$; $a \neq 0$ to the ode so that it reduces to a linear having constant co-efficients. Here $a_0 = 1$; $a_1 = -2$; $a_2 = 2$ and so the transformed ode will be:

$$\left(\frac{1}{a^2}\right) \frac{d^2 y}{dz^2} - \frac{3}{a} \frac{dy}{dz} + 2y = e^{3az} \text{ (see equation (6.19))}$$

i, e, $\quad \dfrac{d^2 y}{dz^2} - 3a \dfrac{dy}{dz} + 2a^2 y = a^2 e^{3az} \ (a \neq 0)$

The auxiliary equation of the corresponding reduced ode is:

$$(m^2 - 3am + 2a^2) = 0,$$

obtained by use of trial solution $y = e^{mz}$.

The roots of auxiliary equation being $m = a, 2a$, the c.f. of the above ode is:

$$y_c = (C_1' e^{az} + C_2' e^{2az}) = (C_1' x + C_2' x^2)$$

y_p can be computed via the operational methods developed earlier :

$$y_p = \left(\frac{1}{D^2 - 3aD + 2a^2}\right)\left(a^2 e^{3az}\right), \text{ where } D \equiv \frac{d}{dz}$$

$$= \frac{1}{(D-a)(D-2a)}\left(a^2 e^{3az}\right)$$

$$= \frac{a^2 e^{3az}}{(3a-a)(3a-2a)} = \frac{1}{2}e^{3az} = \frac{1}{2}x^3$$

Hence the general solution of the given ode is:

$$y = y_c + y_p = \left(C_1' x + C_2' x^2 + \frac{1}{2}x^3\right)$$

We now digress a little and concentrate for the time being on the artifices of handling Cauchy-Euler type odes.

Let's introduce a new operator $\theta \equiv \left(x\frac{d}{dx}\right)$ so that $\theta y = x\frac{dy}{dx}$

$$\therefore \ \theta^2 y = \theta(\theta y) = x\frac{d}{dx}\left(x\frac{dy}{dx}\right) = x\left[\frac{dy}{dx} + x\frac{d^2 y}{dx^2}\right]$$

$$= \theta y + x^2\frac{d^2 y}{dx^2}$$

$$\text{i, e,} \quad x^2\frac{d^2 y}{dx^2} = \theta(\theta - 1)y$$

With this inception of operator technique, we observe that the second order Cauchy-Euler (6.18) ode becomes:

$$\{a_0\,\theta^2 + (a_1 - a_0)\,\theta + a_2\}y = 0 \qquad (6.20)$$

which is in letter and spirit same as the ode (6.19) with the exception that no term involves a. Thus to put up a concordance between this operator method and the method of transformation of a second order ode by changing the independent variable, we may opt for $a = 1$.

Again we may look into (6.20) as a second order linear homogeneous ode with constant co-efficients provided we treat $\theta = x\frac{d}{dx}$ as similar to $D \equiv \frac{d}{dx}$ in usual case. The rest of the matter is just a repetition of the treatment we handed out to the case of odes having constant co-efficients. Further we observe that $\theta \equiv x\frac{d}{dx} = \frac{d}{dz}$.

Example (13) : Solve the ode : $x^2\frac{d^2 y}{dx^2} + 3x\frac{dy}{dx} + y = \frac{1}{(1-x)^2}$

Use $\theta \equiv x\frac{d}{dx}$ so as to reduce the corresponding linear homogeneous ode in the form

$$0 = \{\theta(\theta - 1) + 3\theta + 1\}y = (\theta + 1)^2 y,$$

the solution being

$$y_c = (a + bz)e^{-z} = \frac{(a + b\ln x)}{x},$$

a and b being arbitrary constants of integration. The particular integral will be y_p, where

$$y_p = \frac{1}{(\theta + 1)^2}(1 - x)^{-2} = \frac{1}{(\theta + 1)}\left[\frac{1}{(\theta + 1)}(1 - x)^{-2}\right]$$

We now write $\dfrac{1}{(\theta + 1)}(1 - x)^{-2}$ as u so that $y_p = \dfrac{1}{(\theta + 1)}(u).$

Again from $\quad u = \dfrac{1}{(\theta + 1)}(1 - x)^{-2}\quad$ it follows that

$$(\theta + 1)u = (1 - x)^{-2}$$

i,e, $\quad \dfrac{d}{dx}(xu) = (1 - x)^{-2}, \quad \left[\because \theta \equiv x\dfrac{d}{dx}\right]$

which on integration w.r. to x yields : $\quad xu = \displaystyle\int \frac{dx}{(1 - x)^2} = \frac{1}{x(1 - x)},$

where integration constant has been deliberately omitted as it would correspond to some term present in the complementary function y_c.

Hence $\quad y_p = \dfrac{1}{(\theta + 1)}\left(\dfrac{1}{x(1 - x)}\right).$

Operating $(\theta + 1)$ on both sides we get

$$\frac{1}{x(1 - x)} = (\theta + 1)y_p = \frac{d}{dx}(xy_p)$$

Hence $\quad xy_p = \displaystyle\int \frac{dx}{x(1 - x)} = \ln x - \ln(1 - x),$

the constant of integration being dropped on the basis of the reason stated earlier.

$$\therefore \; y_p = \frac{1}{x}\ln\left(\frac{x}{1 - x}\right)$$

$\therefore$ General solution is : $y = y_c + y_p = \left(\dfrac{a + b\ln x}{x}\right) + \dfrac{1}{x}\ln\left(\dfrac{x}{1-x}\right).$

Example (14) : Solve $x^2\dfrac{d^2y}{dx^2} - x\dfrac{dy}{dx} - 3y = x^2\ln x$

We use the operator $\theta \equiv x\frac{d}{dx}$ so that the above ode becomes:

$$\{\theta(\theta - 1) - \theta - 3\}y = x^2\ln x$$
$$\text{or,} \quad (\theta^2 - 2\theta - 3)y = x^2\ln x$$
$$\text{i, e,} \quad (\theta - 3)(\theta + 1)y = x^2\ln x$$

The complementary function of this ode can be written as:

$$y_c = \frac{C_1}{x} + C_2 x^3$$

while the particular integral is given by

$$
\begin{aligned}
y_p &= \frac{1}{(\theta - 3)(\theta + 1)}\left(x^2\ln x\right) \\
&= \frac{1}{(\theta - 3)(\theta + 1)}\left(ze^{2z}\right) \quad \left(\because x\frac{d}{dx} \equiv \frac{d}{dz} \Rightarrow x = e^z\right) \\
&= e^{2z}.\frac{1}{(\theta + 2 - 3)(\theta + 2 + 1)}(z) \\
&= e^{2z}.\frac{1}{(\theta - 1)(\theta + 3)}(z)
\end{aligned}
$$

We substitute $u \equiv \dfrac{1}{(\theta + 3)}(z)$ so that $y_p = e^{2z}.\dfrac{1}{\theta - 1}(u)$

Therefore, $(\theta + 3)u = z$ i, e, $\dfrac{du}{dz} + 3u = z,$

the general solution of this ode is $u = \frac{1}{3}\left(z - \frac{1}{3}\right)$, where we have dropped the constant of integration as that will be absorbable in the complementary function y_c determined earlier.

$\therefore$ $$y_p = \frac{e^{2z}}{3}.\frac{1}{\theta - 1}\left(z - \frac{1}{3}\right).$$

We now agree to write $\dfrac{1}{\theta - 1}\left(z - \dfrac{1}{3}\right)$ as v so that $(\theta - 1)v = \left(z - \dfrac{1}{3}\right)$

i.e., $\dfrac{dv}{dz} - v = z - \dfrac{1}{3},$

The general solution being given by

$$ve^{-z} = \int e^{-z}\left(z - \frac{1}{3}\right) dz = -\left(z + \frac{2}{3}\right)e^{-z}$$

$$\therefore \quad v = -\frac{1}{3}(3z + 2) \quad \text{(integration constant ignored)}$$

$$\text{Thus} \quad y_p = -\frac{e^{2z}}{9}(3z + 2) = -\frac{1}{9}x^2(3\ln x + 2)$$

Hence the general solution of the given ode is

$$y = y_c + y_p = \frac{C_1}{x} + C_2 x^3 - \frac{1}{9}x^2(3\ln x + 2)$$

Example (15) :   Solve $(1 + x)^2 \dfrac{d^2 y}{dx^2} + (1 + x)\dfrac{dy}{dx} + y = 4\cos\ln(1 + x)$

Here we use the transformation $u = (1 + x)$ so that the ode becomes

$$u^2\frac{d^2 y}{du^2} + u\frac{dy}{du} + y = 4\cos\ln u,$$

which is second order Cauchy-Euler type ode in independent variable u and dependent variable y. If we put $\ln u = z$, then $u = e^z$ so that the transformed ode is:

$$\frac{d^2 y}{dz^2} + y = 4\cos z.$$

Thus the transformation $z = \ln(1 + x)$ has transformed the given ode to a normal form having constant co-efficients.

The complementary function is : $y_c = C_1 \cos z + C_2 \sin z$

The particular integral is given by

$$y_p = \frac{4}{(D^2 + 1)}(\cos z) = 4Re\left[\frac{1}{D^2 + 1}e^{iz}\right]$$

Now $\dfrac{1}{(D^2 + 1)}\left(e^{iz}\right) = \dfrac{1}{(D - i)(D + i)}(e^{iz}) = \dfrac{1}{2i}\cdot\dfrac{1}{(D - i)}\left(e^{iz}\right)$

$$= \frac{1}{2i}\left(ze^{iz}\right) = \frac{1}{2i}(z\cos z + iz\sin z) = \frac{z}{2}\sin z - i\frac{z}{2}\cos z$$

Thus $y_p = 4\left(\dfrac{z}{2}\sin z\right) = 2z\sin z$

So the complete integral of the given ode is:

$$
\begin{aligned}
y &= y_c + y_p \\
&= C_1 \cos z + C_2 \sin z + 2z \sin z \\
&= C_1 \cos(\ln(1+x)) + C_2 \sin(\ln(1+x)) + 2\ln(1+x).\sin(\ln(1+x))
\end{aligned}
$$

Remark: In course of discussion on Cauchy-Euler type odes we observed that the operational factors that are linear functions of $\theta \equiv x\frac{d}{dx}$ commute with each other, although in general, for odes involving variable co-efficients such operational factors involving θ or the likes do not commute with each other as their order of occurrence is the key factor.

This typical behavior of Cauchy-Euler type odes can be explained once we note that the transformation $x = e^{az}$, $a \neq 0$ reduces it to a linear ode having constant co-efficients and for which the commutativity of the operational factors is guarenteed.

For any differentiable function y, observe that the commutator bracket $\left[x^k, \frac{d}{dx}\right] y \neq 0$ as

$$
\begin{aligned}
[x^k, D]y = \left[x^k, \frac{d}{dx}\right] y &= \left(x^k \frac{d}{dx} - \frac{d}{dx}.x^k\right) y = x^k \frac{dy}{dx} - \frac{d}{dx}(x^k y) \\
&= x^k.\frac{dy}{dx} - x^k.\frac{dy}{dx} - kx^{k-1}y = -kx^{k-1}y
\end{aligned}
$$

We now consider the solution of the following ode to illustrate this point.

$$
3x^2 \frac{d^2 y}{dx^2} + (2 - 6x^2)\frac{dy}{dx} - 4y = 0
$$

If we write $\frac{d}{dx}$ as D, the above ode becomes:

$$
\begin{aligned}
[3x^2 D^2 + (2 - 6x^2)D - 4]y &= 0 \\
\text{i, e,} \quad [(3x^2 D^2 + 2D) - (6x^2 D + 4)]y &= 0 \\
\text{i, e,} \quad [(3x^2 D + 2)D - 2(3x^2 D + 2)]y &= 0 \\
\text{i, e,} \quad (3x^2 D + 2)(D - 2)y &= 0
\end{aligned}
$$

To solve the ode, if we write $(D - 2)y = v$, then general solution is obtainable once we solve $(3x^2 D + 2)v = 0$. However, the solution of $(3x^2 D + 2)v = 0$ is: $v = Ae^{\frac{2}{3x}}$, A being an arbitrary constant.

Therefore, $(D - 2)y = Ae^{\frac{2}{3x}}$, which being a first order linear ode in y has the general solution given by

$$
y = Ae^{2x} \int e^{\left(-2x + \frac{2}{3x}\right)} dx + Be^{2x}
$$

where B is another constant of integration.

Note that, if we had reverted the order of the operational factors, we would have got

$$(D-2)(3x^2D+2)y=0$$

Put $(3x^2D+2)y=u$ so that $(D-2)u=0$.

Now the general solution of $(D-2)u=0$ is $u=C_1.e^{2x}$

$$\therefore, \quad (3x^2D+2)y \;=\; C_1e^{2x}$$

$$\text{i, e,} \quad 3x^2\frac{dy}{dx}+2y \;=\; C_1e^{2x}$$

$$\text{i, e,} \quad \frac{dy}{dx}+\frac{2}{3x^2}y \;=\; \frac{C_1}{3x^2}.e^{2x},$$

the general solution of which is:

$$y.e^{-\frac{2}{3x}} \;=\; \frac{C_1}{3}\int \frac{e^{\left(2x-\frac{2}{3x}\right)}}{x^2}\,dx+C_2$$

$$\text{i, e,} \quad y \;=\; \frac{1}{3}C_1e^{\frac{2}{3x}}\int \frac{e^{\left(2x-\frac{2}{3x}\right)}}{x^2}\,dx+C_2e^{\frac{2}{3x}}$$

This shows the drastic difference between two orderings.

6.5 Transformation of a Second order Ode by changing the Dependent Variable:

Sometimes an ordinary differential equation can be transformed to an integrable type by changing its dependent variable. In reference of the homogeneous ode (6.2)

$$\frac{d^2y}{dx^2}+p_1(x)\frac{dy}{dx}+p_2(x)y=0$$

we start with the equation (6.3), viz,

$$f(x)\frac{d^2v}{dx^2}+\left(2\frac{df}{dx}+p_1(x)f(x)\right)\frac{dv}{dx}+\left(\frac{d^2f}{dx^2}+p_1(x)\frac{df}{dx}+p_2(x)f\right)v=0,$$

obtained in course of the method of reduction of order and choose $f(x)$ in a way that makes co-efficient of $\frac{dv}{dx}$ zero. This peculiar choice yields

$$f(x)=\exp\left[-\frac{1}{2}\int^x p_1(t)dt\right] \tag{6.21}$$

and consequently the co-efficient of $v(x)$ as $p_2(x) - \frac{1}{2} \cdot \frac{dp_1}{dx} - \frac{1}{4}p_1^2(x)$.

Defining $P_2(x) \equiv p_2(x) - \frac{1}{2}\frac{dp_1}{dx} - \frac{1}{4}p_1^2(x)$, we have the new second order ode in dependent variable v:

$$\frac{d^2v}{dx^2} + P_2(x)v(x) = 0, \tag{6.22}$$

The equation (6.22) involves no first order derivative term and is useful in qualitative study of second order odes. It may be remarked in passing that the form (6.22) is often encountered as time-independent Schrodinger wave equation (when potential is a function of position only) in non-relativistic quantum mechanics. Equation of the type (6.22) is commonly called 'normal form'.

Example (16) : Solve the ode

$$\frac{d^2y}{dx^2} - 2\tan x \frac{dy}{dx} + 5y = 0, \quad 0 \le x < \infty$$

As $p_1(x) = -2\tan x$ and $p_2(x) = 5$, $f(x) = \exp\left[\int_0^x \tan t\, dt\right] = \sec x$

Therefore $\quad P_2(x) = p_2(x) - \frac{1}{2}\frac{dp_1}{dx} - \frac{1}{4}\,p_1^2(x) = 6.$

For this problem, (6.22) reads $\frac{d^2v}{dx^2} + 6v = 0$, which can be at once solved to have $v(x) = A\cos(\sqrt{6}x) + B\sin(\sqrt{6}x)$.

The given equation thus has its general solution

$$y(x) = \left[A\cos(\sqrt{6}x) + B\sin(\sqrt{6}x)\right]\sec x.$$

Example (17) : Show by means of above method that $x^{-\frac{1}{2}}\sin x$ and $x^{-\frac{1}{2}}\cos x$ are two linearly independent solutions of the Bessel equation of order half, viz.,

$$x^2\frac{d^2y}{dx^2} + x\frac{dy}{dx} + \left(x^2 - \frac{1}{4}\right)y = 0, \quad x > 0$$

Find also a formula for a particular integral of the equation

$$x^2\frac{d^2y}{dx^2} + x\frac{dy}{dx} + \left(x^2 - \frac{1}{4}\right)y = g(x), \quad x > 0$$

We recast the given equation in the standard form

$$\frac{d^2y}{dx^2} + \frac{1}{x}\frac{dy}{dx} + \left(1 - \frac{1}{4x^2}\right)y = \frac{g(x)}{x^2}, \quad x > 0$$

Here $p_1(x) = \frac{1}{x}$ and $p_2(x) = \left(1 - \frac{1}{4x^2}\right)$

$$\begin{aligned}
\text{Thus} \quad f(x) &= \exp\left[-\frac{1}{2}\int^x p_1(t)dt\right] \\
&= \exp\left[-\frac{1}{2}\int^x \frac{dt}{t}\right] = \exp\left[\ln(x^{-\frac{1}{2}})\right] = x^{-\frac{1}{2}}
\end{aligned}$$

$$\begin{aligned}
\text{Again} \quad P_2(x) &\equiv p_2(x) - \frac{1}{2}\frac{dp_1}{dx} - \frac{1}{4}p_1^2(x) \\
&= \left(1 - \frac{1}{4x^2}\right) - \frac{1}{2}\left(-\frac{1}{x^2}\right) - \frac{1}{4x^2} = 1,
\end{aligned}$$

ensuring that equation (6.22) of this problem becomes

$$\frac{d^2v}{dx^2} + v = 0,$$

whose two linearly independent solutions are $\{\sin x, \cos x\}$. The given Bessel equation has therefore the set $\{x^{-\frac{1}{2}}\sin x, x^{-\frac{1}{2}}\cos x\}$ as the linearly independent solutions.

To find particular integral of the second equation we proceed by method of variation of parameters. Chosen particular integral in the form

$$y_p = (c_1(x)\sin x + c_2(x)\cos x)\, x^{-\frac{1}{2}},$$

where $c_1(x)$ and $c_2(x)$ are differentiable functions whose derivatives are given by the equations in matrix form:

$$\begin{bmatrix} x^{-\frac{1}{2}}\sin x & x^{-\frac{1}{2}}\cos x \\ \frac{d}{dx}\left(x^{-\frac{1}{2}}\sin x\right) & \frac{d}{dx}\left(x^{-\frac{1}{2}}\cos x\right) \end{bmatrix}\begin{bmatrix} c_1'(x) \\ c_2'(x) \end{bmatrix} = \begin{bmatrix} 0 \\ \frac{g(x)}{x^2} \end{bmatrix}$$

$$\therefore \quad \begin{bmatrix} c_1'(x) \\ c_2'(x) \end{bmatrix} = \begin{bmatrix} -x^{-\frac{1}{2}}\sin x - \frac{1}{2}x^{-\frac{3}{2}}\cos x & -x^{-\frac{1}{2}}\cos x \\ -x^{-\frac{1}{2}}\cos x + \frac{1}{2}x^{-\frac{3}{2}}\sin x & x^{-\frac{1}{2}}\sin x \end{bmatrix}\begin{bmatrix} 0 \\ \frac{g(x)}{x^2} \end{bmatrix}$$

$$\text{i, e,} \quad \begin{bmatrix} c_1'(x) \\ c_2'(x) \end{bmatrix} = \begin{bmatrix} x^{-\frac{3}{2}}\cos x\, g(x) \\ -x^{-\frac{3}{2}}\sin x\, g(x) \end{bmatrix}$$

Hence $c_1(x) = \displaystyle\int^x t^{-\frac{3}{2}} \cos t \, g(t) dt$ and $c_2(x) = \displaystyle\int^x -t^{-\frac{3}{2}} \sin t \, g(t) dt$,

$$\therefore \quad y_p = x^{-\frac{1}{2}} \sin x \int^x t^{-\frac{3}{2}} \cos t \, g(t) dt - x^{-\frac{1}{2}} \cos x \int^x t^{-\frac{3}{2}} \sin t g(t) dt$$

$$= x^{-\frac{1}{2}} \int^x t^{-\frac{3}{2}} (\sin x \cos t - \cos x \sin t) g(t) dt$$

$$= x^{-\frac{1}{2}} \int^x \frac{\sin(x-t)g(t)}{t^{\frac{3}{2}}} dt, \quad x > 0$$

This shows again that the kernel involves a factor that is function of $(x-t)$. It is not unexpected as on changing the dependent variable, the homogeneous Bessel equation of order half reduces to a second order ode

$$\frac{d^2 v}{dx^2} + v = 0$$ having constant co-efficients.

Alternatively, we see that due to the transformation applied the inhomogeneous equation reduces to

$$\frac{d^2 v}{dx^2} + v = \frac{g(x)}{x^{\frac{3}{2}}} \, .$$

For this problem Green's function $G(x,t) = \sin(x-t)$ and hence v_p, the particular integral of the new equation in x and v is given by

$$v_p = \int^x \frac{\sin(x-t)}{t^{\frac{3}{2}}} g(t) \, dt,$$

yielding the corresponding y_p as

$$y_p = f(x)v_p = x^{-\frac{1}{2}} \int^x \frac{\sin(x-t)}{t^{\frac{3}{2}}} g(t) \, dt$$

6.6 Qualitative Aspects of Second Order Differential Equations.

The foregoing articles have dealt with only the techniques of solving the second order differential equations. We now digress a little and look into some interesting qualitative aspects of these second order equations. In the following we explore a sufficient condition for a second order ode to have all its solutions bounded over $[0, \infty)$. As a prelude to it let's prove the following lemma, viz, **Gronwall-Bellman Inequality:**

Lemma : If $K > 0$ and $u(x)$, $v(x)$ be two non-negative continuous functions on $[x_0, x_1]$ that satisfy $u(x) \leq K + \int_{x_0}^{x} u(s)v(s)ds \ \forall x \in [x_0, x_1]$, then $u(x) \leq K \exp\left[\int_{x_0}^{x} v(s)ds\right] \ \forall x \in [x_0, x_1]$.

Proof: Define $\omega(x) = \int_{x_0}^{x} u(s)v(s)ds$. Therefore, $\omega(x_0) = 0$ and by hypothesis, $u(x) \leq K + \omega(x)$.

$$\therefore \quad \frac{d\omega}{dx} = u(x)v(x) \leq (K + \omega(x))\, v(x) \quad (\because v(x) \text{ is non-negative})$$

$$\text{i.e.} \quad \frac{d\,(K + \omega(x))}{(K + \omega(x))} \leq v(x)dx$$

Integrating this inequality on both sides between x_0 and x and using the monotonicity property of Riemann integrals, we have:

$$K + \omega(x) \leq K \exp\left[\int_{x_0}^{x} v(s)ds\right]$$

$$\therefore \quad u(x) \leq K + \omega(x) \leq K \exp\left[\int_{x_0}^{x} v(s)ds\right]$$

Theorem 6.1 : If $P_2(x)$ be continuously differentiable function over $[0, \infty)$, the second order homogeneous normal ode $y'' + P_2(x)y = 0$ has all its solutions bounded on $[0, \infty)$ provided $P_2(x) \to \infty$ monotonically as $x \to \infty$.

Proof: Multiply the given normal ode by $2y'$ and integrate the resultant between 0 and x to have

$$y'^2 + \int_0^x P_2(s)\frac{d}{ds}(y^2)ds = c$$

Integrating by parts,

$$y'^2 + P_2(x)y^2 = \int_0^x P_2'(s)y^2(s)ds + c + P_2(0)y^2(0)$$

Since $P_2(x) \to \infty$ monotonically as $x \to \infty$, we may assume that $P_2(x)$ is strictly positive for $x \geq 0$. Moreover since $P_2'(x) \geq 0$,

$$|\,P_2(x)y^2\,| \ \leq \ |\,c + P_2(0)y^2(0)\,| + |\int_0^x P_2'(s)y^2(s)ds\,|$$

$$\leq \ |\,c + P_2(0)y^2(0)\,| + \int_0^x |\,P_2'(s)\,|\,y^2(s)ds$$

$$= \ |\,c + P_2(0)y^2(0)\,| + \int_0^x P_2(s)y^2(s)\frac{P_2'(s)}{P_2(s)}ds$$

If the agree to write $\mid P_2(x)y^2 \mid \equiv u(x)$ and $\frac{P_2'(x)}{P_2(x)} \equiv v(x)$ with

$$K \equiv \mid c + P_2(0)y^2(0) \mid,$$

then the above inequality appears in the form

$$u(x) \leq K + \int_0^x u(s)v(s)ds$$

Hence by Gronwall-Bellmann inequality we get

$$u(x) \leq K \exp\left[\int_0^x v(s)ds\right]$$

$$\text{i.e, } \mid P_2(x)y^2 \mid \leq K \exp\left[\int_0^x \frac{P_2'(s)}{P_2(s)} ds\right] = K \left|\frac{P_2(x)}{P_2(0)}\right|$$

Since $P_2(x) > 0$ for $x \geq 0$, we have $y^2 \leq \frac{K}{P_2(0)}$

Taking square root on both sides we have $\mid y(x) \mid \leq \left|\frac{K}{P_2(0)}\right|^{\frac{1}{2}}$ for all $x \in [0, \infty)$. This asserts the claim.

Note: This condition is sufficient but not necessary. For example, take up the normal ode $\frac{d^2y}{dx^2} + y = 0$. Here general solution is the sinusoidal function, viz, $y = A\cos(x + \beta)$.

Solution is obviously bounded but $P_2(x) = 1$ $\forall x$ and so no question of $P_2(x)$ tending to ∞ monotonically with indefinite increase of x. We now study a few properties of the Wronskian in the following.

Theorem 6.2: If $y_1(x)$ and $y_2(x)$ form a fundamental solution set of the second order differential equation $y''(x) + p(x)y'(x) + q(x)y(x) = 0$, where $p(x)$ and $q(x)$ are continuous functions of x over $[a, b]$, then any solution of the equation is expressed as a linear combination of $y_1(x)$ and $y_2(x)$.

Proof: Since y_1 and y_2 form a fundamental solution set, their Wronskian $W(x) \equiv y_1y_2' - y_1'y_2 \neq 0$ over $[a, b]$. Suppose y_3 be any other solution of the given ode. Hence we have the following relations:

$$\left.\begin{array}{l} y_1'' + p(x)y_1' + q(x)y_1 = 0 \\ y_2'' + p(x)y_2' + q(x)y_2 = 0 \\ y_3'' + p(x)y_3' + q(x)y_3 = 0 \end{array}\right\} \qquad (6.23a)$$

Multiplying the first relation by y_2 and second relation by y_1, we have, after elimination of $q(x)y_1y_2$,

$$\frac{d}{dx}(y_1y_2' - y_1'y_2) + p(x)(y_1y_2' - y_1'y_2) = 0$$

or, $$\frac{dW}{dx} + p(x)W(x) = 0, \text{ where } W(x) \neq 0 \text{ over } [a,b]$$

or, $$W(x) = C_{12} \exp\left[-\int^x p(t)dt\right] \qquad (6.23b)$$

Clearly $C_{12} \neq 0$.

In a similar manner, from the last two pair of relations in (6.23a), we have, on elimination of $q(x)$,

$$\frac{d}{dx}(y_2y_3' - y_2'y_3) + p(x)(y_2y_3' - y_2'y_3) = 0$$

or, $$(y_2y_3' - y_2'y_3) = C_{23}\exp\left[-\int^x p(t)dt\right] \qquad (6.23c)$$

By cyclic symmetry, the first & third relation of (6.23a) produces (after elimination of $q(x)$),

$$(y_3y_1' - y_3'y_1) = C_{31}\exp\left[-\int^x p(t)dt\right] \qquad (6.23d)$$

Multiplying (6.23c) by y_1 and (6.23d) by y_2, add them up to get:

$$-y_3(y_1y_2' - y_1'y_2) = (C_{23}y_1 + C_{31}y_2)\exp\left[-\int^x p(t)dt\right]$$

or, $$-y_3 C_{12}\exp\left[-\int^x p(t)dt\right] = (C_{23}y_1 + C_{31}y_2)\exp\left[-\int^x p(t)dt\right]$$

or, $$y_3 = -\frac{C_{23}}{C_{12}}y_1 - \frac{C_{31}}{C_{12}}y_2 \quad (\because C_{12} \neq 0)$$

Hence the proof.

Note: In case $p(x)$ and $q(x)$ are continuous periodic functions of x, ω being their common period, the general solution is a periodic function of $(x - \omega)$ being its period. The proof is furnished below:-

The conditions $p(x + \omega) = p(x)$ and $q(x + \omega) = q(x)$ are given.

The general solution is denoted by $\phi(x)$, say. Hence we have:

$$\phi''(x) + p(x)\phi'(x) + q(x)\phi(x) = 0 \quad \forall x.$$

Replacing x by $(x+\omega)$ and noting that $p(x+\omega) = p(x)$ & $q(x+\omega) = q(x)$ we have

$$\phi''(x+\omega) + p(x)\phi'(x+\omega) + q(x)\phi(x+\omega) = 0 \ \forall x.$$

Hence $\phi(x+\omega)$ is a solution of the given equation.

The practical utility of this result of periodic linear system is that once the fundamental solution set is determined over an interval of length ω, the fundamental solution set and hence the general solution of the given ode determined over the full real-axis.

Theorem 6.3: If $p(x)$ and $q(x)$ are continuous functions and $y_1(x)$, $y_2(x)$ be any solutions of the second order ode

$$y''(x) + p(x)y'(x) + q(x)y(x) = 0 \qquad (6.24)$$

on $[a, b]$, then the following results hold true.

(a) If $y_1(x)$ and $y_2(x)$ vanish at the same point x_0 in $[a, b]$ then they cannot form a fundamental set of solutions in $[a, b]$.

(b) If $y_1(x)$ and $y_2(x)$ achieve a maximum or a minimum at the same point x_0 in $[a, b]$, they cannot form a fundamental set of solutions in this interval.

(c) If $y_1(x)$ and $y_2(x)$ form a fundamental set of solutions in $[a, b]$, they cannot have a common point of inflexion in $[a, b]$ unless $p(x)$ and $q(x)$ vanish simultaneously at that point.

(d) If $y_1(x)$ and $y_2(x)$ be a fundamental set of solutions of the equation (6.24) on $(-\infty, \infty)$, then show that there exists one and only one zero of $y_1(x)$ between two consecutive zeros of $y_2(x)$ and viceversa.

Proof: (a) As $y_1(x)$ and $y_2(x)$ are solutions of the given ode, we have

$$\left.\begin{array}{l} y_1'' + p(x)y_1' + q(x)y_1 = 0 \\ y_2'' + p(x)y_2' + q(x)y_2 = 0 \end{array}\right\} \qquad (6.24a)$$

wherefrom follows that $W'(x) = p(x)W(x)$, $W(x)$ standing for the Wronskian of y_1 and y_2.

Since $W(x_0) = 0$ (why?), $W(x)$ is zero everywhere on $[a, b]$ (refer to Ostrogradsky-Liouville formula). Hence the solutions $y_1(x)$ and $y_2(x)$ are not linearly independent.

(b) If x_0 be a common point of maxima/minima of $y_1(x)$ and $y_2(x)$ in $[a, b]$, then $y_1'(x_0) = y_2'(x_0) = 0$ and hence

$$W(x_0) = y_1(x_0)y_2'(x_0) - y_1'(x_0)y_2(x_0) = 0.$$

The desired result now follows at once.

(c) Solving equations in (6.24a) we get the unique solutions

$$p(x) = \frac{y_2 y_1'' - y_1 y_2''}{W(x)} \quad \text{and} \quad q(x) = \frac{y_1' y_2'' - y_2' y_1''}{W(x)}$$

If possible, let x_0 be a common point of inflexion of $y_1(x)$ and $y_2(x)$ in $[a, b]$. Both the functions are twice-differentiable and as per the assumption, $y_1''(x_0) = y_2''(x_0) = 0$. Hence $p(x_0) = q(x_0)$, proving the assertion.

(d) Let x_1 and x_2 be consecutive zeros of $y_2(x)$. Thus $y_2(x_1) = y_2(x_2) = 0$, making $W(x_1) = y_1(x_1)y_2'(x_1)$ and $W(x_2) = y_1(x_2)y_2'(x_2)$. Due to Rolle's theorem, $y_2'(x)$ vanishes somewhere in (x_1, x_2) and so $y_2'(x_1)$ and $y_2'(x_2)$ must have opposite sign. Again $W(x_1)$ and $W(x_2)$ are of same sign, because otherwise, due to Bolzano's theorem on continuity, $W(x)$ would vanish at some point in (x_1, x_2)-a result which contradicts the hypothesis that $y_{(x)}$ and $y_2(x)$ form a fundamental set of solutions. All that follows is that $y_1(x_1)$ and $y_1(x_2)$ are of opposite sign, ensuring that $y_1(x)$ must vanish at least once in (x_1, x_2). However, $y_1(x)$ vanishes exactly at one point in (x_1, x_2) because otherwise, we may implement the preceding argument to $y_1(x)$ and draw the conclusion that $y_2(x)$ vanishes somewhere in between x_1 and x_2. This would directly contradict the initial assumption that x_1 and x_2 are consecutive zeros of $y_2(x)$. This interesting result is commonly referred to as '**interlacing of zeros**' of the two linearly independent solutions of any second order linear homogeneous ode.

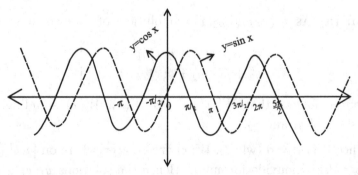

Fig 6.1 : Interlacing of zeros of sine and cosine functions

Very nice illustration has been the 'interlacing of zeros' of $\sin x$ and $\cos x$ over $(-\infty, \infty)$ as $\sin x$ vanishes for integral multiples of π while $\cos x$ vanishes for half-integer multiples of π and they form the fundamental set of solutions of the ode $\frac{d^2y}{dx^2} + y = 0$, obtained as the normal form of Bessel equation of order $\frac{1}{2}$.

Theorem 6.3(d) is known as '**Sturm Separation Theorem**' which tells us that the zeros of any two non-trivial solutions $y_1(x)$ and $y_2(x)$ of equation (6.20) either coincide or occur alternately, depending on whether these solutions are linearly dependent or independent.

Theorem 6.4: If $y(x)$ be any non-trivial solution of the equation

$$y''(x) + p(x)y'(x) + q(x)y(x) = 0$$

over a closed interval $[a, b]$, then $y(x)$ can have atmost a finite number of zeros in this interval.

Proof: We prove the above result by method of contradiction and suppose that $y(x)$ possesses an infinite number of zeros in $[a, b]$. Let's denote the set of zeros of $y(x)$ by S. Since S is an infinite bounded set of real numbers, by virtue of Bolzano-Weierstrass theorem, S has a limit point, say, $x_0 \in [a, b]$. By definition of limit point of a set, there exists a sequence $\{x_n\} \subset S$ that converges to x_0. Since $y(x)$ is continuous and differentiable at x_0,

$$y(x_0) = y\left(\lim_{n\to\infty} x_n\right) = \lim_{n\to\infty} y(x_n) = 0$$

From the definition of derivative, we get

$$y'(x_0) = \lim_{x\to x_0} g(x), \text{ where } g(x) \equiv \frac{y(x) - y(x_0)}{x - x_0}.$$

This implies that the sequence $\{g(x_n)\}$ converges to the limit $y'(x_0)$ for any sequence $\{x_n\}$, $x_n \neq x_0$ converges to x_0. Symbolically,

$$\lim_{x_n \to x_0, x_n \neq x_0} g(x_n) = y'(x_0)$$

However, $\{g(x_n)\}$ being the constant zero sequence, $y'(x_0) = 0$. All these results give us a contradictory conclusion—' $y(x)$ is a trivial solution.' This conclusion was unavoidable as the theorem of uniqueness guarantees that the zero initial conditions, viz, $y(x_0) = y'(x_0) = 0$ are satisfied only by the trivial solution. Hence the theorem.

Remark: Theorems 6.3 and 6.4 jointly furnish very interesting results. If $y_1(x)$ and $y_2(x)$ are two linearly independent solutions of the same differential equation over a preassigned interval $[a, b]$, they oscillate with essentially the same rapidity since the total number of zeros of any of them differs from the total number of zeros of the other by atmost one. You may verify this by drawing graph of any two oscillating functions you like, provided they are linearly independent solutions of one and the same second order ode over same preassigned interval.

Further, as shown in the preceding article, one can always recast equation (6.24) to a form that involves no first order derivative. Stated explicitly, the transformation of the dependent variable y to v given by

$$y(x) = v(x) \exp\left[-\frac{1}{2}\int^x p(t)dt\right]$$

translates (6.24) into the form

$$v''(x) + Q(x)v(x) = 0, \tag{6.25}$$

where $Q(x)$ is given by

$$Q(x) = q(x) - \frac{1}{2}p'(x) - \frac{1}{4}p^2(x) \tag{6.26}$$

This transformation makes life simple but preserves the essential qualitative features of the solutions of (6.24). Only the number of linearly independent solutions of (6.25) will differ from the number of zeros of the corresponding solutions of (6.24) as the extra factor, viz,

$$\exp\left[-\frac{1}{2}\int^x p(t)dt\right]$$

may account for more zeros. Hence without loss of generality one may take a second order ode in the form

$$y''(x) + Q(x)y(x) = 0 \tag{6.26a}$$

If over an interval $[a, b]$, $y(x)$ and $z(x)$ be non-trivial solutions of the second order equations $y''(x) + Q(x)y(x) = 0$ and $y''(x) + R(x)y(x) = 0$ respectively, where $Q(x)$ and $R(x)$ are strictly positive functions such that for each $x \in [a, b]$, $Q(x) > R(x)$ holds true, then $y(x)$ vanishes at least once between two successive zeros of $z(x)$. This result is often referred to as "Sturm's comparison theorem" and it shows how with the increase of $Q(x)$ or the like, the solutions of (6.25a) and hence of (6.24) oscillate more and more rapidly within a given interval.

Theorem 6.5: If for the differential equation $y''(x) + Q(x)y(x) = 0$, $Q(x) > 0$ for all $x > 0$, and $\int_a^\infty Q(x)dx = \infty$, then any non-trivial solution $y(x)$ has infinitely many zeros in $\mathbb{R}^+$, 'a' being some $+ve$ number.

Proof : We shall prove this result by method of contradiction. If possible, let $y(x)$ have finitely many zeros in $\mathbb{R}^+$. Then $\exists$ a point $c > a$ such that for all $x \geqslant c$, $y(x) \neq 0$, So $\forall x \geqslant c$, $y(x)$ maintains same sign. Without loss of generality we presume that $\forall x \geqslant c$, $y(x) > 0$. If one writes $z(x) \equiv \frac{-y'(x)}{y(x)}$, then standard workout yields from the given ode :

$$\frac{dz}{dx} = Q(x) + (z(x))^2 \ \forall \ x \geqslant c$$

Integrating with respect to x over $[c, x]$ we get :

$$z(x) - z(c) = \int_c^x Q(x)dx + \int_c^z (z(x))^2 dx$$

Since $\int_a^x Q(x)dx = \infty$ and $\int_c^x (z(x))^2 dx > 0$, it follows that for sufficiently large $x, z(x)$ is always $+ve$. As by assumption, $y(x) > 0 \ \forall \ x \geqslant c$, it follows that $\forall \ x \geqslant c$, $y'(x) < 0$. Again from the original ode, viz, $y''(x) + Q(x)y(x) = 0$, it follows that for all $x \geqslant c$, $y''(x) < 0$, i,e, in otherwords, $y'(x)$ is a monotone decreasing function. This in turn leads to the existence of an additional zero to the right of c — a direct contradiction to our hypothesis. Hence we assert, by method of contradiction, that the solution $y(x)$ has infinitely many zeros in $\mathbb{R}^+$.

To review what this theorem dishes out for us, lets go back to example (17) of §6.5 dealing with Bessel equation of order half. This equation, as shown earlier, can be cast to the normal form $y''(x) + y(x) = 0$ by the transformation $f(x) = \frac{1}{\sqrt{x}}$ and hence with the aid of the above theorem we infer that the general solution $c_1 \cos x + c_2 \sin x$ of this normal ode has infinitely many zeros, in $\mathbb{R}^+$, their common spacing being π. We further observe that the transformation $f(x) = \exp\left[-\frac{1}{2}\int^x p_1(t)dt\right]$ by which a second order ode of the form (6.21) is cast to the normal form (6.22) has no effect on the zeros of the solution since $f(x)$ never vanishes. So the fundamental solutions $x^{-\frac{1}{2}}\sin x$ and $x^{-\frac{1}{2}}\cos x$ of Bessel equation of order half have also infinitely many zeros lying at intervals of π in $\mathbb{R}^+$ and the zeros of $x^{-\frac{1}{2}}\sin x$ and those of $x^{-\frac{1}{2}}\cos x$ interlace. Later on, we shall see the generalisation of this result.

6.7 Exact Second Order Differential Equations

The concept of exactness that was discussed in Chapter 2 for the first order differential equations can be extended to the second order linear differential equations. Formally, we say that the equation

$$a_0(x)y'' + a_1(x)y' + a_2(x)y = 0 \tag{6.27}$$

is **exact** if it can be recast in the form

$$(a_0(x)y')' + (g(x)y)' = 0 \tag{6.27a}$$

where $g(x)$ is a differentiable function that is to be determined in terms of $a_0(x)$, $a_1(x)$ and $a_2(x)$. Although this definition is given in many texts, we shall adopt a novel approach to define it. D'Alembert's method of reduction of order discussed earlier in this chapter provides us the clue in this case. In that method it was observed that when $y = f(x)$ is a solution of (6.27), it can be reduced to the first order linear ode

$$a_0(x)f(x)u'(x) + \big(a_1(x)f(x) + 2a_0(x)f'(x)\big)\,u(x) = 0 \tag{6.28}$$

the (c.f. derivation of (6.4)).

We now define equation (6.27) to be 'exact' if the associated equation (6.27a) be 'exact' in the usual sense. As is known to us, the necessary and sufficient condition for equation (6.28) to be 'exact' is

$$\frac{d}{dx}(a_0(x)f(x)) = \frac{\partial}{\partial u}\big((a_1(x)f(x) + 2a_0(x)f'(x))u(x)\big),$$

which can be simplified to

$$\frac{f'(x)}{f(x)} = \frac{a_0'(x) - a_1(x)}{a_0(x)}, \quad a_0(x) \neq 0. \tag{6.29}$$

On integration, (6.29) gives

$$f(x) = \exp\left[\int \frac{a_0'(x) - a_1(x)}{a_0(x)}dx\right],$$

wherefrom follows that

$$f'(x) = \left(\frac{a_0'(x) - a_1(x)}{a_0(x)}\right) \exp\left[\int \frac{a_0'(x) - a_1(x)}{a_0(x)} dx\right]$$

$$f''(x) = \left[\left(\frac{a_0'(x) - a_1(x)}{a_0(x)}\right)^2 + \left(\frac{a_0'(x) - a_1(x)}{a_0(x)}\right)'\right] \times$$

$$\exp\left[\int \frac{a_0'(x) - a_1(x)}{a_0(x)} dx\right].$$

Using these expressions into the relation

$$a_0(x)f''(x) + a_1(x)f'(x) + a_2(x)f(x) = 0, \quad \text{we have,}$$

$$a_0(x)\left[\left(\frac{a_0'(x) - a_1(x)}{a_0(x)}\right)^2 + \left(\frac{a_0'(x) - a_1(x)}{a_0(x)}\right)'\right]$$

$$+a_1(x)\left(\frac{a_0'(x) - a_1(x)}{a_0(x)}\right) + a_2(x) = 0,$$

which can be simplified to yield

$$a_0''(x) - a_1'(x) + a_2(x) = 0 \qquad (6.30)$$

Substituting (6.30) back into (6.27) we get back an equation of the form (6.27a) with $g(x) \equiv a_1(x) - a_0'(x)$. If (6.27) satisfies the condition (6.30), we see that (6.27) is exact. If we have (6.27) cast in the form (6.27a), solving the equation is no problem as integrating once w.r. to x, (6.24a) gives a first order ode that can be handled with methods in vogue.

Observe that the equation (6.27a) becomes

$$(a_0(x)y')' + ((a_1(x) - a_0'(x))y)' = 0$$

i, e, $$[a_0(x)y' + (a_1(x) - a_0'(x))y]' = 0,$$

which on integration produces the first integral:

$$a_0(x)y' + (a_1(x) - a_0'(x))y = \text{constant } (c_1);$$

This can be solved by standard methods of solving first order odes.

Remark (a): If instead of D'Alembert's method of order-reduction we had employed the Riccati substitution $u(x) = -\frac{y'(x)}{y(x)}$ to equation (6.27), we would get the non-linear ode (in x and u)

$$a_0(x)u'(x) = a_0(x)u^2(x) - a_1(x)u(x) + a_2(x) \qquad (6.31)$$

Now (6.31) is exact iff

$$\frac{\partial}{\partial u}\left[a_0(x)u^2(x) - a_1(x)u(x) + a_2(x)\right] = -a_0'(x)$$

i, e, iff $\qquad u(x) = \dfrac{a_1(x) - a_0'(x)}{2a_0(x)}, \quad a_0(x) \neq 0$

$$\therefore \quad y(x) = \exp\left[-\int u(x)dx\right] = \exp\left[\int \frac{a_0'(x) - a_1(x)}{2a_0(x)}\, dx\right] \qquad (6.32)$$

Observe that the numerical factor 2 appearing in the denominator of the integrand makes all the difference between (6.32) and the expression of $f(x)$ determined immediately after equation (6.29). So it transpires that on direct substitution, (6.32) can never satisfy (6.30). This exercise envisages that once linearity inherent in the odes is lost due to some transformation, (e,g, Riccati substitution), exactness, if at all were present, is also lost. D'Alembert's method had the advantage of reducing a linear second order ode to a first order linear ode, thereby preserving exactness or non-exactness of the ode. In contrast, Riccati substitution reduced the order at the cost of linearity of the ode and that is why failed to preserve the exactness or non-exactness of the parent equation.

(b) D'Alembert's method is not confined to just second order linear homogeneous differential equations but is equally applicable to any higher order differential equation of the same type. This enthuses us to employ the method of order-reduction in finding the necessary and sufficient condition for exactness of the latter. Intuitively, it is clear that if one applies D'Alembert's method $(n-1)$ times successively, any homogeneous linear ode of the form.

$$a_0(x)y^{(n)}(x) + a_1(x)y^{(n-1)}(x) + \cdots\cdots + a_{n-1}(x)y'(x) + a_n(x)y(x) = 0$$

is reduced to a first order linear ode. The necessary and sufficient condition for exactness of this first order equation ultimately gives back the required condition for exactness of the original nth order ode as:

$$\sum_{k=0}^{n}(-1)^k a_k^{(n-k)}(x) = 0.$$

We shall skip its lengthy proof but may verify its truth for lower order equations, of order 3 or 4 say. Here we shall substitute $y(x) = f(x)v(x)$

in the given linear ode, use Leibnitz's rule of successive differentiation, make use of the fact that $f(x)$ is a known integral of the given equation and finally write $\frac{dv}{dx} \equiv u(x)$ to have an order-reduced linear-ode. If it be of second order, its exactness condition will give us the required relation between the co-efficients of the parent ode.

Example (18) : Determine whether the following differential equations are exact or not. If so, find their solutions.

$$\text{(a)} \quad xy'' - (\cos x)y' + (\sin x)y = 0, \ x > 0$$
$$\text{(b)} \quad (\cot x)y'' - 2(\cot x \ \text{cosec}^2 x)y = 0$$

For the problem (a), $a_0(x) = x$; $a_1(x) = -\cos x$ and $a_2(x) = \sin x$. All these are differentiable functions of x and moreover, $a_0(x) \neq 0$ in the domain $\mathbb{R}^+$. Moreover, $a_0''(x) - a_1'(x) + a_2(x) = 0$. Hence equation is exact. Thus it can be cast in the form (6.27a):

$$(xy')' - ((\cos x)y)' = 0,$$

which on integrating yields

$$xy' - (\cos x)y = c_1, \quad c_1 \text{ being some constant.}$$

This first order linear differential equation has an integrating factor

$$\mu(x) = \exp\left[-\int^x \left(\frac{1}{t} + \frac{\cos t}{t}\right) dt\right]$$

and the solution is

$$y(x) = \frac{1}{x\mu(x)}\left[c_1 \int^x \mu(t)dt + c_2\right], \quad c_1 \text{ and } c_2 \text{ being constants.}$$

For the problem (b), $a_0(x) = \cot x$; $a_2(x) = -2\cot x \ \text{cosec}^2 x$ and $a_1(x) = 0$. It is exact as the necessary & sufficient condition is satisfied here. This equation, when integrated once with respect to x, becomes

$$(\cot x)y' + (\text{cosec}^2 x)y = c_1,$$

which is not exact as $\frac{\partial}{\partial y}(y \ \text{cosec}^2 x) \neq \frac{\partial}{\partial x}(\cot x)$. The integrating factor is given by $\mu(x) = \exp\left[2\int \frac{\text{cosec}^2 x}{\cot x} dx\right] = \tan^2 x$. When multiplied by

$\tan^2 x$, we have the exact equation $d(y \tan x) = c_1 \tan^2 x\, dx$. Integrating once more w.r. to x,

$$y \tan x = c_1(\tan x - x) + c_2$$
$$\text{i, e,} \quad y(x) = c_1(1 - x \cot x) + c_2 \cot x,$$

c_1 and c_2 being arbitrary constants.

Example (19) : Solve $x\frac{d^2y}{dx^2} + (1 - x)\frac{dy}{dx} - y = e^x$ by checking the exactness of the ode.

Comparing with the standard form $a_0(x)\frac{d^2y}{dx^2} + a_1(x)\frac{dy}{dx} + a_2(x)y = X$, we get: $a_0(x); \ a_1(x) = (1 - x); \ a_2(x) = -1; \ X = e^x$.

We verify that the given ode is exact as

$$a_2 - a_1' - a_0'' = -1 - \frac{d}{dx}(1 - x) - \frac{d^2}{dx^2}(x) = 0.$$

The first integral is $\dfrac{dy}{dx} - y = \dfrac{e^x}{x} + \dfrac{c_1}{x}.$

The first order ode has I.F. e^{-x} and so the general solution of this ode will be given by

$$ye^{-x} = \int \left(\frac{1}{x} + \frac{c_1 e^{-x}}{x}\right) dx$$
$$\therefore \quad y = e^x \ln x + c_1 e^x \int \frac{e^{-x}}{x} dx + c_2.$$

This indeed is the general solution of the given second order ode.

Example (20) : Solve the ode $(2x^2+3x)\frac{d^2y}{dx^2}+(6x+3)\frac{dy}{dx}+2y = (x+1)e^x$

by checking exactness of the ode.

Here by comparison with the standard form we have:

$$a_0(x) = 2x^2 + 3x; \ a_1(x) = 6x + 3; \ a_2(x) = 2; \ X = (x + 1)e^x$$

We observe that $a_2 - a_1' + a_0'' = 0$ and so the first integral is:

$$a_0(x)\frac{dy}{dx} + (a_1 - a_0')y = \int X dx$$
$$\text{i, e,} \quad (2x^2 + 3x)\frac{dy}{dx} + 2xy = \int (x + 1)e^x = xe^x + c_1;$$

$$(c_1 \text{ being arbitrary constant})$$

$$\therefore \quad x\left[(2x+3)\frac{dy}{dx}+2y\right]=xe^x+c_1$$

i, e, $\quad \dfrac{d}{dx}(y(2x+3))=e^x+\dfrac{c_1}{x}$

On integrating w.r.to x we get:

$$y(2x+3)=\int\left(e^x+\frac{c_1}{x}\right)dx=e^x+c_1\ln x+c_2$$

This gives the general solution of the given ode.

6.8   Adjoint Equation and Self-adjoint Odes

In case the second order ode (6.27) fails to be exact, there can be found an integrating factor so that the resultant equation becomes exact.

Here again our play is to find an integrating factor of the equation (6.28). When (6.28) is written in the differential form, we have

$$\left(a_1(x)f(x)+2a_0(x)f'(x)\right)u(x)dx+a_0(x)f(x)du=0 \qquad (6.32a)$$

Equation (6.32a) has integrating factor

$$\mu(x)=\frac{f(x)}{a_0(x)}\ \exp\left[\int\frac{a_1(x)}{a_0(x)}\ dx\right]$$

$$\therefore f(x)=a_0(x)\mu(x)\exp\left[-\int\frac{a_1(x)}{a_0(x)}\ dx\right]$$

Differentiation of $f(x)$ gives :

$$f'(x)=\left(a_0'(x)\mu(x)+a_0(x)\mu'(x)-a_1(x)\mu(x)\right)\exp\left[-\int\frac{a_1(x)}{a_0(x)}\ dx\right]$$

$$f''(x)=\left[a_0''(x)\mu(x)+2a_0'(x)\mu'(x)+\mu''(x)a_0(x)-a_1'(x)\mu(x)-a_1(x)\mu'(x)\right.$$
$$\left.-\frac{a_1(x)}{a_0(x)}\left(a_0'(x)\mu(x)+\mu'(x)a_0(x)-a_1(x)\mu(x)\right)\right]\exp\left[-\int\frac{a_1(x)}{a_0(x)}\ dx\right]$$

Since $f(x)$ is a known integral of (6.27), the relation

$$a_0(x)f''(x)+a_1(x)f'(x)+a_2(x)f(x)=0,$$

Differential Equations: A linear Algebra Approach

is true. On direct substitution of $f'(x)$ and $f''(x)$ into this relation and making a bit of simplification one gets the following ode:

$$a_0(x)\mu''(x) + (2a_0'(x) - a_1(x))\mu'(x) + (a_0''(x) - a_1'(x) + a_2(x))\mu(x) = 0$$
$$(6.33)$$

Equation (6.33) is the second order ode satisfied by the integrating factor $\mu(x)$ of (6.29) and hence that of (6.28). Equation (6.33) is known as **Adjoint equation** of (6.28). Very interesting fact about this adjoint equation (6.33) is that its adjoint equation is the original equation (6.27) itself. It brings to light the well-known result of operator theory that adjoint of any operator, if exists, satisfies the relation 'Adjoint of the adjoint of an operator is the operator itself.' You may directly verify this proposition for the pair of equations (6.28) and (6.33). In case they be identical, we say that the differential equation is **self-adjoint**. In the following we derive the necessary condition for self-adjointness of equation (6.28). Observe that comparing co-efficients of $y'(x)$ and $y(x)$ in (6.27)and (6.33) one finds that

$$a_0'(x) = a_1(x) \quad \text{and} \quad a_0''(x) - a_1'(x) = 0$$

But the second condition is nothing new as

$$0 = a_0''(x) - a_1'(x) = \frac{d}{dx}\left(a_0'(x) - a_1(x)\right)$$

gives back the first one on integration.

It is a routine exercise to verify that Legendre equation of order β given by

$$(1 - x^2)y'' - 2xy' + \beta(\beta + 1)y = 0$$

is self-adjoint as $a_0'(x) = \frac{d}{dx}(1 - x^2) = -2x = a_1(x)$. Similarly one can check that the Bessel equation of order ν given by

$$x^2 y'' + xy' + (x^2 - \nu^2)y = 0$$

is not self-adjoint.

Example (21) : Solve the ode $\dfrac{d^2 y}{dx^2} - \left(2x + \dfrac{3}{x}\right)\dfrac{dy}{dx} - 4y = 0$ by finding

solution of the adjoint ode by method of inspection.

Comparing the given second order ode with the standard one, viz,

$$a_0(x)\frac{d^2y}{dx^2} + a_1(x)\frac{dy}{dx} + a_2(x)y = 0, \text{ we get :}$$

$$a_0(x) = 1; \ a_1(x) = -\left(2x + \frac{3}{x}\right); \ a_2(x) = -4$$

Hence the adjoint ode is:

$$\frac{d^2y}{dx^2} + \left(2x + \frac{3}{x}\right)\frac{dy}{dx} - \left(2 + \frac{3}{x^2}\right)y = 0 \qquad (i)$$

By inspection, one finds that $y = x$ is a solution of this adjoint ode and hence on multiplying the original ode by x, it becomes exact.

Therefore, $x\dfrac{d^2y}{dx^2} - (2x^2+3)\dfrac{dy}{dx} - 4xy = 0$ is exact and can be written as :

$$\frac{d}{dx}\left(x\frac{dy}{dx}\right) - \frac{d}{dx}\left((2x^2 + 3)y\right) = 0 \qquad (ii)$$

The first integral of (ii) is:

$$x\frac{dy}{dx} - (2x^2 + 3)y = \text{constant} = a, \text{ say.}$$

$$\therefore \ \frac{dy}{dx} - \left(2x + \frac{3}{x}\right)y = \frac{a}{x} \qquad (iii)$$

Equation (iii) being a first order linear ode in y, its I.F.

$$\mu = \exp\left[-\int\left(2x + \frac{3}{x}\right)dx\right] = \frac{e^{-x^2}}{x^3}$$

The general solution of original ode is : $y = x^3 e^{x^2}\left[\displaystyle\int \frac{a}{x^4}e^{-x^2}\,dx + b\right]$.

We now prove a very powerful theoretical result in context of self-adjointness of a second order ode. It may be enunciated as follows:

Theorem 6.6 A second order ode $a_0(x)\dfrac{d^2y}{dx^2} + a_1(x)\dfrac{dy}{dx} + a_2(x)y = 0$ is

either self-adjoint or it can be transformed to a self-adjoint second order ode by multiplication throughout by a function $w(x)$, where

$$w(x) = \frac{1}{a_0(x)}\exp\left[\int\frac{a_1(x)}{a_0(x)}dx\right],$$

where $a_0(x)$ never vanishes in a specific interval $[a, b]$. This $\omega(x)$ is known as **weight** function. $a_0(x)$, $a_1(x)$, $a_2(x)$ are all continuous in $[a, b]$.

Proof: Let's multiply the given ode by $v(x)$ so that the new ode, viz,

$$(a_0(x)v(x))\frac{d^2y}{dx^2} + (a_1(x)v(x))\frac{dy}{dx} + (a_2(x)v(x))y = 0 \qquad (6.34)$$

becomes self-adjoint.

The adjoint equation of (6.34) is:

$$a_0(x)v(x)\frac{d^2y}{dx^2} + \{2a_0'(x)v'(x) + 2a_0(x)v'(x) - a_1(x)v(x)\}\frac{dy}{dx}$$
$$+\{a_0''(x)v(x) + 2a_0'(x)v'(x) + a_0(x)v''(x) - a_1'(x)v(x)$$
$$-a_1(x)v'(x) + a_2(x)v(x)\}y = 0 \qquad (6.35)$$

Because of self-adjointness, the ode (6.34), must be identical with the ode (6.35).

Therefore, $a_0'(x)v(x) + a_0(x)v'(x) = a_1(x)v(x)$

i, e, $\dfrac{v'(x)}{v(x)} = \dfrac{a_1(x) - a_0'(x)}{a_0(x)}$

Integrating w.t. to x: $\ln v = \displaystyle\int \left(\frac{a_1 - a_0'}{a_0}\right) dx$

$= \displaystyle\int \frac{a_1}{a_0}dx - \ln(a_0) + \text{constant}$

i.e., $\ln(va_0) = \text{constant} + \displaystyle\int \frac{a_1}{a_0}dx$

$\therefore \quad v(x) = \dfrac{\text{constant}}{a_0(x)} \exp\left[\displaystyle\int \frac{a_1(x)}{a_0(x)}dx\right]$

Thus, dropping the constant (because for a homogeneous linear ode this constant bears no extra significance) we see that weight function is:

$$w(x) = \frac{1}{a_0(x)}\left[\int \frac{a_1(x)}{a_0(x)}dx\right]$$

Using this function, we have:

$$\bar{a}_0(x) \equiv a_0(x)w(x) = \exp\left[\int \frac{a_1(x)}{a_0(x)}dx\right]$$

$$\bar{a}_1(x) \equiv a_1(x)\omega(x) = \frac{a_1(x)}{a_0(x)} \exp\left[\int \frac{a_1(x)}{a_0(x)} dx\right]$$

$$\bar{a}_2(x) \equiv a_2(x)\omega(x) = \frac{a_2(x)}{a_0(x)} \exp\left[\int \frac{a_1(x)}{a_0(x)} dx\right]$$

It can be easily checked that $\dfrac{d}{dx}(\bar{a}_0(x)) = \bar{a}_1(x)$

Hence the recast ode is:

$$\frac{d}{dx}\left[\bar{a}_0(x)\frac{dy}{dx}\right] + \bar{a}_2(x)y = 0 \tag{6.36}$$

Hence we observe that every second order linear ode can be always transformed to a self-adjoint ode and that it can be put in the form (6.36) which is typical pattern of a Sturm-Liouville problem.

As claimed before, let's now verify that the adjoint of the adjoint of the differential equation

$$a_0(x)\frac{d^2y}{dx^2} + a_1(x)\frac{dy}{dx} + a_2(x)y = 0 \tag{6.37a}$$

is the original equation (6.27) itself.

Equation (6.27) being given, its adjoint equation is given by

$$a_0(x)\frac{d^2y}{dx^2} + (2a_0'(x) - a_1(x))\frac{dy}{dx} + (a_0''(x) - a_1'(x) + a_2(x))y = 0 \tag{6.37b}$$

The adjoint equation of (6.37b) is

$$a_0(x)\frac{d^2y}{dx^2} + (2a_0'(x) - 2a_0'(x) + a_1(x))\frac{dy}{dx} + (a_0''(x) - 2a_0''(x)$$
$$+ a_1'(x) + a_0''(x) - a_1'(x) + a_2(x))y = 0 \tag{6.37c}$$

However, (6.37c) being identical with (6.37a), our conclusion is true.

Example (22) : Verify whether the following ode is self-adjoint or not. If yes, cast it into the Sturm-Liouville form (6.36), and if not, convert it to self-adjoint ode and then cast it to Sturm-Liouville's form.

$$(x^4 + x^2)\frac{d^2y}{dx^2} + 2x^3\frac{dy}{dx} + 3y = 0$$

Comparison with the standard form of second order ode yields:

$$a_0(x) = x^4 + x^2; \ a_1(x) = 2x^3; \ a_2(x) = 3.$$

Since $a'_0(x) = 4x^3 + 2x \neq a_1(x)$, we conclude that the given ode is not self-adjoint.

In the preceding theorem (i.e, Theorem 6.6) we found that the weight function required to render self-adjointness to this ode is given by

$$
\begin{aligned}
\omega(x) &= \frac{1}{(x^4 + x^2)} \exp\left[\int \frac{2x^3 dx}{x^2(x^2+1)}\right] \\
&= \frac{1}{x^2(x^2+1)} \exp\left[\int \frac{d(x^2+1)}{(x^2+1)}\right] = \frac{1}{x^2}
\end{aligned}
$$

The transformed ode will be of the form (6.36) i,e,

$$
\frac{d}{dx}\left[\bar{a}_0(x)\frac{dy}{dx}\right] + \bar{a}_2(x)y = 0,
$$

where $\left.\begin{aligned} \bar{a}_0(x) &= a_0(x)\omega(x) = (x^2+1) \\ \bar{a}_2(x) &= a_2(x)\omega(x) = \frac{3}{x^2} \end{aligned}\right\}$

Hence the newlook self-adjoint ode is:

$$
\frac{d}{dx}\left[(x^2+1)\frac{dy}{dx}\right] + \frac{3}{x^2}y = 0.
$$

We now once again address the task of finding the formal adjoint of the differential operator

$$
L = \left(a_0(x)\frac{d^2}{dx^2} + a_1(x)\frac{d}{dx} + a_2(x)\right).
$$

To fulfil our aim, let's consider along with L, two sufficiently differentiable functions $f(x)$ and $g(x)$ defined over some interval $[a, b]$.

Define $I = \int_a^b g(x)(Lf(x))dx$ and try to compute it by method of by parts integration.

Therefore, $\displaystyle I = \int_a^b g(x)(Lf(x))dx$

$$
= \int_a^b g(x)\left(a_0(x)\frac{d^2 f}{dx^2} + a_1(x)\frac{df}{dx} + a_2(x)f\right)dx
$$

$$
= \int_a^b g(x)a_0(x)\frac{d^2 f}{dx^2}dx + \int_a^b g(x)a_1(x)\frac{df}{dx}dx
$$

$$
+ \int_a^b g(x)a_2(x)f(x)dx
$$

$$1\text{st term} = \left[\frac{df}{dx}a_0(x)g(x)\right]\Big|_{x=a}^{x=b} - \int_a^b \frac{d}{dx}(a_0(x)g(x))\frac{df}{dx}dx$$

$$= \left[\frac{df}{dx}.a_0(x)g(x)\right]\Big|_{x=a}^{x=b} - \left[\frac{d}{dx}(a_0(x)g(x))f(x)\right]\Big|_{x=a}^{x=b}$$

$$+ \int_a^b \frac{d^2}{dx^2}(a_0(x)g(x)).f(x)dx$$

$$= \left[\frac{df}{dx}.a_0(x)g(x) - \frac{d}{dx}(a_0(x)g(x))f(x)\right]\Big|_{x=a}^{x=b}$$

$$+ \int_a^b f(x)\left[a_0''(x)g(x) + 2a_0'(x)g'(x) + a_0(x)g''(x)\right]dx$$

$$2\text{nd term} = \int_a^b g(x)a_1(x)\frac{df}{dx}dx$$

$$= [(a_1(x)g(x))f(x)]\Big|_{x=a}^{x=b} - \int_a^b \frac{d}{dx}(a_1(x)g(x)).f(x)dx$$

$$= [(a_1(x)g(x))f(x)]\Big|_{x=a}^{x=b} - \int_a^b (a_1'(x)g(x) + a_1(x)g'(x)) \times f(x)dx$$

$$\text{Hence } I = \left[\frac{df}{dx}.a_0(x)g(x) - \frac{d}{dx}(a_0(x)g(x))f(x) + a_1(x)g(x)f(x)\right]\Big|_{x=a}^{x=b}$$

$$+ \int_a^b f(x)\left[a_0''(x)g(x) + 2a_0'(x)g'(x) + a_0(x)g''(x)\right.$$

$$\left. - a_1'(x)g(x) - a_1(x)g'(x) + a_2(x)g(x)\right]dx$$

$$= \int_a^b f(x)\left[a_0\frac{d^2}{dx^2} + (2a_0' - a_1)\frac{d}{dx} + (a_2 - a_1' + a_0'')\right]g(x)dx$$

$$+ \left[\frac{df}{dx}a_0(x)g(x) - \frac{d}{dx}(a_0(x)g(x))f(x) + a_1(x)g(x)f(x)\right]\Big|_{x=a}^{x=b}$$

$$\equiv \int_a^b f(x)\left(L^\dagger g(x)\right)dx$$

$$+ \left[\frac{df}{dx}a_0(x)g(x) - \frac{d}{dx}(a_0(x)g(x))f(x) + a_1(x)g(x)f(x)\right]\Big|_{x=a}^{x=b}$$

$$\text{where } L^\dagger \equiv a_0\frac{d^2}{dx^2} + (2a_0' - a_1)\frac{d}{dx} + (a_2 - a_1' + a_0'')$$

Therefore, $I = \displaystyle\int_a^b f(x)\left(L^\dagger g(x)\right) dx$

$$+ \left[a_0(b)\left(\left.\frac{df}{dx}\right|_{x=b}\cdot g(b) - \left.\frac{dg}{dx}\right|_{x=b} f(b)\right)\right.$$

$$+ \left(a_1(b) - \left.\frac{da_0}{dx}\right|_{x=b}\right) f(b)g(b) - a_0(a)\left(\left.\frac{df}{dx}\right|_{x=a} g(a)\right.$$

$$\left.\left. - \left.\frac{dg}{dx}\right|_{x=a} f(a)\right) - \left(a_1(a) - \left.\frac{da_0}{dx}\right|_{x=a}\right) f(a)g(a)\right]$$

If we now demand the differential operator L to be self-adjoint, then

$$a_1(x) = \frac{da_0}{dx} \ \forall x \in [a,b] \text{ and so}$$

$$I = \int_a^b f(x)\left(L^\dagger g(x)\right) dx + \left[a_0(b)\left(\left.\frac{df}{dx}\right|_{x=b}\cdot g(b) - \left.\frac{dg}{dx}\right|_{x=b} f(b)\right)\right.$$

$$\left. - a_0(a)\left(\left.\frac{df}{dx}\right|_{x=a}\cdot g(a) - \left.\frac{dg}{dx}\right|_{x=a}\cdot f(a)\right)\right] \qquad (6.38)$$

We now want to make this second term (popularly known as surface term in Green's generalised identity) of (6.38) zero so that we have the following relation:

$$\int_a^b g(x)\left(Lf(x)\right) dx = \int_a^b f(x)(L^\dagger g(x))dx,$$

for any functions f and g that are sufficiently differentiable over $[a,b]$. One may perfectly choose $f, g \in C^2[a,b]$.

To fulfil our mission of making the above mentioned surface term zero, we may impose either Dirichlet's conditions, i,e, $f(a) = f(b) = 0$; or Neumann's conditions, viz, $\left.\frac{df}{dx}\right|_{x=a} = \left.\frac{df}{dx}\right|_{x=b} = 0$ or a more generalized mixed condition, viz,

$$\alpha f(a) - \left.\frac{df}{dx}\right|_{x=a} = \beta f(b) - \left.\frac{df}{dx}\right|_{x=b} = 0 \ (\alpha, \beta \text{ being real}).$$

We now address these subsidiary boundary conditions separately.

Case-1: Dirichlet's conditions, viz, $f(a) = f(b) = 0$ assumed.

$\therefore$ The surface term boils down to

$$a_0(b).\left.\frac{df}{dx}\right|_{x=b}\cdot g(b) - a_0(a)\left.\frac{df}{dx}\right|_{x=a}\cdot g(a) \qquad (6.38a)$$

However, $\frac{df}{dx}\Big|_{x=a}$ and $\frac{df}{dx}\Big|_{x=b}$ being arbitrary/unspecified, the only way we can make (6.38a) zero is by setting the co-efficients of $\frac{df}{dx}\Big|_{x=a}$ and $\frac{df}{dx}\Big|_{x=b}$ as zeros, i,e, $a_0(a)g(a) = 0 = a_0(b)g(b)$

However, as $a_0(x) \neq 0$ for $x \in [a, b]$, we conclude that $g(a) = g(b) = 0$.

Hence if f satisfies Dirichlet's conditions and L is a self-adjoint second order differential operator, then g should also satisfy the same Dirichlet's conditions.

Case-2: Neumam's conditions, viz, $\frac{df}{dx}\Big|_{x=a} = \frac{df}{dx}\Big|_{x=b} = 0$ assumed.

Therefore the surface term reduces to

$$a_0(a)f(a) \frac{dg}{dx}\Big|_{x=a} - a_0(b)f(b) \frac{dg}{dx}\Big|_{x=b} \tag{6.38b}$$

However, $f(a)$ and $f(b)$ being arbitrary, the only way one can make (6.38b) equal to zero is by setting the co-efficients of $f(a)$ and $f(b)$ as zeros.

$$\therefore \qquad a_0(a). \frac{dg}{dx}\Big|_{x=a} = 0 \quad \text{and} \quad a_0(b). \frac{dg}{dx}\Big|_{x=b} = 0$$

However, $a_0(a) \neq 0$ and $a_0(b) \neq 0$ implies $\frac{dg}{dx}\Big|_{x=a} = 0 = \frac{dg}{dx}\Big|_{x=b}$.

Thus when f satisfies Neumann's conditions and L is a self-adjoint differential operator, then g should also satisfy Neumann's conditions as f in $[a, b]$.

Case-3: General mixed conditions, viz,

$$\alpha f(a) - \frac{df}{dx}\Big|_{x=a} = \beta f(b) - \frac{df}{dx}\Big|_{x=b} = 0 \text{ assumed.}$$

The surface term now reduces to

$$f(b)a_0(b) \left(\beta g(b) - \frac{dg}{dx}\Big|_{x=b} \right) - f(a)a_0(a) \left(\alpha g(a) - \frac{dg}{dx}\Big|_{x=a} \right) \tag{6.38c}$$

However, $f(a)$ and $f(b)$ being arbitrary or unspecified, the only way by which we can make (6.38c) zero is by setting the co-efficients of $f(a)$ and $f(b)$ equal to zeros. This leads us to

$$\alpha g(a) - \frac{dg}{dx}\Big|_{x=a} = 0 = \beta g(b) - \frac{dg}{dx}\Big|_{x=b} \quad (\because a_0(a), a_0(b) \neq 0)$$

Thus when f satisfies general mixed conditions and L is a self-adjoint second order differential operator, then g should also satisfy the same mixed type boundary conditions as f.

So in all the three cases we observe that the operator L is not only self-adjoint but Hermitian also in the sense f and g satisfy identical boundary conditions. Sometimes we phrase this as following definition:

A differential operator L is **Hermitian** iff

(i) L is self-adjoint, i,e, L is identical with its adjoint $L^\dagger$

(ii) Domains of L and $L^\dagger$ are identical in the sense that the boundary conditions satisfied by the functions f and g are identical.

6.9   Sturm-Liouville Problems

The stage is all set for the introduction of a special type of two-point boundary value problem, that consists of

(a) A homogeneous linear second order differential equation of the form

$$\frac{d}{dx}\left[a_0(x)\frac{dy}{dx}\right] + [a_2(x) + \lambda]\, y = 0 \qquad (6.39a)$$

where $a_2(x) \in C[a,b]$; $a_0(x) \in C^1[a,b]$ and it maintains the same sign in $[a,b]$. λ is a parameter independent of x, and

(b) Two supplementary boundary conditions of the type:

$$\left.\begin{array}{rcl} c_1 y(a) + c_2 y'(a) &=& 0 \\ c_3 y(b) + c_4 y'(b) &=& 0 \end{array}\right\} \qquad (6.39b)$$

where c_1, c_2, c_3, c_4 are real constants so that c_1, c_2 are not simultaneously zeros and likewise, c_3, c_4 are not both zeros.

These two-point boundary conditions are called 'homogeneous' as they remain unaltered when y is replaced by ty in them.

The above boundary value problem (b.v.p.) is known as **Regular Sturm-Liouville problem** (briefly SL problem).

The ode appearing in 6.39(a) is self-adjoint and is of the given form

$$Ly + \lambda y = 0, \qquad (6.40)$$

$$\text{where}\quad L \equiv \frac{d}{dx}\left[a_0(x)\frac{d}{dx}\right] + a_2(x)$$

The Sturm-Liouville problem 6.39(a)-(b) is a self-adjoint b.v.p. in the sense that the underlying differential operator L is Hermitian and moreover, the so-called **Green's relation**

$$\int_a^b (uL(v) - vL(u))\ dx = 0 \qquad (6.41)$$

is satisfied for all $u(x), v(x)$ satisfying the homogeneous boundary conditions given in 6.39(b). To assert this claim let's see

$$
\begin{aligned}
& uL(v) - vL(u) \\
=\ & u\frac{d}{dx}\left(a_0(x)\frac{dv}{dx}\right) - v\frac{d}{dx}\left(a_0(x)\frac{du}{dx}\right) \\
=\ & a_0'(x)(uv' - vx') + a_0(x)(uv'' - vu') \\
=\ & a_0'(x)W(u,v)(x) + a_0(x)W'(u,v)(x) \quad [\because W(u,v) = uv' - u'v] \\
=\ & \frac{d}{dx}(a_0(x)W(u,v)(x))
\end{aligned}
$$

Hence $\qquad \displaystyle\int_a^b (uL(v) - vL(u))dx = [a_0(x)W(u,v)(x)]\Big|_{x=a}^{x=b} = 0$

because u and v satisfy the associated homogeneous boundary conditions 6.39(b).

Remark: (i) The supplementary homogeneous boundary conditions appearing in 6.39(b) closely resemble the form of generalised mixed condition quoted in case (3) of the discussion related to the proof of L being a Hermitian operator. Dirichlet and Neumann conditions are special cases of these boundary conditions.

(ii) The Sturm-Liouville problem we addressed in the above can be further generalised by replacing the second order differential equation with the following one:

$$\frac{d}{dx}\left[a_0(x)\frac{dy}{dx}\right] + [a_2(x) + \lambda a_3(x)]y = 0,$$
$$\text{i, e,}\quad Ly + \lambda a_3(x)y = 0 \qquad (6.41a)$$

where $a_0(x)$ and $a_2(x)$ satisfy in letter and spirit the same conditions as before while the new entrant $a_3(x)$ should be of class $C[a, b]$ and should it maintain same sign over $[a, b]$. $a_3(x)$ is known as weight function.

(iii) Sometimes we observe that a particular format of SL type problem is addressed, where $a_2(x) = 0$ and $a_3(x) = \frac{1}{a_0(x)} \forall x \in [a, b]$. In this case if we switch over to the new independent variable z, defined by

$$z = \int \frac{dx}{p(x)},$$

then the ode appearing in Sturm-Liouville problem reduces to a normal form having constant co-efficients.

As an illustration, let's consider the Sturm-Liouville problem:

$$\frac{d}{dx}\left[x\frac{dy}{dx}\right] + \frac{\lambda}{x}y = 0; \quad y(1) = 0; \quad y'(e^{\pi}) = 0$$

We use the transformation $x = e^z$ so that $dx = e^z dz = x dz$.

Hence the supplementary conditions become:

$$y\Big|_{z=0} = 0; \quad \frac{dy}{dz}\Big|_{z=\pi} = 0$$

while the governing second order ode becomes:

$$\frac{d^2y}{dx^2} + \lambda y = 0.$$

(iv) Since every second order linear homogeneous ode is either self-adjoint or can be easily converted to a self-adjoint ode by use of a suitable multiplicative factor, there is no loss of generality if we assume the form of the ode appearing in SL problem to be of the canonical self-adjoint pattern.

We now concentrate on solving a Sturm-Liouville problem. As any solution of the Sturm-Liouville type boundary value problem must satisfy the second order ode together with the homogeneous boundary conditions (c.f. 6.39(a) and 6.39(b) in the defintion of SL problem), zero solution (i,e, trivial solution) exists to its credit. However, from a physical viewpoint, this trivial solution is of no great significance. We therefore focus on the hunt of non-trivial solution of SL type problem. However, existence of such non-trivial solutions depend on the value of the parameter λ appearing in the ode of the SL-type problem.

The values of the parameter λ for which non-zero solutions of the SL problem exist, are called the 'characteristic values' of the problem relevant to the assigned boundary conditions and the corresponding non-trivial solutions are referred to as 'characteristic functions' or 'eigenfunctions' of the problem. So if $\exists$ a number $\lambda = \lambda_0$ and a function $\psi_0(x) \neq 0$

defined over $[a, b]$ such that the ode and the supplementary conditions are satisfied whenever $\lambda = \lambda_0$ and $\psi_0(x) \neq 0$, then λ_0 is a characteristic value/eigenvalue of the SL problem while ψ_0 is its corresponding characteristic function of eigenfunction. We now explore some salient properties of SL-type b.v.p.

(a) The operator L involved in the Sturm-Liouville problem being Hermitian, all its eigenvalues are real. This basic result in linear algebra can be outlined as follows:

If $Ly = -\lambda y$ and $L = L^\dagger$, then $(Ly)^\dagger = (-\lambda y)^\dagger$

$\quad$ i, e, $\quad y^\dagger L = -\lambda^* y^\dagger$ $\;(\because\; L^\dagger = L$ and λ^* is complex conjugate of λ)

$\quad \therefore \quad (y^\dagger L)y = -\lambda^* y^\dagger y$

$\quad$ i, e, $\quad y^\dagger(Ly) = -\lambda^*(y^\dagger y)$

$\quad$ or, $\quad -\lambda(y^\dagger y) = -\lambda^*(y^\dagger y)$

$\quad \therefore \quad \lambda = \lambda^* \;(\because\; y^\dagger y \neq 0)$ $\quad$ i, e, $\quad \lambda$ is real.

(b) If λ and μ be two distinct eigenvalues of L and $\psi_\lambda(x)$, $\psi_\mu(x)$ be its corresponding eigenfunctions, then

$$L\psi_\lambda(x) = -\lambda\psi_\lambda(x); \; L\psi_\mu(x) = -\mu\psi_\mu(x)$$

As L is self-adjoint, we have, $L^\dagger = L$ and so

$$-\lambda\psi_\lambda(x) = L\psi_\lambda(x) = L^\dagger\psi_\lambda(x)$$

$$\text{i, e,} \quad (-\lambda\psi_\lambda(x))^\dagger = \psi_\lambda^\dagger(x)L$$

$$\text{i, e,} \quad -\lambda\psi_\lambda^\dagger(x) = \psi_\lambda^\dagger(x)L$$

$$\therefore \quad -\lambda\psi_\lambda^\dagger(x)\psi_\mu(x) = \psi_\lambda^\dagger(x)L\psi_\mu(x) = -\mu\psi_\lambda^\dagger(x)\psi_\mu(x)$$

$$\text{i, e,} \quad (\lambda - \mu)\psi_\lambda^\dagger(x)\psi_\mu(x) = 0$$

As $\lambda \neq \mu$, it follows that $\psi_\lambda^\dagger(x)\psi_\mu(x) = 0$, proving that eigenfunctions corresponding to distinct eigenvalues are orthogonal. One may also make use of the Green's relation (6.41) to prove this property

Remark:

(a) $\quad$ If λ and μ be distinct eigenvalues of a SL problem (6.39)(a)-(b) defined over $[a, b]$ and ψ_λ, ψ_μ be their respective eigenfunctions, then they are orthogonal over $[a, b]$ w.r. to the weight function 1.

In other words, these eigenfunctions are pairwise orthogonal over $[a, b]$.

If (λ, ψ_λ) be an eigenpair, then we have in notation of (6.40), $L\psi_\lambda + \lambda\psi_\lambda = 0$ with $c_1\psi_\lambda(a) + c_2\psi_\lambda'(a) = 0$; $c_3\psi_\lambda(b) + c_4\psi_\lambda'(b) = 0$. Similarly if (μ, ψ_μ) be another eigenpair $(\mu \neq \lambda)$ then

$$L\psi_\mu + \mu\psi_\mu = 0 \text{ with } c_1\psi_\mu(a) + c_2\psi_\mu'(a) = 0 \text{ ; } c_3\psi_\mu(b) + c_4\psi_\mu'(b) = 0$$

By (6.41) now it follows that

$$(\lambda - \mu) \int_a^b \psi_\lambda(x) \, \psi_\mu(x) \, dx = \int_a^b (\psi_\mu \, L\psi_\lambda - \psi_\lambda \, L\psi_\mu) \, dx = 0$$

As $\lambda \neq \mu$, it follows that $\displaystyle\int_a^b \psi_\lambda(x) \, \psi_\mu(x) \, dx = 0$

It is worth mentioning at this point that if the differential equation of the Sturm-Liouville problem were taken

$$Ly + \lambda a_3(x)y = 0 \qquad\qquad \cdots\cdots (6.40 \; (a))$$

with the supplementary homogeneous boundary conditions remaining unaltered, then also the above orthogonality would hold, only change being that in lieu of

$\displaystyle\int_a^b \psi_\lambda(x)\psi_\mu(x)dx = 0$ it would read $\displaystyle\int_a^b \psi_\lambda(x)\psi_\mu(x)a_3(x)dx = 0$,

i,e, $\psi_\lambda(x)$ and $\psi_\mu(x)$ would now be orthogonal in [a, b] w.r. to the weight function $a_3(x)$.

(c) For a given Sturm-Liouville problem there exists a countably infinite number of eigenvalues that can always be arranged in the form of a strictly monotone increasing sequence $\{\lambda_n\}$. However, $\{\lambda_n\}$ is unbounded and so by monotone convergence theorem in $\mathbb{R}$, $\{\lambda_n\}$ is divergent. We shall observe this in the examples coming up in the following.

(d) The eigenfunction ψ_n corresponding to each eigenvalue $\lambda_n (n = 1, 2, \cdots \infty)$ has <u>exactly</u> $(n - 1)$ zeros in (a, b).

(e) The eigenvalues of a regular SL-problem are simple, i,e, corresponding to each eigenvalue λ_n, $\exists$ one and only one linearly independent eigenfunction, say ψ_n. So the eigenfunctions associated with a given eigenvalue form a one-dimensional subspace of $C[a, b]$. The proof is done in the following by method of contradiction.

If possible, let there exist two linearly independent solutions $\psi_n(x)$ and $\phi_n(x)$, corresponding to the same eigenvalue λ_n. Define

$$W(\phi_n, \psi_n)(x) = \phi_n(x)\, \psi'_n(x) - \psi_n(x)\, \phi'_n(x) \ \forall \ x \ \in [\, a, b\,]$$

As ϕ_n and ψ_n both satisfy boundary conditions 6.39(b), we have

$$\left. \begin{array}{ll} c_1\, \phi_n(a) + c_2\, \phi'_n(a) = 0 \quad ; \quad c_3\, \phi_n(b) + c_4\, \phi'_n(b) = 0 \\ c_1\, \psi_n(a) + c_2\, \psi'_n(a) = 0 \quad ; \quad c_3\, \psi_n(b) + c_4\, \psi'_n(b) = 0 \end{array} \right\} \quad (6.42)$$

Since c_1, c_2 are not simultaneously zeros and the first and third relations in (6.42) can be written in the matrix form as

$$\begin{pmatrix} \phi_n(a) & \phi'_n(a) \\ \psi_n(a) & \psi'_n(a) \end{pmatrix} \begin{pmatrix} c_1 \\ c_2 \end{pmatrix} = \begin{pmatrix} 0 \\ 0 \end{pmatrix},$$

it follows that the co-efficient matrix has to be singular, i,e,

$$W(a) = \psi'_n(a)\, \phi_n(a) - \phi'_n(a)\, \psi_n(a) = 0.$$

(Similarly had we worked with the second and fourth relations in (6.42), we would get $W(b) = 0$). So by property of Wronskian it follows that $W(x) = 0 \ \forall x \epsilon [a, b]$. This implies that $\phi_n(x)$ and $\psi_n(x)$ are dependent, a direct contradiction to our hypothesis. Hence the conclusion.

Remark: (i) This result (e) depends directly on homogeneity of the separated boundary conditions (6.39)(b). If in lieu of these conditions we had our periodic boundary conditions like $y(a) = y(b)$; $y'(a) = y'(b)$, the eigenvalues of the SL-problem would not be simple as from the given conditions we would be led to $W(a) = W(b)$ type relation that makes no conclusive yield.

(ii) The discrete nature of the eigenspectrum of SL problem (6.39) (a)-(b) would be violated if the continuity conditions of at least one $a_0(x), a_2(x), a_3(x)$ get violated in [a,b]. The following example is really an eye-opener in this sense.

Example (23) :   Solve the eigenvalue problem :

$$x^2 y'' - xy' + xy = 0 \ ; \ 0 \leqslant x \leqslant 1 \text{ subject to } y(0) = y(1) = 0.$$

This being a Cauchy-Euler type ode we may rewrite it as

$$\left\{ (\theta - 1)^2 + (\lambda - 1) \right\} y = 0$$

provided we agree to write $\theta \equiv x\frac{d}{dx}$. The general solution is :

$$y(x) = c_1 x^{(1 + \sqrt{1 - \lambda})} + c_2 x^{(1 - \sqrt{1 - \lambda})}$$

As $y(0) = y(1) = 0$ it follows that

$$y(x) = c_1 \left[x^{(1 + \sqrt{1 + \lambda})} + x^{1 - \sqrt{1 - \lambda}} \right]$$

For real solutions, $\lambda < 1$ is a must. On writing $1 - \lambda = \mu^2$, we have

$$y(x) = c_1 x^{(1 + \mu)} + c_2 x^{1 - \mu}$$

If $\lambda = 0$, then $\mu = 0$ and so $y = c_1 x^2 + c_2$.

However, boundary conditions show that there does not exist any non-zero solution. Similarly $\lambda < 0$ is ruled out as $\lambda = -k^2$ makes

$$y(x) = c_1 x^{(1 + \sqrt{1 + k})} + c_2 x^{(1 - \sqrt{1 + k^2})}$$

$$\therefore \quad 0 = y(1) = c_1 + c_2, \text{ so that}$$

$$y(x) = c_1 \left[x^{(1 + \sqrt{1 + k^2})} - x^{(1 + \sqrt{1 + k^2})} \right]$$

blows out when is to be satisfied.

Hence $\lambda > 0$ is the only option, ensuring that the range or eigenvalues is $(0, 1)$, a continuum. But why this departure? But why this departure? The answer lies in the fact that when the given ode is put in the SL form, viz,

$$\frac{d}{dx} \left(\frac{y'}{x} \right) + \frac{\lambda}{x^3} \, y = 0,$$

then $a_3(x) = \frac{1}{x^3}$ had infinite discontinuity at $x = 0$, a violation to the basic conditions of any SL-problem.

However, if $\lambda = 1$, the roots of the auxiliary equations are same and hence

$$y(x) = (c_1 + c_2 \, ln \, x)x$$

As $y(1) = 0 = c_1$, we would get $y(x) = c_2 \, x \, ln \, x$, so that for $x \to 0+$, $y \to 0$. Hence the continuous range of eigenvalues is the open-closed interval $(0, 1]$.

In the following we present a variety of illustrative examples related to Sturm-Liouville problem.

Example (24) :   Solve the eigenvalue problem:

$$\frac{d^2y}{dx^2} + \lambda y = 0 \ \text{ subject to } y(0) = 0 \text{ and } y'(\pi) = 0$$

Since the given eigenvalue problem is a Sturm-Liouville problem, $\lambda \in \mathbb{R}$ and so by law of trichotomy, $\exists$ three possibilities, viz, $\lambda > 0$, $\lambda = 0$ and $\lambda < 0$.

We shall deal with these three cases separately.

Case-1: If $\lambda > 0$, we may without loss of generality write $\lambda = \omega^2$ so that the bvp becomes:

$$\frac{d^2y}{dx^2} + \omega^2 y = 0 \ \text{ subject to } \ y(0) = 0 \ \text{ and } y'(\pi) = 0$$

The general solution of this ode is:

$$y = c_1 \cos \omega x + c_2 \sin \omega x, \ (c_1 \text{ and } c_2 \text{ being arbitrary constants})$$

Now for the boundary value problem,

$$y(0) = 0 \Rightarrow 0 = c_1$$
$$\therefore \quad y = c_2 \sin \omega x$$
$$\text{But} \quad 0 = y'(\pi) = \omega c_2 \cos(\omega \pi);$$
$$\text{Thus} \quad \cos(\omega \pi) = 0 \ (\because \ c_2 = 0 \text{ leads us to zero solutions})$$
$$\text{i, e,} \quad \omega = \left(n + \frac{1}{2}\right); \ n \in Z \Rightarrow \lambda = \omega^2 = \left(n + \frac{1}{2}\right)^2; \ n \in Z$$

As λ remains same under the substitution $n \to -(n+1)$, it transpires that one can write $\lambda = \lambda_n = \left(n - \frac{1}{2}\right)^2$ with $n \in N$. Thus we have a strictly monotone increasing sequence of eigenvalues for the b.v.p when $\lambda > 0$.

Further, if $\psi_n(x)$ be the eigenfunction for a particular λ_n, then

$$\psi_n(x) = c_2 \sin \left(\left(n - \frac{1}{2}\right) x\right) \ \text{ with } c_2 \text{ being arbitrary.}$$

Without loss of generality we set $c_2 = 1$ so that $\psi_n(x) = \sin \left(\left(n - \frac{1}{2}\right) x\right)$

Case-2: If $\lambda = 0$, then b.v.p given reduces to:

$$\frac{d^2y}{dx^2} = 0 \quad \text{subject to} \quad y(0) = 0 \text{ and } y'(\pi) = 0$$

The general solution being $y = c_1 + c_2 x$, (c_1 and c_2 being arbitrary) we have $y(0) = 0 = c_1$, so that $y = c_2 x$.

Again $y'(\pi) = c_2 \pi = 0 \Rightarrow c_2 = 0$

So for $\lambda = 0$, $\not\exists$ any non-trivial solution to the b.v.p.

Case-3: If $\lambda < 0$, then $\lambda = -k^2$ and hence given b.v.p reads:

$$\frac{d^2y}{dx^2} - k^2 y = 0 \text{ subject to } y(0) = 0 = y'(\pi).$$

The general solution of the b.v.p is: $y = c_1 e^{kx} + c_2 e^{-kx}$,

where $y(0) = 0 = c_1 + c_2$ and $0 = y'(\pi) = k\left(c_1 e^{k\pi} - c_2 e^{-k\pi}\right)$

Thus $0 = c_1\left(e^{k\pi} + e^{-k\pi}\right) = 2c_1 \cosh(k\pi) \Rightarrow c_1 = 0 = c_2$

Hence for $\lambda < 0$ case we have no non-trivial solution.

So we conclude that only for $\lambda > 0$, $\exists$ a sequence of non-zero solutions i,e, eigenfunctions $\left\{ \left(\sin\left(n - \frac{1}{2}\right)\right) x \mid x \in [0, \pi]\right\}$, the corresponding eigenvalues being $\left\{\left(n - \frac{1}{2}\right)^2\right\}$.

Remark (a): We now observe that if m, n are distinct natural numbers, then the eigenfunctions $\psi_m(x)$ and $\psi_n(x)$ associated with eigenvalues $\left(m - \frac{1}{2}\right)^2$ and $\left(n - \frac{1}{2}\right)^2$ are orthogonal as

$$\int_0^\pi \psi_m^*(x)\psi_n(x)dx$$

$$= \int_0^\pi \sin\left(\left(m - \frac{1}{2}\right)x\right)\sin\left(\left(n - \frac{1}{2}\right)x\right)dx$$

$$= \frac{1}{2}\int_0^\pi \{\cos((m-n)x) - \cos((m+n-1)x)\}\, dx$$

$$= \frac{1}{2}\left[\frac{\sin(m-n)x}{(m-n)} - \frac{\sin(m+n-1)x}{(m+n-1)}\right]_0^\pi = 0 \text{ provided } m \neq n.$$

Further if $m = n \in N$, then

$$\int_0^\pi \psi_m^*(x)\psi_n(x)dx = \int_0^\pi \psi_n^*(x)\psi_n(x)dx = \int_0^\pi |\psi_n(x)|^2\, dx$$

$$= \frac{1}{2}\cdot\int_0^\pi 2\sin^2\left(\left(n - \frac{1}{2}\right)x\right)dx = \frac{1}{2}\int_0^\pi \{1 - \cos(2n-1)x\}dx = \frac{\pi}{2}$$

Hence $\left\{ \sqrt{\dfrac{2}{\pi}} \sin\left(\left(n - \dfrac{1}{2}\right)x\right) \bigg| \, x \in [a,b] \right\}$ is a sequence of orthonormal

eigenfunctions of the b.v.p.

(b) We in our earlier discussion found that the b.v.p we are presently dealing with was obtained from the Sturm-Liouville problem

$$\frac{d}{dx}\left[x\frac{dy}{dx}\right] + \frac{\lambda}{x}y = 0; \ y(1) = 0 \ ; \ y'(e^\pi) = 0$$

via a transformation of the type $x = e^z$.

Hence the eigenfunctions of the original b.v.p are:

$$\overline{\psi}_n(x) = \sin\left(\left(n - \frac{1}{2}\right)\ln x\right)$$

and the corresponding eigenvalues being $\lambda_n = \left(n - \frac{1}{2}\right)^2$

Example (25) : Consider the eigenvalue problem:

$$\frac{d^2y}{dx^2} + \lambda y = 0 \ \text{ subject to } \ y'(0) = 0 \ \text{ and } \ y'(2\pi) = 0; \ n \in N$$

Case-1: $\lambda > 0$. If we agree to write $\lambda = \omega^2$, then the general solution of the b.v.p will be given by

$$y = c_1 \cos \omega x + c_2 \sin \omega x,$$

where $y'(0) = c_2\omega$ and $y'(2\pi) = 0 = -\omega c_1 \sin(2\pi\omega) + \omega c_2 \cos(2\pi\omega)$

But this implies $c_2 = 0$ and $0 = \sin(2\pi\omega)$ as otherwise one can't have a non-trivial solution for $\lambda > 0$.

Thus $\quad 2\pi\omega = n\pi \ (n \in z)$ i.e., $\lambda = \omega^2 = \dfrac{n^2}{4}$; $(n \in Z)$

However, it can be put as $\lambda_n = \dfrac{n^2}{4}$ with $n \in N$

The eigenfunctions corresponding to λ_n is : $\ \psi_n(x) = C_1 \cos\left(\dfrac{n}{2}x\right)$

Again if we drop C_1, the arbitrary multiplicative constants, then we may consider $\cos\left(\dfrac{n}{2}x\right)$; $x \in [0, 2\pi]$ as eigenfunction corresponding to

$\lambda_n = \frac{n^2}{4}$; $n \in N$.

Case-2: $\lambda = 0$. Here the general solution of the b.v.p is: $y = a + bx$, where $y'(0) = b = 0$ and $y'(2\pi) = b = 0$. Thus $y = a$, with a being arbitrary. There is enough justification in presuming that $a = 1$, and so $\lambda = 0$ is an eigenvalue of the b.v.p with the constant function 1 as its eigenvector.

Case-3: $\lambda < 0$. Without loss of generality presume $\lambda = -k^2$. Here the b.v.p given reduces to

$$\frac{d^2y}{dx^2} - k^2 y = 0 \quad \text{subject to} \quad y'(0) = 0 \text{ and } y'(2\pi) = 0.$$

The general solution of this b.v.p is: $y = C_1 e^{kx} + C_2 e^{-kx}$ subject to $0 = y'(0) = k(C_1 - C_2)$ and $0 = y'(2\pi) = k\left(C_1 e^{2k\pi} - C_2 e^{-2k\pi}\right)$.

Herefrom only possibility that opens up is: $C_1 = C_2 = 0$, leading to the conclusion that for $\lambda < 0$, $\nexists$ a non-zero solution of the b.v.p.

To sum up all cases, we can infer that the b.v.p has eigenvalues given by $\lambda = 0$ and $\lambda = \frac{n^2}{4}\Big|$ $n \in N$.

The corresponding eigenvalues are : 1 and $\left\{ \cos\left(\frac{n}{2}x\right)\Big| x \in [0, 2\pi]\right\}$

Remark: A comparative study between Example-24 and Example-25 shows that mere change of supplementary boundary conditions in a SL type b.v.p drastically changes the anatomy of the problem. The boundary conditions prescribed in Example-25 is of Neumann type.

Example (26) : Solve the eigenvalue problem :

$$\frac{d^2y}{dx^2} + \lambda y = 0 \; ; \; y(0) = y(l) \text{ and } y'(0) = y'(l).$$

By law of trichotomy, parameter λ is either zero, negative or positive.

Case I $\lambda = 0$. Here $y(x) = c_1 x + c_2$

Since $y(0) = y(l)$, we have $c_1 = 0$ so that $y(x) = c_2$, indicating that $y = 1$ serves as an eigenfunction.

Case II $\lambda < 0$, We write $\lambda = -\mu^2$ so that the general solution can be written as $y(x) = ae^{\mu x} + be^{-\mu x}$. However, the boundary condition $y(0) = y(l)$ implies $a + b = ae^{\mu l} + be^{-\mu l}$ while the boundary condition $y'(0) = y'(l)$ yields, $\mu(a - b) = (ae^{\mu l} - be^{-\mu l})\mu$. These relations can be written in the matrix form

$$\begin{pmatrix} (1 - e^{\mu l}) & (1 - e^{-\mu l}) \\ \mu(1 - e^{\mu l}) & -\mu(1 - e^{-\mu l}) \end{pmatrix} \begin{pmatrix} a \\ b \end{pmatrix} = \begin{pmatrix} 0 \\ 0 \end{pmatrix}$$

If a and b be not zeros simultaneously, we have to make the co-efficient matrix singular. However, the determinant of the co-efficient matrix cannot be zero as its value is 2μ (cosh $\mu l - 1$) where $\mu \neq 0$ and cosh $\mu l > 1$. So $\nexists$ any non-zero solution to this b.v.p for $\lambda < 0$.

Case III   If $\lambda > 0$, one can write $\lambda = \mu^2$ and so

$$\frac{d^2 y}{dx^2} + \mu^2 y = 0 \quad \text{gives} \quad y(x) = A \cos \mu x + B \sin \mu x,$$

where A and B have to satisfy the relations.

$$A(-1 + \cos \mu l) + B \sin \mu l = 0 \text{ and } A\mu \sin \mu l + B \, \mu(1 - \cos \mu l) = 0$$

In matrix form they can be written as :

$$\begin{pmatrix} \cos \mu - 1 & \sin \mu \\ \mu \sin \mu l & -\mu(1 - \cos \mu l) \end{pmatrix} \begin{pmatrix} A \\ B \end{pmatrix} = \begin{pmatrix} 0 \\ 0 \end{pmatrix}$$

However, A, B are not simultaneously zeros if the co-efficient matrix is singular, i.e, in other words,

$$\mu(\cos \mu l - 1)^2 + \mu \sin^2 \mu l = 0$$

As $\mu > 0$, it follows that $\cos \mu l = 1$ is to be satisfied.

But $\cos \mu l = 1$ implies $\mu = \dfrac{2n\pi}{l} \, ; \, n \in Z$

So $\lambda = \mu^2 = \left(\frac{2n\pi}{l}\right)^2$; $n \in Z$, indicating that the eigenvalues form an increasing an sequence. If $\{\lambda_n\}$ be the sequence of eigenvalues then the corresponding eigenfunctions are given as :

$$y_n(x) = A_n \cos \left(\frac{2n\pi}{l} x\right) + B_n \sin \left(\frac{2n\pi}{l} x\right) \, ; \, n \in N$$

This shows that the eigenvalues of this b.v.p are not simple, a direct contradiction to the property of having simple eigen values of a SL-problem. This departure is because of the fact that boundary conditions $y(0) = y(l)$ and $y'(0) = y'(l)$ used in the b.v.p are periodic and not of the separated type like (6.39)(b).

Example (27) : Find the eigenvalues and eigenfunctions of the following Sturm-Liouville problem:

$$\frac{d^2y}{dx^2} + \lambda y = 0; \ y(0) = 0; \ y(\pi) - y'(\pi) = 0$$

Since $\lambda \in \mathbb{R}$, we have to deal with three possible subcases separately.

Case-1 $\lambda > 0$; writing $\lambda = \omega^2$ for conveniences one may rewrite the SL problem as:

$$\frac{d^2y}{dx^2} + \omega^2 y = 0, \text{ subject to the boundary conditions}$$
$$y(0) = 0; \ y(\pi) - y'(\pi) = 0$$

The general solution of this ode will be of the form

$$y = C_1 \cos \omega x + C_2 \sin \omega x.$$
$$\text{where} \quad y(0) = 0 \Rightarrow C_1 = 0$$
$$\text{and} \quad 0 = y(\pi) - y'(\pi) = C_1 \cos \omega \pi + C_2 \sin \omega \pi - \omega C_2 \cos \omega \pi$$
$$\text{i, e,} \quad 0 = C_2(\sin \omega \pi - \omega \cos \omega \pi)$$

For nontrivial solutions of b.v.p, $C_2 \neq 0$ and so $\tan(\omega \pi) = \omega$.

The general solution of this transcendental equation in ω will be obtainable by numerical methods. However, customarily we may conclude that if ω_n be the positive roots of the above equation, then the sequence of eigenvalues $\{\lambda_n\}$ of the given SL problem will be given by $\{\omega_n^2\}$.

The eigenfunction corresponding to eigenvalue λ_n will be

$$\psi_n(x) = A_n \sin(\omega_n x), \text{ where } n = 1, 2, 3 \cdots \infty,$$

where A_n is some non-zero constant.

Case-2: $\lambda = 0$; The b.v.p now reduces to

$$\frac{d^2y}{dx^2} = 0 \text{ subject to } y(0) = 0; \ y(\pi) - y'(\pi) = 0$$

Hence the general solution of the ode will of the form $y = C_1 + C_2 x$,

where $0 = y(0) = C_1$; $0 = y(\pi) - y'(\pi) = C_1 + C_2\pi - C_2 = C_2(\pi - 1)$

$\therefore \ C_1 = C_2 = 0$, leading to the conclusion that $\nexists$ any non-zero solution of the b.v.p.

Case-3: $\lambda < 0$; In this case also, by routine procedure, it is easy to check that no non-zero solution to this b.v.p. exists.

Hence we conclude that the given Sturm-Liouville problem can have non-zero eigenvalues whenever parameter λ involved in it is positive.

In the context of the examples discussed so far, let's verify the theoretical result that the eigenfunction ψ_n corresponding to each eigenvalue $\lambda_n (n = 1, 2, \cdots \infty)$ has exactly $\overline{n-1}$ zeros in (a, b). provided $[a, b]$ is the interval for which the Sturm-Liouville problem is defined.

For example, let's reconsider the problem:

$$\frac{d}{dx}\left[x\frac{dy}{dx}\right] + \frac{\lambda}{x}y = 0; \ y(1) = 0; \ y'(e^\pi) = 0.$$

The eigenvalues are given by the monotone increasing sequence $\left\{\overline{\lambda}_n = \left(n - \frac{1}{2}\right)^2\right\}$ and the corresponding sequence of eigenfunctions by

$$\left\{\overline{\psi}_n(x) = \sin\left(\left(n - \frac{1}{2}\right)\ln x\right)\right\}$$

Now let's find the total number of zeros of $\overline{\psi}_n(x)$ (n fixed) in $[1, e^\pi]$.

$$\because \sin\left(\left(n - \frac{1}{2}\right)\ln x\right) = 0 \Rightarrow \left(n - \frac{1}{2}\right)\ln x = k\pi; \ k \in Z,$$

we have $x = e^{\frac{k\pi}{(n-\frac{1}{2})}}; \ k \in Z.$

Again $1 < x < e^\pi \Longrightarrow 1 < e^{\frac{k\pi}{(n-\frac{1}{2})}} < e^\pi$

$$\Longleftrightarrow 0 < k\pi < \left(n - \frac{1}{2}\right)\pi \quad \text{Since log is an injection}$$

$$\Longleftrightarrow 0 < k < \left(n - \frac{1}{2}\right)$$

Hence the admissible values of k for which $\overline{\psi}_n(x)$ vanishes in the interval $[1, e^\pi]$ are $1, 2, \cdots, \overline{n-1}$. This verifies the theoretical claim.

As another illustration reconsider the SL problem in Example (24)

$$\frac{d^2y}{dx^2} + \lambda y = 0, \text{ subject to } y'(0) = 0 \text{ and } y'(2\pi) = 0.$$

The eigenvalues were given by $\left\{ \lambda_n = \frac{n^2}{4} : n \in N \right\}$ and the corresponding eigenfunctions were $\left\{ \psi_n(x) = \cos\left(\frac{n}{2}x\right) \right\}; x \in [0, 2\pi]$.

Let's now find zeros of $\psi_n(x)$ in $[0, 2\pi]$.

Observe that $\cos\left(\frac{n}{2}x\right) = 0 \Rightarrow x = (2k+1)\frac{\pi}{n}, \ k \in Z$.

Again $0 < x < 2\pi \Longrightarrow 0 < (2k+1)\frac{\pi}{n} < 2\pi \Longleftrightarrow 0 < k\left(n - \frac{1}{2}\right)$

By the same logic as before, the total number of admissible values for which the eigenfunction $\psi_n(x)$ vanishes in $[0, 2\pi]$ is $(n-1)$.

Example (28) : If λ_k be the eigenvalues and $\psi_k(x)$ be the corresponding eigenfunctions of the Sturm-Liouville operator L defined by

$$L \equiv \frac{d}{dx}\left[a_0(x)\frac{d}{dx}\right] + a_2(x) \text{ with } a_0(x) \in C^1[a, b] \quad \& \ a_2(x) \in C[a, b]$$

relevant to the homogeneous boundary conditions given by

$$c_1 \, y(a) + c_2 \, y'(a) = 0 \quad \text{and} \quad c_3 \, y(b) + c_4 \, y'(b) = 0,$$

then prove that a solution of the inhomogeneous b.v.p

$$Ly + \lambda y = \sum_{k=1}^{n} A_k \, \psi_k(x) \qquad (A_k's \ \text{ being preassigned constants}),$$

subject to the same boundary conditions, is given by

$$y = \sum_{k=1}^{n} \frac{A_k \, \psi_k(x)}{\lambda - \lambda_k}, \text{provided } \lambda \text{ is not an eigenvalue of } L.$$

As $\psi_k(x), k = 1, 2, ..., n$ are eigenfunctions of the Strum-Liouville problem given here, we must have $L\psi_k(x) = -\lambda_k\psi_k(x) \ \forall \ k = 1, 2, ..., n; \ \lambda_k$ being corresponding eigenvalue.

To determine a solution of the given inhomogeneous b.v.p, we assume it to be a finite linear combination of the form

$$y(x) = \alpha_1 \, \psi_1(x) + \alpha_2 \, \psi_2(x) + \cdots\cdots + \alpha_n \, \psi_n(x)$$

where α_i's are to be determined so as to satisfy the ode

$$Ly + \lambda y = \sum_{k=1}^{n} A_k \, \psi_k(x)$$

$$\text{i, e, } (L + \lambda I) \sum_{k=1}^{n} \alpha_k \, \psi_k(x) = \sum_{k=1}^{n} A_k \, \psi_k(x)$$

$$\text{i, e, } \sum_{k=1}^{n} (\lambda - \lambda_k)\alpha_k \, \psi_k(x) = \sum_{k=1}^{n} A_k \, \psi_k(x) \tag{a}$$

Because of the linear independence of ψ_k's we conclude :

$$(\lambda - \lambda_k)\alpha_k = A_k \text{ for } k = 1, 2, ..., n \tag{b}$$

$$\text{i, e, } \quad \alpha_k = \frac{A_k}{\lambda - \lambda_k}, \text{ provided } \lambda \neq \lambda_k \text{ for any } k = 1, 2, ..., n$$

Thus only if λ is not an eigenvalue of L,

$$y(x) = \sum_{k=1}^{n} \alpha_k \, \psi_k(x) = \sum_{k=1}^{n} \frac{A_k}{\lambda - \lambda_k} \psi_k(x)$$

is a solution of the b.v.p. One should note that the boundary conditions are automatically satisfied in this case as the inhomogeneous term is a finite linear combination of the eigenfunctions of L and so lies in the inner product space spanned by them. Infact, while deriving the relation (b) from the relation (a) we used the orthogonality of the eigenfunctions in the inner product space $C[a, b]$.

Example (29) : Find the eigenvalues and corresponding eigenfunctions for the Sturm-Liouville problem comprising of the ode

$$y'' + \lambda y = 0; \ \ 0 \leqslant x \leqslant l$$

and the boundary conditions $y(0) = 0$; $y(l) = 0$.

Use the above eigenvalue problem to solve the b.v.p.

$$y'' + \lambda y = \sum_{k=1}^{n} a_k \sin \left(\frac{k\pi x}{l} \right)$$

subject to the boundary conditions $y(0) = y(l) = 0$, if $\lambda \neq \frac{k^2 \pi^2}{l^2}$

To solve the eigenvalue problem appearing in the first part we take up three possible cases, viz $\lambda < 0$; $\lambda = 0$ and $\lambda > 0$.

Case 1 : $\lambda < 0$. Without loss of generality we take $\lambda = -k^2$ so that the general solution of the b.v.p is of the form

$$y(x) = c_1 e^{kx} + c_2 e^{-kx},$$

where c_1, c_2 must abide by the conditions :

$$y(0) = 0 = c_1 + c_2 \quad \text{and} \quad y(l) = 0 = c_1 e^{kl} + c_2 e^{-kl}$$

However, this opens up the only possibility : $c_1 = c_2 = 0$, leading to the conclusion that the only solution for $\lambda < 0$ is the trivial solution and the numbers $\lambda < 0$ are not eigenvalues.

Case 2 : $\lambda = 0$. The differential equation reduces to $y'' = 0$, which has for its general solution

$y = A + Bx$, where A, B satisfy the conditions

$y(0) = 0 = B$ and $y(l) = 0 = A + Bl$

$\therefore A = B = 0$ and we can have only zero solution corresponding to $\lambda = 0$.

Case 3 : $\lambda > 0$. In this case one may write $\lambda = w^2$ so that the general solution of the b.v.p. is of the form :

$$y = c_1 \cos wx + c_2 \sin wx,$$

where $\quad y(0) = 0 = c_1 \quad$ and $\quad y(l) = 0 = c_1 \cos wl + c_2 \sin wl$

$\therefore \quad \sin wl = 0, \quad$ if $\quad c_2 \neq 0$.

Hence for $w = \frac{n\pi}{l}$; $n \in Z$ corresponds to non-trivial solutions of the b.v.p corresponding to the case $\lambda > 0$. One can write down the eigenvalues as $\lambda_n = \frac{n^2 \pi^2}{l^2}$; $n \in N$

The eigenfunctions corresponding to λ_n will be

$$y = c_n \, \sin\left(\frac{n\pi x}{l}\right) \; ; \; n \in N$$

Dropping the arbitrary constant c_n we may treat $\psi_n(x) \equiv \sin\left(\frac{n\pi x}{l}\right)$ as the eigenfunctions and corresponding eigenvalues as $\lambda_n = \frac{n^2\pi^2}{l^2}; n \in N$.

Using the result of the preceding problem, viz, Example- 28, one can straightway write down a solution of the given inhomogeneous b.v.p in the form :

$$y_p(x) = \sum_{k=1}^{n} \frac{a_k \sin\left(\frac{k\pi x}{l}\right)}{\left(\lambda - \frac{k^2\pi^2}{l^2}\right)}, \text{ provided } \lambda \neq \frac{k^2\pi^2}{l^2}$$

The general solution of the b.v.p. is therefore

$$y(x) = \sum_{n=1}^{\infty} A_n \sin\left(\frac{n\pi x}{l}\right) + \sum_{k=1}^{n} \frac{a_k \sin\left(\frac{k\pi x}{l}\right)}{\left(\lambda - \frac{k^2\pi^2}{l^2}\right)}$$

Remark: (a) The above b.v.p. is called inhomogeneous in restricted sense as here the differential equation is inhomogeneous but the boundary conditions are homogeneous. However, there is another type of b.v.p. where both the differential equation and the boundary conditions are inhomogeneous. For example, the boundary value problem comprising of the second order ode

$$Ly \equiv \frac{d}{dx}\left[a_0(x)\frac{dy}{dx}\right] + a_2(x)y = f(x)$$

and the boundary conditions

$$c_1 \, y(a) + c_2 \, y'(a) = A \quad \& \quad c_3 \, y(b) + c_4 \, y'(b) = B$$

where A, B are constants and $f(x), a_2(x) \in C[a, b]$ while $a_0(x) \in C^1[a, b]$, is inhomogeneous in the generalised sense.

It is a routine check that this problem has a unique solution if

$$\Delta = \begin{vmatrix} c_1 \, y(a) + c_2 \, y'(a) & c_1 \, y(b) + c_2 y'(b) \\ c_3 \, y(a) + c_4 \, y'(a) & c_3 \, y(b) + c_4 \, y'(b) \end{vmatrix} \neq 0.$$

Example (30) : Consider the eigenvalue problem : ($\lambda > 0$ case only) $\frac{d^2y}{dx^2} + \lambda y = 0$ subject to the boundary conditions $y(-\pi) + y'(-\pi) = 0$. and $y(\pi) + y'(\pi) = 0$.

Here the associated weight $a_3(x) = 1 > 0$ in and the Hermitian operator $L = \frac{d^2}{dx^2}$. If $\lambda > 0$, then one may write $\lambda = w^2$, so that the bvp becomes :

$$\frac{d^2y}{dx^2} + w^2 y = 0 \;\; \text{subject to} \;\; y(-\pi) + y'(-\pi) = 0$$

The general solution being $y = c_1 \cos(wx) + c_2 \sin(cox)$, we have :

$$\left. \begin{array}{l} 0 = y(-\pi) = c_1 \cos w\pi - c_2 \sin w\pi \\ 0 = y(\pi) = c_1 \cos w\pi + c_2 \sin w\pi \end{array} \right\}$$

Therefore,

$$\left. \begin{array}{l} c_1(\cos w\pi + w \sin w\pi) + c_2(-\sin w\pi + w \cos w\pi) = 0 \\ c_1(\cos w\pi + w \sin w\pi) + c_2(\sin w\pi + w \cos w\pi) = 0 \end{array} \right\}$$

so that

$$\frac{c_1}{c_2} = \frac{\sin w\pi - w \cos w\pi}{w \sin w\pi + \cos w\pi} = \frac{\sin w\pi + w \cos w\pi}{w \sin w\pi - \cos w\pi}$$

$$\therefore \quad \frac{\sin w\pi - w \cos w\pi}{\sin w\pi + w \cos w\pi} = \frac{w \sin w\pi + \cos w\pi}{w \sin w\pi - \cos w\pi}$$

By compodendo & dividendo, $\dfrac{\sin w\pi}{w \cos w\pi} = \dfrac{-w \sin w\pi}{\cos w\pi}$,

$$\tan w\pi = 0 \Rightarrow w = n \in Z$$

As $\lambda_n = w^2 = n^2$, one can choose $n \in N \cup \{0\}$. For $w = n \in N \cup \{0\}$, the eigenfunctions are $\psi_n(x) = c_1 \cos nx + c_2 \sin nx$.

Thus $\{1, \cos x, \sin x, \cos 2x, \sin 2x, \cdots\}$ is an infinite set of eigenfunctions. The corresponding orthonormal set of eigenfunction is

$$\left\{ \frac{1}{\sqrt{2\pi}}, \frac{\cos x}{\pi}, \frac{sinx}{\sqrt{\pi}}, \frac{\cos 2x}{\sqrt{\pi}}, \frac{\sin 2x}{\sqrt{\pi}}, \cdots \right\}$$

If we write $\phi_0(x) \equiv \frac{1}{\sqrt{2\pi}}$, $\phi_n(x) = \frac{1}{\sqrt{\pi}} \cos nx$, $\psi_n(x) = \frac{1}{\pi} \sin nx$ for $n = 1, 2, \cdots \alpha$, then the above orthonormal set will look like $\{\phi_0, \phi_n(x), \psi_n(x)| \; n = 1, 2, \cdots \alpha\}$.

Remark: This problem might seem to be yet another SL problem having a definite pattern of homogeneous boundary conditions. However, this has been worked out keeping in mind the idea of looking into the classical Fourier series expansion of any arbitrary function $f(x)$ as directly related to the above SL problem. For more elaboration, we now pass onto the eigenfunction expansions related to a SL-problem.

Eigenfunction Expansion

The technique of expanding a given function in the form of a power series, specially Taylor series or Maclaurin series, is very common. However, the power series expansion is not valid for most of the continuous functions and so the classical Fourier series expansion in the form of a trigonometric series is a better option. The technique of expanding a function (defined over a closed interval $[-\pi, \pi]$ or the like and satisfying Dirichlet's conditions) in the form of a Fourier series is well-known.

When we expand $f(x)$ defined on $[-\pi, \pi]$ in the form

$$\frac{1}{2}a_0 + \sum_{n=1}^{\infty} (a_n \cos nx + b_n \sin nx),$$

we make use of the orthonormal basis S of $L^2([-\pi, \pi])$ given by

$$S = \left\{ \phi_0(x) \equiv \frac{1}{\sqrt{2\pi}}, \phi_n(x) \equiv \frac{\cos nx}{\pi}, \ \psi_n(x) = \frac{\sin nx}{\sqrt{\pi}} : n\epsilon N \right\}$$

so that the Fourier co-efficients spell out as :

$$a_0 = \sqrt{\frac{2}{\pi}} \int_{-\pi}^{\pi} \phi(x)f(x)dx \ ; \ \ a_n = \frac{1}{\sqrt{\pi}} \int_{-\pi}^{\pi} f(x) \ \phi_n(x)dx$$

$$\text{and} \ \ b_n \ = \frac{1}{\pi} \int_{-pi}^{\pi} f(x) \ \psi_n(x) \ dx. \qquad \forall \ n = 1, 2, \cdots \infty$$

(Mind that the relation $f(x) = \frac{1}{2} a_0 + \sum_{n=1}^{\infty} (a_n \cos nx + b_n \sin nx)$ is valid only at those points where $f(x)$ is continuous). On substituting back these Fourier co-efficients into the Fourier series expanding the function $f(x)$ over $[-\pi, \pi]$, we get :

$$f(x) = \frac{1}{2}a_0 + \sum_{n=1}^{\infty} (a_n \cos nx + b_n \sin nx)$$

$$= \frac{1}{\sqrt{2\pi}} \int_{-\pi}^{+\pi} \phi_0(x')f(x') \ dx' + \sum_{n=1}^{\infty} \left(\int_{-\pi}^{+\pi} f(x')\phi_n(x') \ dx' \right) \phi_n(x)$$

$$+ \sum_{n=1}^{\infty} \left(\int_{-\pi}^{\pi} f(x')\psi_n(x')dx' \right) \psi_n(x)$$

$$= \phi_0\,(x).\int_{-\pi}^{\pi} \phi_0(x')f(x')dx' + \int_{-\pi}^{\pi} dx' f(x') \sum_{n=1}^{\infty} \phi_n(x')\phi_n(x)$$

$$+ \int_{-\pi}^{+\pi} dx' f(x') \sum_{n=1}^{\infty} \psi_n(x')\psi_n(x)$$

$$= \int_{-\pi}^{+\pi} dx' f(x') \left[\phi_0(x')\phi_0(x) + \sum_{n=1}^{\infty} \phi_n(x')\phi_n(x) + \sum_{n=1}^{\infty} \psi_n(x')\psi_n(x) \right]$$

$$\left(\text{where } \phi_0(x) = \frac{1}{\sqrt{2\pi}} \right)$$

$$\therefore \quad \delta(x - x') = \phi_0(x)\phi_0(x') + \sum_{n=1}^{\infty} \phi_n(x')\phi_n(x) + \sum_{n=1}^{\infty} \psi_n(x')\psi_n(x)$$

$$(6.43)$$

This shows that the orthonormal set S of eigenfunctions of Sturm-Liouville problem given in example (30) satisfies the **completeness relation** (6.43) or the so-called **closure property**. One can have a feel from the above workout how each SL-problem gives us an opening of expanding a function in terms of its eigenfunctions.

The stage is all set for us to address now the problem of eigenfunction expansion for the general SL- problem consisting of the homogeneous second order ode

$$0 = Ly + \lambda a_3(x)y \equiv \frac{d}{dx}\left[a_0(x)\frac{dy}{dx} \right] + (a_2(x) + \lambda a_3(x))y = 0 \quad (6.40\text{(a)})$$

and the homogeneous boundary conditions:

$$c_1\, y(a) + c_2\, y'(a) = 0 \ \text{ and } \ c_3\, y(b) + c_4 y'(b) = 0 \qquad (6.39\text{(b)})$$

If $\{\psi_n(x)|n = 1, 2, \cdots \infty\}$ be a set of orthogonal eigenfunction of this SL-problem, then $\int_a^b \psi(x)\psi_\mu(x)a_3(x)dx = 0$ if $\lambda \neq \mu$. If f be a function which is continuous or piecewise continuous together with its first order derivative in $[a, b]$, one can write $f(x)$ as :

$$f(x) = \sum_{n=1}^{\infty} t_n\psi_n(x) \ ; \ t_n \in \mathbb{R} \qquad (6.44)$$

Multiplying (6.44) by $a_3(x)\psi_m(x)$ and integrating over $[a, b]$ we have :

$$\int_a^b a_3(x)\psi_m(x)f(x)dx = \int_a^b a_3(x)\psi_m(x)\sum_n t_n\psi_n(x)dx \quad \text{(using(6.44))}$$

$$= \sum_n t_n \int_a^b a_3(x)\psi_m(x)\psi_n(x)dx$$

$$= \sum_n t_n\delta_{mn} = t_m \tag{6.45}$$

So, $f(x) = \sum_n c_n\psi_n(x) = \sum_n \psi_n(x) \int_a^b a_3(x')\psi_n(x')f(x')dx'$ (using(6.45))

i, e, $\quad \int_a^b f(x')\delta(x - x')dx' = \int_a^b dx' f(x')a_3(x') \sum_n \psi_n(x)\psi_n(x')$

$$\therefore \qquad \delta(x - x') = a_3(x')\sum_{n=1}^\infty \psi_n(x)\psi_n(x') \tag{6.46}$$

The relation (6.46) is usually known as **completeness** relation of the eigenfunctions of the SL problem (6.40(a)) and (6.39(b)). We therefore conclude that it is possible to expand any function f obeying appropriate conditions in terms of a complete set of eigenfunctions of the Hermitian operator L appearing in (6.40(a)). The following couple of illustrations are expected to boost the level of understanding.

Example (31) : Find formal expansion of the function f defined by $f(x) = 1$, $x \in [1, e^\pi]$ in a series of orthonormal eigenfunctions of the SL-problem :

$$\frac{d}{dx}\left(x\frac{dy}{dx}\right) + \frac{\lambda}{x}y = 0 \quad \text{subject to} \quad y(1) = 0 = y'(e^\pi)$$

From the remark(b) following example(24) it is clear that the eigenvalues of this SL-problem are $\{(n - \frac{1}{2})^2 \; ; \; n \in N\}$ while the complete set of orthonormal eigenfunctions are :

$$\psi_n(x) = \sqrt{\frac{2}{\pi}}\sin\left(\left(n - \frac{1}{2}ln\ x\right)\right) \; ; \; n \in N$$

because one can check that

$$\int_1^{e^\pi} a_3(x)\psi_n^2(x)dx = \int_0^\pi \sin^2\left(\left(n - \frac{1}{2}\right)ln\ x\right)d(ln\ x) \quad \left[\because a_3(x) = \frac{1}{x}\right]$$

$$= \int_0^\pi \sin^2\left(\left(n - \frac{1}{2}\right)z\right)dz = \frac{\pi}{2}.$$

In the light of (6.44), one can write

$$1 = f(x) = \sum_{n=1}^{\infty} t_n \psi_n(x); \; t \in \mathbb{R},$$

where t_n's are given by (6.45) as for each $n \in N$,

$$t_n = \int_1^{e^\pi} a_3(x) \psi_n(x) f(x) dx$$

$$= \sqrt{\frac{2}{\pi}} \int_0^\pi \sin\left(\left(n - \frac{1}{2}\right) \ln x\right) d(\ln x)$$

$$= \sqrt{\frac{2}{\pi}} \int_0^\pi \sin\left(\left(n - \frac{1}{2}\right) u\right) du = \frac{2}{(2n-1)} \left(1 - (-1)^2\right) \cdot \sqrt{\frac{2}{\pi}}$$

$$\therefore \; 1 = f(x) = \sum_{n=1}^{\infty} \frac{2}{(2n-1)} \sqrt{\frac{2}{\pi}} \{1 - (-1)^n\} \cdot \sqrt{\frac{2}{\pi}} \sin\left(\left(n - \frac{1}{2}\right) \ln x\right)$$

$$= \frac{4}{\pi} \sum_{n=1}^{\infty} \frac{(1 - (-1)^n)}{(2n-1)} \sin\left(\left(n - \frac{1}{2}\right) \ln x\right); \; x \in [1, e^\pi]$$

Example (32) : Find the formal expansion of the function $f(x) = x^2$ defined over the interval $[0, l]$ using orthonormal set of eigenfunctions of the SL-problem:

$$\frac{d^2 y}{dx^2} + \lambda y = 0 \; ; \; 0 \leqslant x \leqslant l \; ; \; y(0) = y(l) = 0$$

From the workout of Example (29) it follows that the eigenvalues of this SL-problem are $\lambda_n = \frac{n^2 \pi^2}{l^2}$ and corresponding orthonormal eigenfunctions are $c_n \sin\left(\frac{n\pi x}{l}\right); \; n \in N$, where c_n^2's are determined from the relation of orthonormality :

$$\int_0^l c_n^2 \sin^2\left(\frac{n\pi x}{l}\right) = 1 \; [\because \text{ weight function } a_3(x) = 1]$$

Little try by the reader will yield $c_n = \frac{1}{\sqrt{l}}$ so that the orthonormal set of eigenfunctions will be given as :

$$\left\{ \psi_n(x) = \frac{1}{\sqrt{l}} \sin\left(\frac{n\pi x}{\sqrt{l}}\right) \; ; \; n \in N \right\}$$

In the light of (6.44) one can write

$$f(x) = x^2 = \sum_{n=1}^{\infty} t_n \, \psi_n(x) \; ; \; t_n \epsilon \mathbb{R}, \qquad\qquad \text{where}$$

$$t_n = \int_0^l \frac{1}{\sqrt{l}} \sin\left(\frac{n\pi x}{l}\right) x^2 dx = \frac{1}{\sqrt{l}} \left(\frac{l}{n\pi}\right)^3 \int_0^{n\pi} z^2 \sin z \, dz \; \left(\text{if } z = \frac{n\pi x}{l}\right)$$

$$= \left(\frac{l}{n\pi}\right)^3 \frac{1}{\sqrt{l}} \left\{(-1)^n \, (2 - n^2\pi^2) - 2\right\}$$

Hence $\quad x^2 = \displaystyle\sum_{n=1}^{\infty} \{(-1)^n(2 - n^2\pi^2) - 2\} \frac{1}{\sqrt{l}} \left(\frac{l}{n\pi}\right)^3 \frac{n\pi x}{l} \sin\left(\frac{n\pi x}{l}\right)$

$$= \frac{l^2}{\pi^3} \sum_{(n=1)}^{\infty} \left\{ \frac{(-1)^n(2 - n^2\pi^2) - 2}{n}^3 \right\} \sin\left(\frac{n\pi x}{l}\right)$$

Let us now digress for the time being from our main discussion and look into the physical aspect of the problem given in the following.

Example (33) : In fact this example is mathematically modeled form of the physical problem of buckling of a slender column having length l.(i,e, a long thin bar of homogeneous material and uniform rectangular cross section) that is hinged at the bottom and constrained from rotating

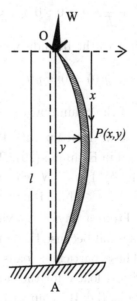

Fig 6.2 : Buckling of a slender column fastened at A

at the top. For purpose of analysis we may conceive a slender column as composed of a bundle of fibres that undergo changes in length under the action of the forces working on it.

In the adjoining diagram, W is the applied load, x is the distance from the top of the column of the point P where the deflection of the column is y. If E is the elastic modulus, I is the moment of inertia of the cross section through the point P w.r.to the neutral axis and p is the radius of curvature of the axis of the buckled column at the point P, then the resisting moment of the stresses across the cross-section through P w.r to the neutral axis will be $\frac{EI}{\rho}$, the quantity EI being known as 'flexural rigidity' of the column. Again this resisting moment must be equal to the algebraic sum of moments of the external forces w.r.to that section.

$$\therefore \qquad -Wy = \frac{EI}{P} = \frac{EIy''}{(1+y'^2)^{\frac{3}{2}}} \approx EIy''$$

(as the curvature of the column is very small, the slope y' is small and so one may ignore y'^2 w.r. to 1 in the above)

Thus we are led to well-known differential equation governing the buckling of column under small deflections :

$$y'' + \frac{W}{EI}\, y = 0 \ ; \ 0 \leqslant x \leqslant l$$

that is to be solved subject to the boundary conditions

$$y(0) = 0 = y(l)$$

($\because$ there is no deflection of the column at the ends)

If we replace $\frac{W}{EI}$ by λ, we land up with the case(3) of the Sturm-Liouville problem discussed in Example-(29). There we found the eigenfunctions as $\psi_n(x) \equiv \sin\left(\frac{n\pi x}{l}\right); n \in N$ with the corresponding eigenvalues being $\lambda_n = \frac{n^2\pi^2}{l^2}$; $n \in N$. Putting back $\lambda = \frac{W}{EI}$, we get $W = \frac{n^2\pi^2}{l^2}EI$; $n \in N$. From a physicist's viewpoint the values of W generated by λ are very special because these are the only applied loads which cause buckling. These particular loads $W = W_n = \frac{n^2\pi^2}{l^2}EI$ are conventionally called 'critical loads' for the slender column. The eigenfunction corresponding to λ_n (i,e, W_n equivalently) is $\psi_n(x) \equiv \sin\left(\frac{n\pi x}{l}\right)$, which provides the buckled shape of the column under this load and n

denotes the mode of buckling. One more point should not be ignored. Although from the viewpoint of solving a SL problem, the subcase $c_2 = 0$ bears no importance, from the angle of a physical problem it is of great interest as it leads to deflection $y = 0$ i,e, no buckling when the applied load is not any of the critical loads W_n given above. Finally we see that the minimum critical load is $W_1 = \frac{\pi^2}{l^2}EI$, i,e, for no applied load W less than W_1 buckling takes place. So to some extent we can explore physics behind the buckling of columns in the light of SL problem.

6.10   Green's Function Approach to IVP

We now consider the second order linear inhomogeneous ode

$$\frac{d^2y}{dx^2} + p_1(x)\frac{dy}{dy} + p_2(x)y = q(x) \tag{6.47}$$

where $p_1(x), p_2(x), q(x) \in C[a,b]$

On multiplying both sides of (6.47) by $a_0(x) = exp[\int p_1(x)dx]$, we get the modified self-adjoint ode given by

$$\frac{d}{dx}\left[a_0(x)\frac{dy}{dx}\right] + a_2(x)y = F(x) \tag{6.48}$$

where $a_2(x) = a_0(x)p_2(x)$ and $F(x) = a_0(x)q(x)$.

If $y = u(x)$ and $y = v(x)$ be two linearly independent solutions of the homogenous ode corresponding to (6.48), then the general solution of (6.48) is found to be

$$y = c_1(x)u(x) + c_2(x)v(x),$$

where $\qquad c_1'(x) = \dfrac{-v(x)F(x)}{a_0(x)W(x)}$ and $c_2'(x) = \dfrac{-u(x)F(x)}{a_0(x)W(x)}$

with $W(x) = u(x)v'(x) - u'(x)v(x)$ being Wronskian of the two functions $u(x)$ and $v(x); x \in [a,b]$. It is a routine mater to check that for any $x \in [a,b]$,

$$a_0(x)W(x) = \text{constant} = A, \text{ say.}$$

Therefore, $c_1(x) = -\displaystyle\int_a^x \dfrac{v(x')F(x')}{A}dx'; \; c_2(x) = -\displaystyle\int_a^x \dfrac{u(x')F(x')}{A}dx$

So a particular solution y_p of (6.47) is given by

$$
\begin{aligned}
y_p &= c_1(x)u(x) + c_2(x)v(x) \\
&= \int_a^x \{v(x)u(x') - u(x)v(x')\}\frac{F(x')}{A}dx' \\
&\equiv \int_a^a G(x, x')F(x')dx', \qquad\qquad (6.49)
\end{aligned}
$$

where we have introduced a new function $G(.,.)$ defined by

$$
G(x, x') = \frac{v(x)u(x') - u(x)v(x')}{A}; \quad x, x' \in [a, b]
$$

Observe that $G(x, x')$ defined above has the simple properties

(i) $G(x, x) = 0 \quad \forall x \in [a, b]$

and (ii) $G(x, x') = -G(x', x) \quad \forall x \in [a, b]$

Since y_p is a particular solution of 6.47, we have

$$
\frac{d}{dx}[a_0(x)\frac{dy_p}{dx}] + a_2(x)y_p = F(x)
$$

i.e., $\dfrac{d}{dx}[a_0(x)\dfrac{d}{dx}\displaystyle\int_a^x G(x, x')F(x')dx'] + a_2(x)\displaystyle\int_a^x G(x, x')F(x')dx' = F(x)$

or, $\displaystyle\int_a^x \left(\frac{\partial}{\partial x}\left(a_0(x)\frac{\partial G}{\partial x}(x, x')\right) + a_2(x)G(x, x')\right)F(x')dx'$

$$
= \int_a^x F(x')\delta(x - x')dx
$$

Since the above result holds true for any arbitrary inhomogeneous term $F(x)$, it follows that

$$
\frac{\partial}{\partial x}(a_0(x)\frac{\partial G}{\partial x}(x, x')) + a_2(x)G(x, x') = \delta(x - x')
$$

Hence for a fixed $x' \in [a, b]$, we observe that $G(x, x')$ satisfies the ode

$$
\frac{d}{dx}\left(a_0(x)\frac{dG}{dx}(x, x')\right) + a_2(x)G(x, x') = \delta(x - x') \qquad (6.50)
$$

i.e., $\qquad\qquad LG(x, x) = \delta(x - x')$

where we have defined the operator L as

$$
L \equiv \frac{d}{dx}\left(a(x)\frac{d}{dx}\right) + a_2(x)
$$

Since,
$$G(x, x') = \frac{1}{A}(v(x)u(x') - v(x')u(x)),$$

$$\left.\frac{\partial G}{\partial x}(x, x')\right|_{x=x'} = \frac{1}{A}\left[v'(x)u(x') - v(x')u'(x)\right]\Big|_{x=x'}$$

$$= \frac{v'(x')u(x') - v(x')u'(x')}{A}$$

$$= \frac{W(x')}{a_0(x')W(x')} = \frac{1}{a_0(x')}$$

Thus for a fixed value of $x' \in [a, b]$, $G(x, x')$ is completely characterized to be the solution of the I.V.P.

$$Ly \equiv \frac{d}{dx}\left(a_0(x)\frac{dy}{dx}\right) + a_2(x)y = \delta(x - x')$$

$$y(x') = 0 \quad \text{and} \quad y'(x') = \frac{1}{a_0(x')}$$

Infact, $G(x, x')$ describes the impact of a concentrated disturbance at the given point x' on the value of y at x —x and x' both being in $[a, b]$.

This function $G(\cdot, \cdot)$ is called **Causal Green's function**, and is also known by the name **propagator** as it transmits a one-point disturbance to the remaining whole:

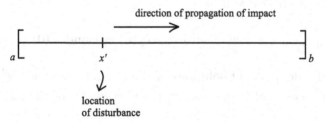

Fig 6.3 : Causal Green's function

Note: Had we defined the Green's function $G(x, x')$ by

$$G(x, x') = \frac{v(x)u(x') - u(x)v(x')}{W(x)},$$

then (6.49) would have been replaced by

$$y_1 = \int_a^x G(x, x')q(x')dx' \qquad (6.49a)$$

and moreover, $G(\cdot, \cdot)$ would satisfy the equation

$$\frac{d^2G}{dx^2} + p_1(x)\frac{dG}{dx} + p_2(x)G = \delta(x - x') \qquad (6.50a)$$

together with the conditions $G(x, x') \big|_{x=x'} = 0$ and $\frac{dG}{dx}(x, x')\big|_{x=x'} = 1$.

i.e., in other words, for a fixed value of x' in $[a, b]$, $G(x, x')$ is completely characterized to be the solution of the IVP

$$\begin{cases} \frac{d^2y}{dx^2} + p_1(x)\frac{dy}{dx} + p_2(x)y = \delta(x - x') \\ y(x') = 0 \;\; \text{and} \;\; y'(x) = 1 \end{cases}$$

Although $G(x, x) = 0$, we lose the very important feature of anti-symmetry as $G(x, x') \neq -G(x', x)$ now.

This is why we should carry on with the choice of G given earlier in connexion with the self-adjoint form of the ode, viz (6.47).

Example (34) : Consider $y'' + y = q(x)$, subject to $y(0) = 1$ and $y'(0) = -1$ for complete solution.

If we compare this ode with (6.48), we get $p_1(x) = 0$ and $p_2(x) = 1$ so that $a_0(x) = 1$ and $a_2(x) = 1$; $F(x) = q(x)$. The obvious self adjoint form of this given IVP will be

$$\frac{d}{dx}\left(\frac{dy}{dx}\right) + y = q(x) \text{ subject to } y(0) = 1 \text{ and } y'(0) = -1$$

Two linearly independent solutions of the homogeneous ode $y'' + y = 0$ being $u(x) = \cos x$ and $v(x) = \sin x$, the C.F will be of the form

$$y_c = c_1 \cos x + c_2 \sin x$$

and the Green's function $G(x, x')$ will be

$$\begin{aligned} G(x, x') &= \frac{1}{A}(v(x)u(x') - u(x)v(x')) \\ &= \sin(x - x') \quad (\because A = a_0(x)W(x) = 1) \\ \therefore \quad y_p &= \int_0^x G(x, x')F(x')dx' = \int_0^x \sin(x - x')q(x')dx' \end{aligned}$$

and the general solution of the inhomogeneous ode is:

$$y = y_c + y_p = c_1 \cos x + c_2 \sin x + \int_0^x \sin(x - x')q(x')dx'.$$

As $y(0) = 1$, we get $c_1 = 1$.

Again $y'(x) = -\sin x + c_2 \cos x + \int_0^x \cos(x - x')q(x')dx'$

$$\Rightarrow y'(0) = -1 = C_2$$

Hence we get the final form of the solution of given IVP as

$$y(x) = \cos x - \sin x + \int_0^x \sin(x - x')q(x')dx'$$

Example (35) : Solve $(D^2 + 3D + 2)y = e^{-2x} \sin x$ subject to $y(0) = y'(0) = 0$

Comparing with $\quad \dfrac{d^2y}{dx^2} + p_1(x)\dfrac{dy}{dx} + p_2(x)y = q(x)$, we get

$$p_1(x) = 3; \ p_2(x) = 2; \ q(x) = e^{-2x} \sin x.$$

Therefore, $\quad a_0(x) = \exp\left[\int p_1(x)dx\right] = e^{3x}$

$$a_2(x) = a_0(x)p_2(x) = 2e^{3x}$$

$$F(x) = e^{3x} \cdot e^{-2x} \sin x = e^x \sin x$$

Thus the self-adjoint ode to which the given ode can be recast is:

$$\frac{d}{dx}\left(e^{3x}\frac{dy}{dx}\right) + 2e^{3x}y = e^x \sin x$$

The two linearly independent solutions of the reduced ode corresponding to the above ode are $u(x) = e^{-x}$ and $v(x) = e^{-2x}$ whose Wronskian $W(x)$ is given by e^{-3x}. Hence $A = a_0(x)W(x) = -1$.

Thus Green's function $G(x, x')$ of this self adjoint ode is given by

$$G(x, x') = e^{-(x+2x')} - e^{(-x'+2x)}$$

$$\therefore y_p = \int_0^x G(x, x')F(x')dx' = \int \left\{e^{-(x+2x')} - e^{-(x'+2x)}\right\} e^{x'} \sin x'dx'$$

$$= \int_0^x e^{-2x'} \sin x' \left(e^{(x-x')} - 1\right) dx'$$

The general solution is of the form

$$y = c_1e^{-x} + c_2e^{-2x} + \int_0^x e^{-2x'} \sin x'(e^{(x-x')} - 1)dx'$$

Differentiating , $y'(x) = -c_1e^{-x} - 2c_2e^{-2x} + \int_o^x e^{-2x'} \sin x'e^{(x-x')}dx'$

$\therefore \ 0 = y(0) = c_1 + c_2$ and $0 = y'(0) = -c_1 - 2c_2$, implying $c_1 = c_2 = 0$.

Therefore, $y(x) = \displaystyle\int_0^x e^{-2x'} \sin x' (e^{x-x'} - 1) dx'$

6.11   Green's Function Approach to BVP

Consider a boundary value problem for a second order linear differential equation on an interval $[a, b]$, say,

$$\frac{d}{dx}\left(a_0(x)\frac{dy}{dx}\right) + a_2(x)y = F(x); \quad a \le x \le b \tag{6.51}$$

subject to boundary conditions $y(a) = 0$ and $y(b) = 0$. (assumed that $a_0(x) > 0 \ \forall\, x \in [a, b]$)

If $u(x)$ and $v(x)$ are linearly independent solutions of the equation

$$\frac{d}{dx}\left(a_0(x)\frac{dy}{dx}\right) + a_2(x)y = 0 \ \text{ subject to } \ y(a) = 0 \text{ and } y(b) = 0,$$

then by using the method of variation of parameters the general solution of (6.51) will be

$$y(x) = c_1 u(x) + c_2 v(x) + \int_a^x \frac{u(x')v(x) - u(x)v(x')}{a_0(x')W(x')} F(x') dx',$$

where $W(x)$ is the Wronskian of $u(x)$ and $v(x)$.

The two boundary conditions entailed are:

$0 = y(a) = c_1 u(a) + c_2 v(a)$

$0 = y(b) = c_1 u(b) + c_2 v(b) + \dfrac{1}{A}\displaystyle\int_a^b \{u(x')v(b) - u(b)v(x')\}\, F(x') dx'$

i.e., $c_1 = \dfrac{-v(a)}{A(u(a)v(b) - u(b)v(a))}\displaystyle\int_a^b \{u(b)v(x') - u(x')v(b)\}F(x')dx'$

$\quad\ \ c_2 = \dfrac{u(a)}{A(u(a)v(b) - u(b)v(a))}\displaystyle\int_a^b \{u(b)v(x') - u(x')v(b)\}F(x')dx'$

$\therefore \ y(x) = \dfrac{1}{A}\displaystyle\int_a^x \{u(x')v(x) - u(x)v(x')\}F(x')dx'$

$\quad + \dfrac{(u(a)v(x) - v(a)u(x))}{A(u(a)v(b) - u(b)v(a))}\displaystyle\int_a^b \{u(b)v(x') - u(x')v(b)\}F(x')dx$

i.e, $y(x) = \int_x^b \frac{(u(a)v(x) - v(a)u(x))(u(b)v(x') - u(x')v(b))}{A(u(a)v(b) - u(b)v(a))} F(x')dx'$

$\qquad + \int_a^x \frac{(-u(x)v(b) + v(x)u(b))(-u(x')v(a) + u(a)v(x'))}{A((u(a)v(b) - u(b)v(a)))} F(x')dx'$

At this point we introduce two functions $g_1(x)$ and $g_2(x)$ defined by

$$\left. \begin{array}{l} g_1(x) = u(a)v(x) - v(a)u(x) \\ g_2(x) = u(b)v(x) - u(x)v(b) \end{array} \right\}$$

so that we get

$y(x) = \int_a^x \frac{g_1(x')g_2(x)}{A(u(a)v(b) - u(b)v(a))} F(x')dx'$

$\qquad + \int_x^b \frac{g_1(x)g_2(x')}{A(u(a)v(b) - u(b)v(a))} F(x')dx'$

$\qquad = \int_a^x \frac{g_1(x')g_2(x)F(x')dx'}{a_0(x')(u(a)v(b) - u(b)v(a))W(x')}$

$\qquad + \int_x^b \frac{g_1(x)g_2(x')F(x')dx'}{a_0(x)(u(a)v(b) - u(b)v(a))W(x)}$

It is easy to check that $y = g_1(x)$ and $y = g_2(x)$ are solutions of the homogeneous ode with $g_1(a) = 0$ and $g_2(b) = 0$ and moreover, Wronskian of $g_1(x)$ and $g_2(x)$ is given by

$W(g_1, g_2)(x) = g_1(x)g_2'(x) - g_2(x)g_1'(x)$

$\qquad\qquad\quad = (u(a)v(b) - u(b)v(a))W(x)$

Therefore, $y(x) = \int_a^x \frac{g_1(x')g_2(x)F(x')}{a_0(x')W(g_1,g_2)(x')}dx' + \int_x^b \frac{g_1(x)g_2(x')F(x')}{a_0(x)W(g_1,g_2)(x)}dx'$

$\qquad\qquad\quad = \int_a^b G(x,x')F(x')dx',$

where $\qquad G(x,x') = \begin{cases} \frac{g_1(x')g_2(x)}{a_0(x')W(g_1,g_2)(x')}, & \text{if } a \leq x' < x \\[2mm] \frac{g_1(x)g_2(x')}{a_0(x)W(g_1,g_2)(x)}, & \text{if } x \leq x' < b \end{cases}$

Observe that $G(x,x')$ is independent of inhomogeneous term $F(x)$ and moreover,

(i) $\quad G(x,x-0) = G(x,x+0)$

(ii) $G(x, x') = G(x', x)$

(iii) $G(x, x') \mid_{x'=a} = 0 = G(x, x') \mid_{x'=b}$

(iv) $\dfrac{dG}{dx'} \mid_{x'=x+0} - \dfrac{dG}{dx'} \mid_{x'=x-0} = \dfrac{g_1'(x)g_2(x) - g_1(x)g_2'(x)}{a_0(x)W(g_1, g_2)(x)}$

$$= \frac{W(g_1, g_2)(x)}{a_0(x)W(g_1, g_2)(x)} = \frac{1}{a_0(x)}$$

For every fixed $x \in [a, b]$, the condition (i) implies that $G(x, x')$ is continuous at $x' = x$. The condition (ii) implies that $G(\cdot, \cdot)$ is symmetric in its arguments. The condition (iii) implies that $G(\cdot, \cdot)$ has to satisfy some end-on condition, viz, $G(x, a) = G(x, b) = 0$. The condition (iv) implies that $\frac{\partial G}{\partial x'}$ has a jump discontinuity of magnitude $-\frac{1}{a_0(x)}$ at $x' = x$. Two conditions (i) and (iv) are conventionally called 'matching conditions'.

Finally, observe that $y(x) = \int_a^b G(x, x')F(x')dx'$ is a solution of (6.51),

$$\text{viz,} \quad Ly \equiv \frac{d}{dx}\left(a_0(x)\frac{dy}{dx}\right) + a_2(x)y = F(x) \text{ in } [a, b],$$

$$\therefore \frac{d}{dx}\left(a_0(x)\frac{d}{dx}\int_a^b G(x, x')F(x')dx'\right) + a_2(x)\int_a^b G(x, x')F(x')dx' = F(x)$$

$$\text{i.e., } \frac{d}{dx}\left(a_0(x)\int_a^b \frac{dG}{dx}(x, x')F(x')dx'\right) + a_2(x)\int_a^b G(x, x')F(x')dx' = F(x)$$

$$\text{i.e., } \int_a^b \frac{d}{dx}\left(a_0(x)\frac{dG}{dx}\right)F(x')dx' + \int_a^b a_2(x)G(x, x')F(x')dx' = F(x)$$

$$\text{i.e., } \int_a^b \left\{\frac{d}{dx}\left(a_0(x)\frac{dG}{dx}\right) + a_2(x)G\right\}F(x')dx' = \int_a^b F(x')\delta(x - x')dx'$$

i.e., $L\,G(x, x') = \delta(x, x')$ $(\because F(\cdot)$ is arbitrary$)$

Hence, Green's function $G(x, x')$ satisfies for all $x' \in [a, b]$ the following auxiliary boundary value problem:

$$LG(x, x') \equiv \frac{d}{dx}\left(a_0(x)\frac{dG}{dx}\right) + a_2(x)G(x, x') = \delta(x - x') \qquad (6.52)$$

together with the boundary conditions

(i) $G(x, x - 0) = G(x, x + 0)$

(ii) $\dfrac{dG}{dx}\bigg|_{x=x'+0} - \dfrac{dG}{dx}\bigg|_{x=x'-0} = \dfrac{1}{a_0(x')}$ for each fixed $x' \in [a, b]$ (6.53)

We are now in a position to explain how the Green's function for a second order ode can be constructed. It follows from the equation $LG(x, x') = \delta(x - x')$ that with the exception of the point $x = x'$, Green's function $G(x, x')$ satisfies the homogeneous ode $LG(x, x') = 0$ in the whole interval $[a, b]$. If $u(x)$ and $v(x)$ be the fundamental solutions of $LG(x, x') = 0$, then one can always express its most general solution in the form

$$G^-(c_1, c_2; x) = c_1 u(x) + c_2 v(x) \quad \text{for } a \leq x < x'$$
$$G^+(d_1, d_2; x) = d_1 u(x) + d_2 v(x) \quad \text{for } x' < x \leq b$$

These two functions can be pieced together to formulate the desired Green's function

$$G(x, x') = \begin{cases} G^-(c_1, c_2; x) & \text{for } a \leqslant x < x' \\ G^+(d_1, d_2; x) & \text{for } x' < x \leqslant b. \end{cases}$$

The four unknown constants, viz., c_1, c_2, d_1, d_2 can be determined from the original boundary conditions together with boundary conditions (6.53) involved in the auxiliary boundary-value problems mentioned before.

Thus
$$\begin{cases} G(a, x') = 0 \; ; \; G(b, x') = 0 \\ \frac{dG}{dx}\big|_{x=x'+0} - \frac{dG}{dx}\big|_{x=x'-0} = -\frac{1}{a_0(x)} \text{ for every fixed } x' \in [a, b] \end{cases}$$

Summing up,
$$\left.\begin{array}{rl} c_1 u(a) + c_2 v(a) & = 0 \\ d_1 u(b) + d_2 v(b) & = 0 \\ c_1 u(x') + c_2 v(x') - d_1 u(x') - d_2 v(x') & = 0 \\ c_1 u'(x') + c_2 v'(x') - d_1 u'(x') - d_2 v'(x') & = \frac{1}{a_0(x)} \end{array}\right\}$$

In practice, c_1, c_2, d_1, d_2 can be determined from these linear equations.

Remark: The original boundary conditions $y(a) = 0 = y(b)$ are simple but of no special importance here. The more generalized form of the homogeneous boundary conditions is

$$\lambda_1 y(a) + \lambda_2 y'(a) = 0; \quad \lambda_3 y(b) + \lambda_4 y'(b) = 0$$

where λ_1, λ_2 are not simultaneously zeros and similarly λ_3, λ_4 are not both zeros.

Alternative way of constructing Green's function for the boundary value problem discussed so far in the present article can be given in the following way. For this purpose we explicitly spell out that $a_0(x), a_2(x)$ and $F(x)$ belong to the infinite dimensional inner product space $C[a, b]$ with inner product as

$$(f, g) = \int_a^b f(x)g(x)dx$$

We seek the solution of the b.v.p. in the form

$$y(x) = \int_a^b G(x, x')f(x')dx'$$

For our convenience we introduce the notation $G_x(x')$ for $G(x, x')$ to emphasize the dependence of G specifically on variable x', thinking of x as fixed and rewrite $y(x)$ as

$$y(x) = \int_a^b G_x(x')f(x')dx' = (G_x, f)$$

(in accordance with the notation of inner product that is in vogue)

As $L \equiv \frac{d}{dx}\left(a_0(x)\frac{d}{dx}\right) + a_2(x)$ and $y(x)$ is a solution of the self-adjoint ode, we have

$$y(x) = (G_x, f) = (G_x, Ly(x)) = (L\ G_x, y(x)) = (L\ G_x, y(x)) + BT$$

(BT being abbreviation of boundary terms)

The boundary term is

$$\left[a_0 G_x y' - a_0' G_x' y + a_0' G_x y \right]_{z=a}^{z=b}$$
$$= \{a_0(b)G_x(b)y'(b) - a_0(a)G_x(a)y'(a)\}$$
$$- \{a_0'(b)G_x(b)y(b) - a_0'(a)G_x'(a)y(a)\} + \{a_0'(b)G_x(b)y(b) - a_0'(a)G_x(a)y(a)\}$$
$$= a_0(b)G_x(b)y'(b) - a_0(a)G_x(a)y'(a),$$

where we have used the boundary conditions, viz, $y(a) = y(b) = 0$. In addition to these boundary conditions if one imposes the extra conditions, viz, $G_x(a) = G_x(b) = 0$, then we get :

$$y(x) = (LG_x\ , \ y(x)), \quad \text{where } G_x(a) = G_x(b) = 0.$$

From the property of delta function it follows that

$$LG_x(z) = \delta(z - x) \text{ with } G_x(a) = G_x(b) = 0$$

so that we have to solve separately the ode

$$LG_x(z) = \begin{cases} 0 & \text{for} \quad a \leqslant z < x \\ 0 & \text{for} \quad x < z \leqslant b \end{cases}$$

and then glue these two pieces together by the matching conditions. We demand that $G_x(z)$ is continuous at $z = x$, i,e,

$$G_x(x - 0) = G_x(x + 0)$$

$$\text{and} \quad 1 = \int_{x-0}^{x+0} \delta(z - x)dz = \int_{x-0}^{x+0} LG_x(z)dz$$

$$= \int_{x-0}^{x+0} \left\{ \frac{d}{dz}\left(a_0(z)\frac{d}{dz}\right) + a_2(z) \right\} G_x(z)dz$$

$$= \left[a_0(z)\frac{dG_x(z)}{dz} \right]_{x-0}^{x+0} + \int_{x-0}^{x+0} a_2(z)G_x(z)dz$$

$$= a_0(x + 0)\frac{dG_x}{dz}(x + 0) - a_0(x - 0)\frac{dG_x}{dz}(x - 0)$$

(other term vanishes due to continuity of integrand)

$$= a_0(x)\{G_x'(x + 0) - G_x'(x - 0)\} \text{ since } a_0(x) \in C[a, b]$$

$$\therefore \quad G_x'(x + 0) - G_x'(x - 0) = \frac{1}{a_0(x)} \,\forall\, x \in [a, b] \,(\because a_0(x) > 0)$$

This last criterion is the jump condition.

Further from the defining relation $LG_x(z) = \delta(z - x)$ we get :

$$\frac{d}{dx}(a_0(x)\frac{d}{dx})G_x(z) + a_2(x)G_x(z) = \delta(z - x)$$

Interchanging z and x we get :

$$\frac{d}{dz}\left(a_0(z)\frac{d}{dz}\right) G_z(x) + a_2(z)G_z(x) = \delta(x - z) = \delta(z - x)$$

This proves that $G_z(x) = G_x(z)$ i,e, $G(z, x) = G(x, z)$

So $G(z, x)$ is symmetric in its arguments.

Example (36) : Use Green's function method to solve the b.v.p:

$$\frac{d}{dx}\left(x^2\frac{dy}{dx}\right) - 2y = F(x) \,;\, y(0) = y(1) = 0$$

Observe that the differential operator $L \equiv \frac{d}{dx}\left(x^2\frac{d}{dx}\right) - 2$ is self-adjoint and so the Green's function $G_x(z)$ of the problem will satisfy

$$LG_x(z) \equiv \frac{d}{dz}\left(z^2\frac{dG_x}{dz}\right) - 2G_x(z) = \delta(z - x);\ G_x(0) = G_x(1) = 0$$

$$\therefore z^2\frac{d^2}{dz^2}G_x(z) + 2z\frac{dG_x(z)}{dz} - 2G_x(z) = 0 \text{ if } z \neq x \text{ with } G_x(0) = G_x(1) = 0$$

Since this is a second order Cauchy-Euler type ode, we can at once write down the general solution in the form :

$$G_x(z) = \begin{cases} c_1\, z + \frac{c_2}{z^2}, & 0 \leqslant z < x \\ c_3\, z + \frac{c_4}{z^2}, & x < z \leqslant 1 \end{cases}$$

where c_1, c_2, c_3, c_4 are evaluated from the two boundary conditions and the matching conditions, viz,

$$G_x(0) = G_x(1) = 0 \qquad\qquad ......(i)$$

$$G_x(x - 0) = G_x(x + 0) \qquad\qquad ......(ii)$$

$$\frac{dG_x}{dz}(x + 0) - \frac{dG_x}{dz}(x - 0) = \frac{1}{x^2} \qquad\qquad ......(iii)$$

See that $G_x(0) = 0$ compels c_2 to be zero while $G_x(1) = 0$ gives : $c_3 + c_4 = 0$

Again from (ii), $c_1\, x = c_3\, x + \frac{c_4}{x^2} = c_3\left(x - \frac{1}{x^2}\right)$ $(\because c_3 = -c_4)$

$\therefore c_1 = c_3\left(1 - \frac{1}{x^3}\right)$

Finally from (iii),

$$\frac{1}{x^2} = c_3 - \frac{2c_4}{x^3} - c_1 = c_3 + \frac{2c_3}{x^3} - c_3\left(1 - \frac{1}{x^3}\right) = \frac{3c_3}{x^3}$$

$$\therefore \quad c_3 = \frac{1}{3}\, x \ ;\ c_1 = \frac{1}{3}\, x\left(1 - \frac{1}{x^3}\right) = \frac{1}{3}\left(x - \frac{1}{x^2}\right)$$

So the desired Green's function of the b.v.p will be :

$$G_x(z) = \begin{cases} \frac{1}{3}\left(x - \frac{1}{x^2}\right)z, & \text{if}\ \ 0 \leqslant z < x \\ \frac{1}{3}x\left(z - \frac{1}{z^2}\right), & \text{if}\ \ x < z \leqslant 1 \end{cases}$$

The solution of the b.v.p is :

$$y(x) = \int_0^1 G_x(z)F(z)dz = \int_0^x \frac{1}{3}\left(x - \frac{1}{x^2}\right)zF(z)dz$$

$$+ \int_x^1 \frac{1}{3}\left(z - \frac{1}{z^2}\right)xF(z)dz$$

Example (37) : Solve the given b.v.p by determining the appropriate Green's function and expressing solution as a definite integral :

$$y'' = F(x) ; \quad y(0) = 0 ; \quad y(1) + y'(1) = 0$$

The operator $L \equiv \frac{d^2}{dx^2}$ is self-adjoint and so the Green's-function $G_x(z)$ will be given by

$$LG_x(z) \equiv \frac{d^2}{dz^2}G_x(z) = \delta(z - x) \qquad \text{......(i)}$$

$$G_x(0) = 0 ; \quad G_x(1) + G'_x(1) = 0 \qquad \text{......(ii)}$$

So the general solution of (i) will be of the form :

$$G_x(z) = \begin{cases} c_1 z + c_2, & \text{if} \quad 0 \leqslant z < x \\ c_3 z + c_4, & \text{if} \quad x < z \leqslant 1 \end{cases}$$

$$G_x(0) = 0 \Rightarrow c_2 = 0$$

As $\quad G'_x(z) = c_3 \quad$ for $\quad x < z \leqslant 1,$

$$0 = G_x(1) + G'_x(1) = 2c_3 + c_4 \qquad \text{......(iii)}$$

From continuity condition of Green's function, $G_x(x + 0) = G_x(x - 0)$

$$c_4 = (c_1 - c_3)x \qquad \text{......(iv)}$$

Again from jump conditions,

$$1 = \frac{dG_x}{dz}(x + 0) - \frac{dG_x}{dz}(x - 0) = c_3 - c_1$$

$$\therefore \quad x = (c_3 - c_1)x = -c_4$$

Hence $\quad 2c_3 = -c_4 = x - c_1 + c_3 = 1$

$$c_3 = \frac{1}{2}x ; \quad c_2 = 0 ; \quad c_1 = c_3 - 1 = \frac{1}{2}(x - 2) ; \quad c_4 = -x$$

So the Green's function will be :

$$G_x(z) = \frac{1}{2}(x - 2)z, \quad \text{if} \quad 0 \leqslant z < x$$

$$= \frac{x}{2}(z - 2), \quad \text{if} \quad x < z \leqslant 1$$

The solution of the boundary value problem is :

$$y(x) = \int_0^1 G_x(z)F(z)dz = \frac{1}{2}(x - 2)\int_0^x zF(z)dz + \frac{x}{2}\int_x^1 (z - 2)F(z)dz$$

Example 38 : Solve the b.v.p $y'' + y = F(x)$; $y'(0) = 0$; $y(1) = 0$ by using Green's function technique.

The operator $L \equiv y'' + y$ being self-adjoint, the Green's function $G_x(z)$ will satisfy :

$$LG_x(z) = \delta(z - x) \; ; \; G_x'(0) = 0 \; ; \; G_x(1) = 0$$

Thus $\quad \dfrac{d^2 G_x}{dz^2} + G_x(z) = 0$ if $z \neq x \quad$ with $\;\; G_x'(0) = 0 \; ; \; G_x'(1) = 0$

The general solution can be written as :

$$G_x(z) = \begin{cases} c_1 \cos(z + c_2), & \text{if } \; 0 \leqslant z < x \\ c_3 \cos(z + c_4), & \text{if } \; x < z \leqslant 1 \end{cases}$$

where the arbitraty constants c_1, c_2, c_3, c_4 can be evaluated from the boundary conditions and the matching conditions.

$$G_x(z) = \left. \begin{array}{l} 0 = G_x(1) = c_3 \cos(1 + c_4) \\ 0 = G_x'(0) = -c_1 \sin c_2 \end{array} \right\}$$

Since $\quad c_1, c_3 \neq 0$, $\sin c_2 = 0$ and $\cos(1 + c_4) = 0$

i, e, $\quad c_2 = 0$ and $c_4 = \dfrac{\pi}{2} - 1$

Again $\quad c_1 \cos x = c_3 \cos(x + \dfrac{\pi}{2} - 1) = -c_3 \sin(x - 1) = c_3 \sin(1 - x)$

Lastly, $\quad \dfrac{dG_x}{dz}(x + 0) - \dfrac{dG_x}{dz}(x - 0) = 1$

i, e, $\quad -c_3 \sin(x + c_4) + c_1 \sin x = 1$

i, e, $\quad -c_3 \sin(x + \dfrac{\pi}{2} - 1) + c_1 \sin x = 1$

i, e, $\quad -c_3 \cos(1 - x) + c_1 \sin x = 1 \qquad \qquad ......(ii)$

From (i) and (ii) we get via cross multiplication,

$$c_1 = \dfrac{-\sin(1 - x)}{\cos 1} \; ; \; c_3 = \dfrac{-\cos x}{\cos 1}$$

so that the Green's function becomes :

$$G_x(z) = \begin{cases} \dfrac{-\sin(1-x)}{\cos 1} \cos z, & \text{if } \; 0 \leqslant z < x \\ \dfrac{-\cos x}{\cos 1} \sin(1 - z), & \text{if } \; x < z \leqslant 1 \end{cases}$$

Green's function is visibly symmetric in arguments z and x. Moreover, the formal solution is given by :

$$
\begin{aligned}
y(x) &= \int_0^1 G_x(z)F(z)dz \\
&= -\left[\int_0^x \frac{\sin(1-x)}{\cos 1}\cos z \cdot F(z)dz + \int_x^1 \frac{\cos x}{\cos 1}\sin(1-z)F(z)dz\right] \\
&= \frac{-[\sin(1-x)\int_0^x \cos z \cdot F(z)dz + \cos x \int_x^1 \sin(1-z)F(z)dz]}{\cos 1}
\end{aligned}
$$

Example (39) : Solve the b.v.p $\frac{d^2y}{dx^2} + y = x$ subject to the condition $y(0) = 0 = y'(\frac{\pi}{4})$ by using Green's function method.

The operator $L \equiv \frac{d^2}{dx^2} + 1$ is self adjoint and so the Green's function $G_x(z)$ of this b.v.p will satisfy

$$
\left.\begin{array}{l}
L\, G_x(z) = \delta(z - x) \\
G_x(0) = 0 \ ; \ G_x'(\frac{\pi}{4}) = 0
\end{array}\right\} \qquad \cdots\cdots \text{(i)}
$$

$$
\therefore \frac{d^2 G_x}{dz^2} + G_x(z) = \begin{cases} 0, & \text{if } \ 0 \leqslant z < x \\ 0, & \text{if } \ x < z \leqslant \frac{\pi}{4} \end{cases}
$$

i,e, the general solution of (i) is of the form :

$$
G_x(z) = \begin{cases} c_1 \cos z + c_2 \sin z, & \text{if } \ 0 \leqslant z < x \\ c_3 \cos z + c_4 \sin z, & \text{if } \ x < z \leqslant \frac{\pi}{4} \end{cases}
$$

$$
G_x(0) = 0 \Rightarrow c_1 = 0
$$

$$
G_x'(\frac{\pi}{4}) = [-c_3 \sin z + c_4 \cos z]_{z=\frac{\pi}{4}} = \frac{1}{2}(c_4 - c_3) = 0
$$

$\therefore c_4 = c_3$ and hence

$$
G_x(z) = \begin{cases} c_2 \sin z, & 0 \leqslant z < x \\ c_3(\cos z + \sin z), & z < x \leqslant \frac{\pi}{4} \end{cases} \qquad \cdots \text{(ii)}
$$

Again from continuity condition, $G_x(x+0) = G_x(x-0)$

i,e, $\qquad c_2 \sin x - c_3(\sin x + \cos x) = 0 \qquad \qquad \text{......(iii)}$

Now $\quad \dfrac{dG_x}{dz}(x+0) - \dfrac{dG_x}{dz}(x-0) = 1$

or, $\quad -c_3(\cos x - \sin x) + c_2 \cos x + 1 = 0 \qquad \qquad \text{......(iv)}$

To find c_2, c_3 we must employ cross multiplication in (iii) & (iv) to get :

$$c_2 = -(\sin x + \cos x); \quad c_3 = -\sin x$$

$\therefore$ The Green's function is given by

$$G_x(z) = \begin{cases} c_2 \sin z = -\sin z(\sin x + \cos x), & \text{if } 0 \leqslant z < x \\ c_3(\cos z + \sin z) = -\sin z + \cos z), & \text{if } x < z \leqslant \frac{\pi}{4} \end{cases}$$

The solution of the b.v.p is given by

$$y(x) = \int_0^{\frac{\pi}{4}} G_x(z)z \, dz = \int_0^x -z \sin z(\sin x + \cos x)dz$$

$$- \int_x^{\frac{\pi}{4}} \sin x(\sin z + \cos z)z \, dz$$

$$= -(\sin x + \cos x) \int_0^x z \sin z dz - (\sin x) \int_x^{\frac{\pi}{4}} z(\sin z + \cos z)dz$$

Now $\quad - \int_0^x z \sin z \, dz = x \cos x - \sin x \quad$ and

$$- \int_x^{\frac{\pi}{4}} z(\sin z + \cos z)dz = x(\sin x - \cos x) + (\sin x + \cos x) - \sqrt{2}$$

$$y(x) = (x \cos x - \sin x)(\sin x + \cos x) + \sin x\{x(\sin x - \cos x)$$
$$+ (\sin x + \cos x) - \sqrt{2}\}$$

$$= x - \sqrt{2}\sin x$$

Hence solution of the inhomogeneous ode will be : $y(x) = x - \sqrt{2}\sin x$

Remark (a) : The Green's function $G_x(z)$ appearing in the solution of an inhomogeneous b.v.p can be interpreted as a response at the point 'z' to a unit impulse at the point 'x'. The inhomogeneous term appearing in a b.v.p defined over some closed interval can be thought of as a continuous distribution of the impulses of magnitude $F(x)$ at that point. The solution thus obtained in terms of Green's function can be treated in the light of superposition of responses to the set of impulses cumulatively represented by $F(x)$.

(b) Green's function associated with a boundary value problem is independent of the inhomogeneous term $F(x)$. It depends solely on the associated homogeneous (i,e,reduced)ode $Ly = 0$ and the imposed boundary conditions. Once the Green's function is determined, the solution of the b.v.p is obtained by a single integration.

The theoretical discussions made so far on the boundary value problem encountered in this section clarifies that in constructing Green's function, one should have apriori knowledge of the independent solutions of the corresponding reduced ode. But if one does not know the independent solution of the reduced ode, one has to obtain the Green's function in the form of an infinite series. We simply spell out the outline of how Green's function is expressed in the form of such a series.

Let's begin with the inhomogeneous second order ode

$$Ly(x) + \lambda y(x) = f(x),$$

whose reduced ode is (6.39)(a) and which is subject to the set of boundary conditions given in (6.39)(b). If $\{\lambda_n\}$ are the eigenvalues of the Hermitian differential operator L and $\{\psi_n(x)\}$ be the corresponding orthonormal eigenfunctions, then both $y(x)$ and $f(x)$ can be expanded in terms of eigenfunctions.

Let $\quad y(x) = \sum_n t_n \psi_n(x) \; ; \; f(x) = \sum_n s_n \psi_n(x); \; s_n, \; t_n \in \mathbb{R} \; \forall \; n \in N$

$$\therefore \; (L + \lambda I) \left(\sum_n t_n \psi_n(x) \right) = \sum_n s_n \psi_n(x)$$

i, e, $\quad L \left(\sum_n t_n \psi_n(x) \right) + \lambda \left(\sum_n t_n \psi_n(x) \right) = \sum_n s_n \, \psi_n \, (x)$

i, e, $\quad \sum_n t_n L \psi_n(x) + \lambda \sum_n t_n \psi_n(x) = \sum_n s_n \psi_n(x)$

i, e, $\quad \sum_n \{ (\lambda_n + \lambda) t_n - s_n \} \psi_n(x) = 0$

Multiplying both sides by $\psi_m(x)$ and integrate on both sides w.r.to x over $[a, b]$ to have

$$\sum_n \int_a^b \{ (\lambda_n + \lambda) t_n - s_n \} \psi_n(x) \psi_m(x) dx = 0$$

i, e, $\sum_n \{ (\lambda_n + \lambda) t_n - s_n \} \lambda_{mn} = 0$ (using orthogonality of eigenfunctions)

i, e, $\quad (\lambda_m + \lambda) t_m = s_m$

Again $\quad f(x) = \sum_n s_n \psi_n(x)$ implies

$$\int_a^b f(x)\psi_m(x)dx = \sum_n s_n \int_a^b \psi_n(x)\psi_m(x)dx = s_m$$

so that $\quad t_m = \dfrac{s_m}{\lambda_m + \lambda} = \dfrac{1}{(\lambda_m + \lambda)} \int_a^b f(x)\psi_m(x)dx$

$$
\begin{aligned}
\therefore \quad y(x) &= \sum_n t_n \psi_n(x) \\
&= \sum_n \frac{1}{(\lambda_n + \lambda)} \left(\int_a^b f(x')\psi_m(x') \right) \psi_n(x) \\
&= \int_a^b dx' f(x') \sum_n \frac{1}{(\lambda_n + \lambda)} \psi_n(x')\psi_n(x) \\
&= \int_a^b dx' f(x') G(x, x')
\end{aligned}
$$

where $\quad G(x, x') = \sum_n \dfrac{1}{(\lambda_n + \lambda)} \psi_n(x')\psi_n(x) \qquad\qquad (6.54)$

So we have obtained a Green's function in the form of an infinite series. It is the price one has to pay for not knowing the solutions of the associated reduced ode. As an instance, reader may check that for example (29),

$$G(x, x') = \sum_n \frac{1}{n^2\pi^2} \sin\left(\frac{n\pi x}{l}\right) \sin\left(\frac{n\pi x'}{l}\right).$$

Exercise 6A

1. Use the method of inspection to show that $y = x$ is a solution of the Legendre equation

$$(1 - x^2)y'' - 2xy' + 2y = 0, \qquad |x| < 1$$

   and hence find a second linearly independent solution.

2. Verify that $y = x^{-\frac{1}{2}} \sin x$ is one solution of the Bessel equation

$$x^2 y'' + xy' + \left(x^2 - \frac{1}{4}\right)y = 0$$

   and determine a second linearly independent solution by order reduction method. (consider the interval $0 < x < \infty$).

3. If $y = x$ be a solution of $(x^2 - x + 1)\frac{d^2y}{dx^2} - (x^2 + x)\frac{dy}{dx} + (x+1)y = 0$, find a linearly independent solution by reducing the order. Find also the general solution.

4. Solve the ode $x\frac{d^2y}{dx^2} - \frac{dy}{dx} - 4x^3y = -4x^5$, given that $y = e^{x^2}$ is a solution of the corresponding homogeneous ode.

5. Use the method of inspection to prove that $y = e^{-2x}$ is a solution of the ode $(1+x)\frac{d^2y}{dx^2} + (4x+5)\frac{dy}{dx} + (4x+6)y = 0$ and hence solve

$$(1+x)\frac{d^2y}{dx^2} + (4x+5)\frac{dy}{dx} + (4x+6)y = e^{-2x}$$

6. Determine whether the equation $xy'' + (x^2 - 1)y' + x^3y = 0$, $0 < x < \infty$ can be transformed into an equation having constant co-efficients by change of independent variables. If possible, find the general solution.

7. Try the same method to the differential equation

$$x^2\frac{d^2y}{dx^2} - \frac{dy}{dx} + 4x^3y = 8x^3\sin\left(x^2\right).$$

What will be the outcome?

8. Solve the following differential equations by suitably changing the independent variable:

   (i) $\frac{d^2y}{dx^2} + (3\sin x - \cot x)\frac{dy}{dx} + 2y\sin^2 x = e^{-\cos x}\sin^2 x$

   (ii) $x\frac{d^2y}{dx^2} + (4x^2 - 1)\frac{dy}{dx} + 4x^3y = 2x^3$.

   (iii) $\frac{d^2y}{dx^2} + (\tan x - 1)^2\frac{dy}{dx} - n(n-1)y\sec^4 x = 0$.

9. Solve the following differential equations:

   (i) $x^2\frac{d^2y}{dx^2} - x\frac{dy}{dx} + y = \log x$

   (ii) $x^2\frac{d^2y}{dx^2} + 5x\frac{dy}{dx} + 4y = x\log x$

   (iii) $x^2\frac{d^2y}{dx^2} + 4x\frac{dy}{dx} + 2y = e^x$

   (iv) $(5+2x)^2\frac{d^2y}{dx^2} - 6(5+2x)\frac{dy}{dx} + 8y = 0$

   (v) $x^2\frac{d^2y}{dx^2} - 2x\frac{dy}{dx} - 4y = x^2 + 2\log x$.

10. Solve the following odes by the method of operational factors:

    (i) $\{xD^2 + (x-1)D - 1\}y = x^2$

    (ii) $\{(x+2)D^2 - (2x+5)D + 2\}y = (1+x)e^x$

    (iii) $\left(x^2D^2 + 2xD + 1\right)y = (1-x)^{-2}$, $0 < x < 1$.

[In all the problems, $D \equiv \frac{d}{dx}$]. Further, check if the order of the factors can be reversed or not.

11. Solve the following ode by removing the first derivative:

    (i) $x^2\dfrac{d^2y}{dx^2} - 2(x^2+x)\dfrac{dy}{dx} + (x^2+2x+2)y = 0$

    (ii) $\dfrac{d^2y}{dx^2} - 4x\dfrac{dy}{dx} + (4x^2-1)y = -3e^{x^2}\sin x$

    (iii) $\dfrac{d^2y}{dx^2} + 2\tan x\dfrac{dy}{dx} + (1+\tan^2 x)y = \sec x\tan x$

    (iv) $x\dfrac{d}{dx}\left(x\dfrac{dy}{dx} - y\right) - 2x\dfrac{dy}{dx} + (x^2+2)y = 0$

12. Solve by any method you like:

    (i) $x^2\dfrac{d^2y}{dx^2} + x\dfrac{dy}{dx} - y = 0$

    (ii) $(\sin x - x\cos x)\dfrac{d^2y}{dx^2} - (x\sin x)\dfrac{dy}{dx} + y\sin x = 0$

    (iii) $x^2\dfrac{d^2x}{dx^2} - 2x(3x-2)\dfrac{dy}{dx} + 3x(3x-4)y = e^{3x}$

As a challenge, try all the methods you have learnt in this chapter, to solve each of the above three odes.

13. Determine whether the following odes are exact or not — if exact, find their first integral and if not, find an integrating factor.

    (i) $2\sin x\dfrac{d^2y}{dx^2} + (2\cos x + 2\sin x)\dfrac{dy}{dx} + 2y\cos x = \cos x$

    (ii) $x^3\dfrac{d^2y}{dx^2} + 2x\dfrac{dy}{dx} - 2y = 1$

    (iii) $\sin^2 x\dfrac{d^2y}{dx^2} = 2y$

    (iv) $x\dfrac{d^2y}{dx^2} + (1-x)\dfrac{dy}{dx} + py = 0$

14. (a) Find the values of n for which the following Cauchy-Euler ode is exact (n being positive integer):

$$x^2 \frac{d^2y}{dx^2} + x \frac{dy}{dx} - n^2 y = 0$$

   (b) Find the adjoint equation for each of the following odes and verify that adjoint of the adjoint is just the original ode.

   (i) $y'' - 2xy' + 2py = 0$

   (ii) $x(1-x)y' + py = 0$

   (iii) $(3-x)y'' + (4x-9)y' + (6-3x)y = 0$,

   where primes denotes derivatives of y w.r. to x.

15. Find the adjoint equation of each of the following odes:

   (i) $x^2 \frac{d^2y}{dx^2} + (2x^3 + 7x)\frac{dy}{dx} + 8(x^2+1)y = 0$

   (ii) $\frac{d^2y}{dx^2} - \left(2x + \frac{3}{x}\right)\frac{dy}{dx} - 4y = 0$

   (iii) $(2x+1)\frac{d^2y}{dx^2} + x^3 \frac{dy}{dx} + x = 0$

   (iv) $(1-x^2)\frac{d^2y}{dx^2} - 2x\frac{dy}{dx} + 2y = 0$, $|x| < 1$.

If possible, solve the above by finding a solution of the adjoint ode either by methods of inspection or any standard method in vogue.

Exercise 6B

1. Find the eigenvalues and corresponding eigenfunctions of the following Sturm-Liouville problems:

   (a) $\frac{d^2y}{dx^2} + \lambda y = 0$; $y(0) = 0 = y(\pi/2)$

   (b) $\frac{d^2y}{dx^2} + \lambda y = 0$; $y(-L) = y(L) = 0$, $L > 0$

   (c) $\frac{d^2y}{dx^2} + \lambda y = 0$; $y'(0) = 0 = y'(L) = 0$; $L > 0$

(d) $\dfrac{d^2y}{dx^2} + \lambda y = 0$; $y(0) - y'(0) = 0$; $y(\pi) - y'(\pi) = 0$

(e) $\dfrac{d}{dx}\left[\dfrac{1}{(3x^2+1)}\dfrac{dy}{dx}\right] + \lambda(3x^2+1)y = 0$; $y(0) = 0 = y(\pi)$

2. Prove that the eigenfunctions corresponding to distinct eigenvalues of each of the following Sturm-Liouville problems are orthogonal over the respective domains:

(a) $\dfrac{d^2y}{dx^2} + \lambda y = 0$; $y(0) = 0 = y(\pi)$

(b) $\dfrac{d^2y}{dx^2} + \lambda y = 0$; $y'(0) = 0 = y'(\pi)$

Find zeros of the eigenfunctions over the respective domains. If possible, find orthonormal set of eigenfunctions of these SL problems.

3. Can Legendre equation be converted to a Sturm-Liouville problem? If yes, find the eigenfunctions of the same and find its zeros.

4. Consider the set of functions $\{\psi_n\}$, where

$$\psi_1(x) = \dfrac{1}{\sqrt{\pi}}; \quad \psi_{n+1}(x) = \sqrt{\dfrac{2}{\pi}}\cos(nx); \quad n = 1, 2, \cdots \infty$$

on the interval $[0, \pi]$. Show that for m, n being distinct positive integers,

$$\int_0^\pi \psi_m(x)\psi_n(x)dx = 1.$$

Can you formulate the Sturm-Liouville problem for which these functions are eigenfunctions.

5. For the ode $y'' + \lambda y = 0$, $0 \leqslant x \leqslant 1$, find the eigenvalues and the corresponding solutions to the following problems and confirm, in each case, that the eigenvalues are real, ordered and simple.

(i) $y(0) = y'(1) = 0$; (ii) $y'(0) = y(1) = 0$; (iii) $y'(0) = y'(1) = 0$

6. (a) Obtain the formal expansion of the function f defined by $f(x) = x$ $(0 \leqslant x \leqslant \pi)$ in a series of orthonormal eigenfunction $\{\phi_n\}$ of the SL-problem $y'' + \lambda y = 0$; $y(0) = 0$; $y(\pi) = 0$

(b) Expand $f(x)$ in the series of orthonormal eigenfunctions $\{\psi_n\}$ of the problem :

$$x^2 y'' + 3xy' + \lambda y = 0 , \quad 1 \leqslant x \leqslant e \text{with } y(1) = y(e) = 1,$$

where $\quad f(x) = e^x ; \ 1 \leqslant x \leqslant \pi$ and $f(x) = e \quad$ if $\pi < x \leqslant e$

7. Use Green's function technique to solve the b.v.p's :

(a) $\dfrac{x^2 d^2 y}{dx^2} + 2x\dfrac{dy}{dx} - 2y = f(x), 0 < x < 1 ; \ y(0) = y(1) = 0$

(b) $\dfrac{d^2 y}{dx^2} = f(x), \ 0 < x < L; \ y(0) = y(L) = 0$

(c) $\dfrac{d^2 y}{dx^2} - 5\dfrac{dy}{dx} + 6y = 0 ; \ y(0) = 3 \ \& \ y(lnz) = 8$

(d) $\dfrac{d^2 y}{dx^2} - 3\dfrac{dy}{dx} + 2y = e^{-x} ; \ y(0) = 1 \ \text{and} \ y'(1) = 3$

(e) $\dfrac{d^2 y}{dx^2} + y \cos x = 0 ; \ y(0) = 3 \ \text{and} \ y'(1) = 1$

(f) $\dfrac{x^2 d^2 y}{dx^2} - 2x(1+x)\dfrac{dy}{dx} + 2(1+x)y = x^3 :$

$$y'(0) = 1; \ 4y(1) : \ -y'(1) = 2$$

8. Solve the following I.V.P's by Green's function method :

(a) $\dfrac{d^2 y}{dx^2} + \dfrac{dy}{dx} + y = 0 ; \ y(0) = 1 \ \& \ y'(0) = 0$

(b) $(1 + x^2)\dfrac{d^2 y}{dx^2} + 3x\dfrac{dy}{dx} + y = f(x) ; \ y(0) = 1 \ \text{and} \ y'(0) = 1$

Chapter 7

Laplace Transformations in Ordinary Differential Equations

7.1 Introduction

In solving linear ordinary differential equations having constant co-efficients, Laplace transformation is a very efficient tool. It is especially useful when the inhomogeneous term involved in such an ode is discontinuous or periodic. The method of Laplace transform is applicable to solve an initial value problem and obtain the specific solution of the inhomogeneous ode directly. Laplace transform is an integral transform that is linear by nature and converts a differential equation to an algebraic equation. Our mission is to solve this resultant algebraic equation and then finally solving the inverse problem to retrieve the desired solution of the orginal ode. In the next couple of articles we shall give a brief account of the 'Laplace Transform' and its rules for derivatives and integrals and then shall pass onto its application in ordinary differential equations.

7.2 Definition and Anatomy of Laplace Transform

Let $f : \mathbb{R}^+ \bigcup \{0\} \to \mathbb{R}$ be a function and s be a real variable. If one defines a new function $F(\cdot)$ by the rule

$$F(s) = \int_0^\infty e^{-sx} f(x) dx, \quad s > 0 \tag{7.1}$$

for all those values of x for which the improper integral converges. The new function $F(s)$ is called the **Laplace transform** of $f(x)$. As per traditions $F(s)$ is also denoted by $\mathcal{L}(f(x))$, where we call script $\mathcal{L}$ to be the 'Laplace Transform Operator'.

There are mainly two points to address about the improper integral appearing in (7.1): (i) The integrand becomes infinite at some point

$c \in (a, b)$. (ii) The improper integral in (7.1) may not exist. Mathematically this amounts to

$$\lim_{x \to c} e^{-sx} f(x) = \infty \text{ and } \int_0^{\infty} f(x) e^{-sx} dx = \infty$$

However, it is easy to see that for any s and any $x > 0$, $e^{-sx} < 1$ and so blowing up of $e^{-sx} f(x)$ is feasible only due to blowing up of $f(x)$. This threat to (7.1) can be avoided if we presume $f(x)$ to be 'sectionally continuous' on $[0, \infty]$. A function $f(x)$ defined on $[a, b]$ is said to be **sectionally continuous** or **piecewise continuous** over $[a, b]$ if $\exists$ a partition $P \equiv \{a = x_0, x_1, \cdots x_{r-1}, x_r \cdots, x_{n-1}, x_n = b\}$ of $[a, b]$ such that f is continuous in the open subintervals (x_{r-1}, x_r) $\forall r = 2, 3, \cdots, \overline{n-1}$ and f is discontinuous at the endpoints of each of these subintervals, i.e.,

$$\lim_{x \to x_r - 0} f(x) \neq \lim_{x \to x_r + 0} f(x) \quad \forall r = 2, 3, \cdots, \overline{n-1}$$

together with $\quad \lim_{x \to a + 0} f(x) = f(a) \quad$ and $\quad \lim_{x \to b - 0} f(x) = f(b)$

A function f defined on $[0, +\infty]$ is said to be piecewise continuous over $[0, \infty)$ if f for every finite $b > 0$, f is piecewise continuous in the above sense.

We must note that at the partition points x_r, $r = 2, 3, \cdots, \overline{n-1}$, $f(x)$ has a jump discontinuity since at each of these points, the left hand limit and the right hand limit of $f(x)$ both exist but are unequal. If f is continuous in $[0, \infty)$, then it is continuous in $[a, b)$ for every $b > 0$. Again continuity of f in $[0, b]$ implies piecewise continuity of f in $[0, b]$ and hence f is Riemann-intergrable in $[0, b]$ for every $b > 0$.

To avoid the pitfall of divergence of the improper integral $\int_0^{\infty} e^{-sx} f(x) dx$, we presume that the function $f(x)$ is of exponential order in the sense that $\exists$ some constants α and $M(> 0)$ such that

$$|f(x)| \leq M e^{\alpha x} \quad \forall \, x \geq 0$$

This assumption is tantamount to the boundedness of $e^{-\alpha x} f(x)$ for all sufficiently large values of x. We also observe that

$$e^{-sx} |f(x)| \leq M e^{-(s-a)x} \quad \forall \, x \geq 0.$$

Observe that

$$\int_0^\infty M e^{-(s-\alpha)x} dx = \lim_{B\to\infty} \int_0^b M e^{-(s-\alpha)x} dx$$

$$= \lim_{B\to\infty} \left(\frac{M}{s-\alpha} \right) \left(1 - e^{-(s-\alpha)B} \right) = \frac{M}{s-\alpha}, \quad \text{if } s > \alpha$$

By comparison test for improper integrals, it therefore follows that $\int_0^\infty e^{-sx} |f(x)| dx < \infty$ for $s > \alpha$. This in turn leads to the conclusion that $\int_0^\infty e^{-sx} f(x) dx < \alpha$ for $s > x$. This result is some kind of existence theorem. Further,

$$\left| \int_0^\infty e^{-sx} f(x) dx \right| \le \int_0^\infty e^{-sx} |f(x)| dx \le M \int_0^\infty e^{-(s-\alpha)x} dx$$

$$= \frac{M}{s-\alpha} \text{ for } s > \alpha$$

i.e., $$\left| F(s) \right| \le \frac{M}{s-\alpha} \text{ for } s > \alpha$$

Therefore, $$\lim_{s\to\alpha} |F(s)| \le \lim_{s\to\alpha} \frac{M}{s-\alpha} = 0 \quad \text{i.e.,} \quad \lim_{s\to\alpha} F(s) = 0$$

Moreover, $$|sF(s)| \le \frac{Ms}{s-\alpha} = \frac{M(s-\alpha+\alpha)}{(s-\alpha)} = M + \frac{\alpha}{s-\alpha} \longrightarrow M$$

as $s \to \infty$, indicating that $sF(s)$ is bounded whenever s is very large.

For making our discussion self-complete, we illustrate the concepts of sectionally continuous functions in $\mathbb{R}$.

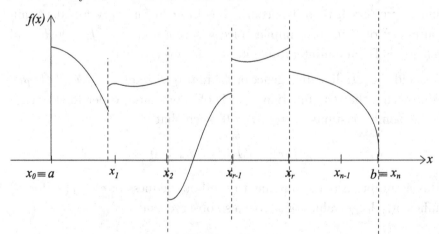

Fig 7.1 : Sectionally continuous function

The most popular example of a sectionally continuous function is the greatest integer function $[x]$ defined by

$$[x] = \begin{array}{c} \sup n \\ n \le x; x \in N \end{array} = \begin{cases} \text{integral part of } x, & \text{if } x \ge 0 \\ \text{integral part of } (x-1), & \text{if } x < 0 \end{cases}$$

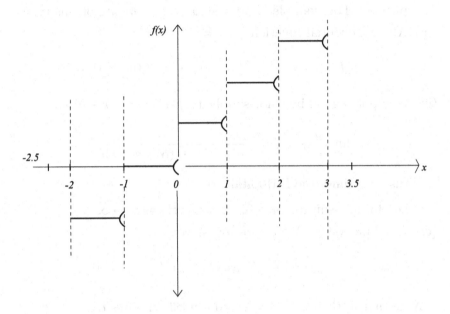

Fig 7.2 : Sectional continuity of the function $f(x) = [x]$

The graph of $f(x) = [x]$ shows that it is sectionally continuous in any finite interval $[a, b]$, the points of jump discontinuity being the integer points in that interval.

We now give a few illustrative examples of functions of exponential order. We recall that a function $f(x)$ is said to be of exponential order α if $\exists$ a constant $M > 0$ such that

$$|f(x)| \le M e^{\alpha x} \quad \forall \ x \ge 0$$

Thus if $f(x)$ is of exponential order α and the values $f(x)$ assumed by f become infinite as $x \to \infty$, these function values cannot grow more rapidly then a multiple M of the corresponding values of $e^{\alpha x}$. Mathematically this is expressed in the form

$$\lim_{x \to \infty} \frac{f(x)}{e^{\alpha x}} = \text{a finite quantity.}$$

or more succinctly, $f(x) = O(Me^{\alpha x})$ for some $M > 0$.

The simplest examples of functions of exponential order are :

(i) Any bounded function is of exponential order since $f(x)$ is bounded $\Rightarrow \exists\ M > 0$ such that $|f(x)| \leq M = Me^{0 \cdot x}$. So $\alpha = 0$ serves our purpose. For example, $f(x) = \sin x$; $\cos x$ etc are of this type.

(ii) Any exponential function $f(x) = e^{\beta x}$ since

$$|f(x)| = e^{\beta x} \leq e^{\alpha x} \text{ for any } \alpha \geq \beta \text{ (here } M = 1)$$

(iii) Any polynomial in x, for simplicity $f(x) = x^n$, $n \in N$ as

$$\lim_{x \to \infty} \frac{x^n}{e^{\alpha x}} = \lim_{x \to \infty} \frac{n!}{\alpha(\alpha - 1) \cdots (\alpha - n + 1)e^{\alpha x}} = 0$$

(using n times the L Hopital's rule)

i.e., for any suitable $\alpha > 0$, $x^n = O(e^{\alpha x})$ as $x \to \infty$.

(iv) Any function $f(x) = e^{ax}\cos bx$, since

$$|f(x)| = e^{\alpha x}\ |\cos bx| \leqslant e^{\alpha x} \quad \forall\ x \geqslant 0$$

Note that in this case $M = 1$ and a itself serves as α .

We have some simple examples of functions that are both sectionally continuous and of exponential order as $x \to \infty$.

(i) Consider $f(x) = xe^{2x}$. This being a continuous function, is sectionally continuous in $[0, b]\ \forall\ b > 0$. It is also of exponential order as

$$\lim_{x \to \infty} \frac{xe^{2x}}{e^{\alpha x}} = \lim_{x \to \infty} \frac{1}{(\alpha - 2)e^{(\alpha - 2)x}} = 0, \text{ whenever } \alpha > 2$$

(ii)   If $f(x) = \begin{cases} (x + 1)^{-1}, & \text{if } 0 < x < 2 \\ 1 & , & \text{if } x > 2 \end{cases}$

then $\displaystyle\lim_{x \to 2-0} f(x) = \lim_{x \to 2-0} \frac{1}{x + 1} = \frac{1}{3}$ and $\displaystyle\lim_{x \to 2+0} f(x) = 1$

Since $f(2 - 0)$ and $f(2 + 0)$ both exist but are unequal, it follows that f has a jump discontinuity at $x = 2$. So in any interval $[0.b]$

with $b > 2$, $f(x)$ is sectionally continuous. Since for $x > 2$, $f(x) = 1$, $f(x) = O(Me^{\alpha x})$ as $x \to \infty$ with $\alpha \geqslant 0$ and $M = 1$.

(iii) If $f(x)$ is sectionally continuous and is of exponential order, then the function $g(\cdot)$ defined by $g(x) = \int_0^x f(y)\,dy$ is also sectionally continuous and also of exponential order as can be established below.

Observe that f is R-integrble in any finite interval and its primitive $g(\cdot)$ is a continuous function of x there at. Again $f(x)$ being of exponential order, $\exists$ constants α and $M(> 0)$ such that $\forall\ x \geqslant 0$,

$$|g(x)| = \left| \int_0^x f(y)dy \right| \leqslant \int_0^x |f(y)|dy \leqslant M \int_0^x e^{\alpha y} dy = \frac{M}{\alpha}(e^{\alpha x} - 1)$$

Therefore, $\quad e^{-\alpha x}|g(x)| = \dfrac{M}{\alpha}(1 - e^{-\alpha x}) < \dfrac{M}{\alpha}$, $\quad$ whenever $\alpha > 0$

i.e, $\qquad\qquad |g(x)| < \dfrac{M}{\alpha}\, e^{\alpha x}$, whenever $\quad \alpha > 0$ and $x \geqslant 0$

i.e, $\quad g(\cdot)$ is of exponential order α.

Some simple properties of Laplace Transforms

(i) If $f(x)$ and $g(x)$ are real-valued functions defined on the half line $[0, \infty]$ such that their Laplace transforms exist for $s > \alpha$ and $s > \beta$ respectively, then for arbitrary real constants c_1 and c_2, Laplace transform of $(c_1 f(x) + c_2 g(x))$ exist for $s > \text{Max}\{\alpha, \beta\}$ and moreover,

$$\mathcal{L}[(c_1 f(x) + c_2 g(x))] = c_1 \mathcal{L}[f(x)] + c_2 \mathcal{L}[g(x)]$$

By definition, $\quad \mathcal{L}(f(x)) \quad = \quad \displaystyle\int_0^\infty e^{-sx} f(x)dx < \infty$ whenever $s > \alpha$

$$\mathcal{L}(g(x)) \quad = \quad \int_0^\infty e^{-sx} g(x)dx < \infty \text{ whenever } s > \beta$$

$$\therefore c_1 \mathcal{L}(f(x)) + c_2 \mathcal{L}(g(x)) = c_1 \int_0^\infty e^{-sx} f(x)dx + c_2 \int_0^\infty e^{-sx} g(x)dx$$

$$= \int_0^\infty e^{-sx}(c_1 f(x) + c_2 g(x))dx$$

$$\text{for } s > \text{Max}\{\alpha, \beta\}$$

But this right hand side defines the Laplace transform of the function $(c_1 f(x) + c_2 g(x))$ and so can be written as $\mathcal{L}(c_1 f(x) + c_2 g(x))$. This proves the linearity property of Laplace transforms.

(ii) If $f(x)$ admits of Laplace transform and $F(s)$ denotes its Laplace transform, then the Laplace transform of $g(x) = e^{\beta x} f(x)$ also exists and is given by $G(s) = F(s - \beta)$. This result can be proved from the definition of Laplace transform itself. Assuming $F(s)$ to be well-defined for $s > \alpha$, we have

$$\int_0^\infty e^{-sx}(e^{\beta x} f(x))dx = \int_0^\infty e^{-(s-\beta)x} f(x)dx < \infty \quad \text{if } s > (\alpha + \beta)$$

Thus $G(s)$ exist for $s > (\alpha + \beta)$ and $G(s) = F(s - \beta)$.

This criterion is the **first shifting property** of Laplace transforms.

(iii) If $f(x)$ admits of Laplace transform and $F(s)$ denotes its Laplace transform, then the Laplace transform of the function $g(x)$ defined by

$$g(x) = \begin{cases} f(x - \beta), & \text{if} \quad x > \beta > 0 \\ 0, & \text{if} \quad 0 < x < \beta \end{cases}$$

then Laplace transform of $g(x)$ exists because if $s > \alpha$,

$$\begin{aligned} \int_0^\infty e^{-sx} g(x)dx &= \int_0^\beta e^{-sx} g(x)dx + \int_\beta^\infty e^{-sx} g(x)dx \\ &= \lim_{A \to \infty} \int_\beta^A e^{-sx} f(x - \beta)dx \\ &= \lim_{A \to \infty} \int_0^{A-\beta} e^{-(s(y+\beta))} f(y)dy \quad [\text{where } x - \beta = y] \\ &= e^{-\beta s} \lim_{A \to \infty} \int_0^{A-\beta} e^{-sy} f(y)dy = e^{-\beta s} F(s) < \infty, \end{aligned}$$

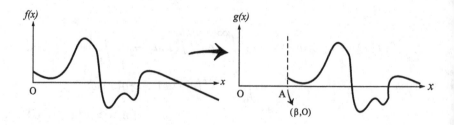

Fig 7.3 : Some signal with a delayed start

assuming that $F(s)$ is well-defined for $s > \alpha$.

Therefore, $G(s) = e^{-\beta s} F(s) < \infty$ if $s > \alpha$.

This is known as **second shifting property** of Laplace transforms.

The first shifting property shows that the substitution $(s - \beta)$ for the variable s in the Laplace transform is tantamount to the multiplication of the object function by the function $e^{\beta x}$. Similarly the second shifting property envisages that if we substitute x by $(x - \beta)$ in the original object function, this corresponds to the multiplication of its Laplace transform by the function $e^{-\beta s}$. One may interpret g as f delayed by β and 'zero padded upto β (See figure 7.3)

(iv) If $F(s)$ denotes the Laplace transform of $f(x)$, then setting $x = \frac{y}{\beta}$ in $F(s)$ we have

$$F(s) = \int_0^\infty e^{-sx} f(x) dx = \lim_{A \to \infty} \int_0^A e^{-sx} f(x) dx$$

$$= \lim_{A \to \infty} \int_0^{A/\beta} e^{-\left(\frac{s}{\beta}\right)y} f\left(\frac{y}{\beta}\right) \frac{dy}{\beta}$$

Therefore, $\quad \beta F(\beta s) = \lim_{A \to \infty} \int_0^{A/\beta} e^{-sy} f\left(\frac{y}{\beta}\right) dy$

$$= \int_0^\infty e^{-sy} f\left(\frac{y}{\beta}\right) dy = \mathcal{L}\left(f\left(\frac{x}{\beta}\right)\right)$$

This result is called **scaling property** of Laplace transforms.

(v) If $f(x)$ and $g(x)$ admit of Laplace transformation with $F(s) = \mathcal{L}(f(x))$ existing for $s > \alpha$ and $G(s) = \mathcal{L}(g(x))$ existing for $s > \beta$, then the convolution $\phi(x)$ of $f(x)$ and $g(x)$ defined by

$$\phi(x) = \int_0^x f(y)\, g(x - y)\, dy = \int_0^x f(x - y)\, g(y)\, dy$$

also admits of Laplace transformation for $s > \text{Max}\{\alpha, \beta\}$ and

$$\Phi(s) = \mathcal{L}(\phi(x)) = F(s) \cdot G(s)$$

Since $F(s)$ exists for $s > \alpha$ and $G(s)$ exists for $s > \beta$, we can assume without loss of generality that

$$f(x) = O(M_1\, e^{\alpha x}) \quad \text{and} \quad g(x) = O(M_2\, e^{\beta x})$$

i.e, $\quad |f(x)| \leqslant M_1\, e^{\alpha x} \quad \text{and} \quad |g(x)| \leqslant M_2\, e^{\beta x} \quad \text{as} \quad x \to \infty$

Again without loss of generality one may assume that $f(x)$ and $g(x)$ are piecewise continuous for a given $x > 0$ and so the integrand $f(y)g(x-y)$ is also a piecewise continuous function of y. Hence $f(y)g(x - y)$ is Riemann integrable over $[0, x]$. Thus $\int_0^x f(y)g(x - y)dy$ is well-defined for $x > 0$ and a continuous function of the upper limit x. Moreover,

$$
\begin{aligned}
|\phi(x)| &= \left| \int_0^x f(y)\, g(x - y)\, dy \right| \\
&\leqslant \int_0^x |f(y)|\, |g(x - y)|\, dy \\
&\leqslant M_1 M_2 \int_0^x e^{\alpha y} \cdot e^{\beta(x-y)}\, dy \\
&= M_1 M_2\, e^{\beta x} \int_0^x e^{(\alpha - \beta)y}\, dy = \frac{M_1 M_2}{(\alpha - \beta)} \left(e^{\alpha x} - e^{\beta x} \right)
\end{aligned}
$$

By law of trichotomy three subcases arise —

(i) $\alpha > \beta$ (ii) $\alpha < \beta$ (iii) $\alpha = \beta$

If $\alpha > \beta$, we may neglect $e^{\beta x}$ to get

$$
|\phi(x)| < M_1 M_2 \cdot \frac{e^{\alpha x}}{\alpha - \beta} \qquad \text{for infinitely large } x.
$$

If $\beta > \alpha$, we may neglect $e^{\alpha x}$ to get

$$
|\phi(x)| < M_1 M_2 \cdot \frac{e^{\beta x}}{\beta - \alpha} \qquad \text{for infinitely large } x.
$$

If $\beta = \alpha$, then for infinitely large x,

$$
|\phi(x)| = \left| \int_0^x f(y)\, g(x - y)\, dy \right| \leqslant M_1 M_2\, e^{\beta x} \int_0^x dy = M_1 M_2\, x\, e^{\beta x}
$$

So for all cases, $\phi(x)$ being of exponential order and we can conclude

$$
\Phi(s) = \mathcal{L}[\phi(x)] = \mathcal{L}\left[\int_0^x f(y)\, g(x - y)\, dy \right] \text{ exists.}
$$

We now set out to prove that $\Phi(s) = F(s) \cdot G(s)$.

$$
\begin{aligned}
\Phi(s) &= \mathcal{L}\,[\,\phi(x)\,] \\
&= \int_0^\infty e^{-sx} \left(\int_0^x f(y)g(x - y)dy \right) dx
\end{aligned}
$$

$$= \lim_{X\to\infty} \int_0^X dx \cdot e^{-sx}\left(\int_0^x f(y)g(x-y)dy\right)$$

$$= \lim_{X\to\infty} \int_0^X dy \int_y^X dx\, e^{-sx}\, f(y)g(x-y)dy$$

$$= \lim_{X\to\infty} \int_0^X dy \int_y^X dx\, e^{-s(x-y)}\cdot e^{-sy} f(y)\, g(x-y)$$

$$= \lim_{X\to\infty} \int_0^X dy\, e^{-sy} f(y) \int_y^X e^{-s(x-y)}g(x-y)dx$$

$$= \lim_{X\to\infty} \int_0^X dy\, e^{-sy} f(y) \int_0^{X-y} e^{-sz}g(z)dz \quad \text{where} \quad z \equiv x - y$$

$$= \left(\int_0^\infty dy\, e^{-sy} f(y)\right)\left(\int_0^\infty e^{-sz}g(z)dz\right) = F(s)\cdot G(s),$$

$$\text{for } s > \text{Max}\{\alpha,\beta\}$$

This ensures the claim. The above result shows that the Laplace transform of the convolution of two functions $f(x)$ and $g(x)$ is equal to the product of their individual Laplace transforms.

Laplace Transform of Elementary Functions

(a)
$$f(x) = \begin{cases} 1, & x \geq 0 \\ 0, & x < 0 \end{cases}$$

∴
$$\mathcal{L}(1) = \int_0^1 e^{-sx}dx = \frac{1}{s} \; ; \; s > 0$$

(b)
$$f(x) = \begin{cases} e^{kx}, & x \geq 0 \\ 0, & x < 0 \end{cases}$$

∴
$$\mathcal{L}(e^{kx}) = \int_0^\infty e^{kx}\cdot e^{-sx}dx = \int_0^\infty e^{-(s-k)x}dx = \frac{1}{s-k} \; ; \; s > k$$

(c)
$$f(x) = \begin{cases} \sin wx, & x \geq 0 \\ 0, & x < 0 \end{cases}$$

∴
$$\mathcal{L}(\sin wx) = \frac{w}{s^2 + w^2} \; ; \; s > 0$$

(d) $$f(x) = \begin{cases} \cos wx, & x \geqslant 0 \\ 0, & x < 0 \end{cases}$$

∴ $$\mathcal{L}(\cos wx) = \frac{s}{s^2 + w^2} \; ; \; s > 0$$

(e) $$f(x) = \begin{cases} e^{\lambda x} \cos wx, & x \geqslant 0 \\ 0, & x < 0 \end{cases}$$

∴ $$\mathcal{L}\left[e^{\lambda x} \cos wx\right] = \frac{(s - \lambda)}{\{(s - \lambda)^2 + w^2\}} \; ; \; s > \lambda$$

(f) $$f(x) = \begin{cases} e^{\lambda x} \sin wx, & x \geqslant 0 \\ 0, & x < 0 \end{cases}$$

∴ $$\mathcal{L}\left[e^{\lambda x} \sin wx\right] = \frac{w}{\{(s - \lambda)^2 + w^2\}} \; ; \; s > \lambda$$

(g) $$f(x) = \begin{cases} x^{\nu}, & \text{if } x \geqslant 0 \; ; \; \nu > 0 \\ 0, & \text{elsewhere} \end{cases}$$

∴ $$\mathcal{L}\left[x^{\nu}\right] = \frac{\Gamma(\nu + 1)}{s^{\nu+1}} \; ; \; s > 0$$

If ν be an integer, say n, then $\mathcal{L}\left[x^n\right] = \frac{\Gamma(n+1)}{s^{n+1}}$ where $s > 0$. In this special case one may deduce the result simply by integration by parts and mathematical induction.

(h) $$f(x) = \begin{cases} x^n e^{kx}, & x \geqslant 0 \; ; \; k > -1 \\ 0, & x < 0 \end{cases}$$

∴ $$\mathcal{L}\left[x^n e^{kx}\right] = \frac{n!}{(s - k)^{n+1}} \; ,$$

n being an integer but k not so. $s > k$ is a must for convergence.

(i) $$f(x) = \begin{cases} \cosh(kx), & x \geqslant 0 \\ 0, & x < 0 \end{cases}$$

∴ $$\mathcal{L}\left[\cosh(kx)\right] = \frac{s}{s^2 - k^2} \quad \text{with} \quad s > |k|$$

(j)
$$f(x) = \begin{cases} \sinh\,(kx), & \text{if } x \geqslant 0 \\ 0, & \text{if } x < 0 \end{cases}$$

$\therefore \quad \mathcal{L}\,[\sinh\,(kx)] = \dfrac{k}{s^2 - k^2}\;;\; s > |k|$

(k)
$$f(x) = \begin{cases} x\,\sin\,wx, & x \geqslant 0 \\ 0, & \text{elsewhere} \end{cases}$$

$\therefore \quad \mathcal{L}\,[x\,\sin\,wx] = \dfrac{2ws}{(s^2 + w^2)^2}$

(l)
$$f(x) = \begin{cases} x\,\cos\,wx, & x \geqslant 0 \\ 0, & \text{elsewhere} \end{cases}$$

$\therefore \quad \mathcal{L}\,[x\,\cos\,wx] = \dfrac{s^2 - w^2}{(s^2 + w^2)^2}$

(m)
$$f(x) = \begin{cases} 0, & \text{if } x < a \\ 1, & \text{if } x \geqslant a \end{cases}$$

$\therefore \quad \mathcal{L}[f(x)] = \displaystyle\int_0^\infty e^{-sx} f(x)\,dx = \int_a^\infty e^{-sx}\,dx$

$$= \lim_{A \to \infty} \int_a^A e^{-sx}\,dx = \lim_{A \to \infty} \left[\frac{e^{-sx}}{-s}\right]_{x=a}^{x=A}$$

$$= \lim_{A \to \infty} \frac{1}{s}\left[e^{-as} - e^{-sA}\right] = \frac{e^{-as}}{s} \quad \text{for} \quad s > 0$$

In particular, if $a = 0$, so that $f(x)$ can be rewritten as :

$$f(x) = \begin{cases} 0, & \text{if } x < 0 \\ 1, & \text{if } x \geqslant 0 \end{cases}$$

then $\mathcal{L}[f(x)]$ would be $\frac{1}{s}$, $s > 0$ and we would get back case (i). The function f defined above is called **Unit step** or **Heaviside function**.

The function $f(x)$ defined by

$$f(x) = \begin{cases} 0, & \text{if } x < a \\ 1, & \text{if } x \geqslant a \end{cases}$$

is called the Unit step function at a and denoted by $U_a(x)$. In particular, $U_0(x)$ is also denoted by $H(x)$ with H standing for Heaviside function.

(n)
$$f(x) = \begin{cases} \frac{1}{a}, & \text{if } 0 < x < a \\ 0, & \text{elsewhere} \end{cases}$$

has its Laplace transform $\mathcal{L}(f(x))$, where

$$\begin{aligned} \mathcal{L}(f(x)) &= \frac{1}{a}\, \mathcal{L}(U_0(x) - U_a(x)) \\ &= \frac{1}{a}\, \{\mathcal{L}(U_0(x)) - \mathcal{L}(U_a(x))\} \text{ (by linearity of Laplace transform)} \\ &= \frac{1}{a}\left(\frac{1}{s} - \frac{e^{-as}}{s}\right) = \frac{1 - e^{-as}}{as} \text{ with } s > 0 \end{aligned}$$

$f(x)$ is the uniform density function used in probability theory.

In particular, if $a \to 0$, $f(x)$ becomes **the Unit Impulse function** $\delta(x)$, also popular as **Dirac delta function**.

Therefore,
$$\begin{aligned} \mathcal{L}(\delta(x)) &= \mathcal{L}\left(\lim_{a \to 0} f(x)\right) \\ &= \lim_{a \to 0} \mathcal{L}(f(x)) \\ &= \lim_{a \to 0} \frac{1 - e^{-as}}{as} = \lim_{a \to 0} \frac{ae^{-as}}{a} = 1 \end{aligned}$$

(in last but one step we used L Hopital's rule). Dirac delta function can be interpreted as containing an impulse at $x = 0$.

$f(x)$ given by following diagram is known as **Square Wave function**.

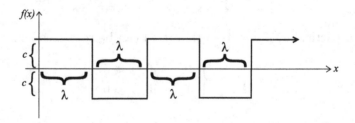

Fig 7.4 : Square Wave function

From the diagram it is easy to identify that f is periodic by nature, the period being 2λ.

Thus $f(x) = c\, u_0(x) - 2c\, u_\lambda(x) + 2c\, u_{2\lambda}(x) - 2c\, u_{3\lambda}(x) + \cdots$

$$= c\,u_0(x) + 2c\sum_{k=1}^{\infty}(-1)^k u_{k\lambda}(x)$$

By linearity property of Laplace transforms,

$$\mathcal{L}\left(f(x)\right) = \mathcal{L}\left(c\,u_0(x) + 2c\sum_{k=1}^{\infty}(-1)^k u_{k\lambda}(x)\right)$$

$$= c\,\mathcal{L}\left(u_0(x)\right) + 2c\sum_{k=1}^{\infty}(-1)^k \mathcal{L}\left(u_{k\lambda}(x)\right) \;(\because\; \mathcal{L}(\cdot) \text{ is linear}).$$

$$= \frac{c}{s}\left[1 + 2\cdot\sum_{k=1}^{\infty}(-1)^k\,\frac{e^{k\lambda s}}{s}\right]$$

$$= \frac{c}{s}\left[1 + \frac{2}{s}\sum_{k=1}^{\infty}(-1)^k\,r^k\right],$$

(where $r \equiv e^{-s\lambda} < 1$ ensures convergence of infinite g.p)

$$= \frac{c}{s}\left(1 - \frac{2r}{1+r}\right) = \frac{c}{s}\left(\frac{1-r}{1+r}\right)$$

$$= \frac{c}{s}\left(\frac{1 - e^{\lambda s}}{1 + e^{-\lambda s}}\right)\;,\; s > 0$$

If $f(x)$ be a periodic function having period λ and f admits of Laplace Transform, then one can write

$$\mathcal{L}(f(x)) = \int_0^{\lambda}\frac{e^{-sy}f(y)dy}{(1 - e^{-\lambda s})}\;;\; s > 0$$

The proof runs as follows.

$$\mathcal{L}(f(x)) = \int_0^{\infty}e^{-sx}f(x)\,dx = \lim_{N\to\infty}\int_0^{\lambda N}e^{-sx}f(x)\,dx$$

$$= \lim_{N\to\infty}\left[\int_0^{\lambda}e^{-sx}f(x)\,dx + \int_{\lambda}^{2\lambda}e^{-sx}f(x)dx\right.$$

$$\left. + \int_{2\lambda}^{3\lambda}e^{-sx}f(x)dx + \cdots + \int_{\lambda(N-1)}^{\lambda N}e^{-sx}f(x)dx\right]$$

$$= \lim_{N\to\infty}\left[\sum_{k=1}^{N}\int_{\lambda(k-1)}^{\lambda k}e^{-sx}f(x)dx\right]$$

$$= \lim_{N\to\infty}\left[\sum_{k=1}^{N}\int_{\lambda(k-1)}^{\lambda k}e^{-sx}f(x - \lambda(k-1))dx\right]$$

(In the last step we used the fact that f is periodic with period λ,

i, e, $\quad f(x) = f(x - \lambda) = f(x - 2\lambda) = \cdots\cdots = f(x - (k-1)\lambda)$

We now compute $\quad \displaystyle\int_{\lambda(k-1)}^{\lambda k} e^{-sx} f(x - (k-1)\lambda) \, dx$

Putting $x - (k-1)\lambda = y$ we have $dx = dy$

and $x = \lambda(k-1) \Rightarrow y = 0$; $x = \lambda k \Rightarrow y = \lambda$ so that

$$\int_{\lambda(k-1)}^{\lambda k} e^{-sx} f(x - (k-1)\lambda)dx = \int_0^\lambda e^{-s(y+(k-1)\lambda)} f(y)dy$$

$$= \int_0^\lambda e^{-(k-1)\lambda} e^{-sy} f(y)dy$$

Hence $\quad \displaystyle \mathcal{L}(f(x)) = \lim_{N\to\infty} \left[\sum_{k=1}^N \int_0^\lambda e^{-sy} f(y)dy \cdot e^{-(k-1)\lambda s} \right]$

$$= \int_0^\lambda e^{-sy} f(y)dy \sum_{k=1}^\infty e^{-(k-1)\lambda s}$$

$$= \int_0^\lambda e^{-sy} f(y)dy \cdot \sum_{k=0}^\infty (e^{-\lambda s})^k$$

Therefore, $\quad \displaystyle \mathcal{L}(f(x)) = \int_0^\lambda \frac{e^{-sy} f(y)dy}{(1 - e^{-\lambda s})}$

(since the series $\displaystyle\sum_{k=0}^\infty (e^{-\lambda s})^k$ is a convergent geometric series with the

common ratio $e^{-\lambda s} < 1$)

As a verification let's consider $f(x)$ defined by (A) & (B) & (C) :

(A) $\qquad f(x) = \begin{cases} \cos wx, & x \geq 0 \\ 0, & x < 0 \end{cases}$

We know that $f(x)$ is periodic function with period $\frac{2\pi}{w}$ and so according to the result established,

$$\mathcal{L}(\cos wx) = \int_0^{2\pi/w} e^{-sy} \cos wy \, dy \left/ \left(1 - e^{\frac{-2\pi s}{w}}\right)\right.$$

From elementary calculus we have :

$$\int_0^{2\pi/w} e^{-sy} \cos wy \; dy = \frac{e^{-sy}(w \sin wy - s \cos wy)}{(s^2 + w^2)} \Big|_0^{2\pi/w}$$

$$= \frac{s(1 - e^{\frac{2\pi}{w} s})}{(s^2 + w^2)}$$

Thus $\quad \mathcal{L}(\cos wx) = \dfrac{s}{s^2 + w^2} \; ; \; s > 0$ $\hspace{2cm}$ (c.f. (d))

(B) $f(x)$ is the square-wave function defined earlier. It is periodic function having period 2λ and so according to the above formula,

$$\mathcal{L}(f(x)) = \int_0^{2\lambda} e^{-sy} f(y) \; dy \; / \; (1 - e^{-2\lambda s})$$

But $f(y) = c$ if $0 \leqslant y \leqslant \lambda$ and $f(y) = -c$ if $\lambda \leqslant y \leqslant 2\lambda$.

Hence $\quad \displaystyle\int_0^{2\lambda} e^{-sy} f(y) dy = c \int_0^{\lambda} e^{-sy} dy - c \int_{\lambda}^{2\lambda} e^{-sy} dy$

$$= \frac{1}{s}\left(1 - e^{-\lambda s}\right) + \frac{1}{s}\left(e^{-2\lambda s} - e^{-\lambda s}\right) = \frac{1}{s}\left(1 - e^{-\lambda s}\right)^2$$

$$\therefore \quad \mathcal{L}(f(x)) = \frac{1}{s} \cdot \frac{\left(1 - e^{-\lambda s}\right)^2}{\left(1 - e^{-2\lambda s}\right)} = \frac{1}{s} \cdot \left(\frac{1 - e^{-\lambda s}}{1 + e^{-\lambda s}}\right),$$

(C) If x is a real number, then one can write $x = [x] + \{x\}$, where $[x]$ is the greatest integer $\leqslant x$ and the fractional part $\{x\}$ is periodic and of period 1. Since $0 \leqslant \{x\} < 1$, it is bounded and so must be of exponential order.$\{x\}$ is a continuous function and so is necessarily sectionally continuous. We are therefore ensured that $\mathcal{L}(\{x\})$ exists. Since $[x] = x - \{x\}$, by linearity property of Laplace transform,

$$\mathcal{L}([x]) = \mathcal{L}(x) - \mathcal{L}(\{x\}) \quad \text{where } x \geqslant 0$$

$$\mathcal{L}[\{x\}] = \frac{\int_0^1 e^{-sy}\{y\}dy}{(1 - e^{-s})} = \frac{\int_0^1 e^{-sy} \; y \; dy}{(1 - e^{-s})}$$

Again $\quad \displaystyle\int_0^1 e^{-sy} \; y \; dy = \left[\left(-\frac{1}{s} e^{-sy}\right) \cdot y\right]_{y=0}^{y=1} + \frac{1}{s} \int_0^1 e^{-sy} dy$

$$= \frac{1}{s^2}(1 - e^{-s}) - \frac{1}{s} e^{-s}$$

Therefore, $\quad \mathcal{L}[\{x\}] = \dfrac{\frac{1}{s^2}(1 - e^{-s}) - \frac{1}{s} e^{-s}}{(1 - e^{-s})} = \dfrac{1}{s^2} - \dfrac{1}{s} \dfrac{e^{-s}}{(1 - e^{-s})}$

Thus
$$\mathcal{L}([x]) = \mathcal{L}(x) - \mathcal{L}(\{x\})$$
$$= \frac{1}{s}\left(\frac{e^{-s}}{1-e^{-s}}\right) \; ; \; s > 0 \; x \geqslant 0$$

(D) As another illustration we take up the saw-tooth function which is a periodic function having period 2. It is formed of the triangular function given by

$$T(x) = \begin{cases} \lambda x , & \text{if } 0 < x < 1 \\ \lambda(2-x) , & \text{if } 1 < x < 2 \end{cases} \; ; \; \lambda > 0$$

The graph of saw tooth function is shown in the following :

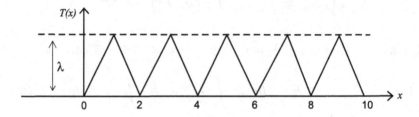

Fig 7.5 : Saw-tooth function, composed of **T(x)**

As $T(x)$ is periodic function, having period 2, its Laplace transform is :

$$\mathcal{L}(T) = \frac{1}{(1-e^{-2s})}\int_0^2 T(x)e^{-sx}dx \; , \text{where}$$

$$\int_0^2 T(x)e^{-sx}dx$$

$$= \lambda\int_0^1 e^{-sx}xdx + \lambda\int_1^2 (2-x)e^{-sx}$$

$$= \lambda\int_0^1 e^{-sx}xdx + \lambda\int_0^1 (1-x)e^{-s}e^{-sx}dx$$

$$= \lambda\int_0^1 e^{-sx}xdx + \lambda e^{-s}\int_0^1 e^{-sx}dx - \lambda e^{-s}\int_0^1 e^{-sx}.x\,dx$$

$$= \lambda(1-e^{-s})\int_0^1 e^{-sx}xdx + \frac{\lambda e^{-s}e^{-sx}}{(-s)}\Big|_0^1$$

$$= \lambda(1-e^{-s})\int_0^1 e^{-sx}xdx + \frac{\lambda e^{-s}}{s}(1-e^{-s}) = \frac{\lambda}{s^2}(1-e^{-s})^2$$

$$\therefore \quad \mathcal{L}(T) = \frac{1}{(1-e^{-2s})}\frac{\lambda}{s^2}(1-e^{-s})^2 = \frac{\lambda}{s^2}\left(\frac{1-e^{-s}}{1+e^{-s}}\right) = \frac{\lambda}{s^2}\tanh(\frac{s}{2})$$

Remark: One may also derive the same from the formula of Laplace transforms of square wave functions and its integrals.

Idea of Inverse Laplace Transforms

Thus far, we have been working on finding the Laplace transforms of a function $f(x)$ that is non-zero only if $x \geqslant$ and is of exponential order for infinitely large x and moreover, piecewise continuous in $[0, \infty)$. One may regard $f(x)$ as signals defined over their respective time domains and the corresponding Laplace transform $F(s)$ as defined over the frequency domains of signals.

We now set for ourselves the inverse problem, i,e, given a function $F(s)$, can we always find a function whose Laplace transform will be $F(s)$? We cannot be sure to provide an affirmative answer to this query because $\exists$ functions that are not Laplace transforms of any function f. Secondly, if for a given F, we assume that the inverse transform f exists, is that f unique? In the strict sense of the term, the inverse Laplace transform is not unique.

If $F(s) \equiv \mathcal{L}(f(x))$ be Laplace transform of $f(x)$, i,e, $\mathcal{L}(f(x)) = F(s)$, then conventionally we use the symbol $\mathcal{L}^{-1}(F(s))$ to denote a function whose Laplace transform is $F(s)$. Since we know that for the function $g(x)$ defined by $g(x) = e^{kx}$ for $x \geq 0$,

$$G(s) \equiv \mathcal{L}(g(x)) = \tfrac{1}{s-k} \ , \ s > k$$

As a matter of fact, $\mathcal{L}^{-1}\left(\tfrac{1}{s-k}\right) = e^{kx}$.

Again consider the function $h(x)$ defined by

$$h(x) = \begin{cases} e^{kx}, & \text{if} \quad x \in (0,1) \ \bigcup(1,3) \ \bigcup(3,\infty) \\ 1, & \text{if} \quad x = 1 \\ 1, & \text{if} \quad x = 3 \end{cases}$$

$$\begin{aligned} \therefore \mathcal{L}(h(x)) &= \int_0^\infty e^{-sx} h(x)dx \\ &= \int_0^1 e^{-sx} h(x)dx + \int_1^3 e^{-sx} h(x)dx + \int_3^\infty e^{-sx} h(x)dx \\ &= \int_0^1 e^{-(s-k)x} dx + \int_1^3 e^{-(s-k)x} dx + \int_3^\infty e^{-(s-k)x} dx \end{aligned}$$

$$= \int_0^\infty e^{-(s-k)x} dx = \mathcal{L}(e^{kx}) \equiv \mathcal{L}(g(x))$$

This illustrates that $\mathcal{L}^{-1}\left(\frac{1}{s-k}\right)$ is not unique as both $g(x)$ and $h(x)$ are equally acceptable. Had we chosen $h(x)$ to be differing from $g(x)$ over a finite set of points or even a countably infinite set of points having measure zero, in $\mathbb{R}^+ \bigcup \{0\}$, then we will get no change in Laplace transform. This non-uniqueness of the inverse Laplace transform of a function $F(s)$ reminds us of the multivaluedness of inverse trigonometric functions, due to their periodicity and other properties. One can check that the non-uniqueness of inverse Laplace transformation can be removed if we restrict ourselves to a special class $\mathbb{A}$ comprising of functions f having the following features —

(i) $f(x)$ is defined for $x \geqslant 0$
(ii) $f(x)$ is of exponential order in the sense $\exists$ constants α and $M(> 0)$ such that $|f(x)| \leqslant Me^{\alpha x}$
(iii) $f(x)$ is piecewise continuous on every bounded subinterval of $[0, \infty)$ and at each point of discontinuity $C \in (0, \infty)$,

$$f(c) = \frac{1}{2}\left(f(c+0) + f(c-0)\right)$$

It is easy to prove that among all fuctions of class $\mathbb{A}$, no two function g and h can have the same Laplace transform for all $s > \alpha$.

Inverse Laplace transform enjoys linearity as shown below.

If $f, g \in \mathbb{A}$; $c_1, c_2 \in \mathbb{R}$, f is of exponential order α and g is of exponential order β, then via linearity of Laplace transforms, we get :

$$c_1\mathcal{L}(f(x)) + c_2\mathcal{L}(g(x)) = \mathcal{L}(c_1 f(x) + c_2 g(x)) \quad \forall \ s > \max\{\alpha, \beta\}$$

Applying $\mathcal{L}^{-1}$ on both sides we have :

$$\mathcal{L}^{-1}(c_1\mathcal{L}(f(x)) + c_2\mathcal{L}(g(x)) = c_1\mathcal{L}(f(x) + c_2\mathcal{L}(g(x)$$

If one agrees to write $\mathcal{L}(f(x)) \equiv F(s)$ and $\mathcal{L}(g(x)) \equiv G(s)$, then $f(x) = \mathcal{L}^{-1}(F(S))$ and $g(x) = \mathcal{L}^{-1}(G(s))$ and consequently we get

$$\mathcal{L}^{-1}(c_1 F(s) + c_2 G(s)) = c_1\mathcal{L}^{-1}(F(s)) + c_2\mathcal{L}^{-1}(G(s)),$$

proving the linearity of inverse Laplace transform.

This linearity property of $\mathcal{L}^{-1}$ and the method of partial fractions are often useful in determining the inverse Laplace transformation of a sizeable subclass of functions in $\mathbb{A}$.

From product of Laplace transforms of two given fuctions f and g, inverse Laplace transformation operator $\mathcal{L}^{-1}$ enables us to retrieve the convolution $f * g$ of f and g.

If $F(s) = \mathcal{L}(f(x))$ and $G(s) = \mathcal{L}(g(x))$, then

$$\mathcal{L}^{-1}(F(s) \cdot G(s)) = \int_0^x f(x-y)g(y) \ dy = (f * g)(x).$$

Example (1) : Using Laplace transform of elementary functions prove:

$$\mathcal{L}^{-1}\left(\frac{5s+3}{(s-1)(s^2+2s+5)}\right) = e^x - e^{-x}\cos 2x + \frac{3}{2} e^{-x}\sin 2x$$

By method of partial fractions, we write

$$\frac{5s+3}{(s-1)(s^2+2s+5)} = \frac{A}{s-1} + \frac{Bs+C}{s^2+2s+5} \ ;$$
$$(A, B \text{ and } C \text{ being constants yet to be determined})$$
$$\therefore \quad 5s+3 = A(s^2+2s+5) + (s-1)(Bs+C)$$

Put $s = 1$ to get $A = 1$
Put $s = 0$ to get $C = 2$
Put $s = -1$ to get $C - B = 3$ so that $B = -1$.

Thus $\dfrac{5s+3}{(s-1)(s^2+(s^2+2s+5))} = \dfrac{1}{(s-1)} + \dfrac{2-s}{(s+1)^2+2^2}$

$$= \frac{1}{(s-1)} + \frac{3-(s+1)}{(s+1)^2+2^2}$$

Since $\mathcal{L}^{-1}$ is a linear operator,

$$\mathcal{L}^{-1}\left(\frac{5s+3}{(s-1)(s^2+2s+5)}\right) = \mathcal{L}^{-1}\left(\frac{1}{s-1}\right) + 3\mathcal{L}^{-1}\left(\frac{1}{(s+1)^2+2^2}\right)$$

$$-\mathcal{L}^{-1}\left(\frac{s+1}{(s+1)^2+2^2}\right)$$

$$= e^x + \frac{3}{2} e^{-x}\sin 2x - e^{-x}\cos 2x$$

where in the last step we have used first shifting property of Laplace transforms, i,e, $\mathcal{L}(e^{\beta x} f(x)) = F(s - \beta)$

Example (2) : Use convolution theorem to prove that

$$\mathcal{L}^{-1}\left(\frac{s^2}{(s^2+4)^2}\right) = \frac{x}{2} \cdot \cos 2x + \frac{1}{4}\sin 2x$$

Now $\mathcal{L}^{-1}\left(\dfrac{s^2}{(s^2+4)^2}\right) = \mathcal{L}^{-1}\left(\dfrac{s}{(s^2+4)} \cdot \dfrac{s}{(s^2+4)}\right)$

$$= \mathcal{L}^{-1}\left(\frac{s}{(s^2+2^2)} \cdot \frac{s}{(s^2+2^2)}\right)$$

$$= \mathcal{L}^{-1}(\ \mathcal{L}\ (\cos\ 2x) \cdot \mathcal{L}\ (\cos 2x))$$

$$= \int_0^x \cos 2y \cdot \cos(2x - 2y)\ dy$$

$$= \int_0^x \cos 2y \cdot \cos 2(x - y)\ dy \qquad \text{(using Convolution Theorem)}$$

$$= \frac{1}{2}\int_0^x \{\cos 2x + \cos 2(x - 2y)\}\ dy$$

$$= \frac{1}{2}x\cos 2x + \frac{1}{2}\int_0^x \cos 2(2y - x)dy$$

$$= \frac{1}{2}x\cos 2x + \frac{1}{4}\sin 2x.$$

This completes our verification.

Example (3) : Verify scaling property of Laplace transforms by finding $\mathcal{L}(f(x))$ in two ways — one by computing $f(x)$ from $\mathcal{L}(f(x))$ and then finding $f(2x)$, followed by Laplace transform; and another by scaling property, if given that

$$\mathcal{L}(f(x)) = \frac{(s^2 - s + 1)}{(2s + 1)^2(s - 1)}$$

$\dfrac{(s^2 - s + 1)}{(2s + 1)^2(s - 1)}$ writing as $\dfrac{A}{(2s + 1)} + \dfrac{B}{(2s + 1)^2} + \dfrac{C}{(s - 1)}$, we get

$$s^2 - s + 1 = A(s - 1)(2s + 1) + B(s - 1) + C(2s + 1)^2$$

$$= A(2s^2 - s - 1) + B(s - 1) + C(4s^2 + 4s + 1)$$

$$= (2A + 4C)s^2 + (4c + B - A)s + (C - A - B)$$

Comparing co-efficients of like powers of s, we get:

$$2A + 4C = 1; \quad A - B - 4C = 1; \quad C - A - B = 1$$

$$\frac{s^2 - s + 1}{(2s + 1)^2(s - 1)} = \frac{5}{18}\frac{1}{(2s + 1)} - \frac{7}{6}\frac{1}{(2s + 1)^2} + \frac{1}{9}\frac{1}{(s - 1)}$$

$$\therefore \quad f(x) = \frac{5}{18}\mathcal{L}^{-1}\left(\frac{1}{2s + 1}\right) - \frac{7}{6}\mathcal{L}^{-1}\left(\frac{1}{(2s + 1)^2}\right) + \frac{1}{9}\mathcal{L}^{-1}\left(\frac{1}{s - 1}\right)$$

$$= \frac{5}{36}\mathcal{L}^{-1}\left(\frac{1}{s + \frac{1}{2}}\right) - \frac{7}{24}\mathcal{L}^{-1}\left(\frac{1}{(s + \frac{1}{2})^2}\right) + \frac{1}{9}\mathcal{L}^{-1}\left(\frac{1}{s - 1}\right)$$

$$= \frac{5}{36}e^{-\frac{x}{2}} - \frac{7}{24}xe^{-\frac{x}{2}} + \frac{1}{9}e^{-x}$$

Therefore, $\quad f(2x) = \dfrac{5}{36}e^{-x} - \dfrac{7}{24}(2x)e^{-x} + \dfrac{1}{9}e^{2x}$

$$\mathcal{L}(f(2x)) = \frac{5}{36}\mathcal{L}(e^{-x}) - \frac{7}{12}\mathcal{L}(xe^{-x}) + \frac{1}{9}\mathcal{L}(e^{2x})$$

$$= \frac{5}{36}\frac{1}{(s + 1)} - \frac{7}{12}\frac{1}{(s + 1)^2} + \frac{1}{9}\frac{1}{(s - 2)}$$

$$= \frac{1}{4}\frac{s^2 - 2s + 4}{(s + 1)^2(s - 2)} = \frac{(\frac{s}{2})^2 - (\frac{s}{2}) + 1}{2(2\frac{s}{2} + 1)^2(\frac{s}{2} - 1)}$$

If we denote $\mathcal{L}(f(x))$ by $F(s)$, then $\mathcal{L}(f(2x)) = \frac{1}{2}F(\frac{s}{2})$

This verifies the scaling property of Laplace transforms.

Example (4) : To find the inverse Laplace transform of $\dfrac{1 + e^{-\pi s}}{(s^2 + 1)}$

Since $\qquad \dfrac{1 + e^{-\pi s}}{(s^2 + 1)} = \dfrac{1}{(s^2 + 1)} + \dfrac{e^{-\pi s}}{(s^2 + 1)}$

and from the known tables of Laplace transforms, it follows that

$$\frac{1}{s^2 + 1} = \mathcal{L}(f(x)), \text{ where } f(x) = \begin{cases} \sin x, & \text{if } x \geqslant 0 \\ 0, & \text{if } x < 0 \end{cases}$$

and $\quad \dfrac{e^{-\pi s}}{(s^2 + 1)} = \mathcal{L}(g(x)), \text{ where } g(x) = \begin{cases} \sin(x - \pi), & \text{if } x \geqslant \pi \\ 0, & \text{if } x < \pi \end{cases}$

we have by linearity of $\mathcal{L}^{-1}$,

$$\mathcal{L}^{-1}\left(\frac{e^{-\pi s}}{s^2 + 1}\right) = f(x) + g(x),$$

where $\quad f(x) + g(x) = \begin{cases} \sin\ x + 0 = \sin\ x, & \text{if } 0 < x < \pi \\ \sin\ x + \sin(x - \pi) = 0, & \text{if } x \geqslant \pi \\ 0 + 0 = 0, & \text{if } x < 0 \end{cases}$

i, e, $\quad f(x) + g(x) = \begin{cases} \sin\ x, & \text{if } 0 < x < \pi \\ 0, & \text{if } x \geqslant \pi \ \text{ or } \ x < 0 \end{cases}$

Observe that in working out this problem, we used second shifting property of Laplace transforms.

Example (5) : If $F(s)$ be the Laplace transform of $f(x)$, then

$$\mathcal{L}\left(\frac{f(x)}{x}\right) = \int_s^\infty F(s)ds, \text{ whenever } s > 0.$$

Infact, $\displaystyle\int_s^\infty F(s)ds = \lim_{B\to\infty} \int_s^B F(s)ds$

$$= \lim_{B\to\infty} \int_s^B \left(\int_0^\infty e^{-sx}f(x)dx\right)ds$$

$$= \lim_{B\to\infty} \int_s^B \left(\lim_{A\to\infty}\int_0^A e^{-sx}f(x)dx\right)ds$$

$$= \lim_{B\to\infty}\lim_{A\to\infty}\int_0^A \left(f(x)\int_s^B e^{-sx}ds\right)dx$$

$$= \lim_{B\to\infty}\lim_{A\to\infty}\int_0^A \frac{f(x)}{x}\left(e^{-sx} - e^{-Bx}\right)dx$$

$$= \int_0^\infty \frac{f(x)}{x}e^{-sx}dx = \mathcal{L}\left(\frac{f(x)}{x}\right) \ (\because \lim_{B\to\infty} e^{-Bx} = 0)$$

This proves the desired result, which in turn can be used for evaluation of few common improper integrals.

For example, we shall use the above result in establishing

(i) $\displaystyle\int_0^\infty \frac{\sin x}{x}dx = \frac{\pi}{2}$ and (ii) $\displaystyle\int_0^\infty \frac{e^{-ax}\sin bx}{x}dx$; $a, b > 0$

(i) $\mathcal{L}\left(\dfrac{\sin x}{x}\right) = \displaystyle\int_0^\infty e^{-sx}\frac{\sin x}{x}dx = \int_s^\infty \mathcal{L}(\sin x)ds$

$\qquad\qquad = \displaystyle\int_s^\infty \frac{ds}{(s^2+1)} = \left[\tan^{-1}s\right]_s^\infty = \frac{\pi}{2} - \tan^{-1}s \ ; \ s > 0$

On substituting $s = 0$ in the limiting form, we get

$$\int_0^\infty \frac{\sin x}{x}\, dx = \lim_{s \to 0}\left(\frac{\pi}{2} - \tan^{-1}s\right) = \frac{\pi}{2} \qquad \text{(q.e.d)}$$

(ii) By the above result,

$$\mathcal{L}\left(\frac{e^{-ax}\sin bx}{x}\right) = \int_s^\infty \mathcal{L}(e^{-ax}\sin bx)\, ds; \quad s > 0$$

i, e, $\displaystyle e^{-sx}\int_0^\infty \left(\frac{e^{-ax}\sin bx}{x}\right) dx = b\int_s^\infty \frac{ds}{(s+a)^2 + b^2}; \quad s > 0$

(Using first shifting property)

$$\therefore \int_0^\infty e^{-sx}\left(\frac{e^{-ax}\cdot \sin bx}{x}\right) dx = \tan^{-1}\left(\frac{s-a}{b}\right)\Big|_s^\infty = \frac{\pi}{2} - \tan^{-1}\left(\frac{s-a}{b}\right)$$

On substituting $s = 0$ in the limiting form we get :

$$\int_0^\infty \frac{e^{-ax}\cdot \sin bx}{x}\, dx = \frac{\pi}{2} - \tan^{-1}(0) = \frac{\pi}{2} \qquad \text{(q.e.d)}$$

(iii) According to the above result,

$$\mathcal{L}\left(\frac{e^{-ax} - e^{-bx}}{x}\right) = \int_s^\infty \mathcal{L}\left(e^{-ax} - e^{-bx}\right) ds$$

i, e, $\displaystyle \int_0^\infty e^{-sx}\left(\frac{e^{-ax} - e^{-bx}}{x}\right) dx = \int_s^\infty \left\{\mathcal{L}\left(e^{-ax}\right) - \mathcal{L}\left(e^{-bx}\right)\right\} ds; a, b > 0$

$$\text{R.H.S} = \int_s^\infty \left(\frac{1}{s+a} - \frac{1}{s+b}\right) ds$$

$$= \lim_{B \to \infty}\int_s^B \left(\frac{1}{s+a} - \frac{1}{s+b}\right) ds$$

$$= \lim_{B \to \infty} ln\left(\frac{s+a}{s+b}\right)\Big|_s^b = \lim_{B \to \infty} ln\left(\frac{1+\frac{a}{b}}{1+\frac{b}{B}}\right) - ln\left(\frac{s+a}{s+b}\right)$$

$$= ln\left(\frac{s+b}{s+a}\right)$$

Thus $\displaystyle \int_0^\infty e^{-sx}\left(\frac{e^{-ax} - e^{-bx}}{x}\right) dx = ln\left(\frac{s+b}{s+a}\right)$

In the limiting form if substituted $s = 0$, we get form above :

$$\int_0^\infty \frac{e^{-ax} - e^{-bx}}{x} dx = ln \left(\frac{b}{a} \right)$$

Note that $b > a$ must hold as otherwise one cannot ensure non-negativity of the function appearing in $\int_s^\infty \mathcal{L} \left(e^{-ax} - e^{-bx} \right) ds$.

Example (6) : Use the above result of Example (5) to prove

$$\mathcal{L} \left(\frac{\sin^2 x}{x} \right) = \frac{1}{4} ln \left(\frac{s^2 + 4}{s^2} \right)$$

We choose $f(x) = \sin^2 x$, so that

$$\mathcal{L} \left(\frac{\sin^2 x}{x} \right) = \mathcal{L} \left(\frac{f(x)}{x} \right) = \int_s^\alpha F(s) ds, \text{ where } F(s) \equiv \mathcal{L}(f(x))$$

Now $\quad \mathcal{L}(f(x)) = \mathcal{L}(\sin^2 x) = \frac{1}{2}\mathcal{L}(1 - \cos 2x) = \frac{1}{2}\{\mathcal{L}(1) - \mathcal{L}(\cos 2x)\}$

$$\therefore \quad F(s) = \frac{1}{2} \left[\frac{1}{s} - \frac{s}{s^2 + 4} \right] = \frac{2}{s(s^2 + 4)}$$

Hence $\quad \displaystyle\int_s^\infty F(s) ds = 2 \int_s^\infty \frac{ds}{s(s^2 + 4)} = \int_{s^2}^\infty \frac{d(s^2)}{s^2(s^2 + 4)}$

$$= \frac{1}{4} \left[\int_{s^2}^\infty \frac{d(s^2)}{s^2} - \int_{s^2}^\infty \frac{d(s^2)}{(s^2 + 4)} \right]$$

$$= \frac{1}{4} \left[-ln(s^2) + ln(s^2 + 4) \right] = \frac{1}{4} ln \left(\frac{s^2 + 4}{s^2} \right) \qquad \text{(q.e.d)}$$

Laplace Transform of the Derivatives and Derivatives of Laplace Transforms

The knowledge of derivatives of the Laplace transform help us in two ways, viz,

(i) finding the Laplace transform of new functions from the knowledge of Laplace transforms of a handful set of functions.

(ii) solving linear ordinary differential equations.

The following result shows how the Laplace transform of new function $f'(x)$ may be computed from that of $f(x)$.

If a function $f(x)$ is continuous and is of exponential order in the sense $\exists\ \alpha$ and $M(>0)$ such that $|f(x)| \leqslant Me^{-\alpha x}$ for infinitely large x and if its derivative $f'(x)$ is piecewise continuous in every finite closed interval, then the Laplace transform of $f'(x)$ exists for $s > \alpha$ and is given by $\mathcal{L}\left(f'(x)\right) = s\mathcal{L}\left(f(x)\right) - f(0)$.

Proof : By definition of Laplace transform,

$$\mathcal{L}\left(f'(x)\right) = \lim_{B \to \infty} \int_0^B f'(x)e^{-sx}dx \ ; \ s > 0$$

provided the improper integral is convergent. Since by assumption $f'(x)$ is piecewise continuous in $[0, B]$, $f'(x)$ has a finite number of jump discontinuities in $[0, B]$ that we may denote by $x_1, x_2, \cdots, x_{n-1}$. So if one agrees to write $0 = x_0$ and $B = x_n$, then we eventually get a partition $P \equiv \{0 = x_0 < x_1 < \cdots\cdots x_{r-1} < x_r < \cdots\cdots < x_n = B\}$ of $[0, B]$ such that $f'(x)$ is right continuous in each of the sub-intervals $[x_{r-1}, x_r]$. We may therefore apply integration by parts to each of them.

Now $\displaystyle\int_0^B e^{-sx}f'(x)dx = \sum_{r=1}^{n} \int_{x_{r-1}}^{x_r} f'(x)e^{-sx}dx$

$$= \sum_{r=1}^{n} s \int_{x_{r-1}}^{x_r} f(x)e^{-sx}dx + \sum_{r=1}^{n} \left[e^{-sx}f(x)\right]_{x_{r-1}}^{x_r}$$

$$= e^{-sB}f(B) - f(0) + s \int_0^B f(x)e^{-sx}dx$$

as $f(x)$ is continuous over this subinterval, we can have

$$\sum_{r=1}^{n} \int_{x_{r-1}}^{x_r} f(x)e^{-sx}dx = \int_0^B f(x)e^{-sx}dx$$

and $f(x_r - 0) = f(x_r + 0)$ for $r = 1, 2, \cdots, n-1$

Finally as f is of exponential order, $|e^{-sB}f(B)| < Me^{-(s-\alpha)B}$ and so for any $B > 0$ and any $s > \alpha$,

$$\lim_{B \to \infty} e^{-sB}f(B) = 0$$

Therefore proceeding to the limit $B \to \infty$, we get

$$\mathcal{L}(f'(x)) = \int_0^\infty f'(x)e^{-sx}dx = \lim_{B \to \infty} \int_0^B f'(x)e^{-sx}dx = s\,\mathcal{L}(f(x)) - f(0)$$

Differential Equations: A linear Algebra Approach

A generalisation of this result concerning the Laplace transform of the nth derivative of the function $f(x)$ is given in the following result:

If $(n-1)$th order derivative $f^{(n-1)}(x)$ of a function $f(x)$ is continuous and its nth order derivative $f^{(n)}(x)$ is sectionally continuous in any finite interval of x and if each of $f(x), f'(x), \cdots\cdots f^{(n-1)}(x)$ be of exponential order as x becomes infinitely large, then the Laplace transform

$$\mathcal{L}\left(f^{(n)}(x)\right) = s^n F(s) - s^{n-1} f(0) + s^{n-2} f'(0) - \cdots\cdots + (-1)^n f^{(n-1)}(0)$$

Infact, this last result is the basic weapon of applying Laplace transforms to solve ordinary linear differential equations. One literally observes that Laplace transformation technique can be applied to only initial-value problems where one has the apriori knowledge of the values of $f(0), f'(0), \cdots, f^{(n-1)}(0)$. Proving this result is nothing special if we show the existence of $\mathcal{L}\left(f^{(n)}(x)\right)$ for $s > \alpha$ and then apply principle of mathematical induction.

We now find successive derivatives of the Laplace transform of a function f that is continuous and is of exponential order α for large x. According to the definition of Laplace transform,

$$F(s) \equiv \mathcal{L}(f(x)) = \int_0^\infty e^{-sx} f(x) dx \ ; \ s > \alpha$$

$$\therefore F'(s) = \frac{d}{ds}\left(\int_0^\infty e^{-sx} f(x) dx\right) = \int_0^\infty (-x) e^{-sx} f(x) dx = -\mathcal{L}(xf(x))$$

since differentiation under the sign of integration is valid here and $xf(x)$ satisfies both the properties assumed for $f(x)$.

In general, for any positive integer r, the function $x^r f(x)$ does obey both these properties of $f(x)$ and so $\mathcal{L}(x^r f(x))$ exists. By principle of induction, one may prove that if $s > \alpha$, then

$$\mathcal{L}(x^n f(x)) = (-1)^n F^{(n)}(s) \text{ for any finite } + \text{ve integer } n \ .$$

Thus through successive differentiation of the Laplace transform of a given function, one can generate the Laplace transform of a new class of functions. This result is useful in solving some differential equations having variable co-efficients.

Integrals of Laplace transforms has been discussed earlier. We therefore give a brief account of the Laplace transform of integrals.

Laplace Transform of Integrals

We come up with the problem of finding the Laplace transform of the primitive of a function $f(x)$ (defined by $g(x) = \int_0^x f(y)dy$) that is sectionally continuous and is of exponential order α for infinitely large x. We have established in earlier part of this section that the primitive $g(x)$ of $f(x)$ is also sectionally continuous and of exponential order α. Thus $\mathcal{L}(f(x))$ and $\mathcal{L}(g(x))$ both exist. Our present aim will be to prove that

$$\mathcal{L}(g(x)) = \frac{1}{s}\mathcal{L}(f(x)) \quad \text{for} \quad s > 0 \cdot$$

Observe that $g'(x) = f(x)$ is piecewise continuous and $g(0) = 0$. So by the basic result of Laplace transform for derivatives we get

$$\mathcal{L}(g'(x)) = s\mathcal{L}(g(x)) - g(0) = s\mathcal{L}(g(x)) \quad (\because g(0) = 0)$$

Therefore, $\quad \mathcal{L}(f(x)) = s\mathcal{L}(g(x)) \quad$ for $\quad s > 0$

or, equivalently $\quad F(s) = sG(s) \quad\quad$ for $\quad s > 0$.

Illustrative examples :

(i) If $f(x) = \begin{cases} 1 & \text{for} \quad x \geqslant 0 \\ 0 & \text{for} \quad x < 0, \end{cases}$ then $F(s) = L(f(x)) = \dfrac{1}{s}$ for $s > 0$

Therefore, $\quad F^{(n)}(s) = \dfrac{d^n}{ds^n}\left(\dfrac{1}{s}\right) = \dfrac{(-1)^n \, n!}{s^{n+1}}$, $s > 0$

$\therefore \qquad\qquad L(x^n) = L(x^n f(x)) = (-1)^n F^{(n)}(s)$

$$= \frac{(-1)^n \cdot (-1)^n n!}{s^{n+1}} = \frac{n!}{s^{n+1}} \; ; \; s > 0$$

(ii) If $f(x) = \begin{cases} e^{kx}, & x \geqslant 0 \\ 0, & \text{elsewhere} \end{cases}$ then it is known that

$$F(s) \equiv \mathcal{L}(f(x)) = \frac{1}{s-k} \quad \text{for} \quad s > k$$

$$\therefore \quad \mathcal{L}(x^n e^{kx}) \;\equiv\; G(s) = (-1)^n\, F^n(s) = (-1)^n \cdot \frac{d^n}{ds^n}\left\{(s-k)^{-1}\right\}$$

$$= \frac{(-1)^n \cdot (-1)^n \cdot n!}{(s-k)^{n+1}} = \frac{n!}{(s-k)^{n+1}} \quad \text{for } s > k$$

(iii)   If $f(x) = \begin{cases} \sin wx, & x \geqslant 0 \\ 0, & x < 0, \end{cases}$ and $g(x) = \begin{cases} x \sin wx, & x \geqslant 0 \\ 0, & x < 0, \end{cases}$

then $\quad \mathcal{L}(f(x)) \equiv F(s) = \dfrac{w}{s^2 + w^2}$; $s > 0$

$$\therefore \quad \mathcal{L}(x \sin wx) = \mathcal{L}(g(x)) = (-1)\frac{d}{ds}F(s) = (-1)\frac{d}{ds}\left(\frac{w}{s^2+w^2}\right)$$

$$\therefore \quad \mathcal{L}(x \sin wx) = \frac{2sw}{(s^2+w^2)^2} \; ; \; s > 0$$

(iv)   If $\quad f(x) = \dfrac{\sin wx - wx \cos wx}{2w^3}$, then we may find $\mathcal{L}(f(x))$ by

use of formula for integrals.

Define $\; h(x) = x \sin wx \;$ and $\; g(x) = \displaystyle\int_0^x h(y)\,dy$

Now $\; \displaystyle\int_0^x h(y)\,dy = \int_0^x (y \sin wy)\,dy = \dfrac{\sin wx - wx \cos wx}{w^2}$

So we observe that $g(x)$ is identified as $2wf(x)$ and also

$$\mathcal{L}(g(x)) \equiv G(s) = \frac{1}{s}\mathcal{L}(h(x)) = \frac{1}{s}\mathcal{L}(x \sin wx) = \frac{2w}{(s^2+w^2)^2} \; ; \quad s > 0$$

Hence $\quad \mathcal{L}\left(\dfrac{\sin wx - wx \cos wx}{2w^3}\right) = \dfrac{1}{(s^2+w^2)^2} \; ; \; s > 0$

7.3   Laplace transformation technique of solving Ordinary Differential Equations

Our main objective of the present article is to solve odes by the Laplace transform method. However, this method applies to the initial-value problems comprising of a nth order linear ode having constant co-efficients:

$$a_0\frac{d^n y}{dx^n} + a_1\frac{d^{n-1} y}{dx^{n-1}} + \cdots\cdots + a_{n-1}\frac{dy}{dx} + a_n\, y = b(x),$$

where a_k's are constants and $b(x)$ is a continuous function over some interval I together with prescribed initial conditions

$$y(0) = c_0 \; ; \; y'(0) = c_1 \; ; \; \cdots\cdots \; ; \; y^{(n-1)}(0) = c_{n-1}$$

From the discussions of Chapter 5 it is assumed that this IVP admits of a unique solution. We aim to explore it. As told earlier, the purpose of Laplace transform technique for solving initial value problems is to translate it into an algebraic equation in an unknown function $Y(s)$(where $Y(s)) \equiv \mathcal{L}(y(x))$. This implies that solving an I.V.P involving a constant co-efficient linear ode in $y(x)$ is tantamount to solving a polynomial equation in $Y(s)$. Once $Y(s)$ is obtained, uniqueness of Laplace inversion enables us to retrieve the desired $y(x)$ with the help of table of standard results of Laplace transforms.

This technique may be employed in solving initial value problems where the involved ode is one with variable co-efficients. We shall deal with a couple of such problems in the next article — one being the Cauchy-Euler type equations and other being Legendre and Bessel equations.

Example (7) : Solve the first order ode by the Laplace transformation method : $y' + 2y = e^{-2x}$ subject to $y(0) = 3$.

Here the inhomogeneous term e^{-2x} is of exponential order and y, y' are also of exponential order. Again y and e^{-2x} being continuous of x, y' follows suit. Hence one can directly apply Laplace transformation technique to solve the initial value problem.

Taking Laplace transform on both sides of the equation and applying the linearity of the transformation, we have:

$$\mathcal{L}(y') + 2\mathcal{L}(y) = \mathcal{L}(e^{-2x})$$

$$\therefore \quad s\mathcal{L}(y) - y(0) + 2\mathcal{L}(y) = \mathcal{L}(e^{-2x})$$

i.e, $$(s+2)Y(s) = 3 + \frac{1}{s+2} \qquad (\because y(0) = 3)$$

Hence $$Y(s) = \frac{3}{(s+2)} + \frac{1}{(s+2)^2}$$

By uniqueness of Laplace Inversion, we get

$$y(x) = \mathcal{L}^{-1}(Y(s)) = \mathcal{L}^1\left(\frac{3}{s+2} + \frac{1}{(s+2)^2}\right) = 3e^{-x} + xe^{-2x}$$

Example (8) : Solve the following initial value problem by Laplace transformation method:

$y'' + 2y' + 2y = 2$ subject to the conditions $y(0) = 0$ and $y'(0) = 1$

For the differential equation, the inhomogeneous term 2 being of exponential order, y', y'' are also of exponential order. Again $(2 - 2y)$ being a continuous function of x, $(y'' + 2y')$ is also so.

Therefore the Laplace transformation technique is applicable. Taking Laplace transform on both sides and using its linearity we have:

$$\mathcal{L}(y'') + 2\mathcal{L}(y') + 2\mathcal{L}(y) = 2\mathcal{L}(1) \tag{a}$$

$$\therefore \quad \{s^2 Y(s) - sy(0) - y'(0)\} + 2\{sY(s) - y(0)\} + 2Y(s) = \frac{2}{s},$$

where $Y(s) = \mathcal{L}(y(x))$

Regrouping the terms and using the given initial conditions,

$$Y(s) = \frac{(s+2)}{s(s^2 + 2s + 2)} \tag{b}$$

Using method of partial fractions, we may write

$$\frac{(s+2)}{s(s^2 + 2s + 2)} = \frac{A}{s} + \frac{B(s+1) + C}{s(s^2 + 2s + 2)},$$

where A, B, C are constants that can be determind as follows :

$$(s + 2) = A(s^2 + 2s + 2) + s\{B(s+1) + C\}$$

Putting $s = 0$, we have $A = 1$ and so $(B+1)(s+1) + C = 0$

Putting $s = -1$, we get $C = 0$ and consequently $B = -1$

$$
\begin{aligned}
\therefore \quad y(x) &= \mathcal{L}^{-1}(Y(s)) = \mathcal{L}^{-1}\left(\frac{1}{s} - \frac{(s+1)}{s(s^2 + 2s + 2)}\right) \\
&= \mathcal{L}^{-1}\left(\frac{1}{s}\right) = \mathcal{L}^{-1}\left(\frac{s+1}{(s+1)^2 + 1}\right) \\
&= \mathcal{L}^{-1}(1) - \mathcal{L}^{-1}\left(e^{-x}\cos x\right) \text{ (using first shifting property)} \\
&= \mathcal{L}^{-1}\left(1 - e^{-x}\cos x\right)
\end{aligned}
$$

$\therefore$ By uniqueness of Laplace inversion, $y(x) = 1 - e^{-x}\cos x$

Example (9) : Solve the ode : $y'' + 2y' + 5y = 3e^{-x}\sin x$ subject to $y(0) = 0$; $y'(0) = 3$

Since the inhomogeneous term $3e^{-x}\sin x$ is continuous and is of exponential order, y, y' and y'' are all continuous and of exponential order. Consequently we can apply Laplace transform method. Taking Laplace transformation and applying its linearity property,

$$\mathcal{L}(y'') + 2\mathcal{L}(y') + 5\mathcal{L}(y) = 3\mathcal{L}(e^{-x}\sin x)$$

$$\therefore \ (s^2 Y(s) - sy(0) - y'(0)) + 2(sY(s) - y(0)) + 5Y(s) = \frac{3}{(s+1)^2 + 1}$$

Regarding the terms we have :

$$(s^2 + 2s + 5)Y(s) = 3 + \frac{3}{s^2 + 2s + 2} \quad (\because y(0) \text{ and } y'(0) = 3)$$

i.e, $\ Y(s) = \dfrac{3}{s^2 + 2s + 5} + \dfrac{3}{(s^2 + 2s + 2)(s^2 + 2s + 5)}$

One may also write $\ Y(s) = \left(\dfrac{1}{s^2 + 2s + 2} + \dfrac{2}{s^2 + 2s + 5} \right)$

Using uniqueness of Laplace inversion, we get

$$\begin{aligned} y(x) &= \mathcal{L}^{-1}(Y(s)) = \mathcal{L}^{-1}\left(\frac{1}{s^2 + 2s + 2} + \frac{2}{s^2 + 2s + 5} \right) \\ &= \mathcal{L}^{-1}\left(\frac{1}{(s+1)^2 + 1} + \frac{2}{(s+1)^2 + 2^2} \right) \\ &= e^{-x}(\sin x + \sin 2x), \end{aligned}$$

where in the last step we used linearity of Laplace inversion operator $\mathcal{L}^{-1}$ and also the first shifting property of Laplace transform.

Example (10) : Solve the ode $(y'' + 2y' + y) = 3xe^{-x}$ subject to the boundary conditions $y(0) = 4$ and $y'(-1) = 2$.

We refrain from detailing the formalities that ensures the applicability of the Laplace transformation technique. However, there is one hiccup present, viz, the fact that given problem is a b.v.p and not an i.v.p as are encountered usually while applying this method. To do away with this difficulty we presume $y'(0) = c$, say, and then evaluate the unknown

c with the help of the unused boundary condition $y'(-1) = 2$. Applying Laplace transformation technique we have :

$$\mathcal{L}(y'') + 2\mathcal{L}(y') + \mathcal{L}(y) = 3\mathcal{L}(xe^{-x}) \tag{a}$$

$$\therefore \ (s^2 Y(s) - sy(0) - y'(0)) + 2(sY(s) - y(0)) + Y(s) = \frac{3}{(s+1)^2} \ ,$$

where we have made use of the rules of Laplace transform for derivatives and the first shifting property. Regrouping the terms we get on simplification,

$$Y(s) = \frac{4(s+2) + c}{(s+1)^2} + \frac{3}{(s+1)^4} = \frac{4}{(s+1)} + \frac{4+c}{(s+1)^2} + \frac{3}{(s+1)^4}$$

Taking the Laplace inversion operator $\mathcal{L}^{-1}$ on both sides and making use of the fact that $\mathcal{L}^{-1}$ is also linear, we get :

$$\begin{aligned}
y(x) &= 4\mathcal{L}^{-1}\left(\frac{1}{(s+1)}\right) + (4+c)\mathcal{L}^{-1}\left(\frac{1}{(s+1)^2}\right) + 3\mathcal{L}^{-1}\left(\frac{1}{(s+1)^4}\right) \\
&= 4e^{-x} + (4+c)(xe^{-x}) + \frac{1}{2}x^3 e^{-x} \\
&= \left\{4 + (4+c)x + \frac{1}{2}x^3\right\}e^{-x} \tag{b}
\end{aligned}$$

Our next job is to evaluate constant c. Differentiating (b) $w.r$ to x,

$$y'(x) = \left\{c - (c+4)x + \frac{3}{2}x^2 - \frac{1}{2}x^3\right\}e^{-x}$$

As $\ y'(-1) = 2$, we get :

$$2 = \left\{c + (c+4) + \frac{3}{2} + \frac{1}{2}\right\}e = 2(c+3)e$$

$$\therefore c = -3 + \frac{1}{e}$$

Putting back this value of c in (b), we reach the desired solution :

$$y(x) = \left\{4 + \left(1 + \frac{1}{e}\right)x + \frac{1}{2}x^3\right\}e^{-x}$$

Example (11) : Solve the initial value problem $y'' + 4y = \sin 2x$, subject to the conditions $y(0) = 1$ and $y'(0) = -2$

Skipping the formalities of proving the legitimacy of Laplace transformation technique to this i.v.p, we straightway use Laplace transformation to the given ode so as to get :

$$\mathcal{L}(y'') + 4\mathcal{L}(y) = \mathcal{L}(\sin 2x)$$

i, e, $(s^2 Y(s) - sy(0) - y'(0)) + 4Y(s) = \dfrac{2}{s^2 + 4}$, where $Y(s) = \mathcal{L}(y(x))$

or $(s^2 + 4)y(s) - (s - 2) = \dfrac{2}{s^2 + 4}$ $(\because y(0) = 1 \, ; \, y'(0) = -2)$

$$\therefore \; y(x) = \mathcal{L}^{-1}\left(\frac{s}{s^2 + 2^2}\right) - \mathcal{L}^{-1}\left(\frac{2}{s^2 + 2^2}\right) + \mathcal{L}^{-1}\left(\frac{2}{(s^2 + 4)^2}\right)$$

$$= \cos 2x - \sin 2x + \mathcal{L}^{-1}\left(\frac{2}{(s^2 + 4)^2}\right)$$

Now $\mathcal{L}^{-1}\left(\dfrac{2}{(s^2 + 4)^2}\right) = \dfrac{1}{2}\mathcal{L}^{-1}\left(\mathcal{L}(\sin 2x)\mathcal{L}(\sin 2x)\right)$

$$= \frac{1}{2}\int_0^x \sin 2(x - y)\sin 2y \, dy \quad \text{(by convolution theorem)}$$

$$= \frac{1}{4}\int_0^x \{\cos(2x - 4y) - \cos 2x\} \, dy$$

$$= \frac{1}{4}\left[\frac{1}{4}\sin(4y - 2x) - y\cos 2x\right]_0^x = \frac{1}{8}(\sin 2x - 2x\cos 2x)$$

Therefore the solution is : $y(x) = \cos 2x - \sin 2x + \frac{1}{8}(\sin\ 2x - \cos 2x)$

Example (12) : Solve the ode

$$y'' + y' - 2y = g(x) = \begin{cases} e^{3x} & , \quad \text{if} \quad 0 \leqslant x < 1 \\ e^{3x} + 1, & \text{if} \quad x \geqslant 1 \end{cases}$$

subject to the initial conditions $y(0) = y'(0) = 0$

Applying Laplace transform to this ode we have :

$$\mathcal{L}(y'') + \mathcal{L}(y') + 2\mathcal{L}(y) = \mathcal{L}(g(x))$$

i.e, $(s^2 Y(s) - sy(0) - y'(0)) + (sY(s) - y(0) - 2Y(s)) = G(s))$

where $Y(s) \equiv \mathcal{L}(y(x))$ and $G(s) \equiv \mathcal{L}(g(x))$

Therefore $Y(s) = \dfrac{1}{(s-1)(s+2)}G(s)$ \hfill (a)

Before computing $G(s)$ let's note that $g(x)$ defined in the problem is piecewise continuous as it has jump discontinuity at $x = 1$ and continuity elsewhere in $[0, \infty)$. Moreover, $g(x)$ satisfies the inequality

$$|g(x)| = \begin{cases} e^{3x} & \leqslant 2e^{3x} \quad \text{if } 0 \leqslant x < 1 \\ e^{3x} + 1 & \leqslant 2e^{3x} \quad \text{if } x \geqslant 1 \end{cases}$$

and so $g(x)$ is of exponential order. Hence $G(s)$ is well defined for $s > 0$.

$$\therefore \; G(s) = \int_0^\infty e^{-sx} g(x)\, dx = \int_0^1 e^{-(s-3)x}\, dx + \int_1^\infty e^{-sx}(e^{3x} + 1)dx$$

$$= \left. \frac{e^{-(s-3)x}}{-(s-3)} \right|_0^1 + \lim_{B \to \infty} \left[\frac{e^{-(s-3)x}}{-(s-3)} - \frac{e^{-sx}}{s} \right]_1^B$$

$$= \frac{1}{s-3} + \frac{e^{-s}}{s} + \lim_{B \to \infty} \left[\frac{-e^{-(s-3)B}}{(s-3)} - \frac{e^{-sB}}{s} \right] = 0$$

From (a) we therefore get :

$$Y(s) = \frac{1}{(s-1)(s+2)} \left[\frac{1}{s-3} + \frac{e^{-s}}{s} \right] \quad \text{for } s > 3$$

$$= \frac{1}{(s-1)(s+2)(s-3)} + \frac{1}{s(s-1)(s+2)} e^{-s} \quad \text{for } s > 3 \quad \text{(b)}$$

We now use method of partial fraction to get

$$\frac{1}{(s-1)(s+2)(s-3)} = -\frac{1}{6} \cdot \frac{1}{(s-1)} + \frac{1}{15} \cdot \frac{1}{(s+2)} + \frac{1}{10} \cdot \frac{1}{(s-3)}$$

Similarly $\quad \dfrac{1}{s(s-1)(s+2)} = -\dfrac{1}{2} \cdot \dfrac{1}{s} + \dfrac{1}{3} \cdot \dfrac{1}{(s-1)} + \dfrac{1}{6} \cdot \dfrac{1}{(s+2)}$

From (b) we now have :

$$Y(s) = -\frac{1}{6} \cdot \frac{1}{(s-1)} + \frac{1}{15} \cdot \frac{1}{(s+2)} + \frac{1}{10} \cdot \frac{1}{(s-3)}$$

$$+ \left\{ -\frac{1}{2} \cdot \frac{1}{s} + \frac{1}{3} \cdot \frac{1}{(s-1)} + \frac{1}{6} \cdot \frac{1}{(s+2)} \right\} e^{-s}$$

$$= \mathcal{L} \left[-\frac{1}{6} e^x + \frac{1}{15} e^{-2x} + \frac{1}{10} e^{3x} \right] + \mathcal{L} \left[-\frac{1}{2} + \frac{1}{3} e^x + \frac{1}{6} e^{-2x} \right] e^{-s} \quad \text{(c)}$$

Using the relation

$$\mathcal{L} \left[-\frac{1}{2} + \frac{1}{3} e^x + \frac{1}{6} e^{-2x} \right] e^{-s} = \begin{cases} 0, & \text{if } 0 \leqslant x < 1 \\ -\frac{1}{2} + \frac{1}{3} e^{(x-1)} + \frac{1}{6} e^{-2(x-1)}, & \\ & \text{if } x \geqslant 1 \end{cases}$$

in (c) we get through uniqueness of Laplace inversion,

$$y(x) = \begin{cases} -\frac{1}{6} e^x + \frac{1}{15} e^{-2x} + \frac{1}{10} e^{3x}, & \text{if } 0 \leqslant x < 1 \\ -\frac{1}{6} e^x + \frac{1}{15} e^{-2x} + \frac{1}{10} e^{3x} - \frac{1}{2} + \frac{1}{3} e^{(x-1)} + \frac{1}{6} e^{-2(x-1)}, & \text{if } x \geqslant 1 \end{cases}$$

Example (13) : Solve $\dfrac{d^2}{dx^2} + 4y = b(x)$ if $y(0) = 2$; $y'(0) = 0$ and

$$b(x) = \begin{cases} 8\pi - 4x, & \text{if } 0 < x < 2\pi \\ 0, & \text{if } x > 2\pi \end{cases}$$

Taking Laplace transform on the given ode we have:

$$\mathcal{L}\left(\frac{d^2y}{dx^2}\right) + 4\mathcal{L}(y) = \mathcal{L}\left(b(x)\right)$$

i.e., $\quad s^2 Y(s) - sy(0) - y'(0) + 4Y(s) = -4 \displaystyle\int_0^{2\pi} e^{-sx}(x - 2\pi)dx$

$\therefore \quad (s^2 + 4)Y(s) - 2s = -4 \displaystyle\int_{-2\pi}^{0} e^{-s(u+2\pi)} u \, du$ (where $x - 2\pi \equiv u$)

$\therefore \quad Y(s) = \dfrac{2s}{s^2 + 4} + \dfrac{4}{s^2 + 4} \displaystyle\int_0^{-2\pi} e^{-us} \cdot e^{-2\pi s} \cdot u \, du$

$\qquad\quad = \dfrac{2s}{s^2 + 4} + \dfrac{4e^{-2\pi s}}{s^2 + 4} \displaystyle\int_0^{-2\pi} e^{-us} u \, du$

$\qquad\quad = \dfrac{2s}{s^2 + 4} + \dfrac{4e^{-2\pi s}}{s^2 + 4} \left[\dfrac{2\pi \, e^{2\pi s}}{s} - \dfrac{1}{s^2}\left(e^{2\pi s} - 1\right)\right]$

$\qquad\quad = \dfrac{2s}{s^2 + 4} + \left(\dfrac{8\pi}{s} - \dfrac{4}{s^2} + \dfrac{4}{s^2} e^{-2\pi s}\right) \dfrac{1}{(s^2 + 4)}$

$\qquad\quad = \dfrac{2s}{s^2 + 4} + \dfrac{8\pi}{s(s^2 + 4)} - \dfrac{4}{s^2(s^2 + 4)} + \dfrac{4}{s^2(s^2 + 4)} e^{-2\pi s}$

Now $\quad \dfrac{2s}{s^2 + 4} = \mathcal{L}(2\cos 2x)$

$$\frac{8\pi}{s(s^2 + 4)} = \frac{(2\pi s) \cdot 4}{s^2(s^2 + 4)} = 2\pi\left(\frac{1}{s} - \frac{s}{s^2 + 4}\right) = 2\pi\mathcal{L}(1 - \cos 2x)$$

$$\frac{4}{s^2(s^2 + 4)} = \left(\frac{1}{s^2} - \frac{1}{s^2 + 4}\right) = L\left(x - \frac{1}{2} \sin 2x\right)$$

$$\frac{4e^{-2\pi s}}{s^2(s^2 + 4)} = \left(\frac{1}{s^2} - \frac{1}{s^2 + 4}\right) e^{-2\pi s}$$

$$= \mathcal{L}\left(x - \frac{1}{2} \sin 2\pi s\right) \cdot e^{-2\pi s} = \mathcal{L}(f(x)),$$

where $\quad f(x) = \begin{cases} (x - 2\pi) - \frac{1}{2}\sin 2(x - 2\pi)\,, & \text{if } x > 2\pi \\ 0\,, & \text{if } 0 < x < 2\pi \end{cases}$

Thus $Y(s) = \begin{cases} \mathcal{L}[(2 - 2\pi)\cos 2x], & \text{if } x > 2\pi \\ \mathcal{L}[(-x + 2\pi) + \frac{1}{2}\sin 2x + (2 - 2\pi)\cos 2x], & \text{if } 0 < x < 2\pi \end{cases}$

By uniqueness of Laplace inversion,

$$y(x) = \begin{cases} (2 - 2\pi)\cos 2x, & \text{if} \quad x > 2\pi \\ (-x + 2\pi) + \frac{1}{2}\sin 2x + (2 - 2\pi)\cos 2x, & \text{if} \quad 0 < x < 2\pi \end{cases}$$

Example (14) : Solve the initial-value problem :

$$\frac{dy}{dx} + 2y = b(x), \quad \text{where } b(x) = \begin{cases} 1, & \text{if } 0 \leqslant x < 2 \\ 0, & \text{if } x \geqslant 2 \end{cases} \quad \text{and } y(0) = 0$$

We observe that the inhomogeneous term $b(x)$ is a piecewise constant function and so sectionally continuous. In this case Laplace transform technique is used to solve it.

If Laplace transformation of $y(x)$ is denoted by $\mathcal{L}(y(x)) = Y(s)$, then by rules of derivatives we get

$$sY(s) - y(0) + 2Y(s) = \mathcal{L}(b(x)) = \int_0^2 e^{-sx} \cdot 1 \; dx = \frac{1}{s}\left(1 - e^{-2s}\right)$$

$$\therefore \quad Y(s) = \frac{1}{s(s + 2)}(1 - e^{-2s}) \quad (\because y(0) = 0)$$

$$= \frac{1}{2}\left(\frac{1}{s} - \frac{1}{s + 2}\right) - \frac{1}{2}\frac{e^{-2s}}{s} + \frac{1}{2}\frac{1}{(s + 2)} \cdot e^{-2s}$$

$$= \frac{1}{2}\left(\frac{1}{s} - \frac{1}{s + 2}\right) - \frac{1}{2}\left(\frac{1}{s} - \frac{1}{s + 2}\right)e^{-2s}$$

$$= \mathcal{L}\left(\frac{1}{2}\left(1 - e^{-2x}\right)\right) - \mathcal{L}(f(x)),$$

where $\quad f(x) \quad = \quad \begin{cases} \frac{1}{2}\left(1 - e^{-2(x-2)}\right), & \text{if } x > 2 \quad \text{(by shifting property)} \\ 0, & \text{if } 0 < x < 2 \end{cases}$

$$= \quad \begin{cases} \frac{1}{2}\left(1 - e^4\, e^{-2x}\right), & \text{if } x \geqslant 2 \\ 0, & \text{if } 0 < x < 2 \end{cases}$$

Hence $Y(s) = \mathcal{L}\left[\frac{1}{2}\left(1 - e^{-2x}\right) - f(x)\right]$, with $f(x)$ defined as above.

$$\therefore \quad y(x) = \begin{cases} \frac{1}{2}\left(1 - e^{-2x}\right) - \frac{1}{2}\left(1 - e^4 e^{-2x}\right), & \text{if } x \geqslant 2 \\ \frac{1}{2}\left(1 - e^{-2x}\right), & \text{if } 0 < x < 2 \end{cases}$$

$$\text{i.e,} \quad y(x) = \begin{cases} \frac{1}{2}\left(e^4 - 1\right)e^{-2x}, & \text{if } x \geqslant 2 \\ \frac{1}{2}\left(1 - e^{-2x}\right), & \text{if } 0 < x < 2 \end{cases}$$

Remark : We now discuss in a nutshell what would be the difficulty faced by us if we try to solve this ode by the usual methods. First observe that the differential equation can be explicitly written as

$$\frac{dy}{dx} + 2y = \begin{cases} 1, & \text{if } 0 \leqslant x < 2 \\ 0, & \text{if } x \geqslant 2 \end{cases}$$

subject to the initial condition $y(0) = 0$.

For the first part, i.e, $0 \leqslant x < 2$, the solution is $y(x) = \frac{1}{2} - \frac{1}{2}e^{-2x}$ and for the second part, i.e, $x \geqslant 2$, the solution is $y(x) = Ae^{-2x}$, where A cannot be determined from $y(0)$. Hence we have to inculcate the physical considerations and presume continuity of the solution $y(x)$. We therefore use the fact

$$y(2) = y(2 - 0) = \frac{1}{2} - \frac{1}{2}\,e^{-4}$$

to evaluate A and finally get

$$y(x) = \frac{1}{2}(e^4 - 1)e^{-2x} \quad \text{for} \quad x \geqslant 2.$$

In case of Laplace transformation technique we need not bother about the physical consideration — infact shifting property of Laplace transform takes care of this connectivity between two parts.

Example (15) : Solve the initial value problem

$$\frac{dy}{dx} + y = b(x), \text{ where } b(x) = \begin{cases} 0, & \text{if } 0 \leqslant x < 1 \\ 1, & \text{if } 1 \leqslant x \leqslant 2 \end{cases} \quad \text{and} \quad y(0) = 2$$

Because $b(x)$ is piecewise constant and is obviously of exponential order, we can apply Laplace transformation technique to solve the given ode.

Applying technique of Laplace transforms we have:

$$\mathcal{L}\left(\frac{dy}{dx}\right) + \mathcal{L}(y(x)) = \mathcal{L}(b(x))$$

i, e, $(sY(s) - y(0)) + Y(s) = \mathcal{L}(b(x))$

$\therefore$ $(s+1)\, Y(s) = 2 + \mathcal{L}(b(x))$

i, e, $Y(s) = \dfrac{2}{s+1} + \dfrac{1}{(s+1)}\, \mathcal{L}(b(x))$

$$= 2\mathcal{L}\left(e^{-x}\right) + \mathcal{L}\left(e^{-x}\right)\mathcal{L}(b(x))$$

$$= 2\mathcal{L}\left(e^{-x}\right) + \mathcal{L}\left\{\int_0^x e^{-(x-y)} b(y)dy\right\}$$

$\therefore$ $y(x) = 2e^{-x} + e^{-x}\displaystyle\int_0^x e^y b(y)dy$

Again $\displaystyle\int_0^x e^y\, b(y)\, dy = \begin{cases} 0, & \text{if } x < 1 \text{ as } b(y) = 0 \text{ for } 0 \leqslant x < 1 \\ \int_1^x e^y dy = (e^x - e) & \text{if } 1 \leqslant x \leqslant 2 \end{cases}$

Therefore, by uniqueness of Laplace inversion it follows that

$$y(x) = \mathcal{L}^{-1}(Y(s)) = \begin{cases} 2e^{-x} + e^{-x} \cdot 0 = 2e^{-x}, & \text{if } 0 \leqslant x < 1 \\ = (2-e)e^{-x} + 1, & \text{if } 1 \leqslant x \leqslant 2 \end{cases}$$

Hence we get the final solution as :

$$y(x) = \begin{cases} 2e^{-x}, & \text{if } 0 \leqslant x < 1 \\ (2-e)e^{-x} + 1, & \text{if } 1 \leqslant x \leqslant 2 \end{cases}$$

Observation : We shall arrive at the same solution by the standard technique (i,e, other than Laplace transform), but we have to take into account physical consideration of the problem as shown in the following:

The given ode can be explicitly written as :

$$\frac{dy}{dx} + y = 0, \quad \text{if } 0 \leqslant x < 1 \; ; \; y(0) = 2$$

and $\dfrac{dy}{dx} + y = 1, \quad \text{if } 1 \leqslant x \leqslant 2$

For the first part, the particular solution is $ye^x = 2$ while for the second part, the general solution is : $ye^x = e^x + c$. However, there is no condition left with us to evaluate c. Thus we presume left continuity of the solution at $x = 1$.

$$\therefore \lim_{x \to 1-0} y(x) = y(1), \quad \text{implying} \quad y(1) = \frac{2}{e}$$

Thus $y(1)e = e + c$, i,e, $c = (2 - e)$ and the solution for $1 \leqslant x \leqslant 2$ is $y\, e^x = (2 - e) + e^x$, same as what we deduced by Laplace technique.

Example (16) : Consider a *LRC* circuit consisting of resistance R, inductance L and capacitance C connected in series. An electromotive force E is impressed across the terminals. If I be the current flowing in the circuit, then by Kirchoff's first law, we get a second order ode :

$$L\frac{d^2 I}{dt^2} + R\frac{dI}{dt} + \frac{1}{c} I = E \; ;$$

Solve this ode by Laplace transformation technique under the initial conditions $I(0) = 0$ and $\frac{dI}{dt} = A$ for three choices of E, viz,

(i) $E(t) = E_0\, U(t)$ (ii) $E(t) = E_0 \delta(t)$ and (iii) $E(t) = E_0 \sin pt$

$(U(t)$ denotes the unit step function).

Applying Laplace transform and using its linearity we get:

$$L\mathcal{L}\left(\frac{d^2 I}{dt^2}\right) + R\mathcal{L}\left(\frac{dI}{dt}\right) + \frac{1}{c}\mathcal{L}(I) = (E)$$

i,e, $\mathcal{L}(s^2 \bar{I}(s) - sI(0) - I'(0)) + R(s\bar{I}(s) - I(0)) + \frac{1}{c}\bar{I}(s) = \mathcal{L}(E)$

i,e, $(Ls^2 + Rs + \frac{1}{c})\bar{I}\,(s) - AL = \mathcal{L}(E)$

$$\therefore \bar{I}(s) = \frac{AL}{(Ls^2 + Rs + \frac{1}{c})} + \frac{\mathcal{L}(E)}{L(s^2 + \frac{R}{L}s + \frac{1}{Lc})}$$

$$= \frac{A}{\{(s + \frac{R}{2L})^2 + w^2\}} + \frac{\mathcal{L}(E)}{\{(s + \frac{R}{2L})^2 + w^2\}} \qquad \text{(a)}$$

Now $\dfrac{A}{(s + \frac{R}{2L})^2 + w^2} = L\left(\dfrac{A}{w}\, e^{\frac{-R}{2L}\cdot t} \sin wt\right)$, where $w^2 \equiv \dfrac{1}{Lc} - \dfrac{R^2}{4L^2}$

(Here we presume that R is sufficiently small)

Again (i) $\mathcal{L}(E) \equiv \mathcal{L}(E_0 U(t)) = E_0\mathcal{L}(U(t)) = E_0\left(\dfrac{1 - e^{-s}}{s}\right)$

(ii) $\mathcal{L}(E) \equiv \mathcal{L}(E_0 \delta(t)) = E_0 \cdot 1 = E_0$

(iii) $\mathcal{L}(E) \equiv \mathcal{L}(E_0 \sin pt) = E_0\, \mathcal{L}(\sin pt)$

Hence we get from (a), (case (iii))

$$\bar{I}(s) = \mathcal{L}\left(\frac{A}{w}e^{-\frac{R}{2L}t}\sin wt\right) + \mathcal{L}\left(\frac{E_0}{Lw}e^{-\frac{R}{2L}t}\sin wt\right).\mathcal{L}(\sin pt)$$

$$= \mathcal{L}\left[\frac{A}{w}e^{-\frac{R}{2L}t}\sin wt + \frac{E_0}{Lw}\int_0^t e^{-\frac{R}{2L}\tau}\sin wt \sin \phi(t-\tau)d\tau\right]$$

$\therefore$ By uniqueness of Laplace inversion,

$$I(t) = \mathcal{L}^{-1}(\bar{I}(s))$$

$$= \frac{A}{w}e^{-\frac{R}{2L}t}\sin wt + \frac{E_0}{wL}\int_0^t e^{-\frac{R}{2L}t}\sin w\tau \sin p(t-\tau)d\tau$$

In case (i), we get

$$I(t) = \mathcal{L}^{-1}(\bar{I}(s))$$

$$= \frac{A}{w}e^{-\frac{R}{2L}t}\sin wt + \frac{E_0}{L}e^{-\frac{R}{2L}t}\sin wt$$

In case (ii), we get

$$I(t) = \mathcal{L}^{-1}(\bar{I}(s))$$

$$= \frac{A}{w}e^{-\frac{R}{2L}t}\sin wt + \frac{E_0}{L}\int_0^t U(t-\tau)e^{-\frac{R}{2L}\tau}\sin w\tau \, d\tau,$$

(In all these cases we used convolution theorem).

Example (17) : Solve the I.V.P :

$$\left.\begin{array}{l}\frac{d^2y}{dx^2} + y = f(x) \\ y(0) = 2 \ ; \ y'(0) = 3\end{array}\right\} \text{ where } f(x) = \begin{cases} x, & 0 < x < \pi \\ \pi, & x > \pi \end{cases}$$

Taking Laplace transform on both sides and using its linearity property we get from the ode :

$$\mathcal{L}\left(\frac{d^2y}{dx^2}\right) + \mathcal{L}(y) = \mathcal{L}(f(x)) \tag{a}$$

Again $\mathcal{L}(f(x)) = \displaystyle\int_0^\pi e^{-sx}\cdot x \, dx + \pi \int_\pi^\infty e^{-sx}\, dx \qquad [s>0]$

$$= \left[\frac{e^{-sx}}{-s}\cdot x\right]_{x=0}^{x=\pi} + \frac{1}{s}\int_0^\pi e^{-sx}\, dx + \pi\int_\pi^\infty e^{-sx}\, dx$$

$$= \frac{-\pi}{s}e^{-s\pi} - \frac{1}{s^2}e^{-\pi s} + \frac{1}{s^2} + \frac{\pi}{s}e^{-s\pi} = \frac{1}{s^2}\left(1 - e^{-\pi s}\right)$$

Using rules of derivatives, (a) therefore becomes :

$$\{s^2 Y(s) - sy(0) - y'(0)\} + Y(s) = \frac{1}{s^2}\left(1 - e^{-\pi s}\right)$$

i, e, $(s^2 + 1)Y(s) - 2s - 3 = \dfrac{1}{s^2}\left(1 - e^{-\pi s}\right)$ $(\because y(0) = 2 \ \& \ y'(0) = 3)$

$$\therefore \ Y(s) = \frac{2s}{s^2 + 1} + \frac{3}{s^2 + 1} + \frac{1}{(s^2 + 1)} \cdot \frac{1}{s^2}\left(1 - e^{-\pi s}\right) \qquad \text{(b)}$$

Taking Laplace inversion, and using its linearity we get :

$$y(x) = \mathcal{L}^{-1}(Y(s)) = 2\mathcal{L}^{-1}\left(\frac{s}{s^2 + 1}\right) + 3\mathcal{L}^{-1}\left(\frac{1}{s^2 + 1}\right)$$

$$+ \mathcal{L}^{-1}\left(\frac{1}{s^2 + 1} \cdot \frac{1}{s^2}(1 - e^{-\pi s})\right)$$

$$= 2\cos x + 3\sin x + \mathcal{L}^{-1}\left(\mathcal{L}(\sin \ s) \cdot \mathcal{L}(f(x))\right)$$

$$= 2\cos x + 3\sin x + \int_0^x \sin(x - y)f(y)dy \ \text{(by convolution theorem)}$$

$$\text{But } \int_0^x \sin(x - y)f(y)dy = \begin{cases} \int_0^x y\sin(x - y)dy, & \text{if } 0 < x < \pi \\[2mm] \int_0^\pi y\sin(x - y)dy + \pi \int_\pi^x \sin(x - y)dy, \\[2mm] \hspace{4cm} \text{if } x > \pi \end{cases}$$

$$\text{where } \int_0^x y\sin(x - y)dy = \left[y\cos(x - y) + \sin(x - y)\right]\Big|_{y=0}^{y=x} = (x - \sin x)$$

$$\int_0^\pi y\sin(x - y)dy = \left[y\cos(x - y) + \sin(x - y)\right]\Big|_{y=0}^{y=\pi} = -\pi\cos x - 2\sin x$$

$$\int_\pi^x \sin(x - y)dy = \cos(x - y)\Big|_{y=\pi}^{y=x} = 1 + \cos x$$

$$\text{Hence } \int_0^x \sin(x - y)f(y)dy = \begin{cases} (x - \sin x), & \text{if } 0 < x < \pi \\ (1 - 2\sin x) + (1 - \pi)\cos x, & \text{if } x > \pi. \end{cases}$$

Piecing together the parts we get the desired solution.

Example (18) : Solve the ode

$$\frac{d^2 y}{dx^2} + 2\frac{dy}{dx} - y = xe^{-x} \ \text{subject to } y(0) = 0 \ ; \ y'(0) = 1$$

The equation $f(x) = xe^{-x}$ appearing as the inhomogeneous term in the given ode is continuous and of exponential order as shown below :

$$
\begin{aligned}
\lim_{x \to \infty} \frac{f(x)}{e^{\alpha x}} &= \lim_{x \to \infty} \frac{xe^{-x}}{e^{\alpha x}} = \lim_{x \to \infty} \frac{x}{e^{(\alpha+1)x}} \\
&= \lim_{x \to \infty} \frac{1}{(\alpha + 1)e^{\alpha+1}x} \\
&= 0, \text{ for any } \alpha > 0
\end{aligned}
$$

Thus $f(x)$ admits of Laplace transform and

$$
\mathcal{L}(f(x)) = \mathcal{L}(xe^{-x}) = (-1)\frac{d}{ds}(\mathcal{L}(e^{-x})) = -\frac{d}{ds}\left(\frac{1}{(s+1)}\right) = \frac{1}{(s+1)^2}
$$

Applying Laplace transformation on both sides of the ode, we get :

$$
\mathcal{L}\left(\frac{d^2 y}{dx^2}\right) + 2\mathcal{L}\left(\frac{dy}{dx}\right) - \mathcal{L}(y) = \mathcal{L}(xe^{-x})
$$

i, e, $(s^2 Y(s) - sy(0) - y'(0)) + 2(sY(s) - y(0)) - Y(s) = \dfrac{1}{(s+1)^2}$

i, e, $(s^2 + 2s - 1)Y(s) = 1 + \dfrac{1}{(s+1)^2} = \dfrac{(s^2 + 2s + 2)}{(s^2 + 2s + 1)}$

Therefore $Y(s) = \dfrac{(s^2 + 2s + 2)}{(s^2 + 2s - 1)(s^2 + 2s + 1)}$

$$
= \frac{3}{2}\left(\frac{1}{s^2 + 2s - 1}\right) - \frac{1}{2}\left(\frac{1}{s^2 + 2s + 1}\right)
$$

$$
= \frac{3}{2\sqrt{2}}\frac{(\sqrt{2})}{(s+1)^2 - (\sqrt{2})^2} - \frac{1}{2} \cdot \frac{1}{(s+1)^2}
$$

Hence $y(x) = \mathcal{L}^{-1}(y(s))$

$$
= \frac{3}{2\sqrt{2}}\mathcal{L}^{-1}\left(\frac{\sqrt{2}}{(s+1)^2 - (\sqrt{2})^2}\right) - \frac{1}{2}\mathcal{L}^{-1}\left(\frac{1}{(s+1)^2}\right)
$$

$$
= \frac{3}{2\sqrt{2}}e^{-x}\sinh(\sqrt{2}x) - \frac{1}{2}\,xe^{-x},
$$

where we have used the shifting property of Laplace transform.

Differential Equations having Variable Co-efficients

Let's now discuss the applicability of Laplace transformation technique to ordinary differential equations with variable co-efficients. The first thing which strikingly differs from the scenario of constant co-efficient odes is the fact that in the variable case co-efficient, the Laplace transformation produces again a differential equation while in constant co-efficient odes we get an algebric equation. In what follows we shall observe that Laplace transformation technique fails miserably if the co-efficients appearing in the ode are polynomials of degree strictly less than the order of the ode. To establish our claim we pick up four instances.

(A) Consider a general second order Cauchy-Euler equation in the form:

$$a_0 \, x^2 \, \frac{d^2 y}{dx^2} + a_1 \, x \frac{dy}{dx} + a_2 \, y = 0$$

subject to initial conditions given by $y(0) = \alpha$; $y'(0) = \beta$.

Let's apply Laplace transformation technique directly to solve this ode. By linearity of Laplace transforms we have :

$$a_0 \, \mathcal{L}\left(x^2 \frac{d^2 y}{dx^2}\right) + a_1 \, \mathcal{L}\left(x \frac{dy}{dx}\right) + a_2 \, \mathcal{L}(y) = 0$$

By rules of derivatives of Laplace transforms we have :

$$a_0 \frac{d^2}{ds^2}\left\{ \mathcal{L}\left(\frac{d^2 y}{dx^2}\right) \right\} - a_1 \frac{d}{ds}\left\{ \mathcal{L}\left(\frac{dy}{dx}\right) \right\} + a_2 \, \mathcal{L}(y) = 0$$

$$\therefore \quad a_0 \frac{d^2}{ds^2}\left\{ s^2 Y(s) - sy(0) - y'(0) \right\} - a_1 \frac{d}{ds}\left\{ sy(s) - y(0) \right\} + a_2 \, y(s) = 0$$

where $Y(s) \equiv L(y(x))$.

Applying Leibnitz's rule for successive derivatives we get :

$$a_0 \left[s^2 \frac{d^2 Y}{ds^2} + 4s \frac{dY}{ds} + 2Y \right] - a_1 \left[s \frac{dY}{ds} + Y \right] + a_2 \, Y = 0$$

i,e, $$a_0 s^2 \frac{d^2 Y}{ds^2} + (4a_0 - a_1)s \frac{dY}{ds} + (2a_0 - a_1 + a_2)Y = 0 \qquad \text{(a)}$$

But (a) is again a Cauchy-Euler type ode!! This form-invariance of Cauchy-Euler type odes under Laplace transformation envisages that direct application of this technique to a Cauchy-Euler type ode is of no use. We have to first convert it into a constant co-efficient type ode by transformation of the type $x = e^{az}$ and then solve it, if required by Laplace transformation technique.

(B) Let's consider the ivp

$$\frac{d^2y}{dx^2} + x^2 y = 0 \qquad \text{subject to } y(0) = \alpha \text{ and } y'(0) = \beta. \tag{b}$$

Applying Laplace transformation technique to this ode we get :

$$\mathcal{L}\left(\frac{d^2y}{dx^2}\right) + \mathcal{L}(x^2 y) = 0$$

i, e, $\{s^2 Y(s) - sy(0) - y'(0)\} + \dfrac{d^2}{ds^2} Y(s) = 0$ where $Y(s) \equiv \mathcal{L}(y(x))$

Therefore, $\quad \dfrac{d^2 Y}{ds^2} + s^2 Y(s) = sy(0) + y'(0) = s\alpha + \beta \tag{c}$

On comparing the ode (c) with the given ode (b) we observe that the Laplace transformation technique failed to make any progress since we have got back an ode of the same pattern as that of the given one. In general, we have again discovered a kind of form-invariance of the ode (b) under Laplace transformation method.

(C) We consider Legendre equation

$$(1 - x^2)\frac{d^2y}{dx^2} - 2x\frac{dy}{dx} + n(n+1)y = 0$$

subject to initial conditions $y(0) = \alpha$ and $y'(0) = \beta$.

Applying Laplace transform to this second order ode, we have, after a number of routine steps, the transformed ode :

$$\frac{s^2 d^2 Y}{ds^2} + 2s\frac{dY}{ds} - \left\{n(n+1) + s^2\right\} Y(s) = 0, \tag{d}$$

which shows the failure of Laplace transform technique to deliver. Unlike the first two cases, the form invariance of ode not obtained. In the next chapter we shall show that the Legendre equation has two regular singular points, viz, $x = \pm 1$ while the transformed second order ode (d) will have just one regular singular point, namely $s = 0$.

(D) Finally we come up with the Bessel equation, viz,

$$x^2\frac{d^2y}{dx^2} + x\frac{dy}{dx} + (x^2 - p^2)y = 0, \ p \text{ being a constant.} \tag{e}$$

If $p \neq 0$, Bessel quation will have the same fate on being tried the Laplace transformation technique. Only if $p = 0$, we can successfully apply Laplace transformation method as it becomes

$$x\frac{d^2y}{dx^2} + \frac{dy}{dx} + xy = 0 \qquad (f)$$

(It is known as **Bessel equation of order zero** and its solutions are called **Bessel functions of order zero**).

Applying Laplace transform to the ode (f) we have :

$$\mathcal{L}\left(x\frac{d^2y}{dx^2}\right) + \mathcal{L}\left(\frac{dy}{dx}\right) + \mathcal{L}(xy) = 0$$

$$\therefore \qquad -\frac{d}{ds}\mathcal{L}\left(\frac{d^2y}{dx^2}\right) + \mathcal{L}\left(\frac{dy}{dx}\right) - \frac{d}{ds}\mathcal{L}(y) = 0$$

i, e, $\qquad -\frac{d}{ds}\left\{s^2Y(s) - sy(0) - y'(0)\right\} + \left\{sY(s) - y(0)\right\} - \frac{dY}{ds} = 0$

i, e, $\qquad -2sY(s) - s^2\frac{dY}{ds} + y(0) + sY(s) - y(0) - \frac{dY}{ds} = 0$

Therefore, $\quad (s^2 + 1)\frac{dY}{ds} + sY(s) = 0$, a first order ode in s and $Y(s)$.

The general solution of this ode can be written as $Y(s) = A \cdot \frac{1}{\sqrt{s^2+1}}$, A being an arbitrary constant.

We may now use Binomial expansion for -ve fractional index to get

$$\frac{1}{\sqrt{s^2+1}} = \frac{1}{s} \cdot \frac{1}{\sqrt{1+\frac{1}{s^2}}} = \frac{1}{s} \cdot \left(1 + \frac{1}{s^2}\right)^{-\frac{1}{2}}$$

$$= \frac{1}{s}\left(1 - \frac{1}{2} \cdot \frac{1}{s^2} + \frac{1}{2!} \cdot \frac{1}{2} \cdot \frac{3}{2} \cdot \frac{1}{s^4} - \frac{1}{3!} \cdot \frac{1}{2} \cdot \frac{3}{2} \cdot \frac{5}{2} \cdot \frac{1}{s^6} + \cdots \infty\right)$$

$$= \frac{1}{s}\sum_{k=0}^{\infty} \frac{(2k)!(-1)^k}{2^{2k}(k!)^2 s^{2k}} = \sum_{k=0}^{\infty} \frac{(-1)^k \cdot (2k)!}{2^{2k}(k!)^2} \mathcal{L}\left(\frac{x^{2k+1}}{\Gamma(2k+1)}\right)$$

where we used the fact that $L(x^\nu) = \frac{\Gamma(\nu+1)}{s^{\nu+1}}$, $s > 0$ in the penultimate step. Further one may note that the series $\sum_{k=0}^{\infty} \frac{(-1)^k \cdot (2k)!}{2^{2k}(k!)^2} \cdot \frac{x^{2k+1}}{\Gamma(2k+1)}$ is uniformly convergent and so term by term integration is feasible. This eventually leads us to the result :

$$\mathcal{L}\left(\sum_{k=0}^{\infty} \frac{(-1)^k \cdot (2k)! \, x^{2k+1}}{2^{2k} \cdot (k!)^2 \, \Gamma(2k+1)}\right) = \sum_{k=0}^{\infty} \frac{(-1)^k (2k)!}{2^{2k} \cdot (k!)^2} \mathcal{L}\left(\frac{x^{2k+1}}{\Gamma(2k+1)}\right)$$

Hence
$$\frac{1}{\sqrt{s^2+1}} = \mathcal{L}\left(\sum_{k=0}^{\infty} \frac{(-1)^k \cdot (2k)! \, x^{2k+1}}{2^{2k} \cdot (k!)^2 \, \Gamma(2k+1)}\right)$$

$$\therefore \qquad Y(s) = \frac{A}{\sqrt{s^2+1}} = \mathcal{L}\left(A \sum_{k=0}^{\infty} \frac{(-1)^k \cdot (2k)! \, x^{2k+1}}{2^{2k} \cdot (k!)^2 \, \Gamma(2k+1)}\right)$$

and consequently by uniqueness of Laplace inversion, we have

$$
\begin{aligned}
y(x) = \mathcal{L}^{-1}(Y(s)) \;=\; & A \sum_{k=0}^{\infty} \frac{(-1)^k \cdot (2k)! \, x^{2k+1}}{2^{2k} \cdot (k!)^2 \cdot \Gamma(2k+1)} \\
=\; & A \sum_{k=0}^{\infty} \frac{(-1)^k \, x^{2k+1}}{2^{2k} \cdot (k!)^2} \quad (\because \Gamma(2k+1) = (2k)!)
\end{aligned}
$$

If one imposes the initial condition $y(0) = 1$, then $A = 1$ so that

$$y(x) = \sum_{k=0}^{\infty} \frac{(-1)^k \cdot x^{2k+1}}{2^{2k}(k!)^2} \equiv J_0(x)$$

Thus one identifies one solution of the Bessel equation of order zero to be $J_0(x)$, the so called Bessel function of order zero.

Example (19) : Solve the ode $\frac{d^2y}{dx^2} + y = xe^{2x}$ by Laplace transform.

Here there is no initial condition. Hence we prescribe $y(0) = \alpha$ and $y'(0) = \beta$ for the formal set up. Later on we shall observe that this α and β will generate arbitrary constants appearing in the solution.

Taking Laplace transform of the given ode and using linearity :

$$\mathcal{L}\left(\frac{d^2y}{dx^2}\right) + \mathcal{L}(y) = \mathcal{L}\left(xe^{2x}\right)$$

Again this is equivalent to

$$\left(s^2 Y(s) - sy(0) - y'(0)\right) + Y(s) = \frac{1}{(s-2)^2}$$

i, e, $$(s^2 + 1)\, Y(s) = (\alpha s + \beta) + \frac{1}{(s-2)^2}$$

i, e, $$
\begin{aligned}
Y(s) &= \frac{\alpha s + \beta}{s^2 + 1} + \frac{1}{(s^2+1)(s-2)^2} \\
&= \frac{\alpha s + \beta}{s^2 + 1} + \frac{\gamma s + \delta}{s^2 + 1} + \frac{A}{(s-2)} + \frac{B}{(s-2)^2}
\end{aligned}
$$

where A, B, γ, δ are determined via method of partial fractions.

$$\therefore Y(s) = \frac{(\alpha+\gamma)s}{(s^2+1)} + \frac{(\beta+\delta)}{(s^2+1)} + \frac{A}{(s-2)} + \frac{B}{(s-2)^2}$$

$$= \frac{Cs}{s^2+1} + \frac{D}{(s^2+1)} + \frac{A}{(s-2)} + \frac{B}{(s-2)^2}$$

$$= C\,\mathcal{L}(\cos x) + D\,\mathcal{L}(\sin x) + A\,\mathcal{L}(e^{-2x}) + B\,\mathcal{L}(x\,e^{-2x})$$

$$= \mathcal{L}\left(C\cos x + D\sin x + Ae^{-2x} + Bxe^{-2x}\right)$$

$$\therefore y(x) = \mathcal{L}^{-1}(Y(s)) = C\cos x + D\sin x + Ae^{-2x} + Bxe^{-2x}$$

If we literally find the co-efficients A, B by method of partial fractions, our y(x) will read

$$y(x) = (C\cos x + D\sin x) + \frac{e^{-2x}}{25}(5x-4)$$

Example (20) : Solve the ode $x\frac{d^2y}{dx^2} + (2x+3)\frac{dy}{dx} + (x+3)y = 3e^{-x}$
Assuming that Laplace transform of y, $\frac{dy}{dx}$ and $\frac{d^2y}{dx^2}$ exist, we apply Laplace transform on both sides and use its linearity to get

$$\mathcal{L}\left(x\frac{d^2y}{dx^2}\right) + \mathcal{L}\left((2x+3)\frac{dy}{dx}\right) + \mathcal{L}((x+3)y) = 3\mathcal{L}(e^{-x})$$

i.e., $\quad \mathcal{L}\left(x\frac{d^2y}{dx^2}\right) + 2\mathcal{L}\left(x\frac{dy}{dx}\right) + 3\mathcal{L}\left(\frac{dy}{dx}\right) + \mathcal{L}(xy) + 3\mathcal{L}(y) = 3\mathcal{L}(e^{-x})$

i.e., $\quad -\frac{d}{ds}\mathcal{L}\left(\frac{d^2y}{dx^2}\right) - 2\frac{d}{ds}\mathcal{L}\left(\frac{dy}{dx}\right) + 3\mathcal{L}\left(\frac{dy}{dx}\right) - \frac{d}{ds}\mathcal{L}(y) + 3\mathcal{L}(y) = 3\mathcal{L}(e^{-x})$

i.e., $\quad -\frac{d}{ds}[s^2Y(s) - sy(o) - y'(0) + 2sY(s) - 2\,y(0) + Y(s)]$

$$+ 3[sY(s) - y(0)] + 3Y(s) = \frac{3}{(s+1)} \quad (\text{where } Y(s) = \mathcal{L}(y(x)))$$

i.e., $\quad -\left[2sY(s) + s^2\frac{dY}{ds} + y(0) + 2Y(s) + 2s\frac{dY}{ds} + \frac{dY}{ds}\right]$

$$+3(s+1)Y(s) - 3y(0) = \frac{3}{(s+1)}$$

i.e., $-[(s^2+2s+1)\frac{dY}{ds} + y(0) + 2(s+1)Y(s)] + 3(s+1)Y(s) - 3y(0) = \frac{3}{(s+1)}$

i.e., $\quad -(s+1)^2\frac{dY}{ds} + (s+1)Y(s) - 4y(0) = \frac{3}{s+1}$

Dividing this equation by $(s+1)^3$, we have:

$$-\frac{1}{(s+1)}\frac{dY}{ds} + \frac{1}{(s+1)^2}Y(s) = \frac{4y(0)}{(s+1)^3} + \frac{3}{(s+1)^4}$$

i.e.,
$$\frac{d}{ds}\{\frac{1}{(s+1)}Y(s)\} = -\frac{4y(o)}{(s+1)^3} - \frac{3}{(s+1)^4}$$

Integrating w.r.to s, we get:

$$-\frac{1}{(s+1)}Y(s) = \frac{2y(0)}{(s+1)^2} + \frac{1}{(s+1)^3} + A$$

$$Y(s) = \frac{2y(0)}{(s+1)} + \frac{1}{(s+1)^2} + A(s+1) \qquad (a)$$

i.e.,
$$\mathcal{L}(y(x)) = 2y(0)\mathcal{L}(e^{-x}) + \mathcal{L}(xe^{-x}) + A(s+1)$$

By uniqueness of Laplace inversion we get:

$$y(x) = 2y(0)e^{-x} + xe^{-x} + A\mathcal{L}^{-1}((s+1)) \qquad (b)$$

Again if $(s+1)$ is the Laplace transforms of $g(x)$, then we may write:

$$g(x) = e^{-x}f(x), \quad \text{where} \quad \mathcal{L}(f(x)) = s$$

Further from the theory of Laplace transform for integrals we get:

$$\mathcal{L}(f(x)) = s = s\,\mathcal{L}(\delta(x))$$

$$\delta(x) = \int_0^x f(y)dy = \int_0^x e^y g(y)dy$$

$$\therefore \qquad y(x) = 2y(0)e^{-x} + xe^{-x} + Ag(x), \qquad (c)$$

where $g(x)$ is given by the integral relation (c)

Thus we see Laplace transform technique does not provide us a simplistic model of answer. We therefore approach anew and rewrite the ode given as:

$$\{xD^2 + (2x+3)D + (x+3)\}y = 3e^{-x}$$

i.e., $\{xD^2 + xD + (x+3)D + (x+3)\}y = 3e^{-x}$

i.e., $\{xD(D+1) + (x+3)(D+1)\}y = 3e^{-x}$

i.e., $(xD + x + 3)(D+1)y = 3e^{-x}$

i.e., $\{x(D+1) + 3\} = 3e^{-x} \qquad (d)$

Assume $(D+1)y \equiv z$ so that (d) is recast as:

$$\{x(D+1) + 3\}z = 3e^{-x}$$

i.e., $$(D+1)z + \frac{3}{x}y = 3e^{-x}$$

i.e., $$\frac{dz}{dx} + (1 + \frac{3}{x})z = \frac{3e^{-x}}{x} \qquad (e)$$

(e) being a linear ode in x and z, integrating factor is given by

$$\mu(x) = exp\left[\int \left(1 + \frac{3}{x}\right)dx\right] = x^3 e^x$$

Therefore, on using $\mu(x)$, equation (e) can be integrated as:

$$z(x^3 e^x) = 3\int \frac{e^{-x}}{x}x^3 e^x dx = \int 3x^2 dx = x^3 + A$$

i.e., $$z = e^{-x} + \frac{A}{x^3}e^{-x}$$

$$\therefore \qquad \frac{dy}{dx} + y = e^{-x} + \frac{A}{x^3}e^{-x} \qquad (\because (D+1)y \equiv z)$$

or, $$\frac{d}{dx}(e^x y) = \left(1 + \frac{A}{x^3}\right)$$

Integrating w.r.to. x : $$ye^x = x - \frac{A}{2x^2} + B$$

i.e., $$y(x) = (x+B)e^{-x} - \frac{A}{2x^2}e^{-x}$$

Remark: In this approach we have managed to get a decent form of the solution and observe that the domain of $y(x)$ must exclude the point $x = 0$.Infact, this might be the reason of the failure of Laplace transforms technique to deliver.

Example (21) : Use Laplace transforms to solve the following I V P:

$$\frac{d^2 y}{dx^2} + 2\frac{dy}{dx} + 5y = 2\delta(x-2) + 1, \quad \text{subject to } y(0) = 2 ; \ y'(0) = -2$$

Solution: Applying the Laplace transform to the initial value problem and using linearity of $\mathcal{L}$ we have:

$$\mathcal{L}\left(\frac{d^2 y}{dx^2}\right) + 2\mathcal{L}\left(\frac{dy}{dx}\right) + 5\mathcal{L}(y) = 2\mathcal{L}(\delta(x-2)) + \mathcal{L}(1)$$

$$\therefore \quad \{s^2 Y(s) - sy(0) - y'(0)\} + 2\{sY(s) - y(0)\} + 5Y(s) = 2e^{-2s} + \frac{1}{s}$$

i.e., $\quad (s^2 + 2s + 5)Y(s) = 2(s+1) + \dfrac{1}{s} + 2e^{-2s}$

$$\therefore \quad Y(s) = \frac{2(s+1)}{s^2 + 2s + 5} + \frac{1}{s(s^2 + 2s + 5)} + \frac{2e^{-2s}}{s^2 + 2s + 5}$$

$$= \frac{2(s+1)}{(s+1)^2 + 2^2} + \frac{1}{s(s^2 + 2s + 5)} + \frac{2e^{-2s}}{(s+1)^2 + 2^2}$$

Since $\quad \dfrac{s}{s^2 + w^2} = \mathcal{L}\,(\cos wx) \;$ and $\; \dfrac{w}{s^2 + w^2} = \mathcal{L}\,(\sin wx)\,,$

$$Y(s) = \mathcal{L}(2e^{-x}\cos 2x) + \frac{1}{s(s^2 + 2s + 5)} + \mathcal{L}(\delta(x-2))\mathcal{L}(e^{-x}\sin 2x)$$

$$= \mathcal{L}(2e^{-x}\cos 2x) + \mathcal{L}\left(\int_0^x \delta(x'-2)e^{-(x-x')}\sin(2(x-x'))dx'\right)$$

$$+\frac{1}{s(s^2 + 2s + 5)}$$

$$\therefore \quad Y(s) = \mathcal{L}\left[2e^{-x}\cos 2x + e^{-(x-2)}\sin(2(x-2)\right] + \frac{1}{s(s^2 + 2s + 5)} \quad \text{(a)}$$

Here we may write by method of partial fractions,

$$\frac{1}{s(s^2 + 2s + 5)} = \frac{A}{s} + \frac{Bs + C}{s^2 + 2s + 5};$$

where $\quad A, B, C$ are yet to be evaluated.

$$\therefore \quad 1 = A(s^2 + 2s + 5) + s(Bs + C) = (A + B)s^2 + (2A + C)s$$

Equating the co-efficients of like powers of s we get:

$$5A \;\; = \;\; 1, \; A + B = 0; \; 2A + C = 0$$

$$\therefore \quad A \;\; = \;\; \frac{1}{5} \;;\; B = -\frac{1}{5} \;;\; C = -\frac{2}{5}$$

Thus $\quad \dfrac{1}{s(s^2 + 2s + 5)} = \dfrac{1}{5s} - \dfrac{(s+2)}{5(s^2 + 2s + 5)}$

$$= \frac{1}{5s} - \frac{(s+1)}{5\{(s+1)^2 + 2^2\}} - \frac{1}{5}\frac{1}{\{(s+1)^2\}}$$

$$= \mathcal{L}[(\frac{1}{5}) - \frac{1}{5}e^{-x}\cos 2x - \frac{1}{10}e^{-x}\sin 2x] \quad \text{(b)}$$

Piecing together (a) and (b) and exploiting the uniqueness Laplace inversion we get the desired solution of IVP as :

$$y(x) = e^{-(x-2)}\sin(2(x-2)) + \frac{1}{5} - \frac{9}{5}e^{-x}\cos 2x - \frac{1}{10}e^{-x}\sin 2x$$

i,e, $y(x) = \dfrac{9}{5}e^{-x}\cos 2x - \dfrac{1}{10}e^{-x}\sin 2x + e^{-(x-2)}\sin(2(x-2)) + \dfrac{1}{5}$

In particular, if the resistance in the LRC circuit is negligible, i,e, $R = 0$, then the ode evolving from Kirchoff's first law reduces to

$$L\frac{d^2Q}{dt^2} + \frac{1}{C}Q = E(t) \tag{c}$$

which has to be solved subject to the initial conditions $Q(0) = 0$ and $Q'(0) = A$. Under this circumstance one gets

$$\bar{Q}(s) = \frac{A}{s^2 + w^2} + \frac{\mathcal{L}(E)}{L(s^2 + w^2)} \qquad \left[\text{where } w^2 = \frac{1}{LC}\right]$$

The initial value problem (c) has a close analogy with the problem of forced vibrations without damping encountered in classical mechanics.

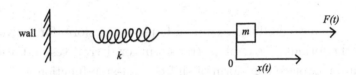

Fig 7.6 : Forced vibration of a block of mass **m** tied to a spring with spring constant **k** and acted on by external force **F(t)**.

The displacement $x(t)$ of mass m at time t satisfies the second order ode:

$$m\frac{d^2x}{dt^2} + kx = F(t)$$

Suppose we solve subject to the initial conditions $x(0) = 0$ and $x'(0) = v_0$ (i,e, we presume the mass to start its motion from the origin and its initial velocity was v_0). If $X(s)$ denotes the Laplace transform of $x(t)$, then in analogy to equation (d), we can write:

$$X(s) = \frac{v_0}{s^2 + w^2} + \frac{\mathcal{L}(F(t))}{m(s^2 + w^2)}$$

where $w^2 = \frac{k}{m}$ denotes the natural frequency of the spring system

Therefore, $x(t) = \dfrac{v_0}{w}\sin wt + \dfrac{1}{mw}\displaystyle\int_0^t \sin w(t - \tau)F(\tau)d\tau$

Depending on the choices of $F(t)$, viz,

(i) $F(t) = F_0$ (some constant)   ; (iii) $F(t) = F_0\delta(t)$

(ii) $F(t) = F_0\mu(t)$; (iv) $F(t) = F_0\sin pt$

we shall eventually get the final expressions of displacement $x(t)$.

Example (22) : Solve the IVP : $y'' + y' + y = g(x)$,

where $y(0) = 1$; $y'(0) = 0$ and $g(x) = \begin{cases} 0, & 0 \leqslant x < 1 \\ 1, & x \geqslant 1 \end{cases}$

What happens if $x \to \infty$?

Applying Laplace transform and its linearity we have :

$$\mathcal{L}(y'') + \mathcal{L}(y') + \mathcal{L}(y) = \mathcal{L}(g(x))$$

i,e, $\{s^2 Y(s) - s\} + \{sY(s) - 1\} + Y(s) = \dfrac{e^{-s}}{s}$,

where we have used the results of Laplace transform for derivatives and also the formula for Laplace transform for derivatives and also the formula for Laplace transform of shifted unit-step function.

After rearrangement of terms we get :

$$Y(s) = \frac{e^{-s}}{s(s^2 + s + 1)} + \frac{s+1}{s^2 + s + 1} ,$$

Now using method of partial fractions we write

$$\frac{1}{s(s^2 + s + 1)} = \frac{A}{s} + \frac{Bs + C}{s^2 + s + 1} ,$$

where one can compute $A = 1$, $B = -1$ and $C = -1$.

$$\therefore \quad Y(s) = e^{-s}\left(\frac{1}{s} - \frac{s+1}{s^2 + s + 1}\right) + \frac{s+1}{\left(s + \frac{1}{2}\right)^2 + \left(\frac{\sqrt{3}}{2}\right)^2}$$

Hence $y(x) = \mathcal{L}^{-1}\left[e^{-s}\left(\dfrac{1}{s} - \dfrac{s+1}{s^2+s+1}\right)\right] + \mathcal{L}^{-1}\left[\dfrac{s+1}{\left(s+\frac{1}{2}\right)^2 + \left(\frac{\sqrt{3}}{2}\right)^2}\right]$

Using result of Laplace transforms one can have :

$$y(x) = U_1(x) \left[1 - e^{-\frac{1}{2}(x-1)} \left(\cos\left(\frac{\sqrt{3}}{2}(x-1) \right) - \sin\left(\frac{\sqrt{3}}{2}(x-1) \right) \right) \right]$$
$$+ e^{-\frac{1}{2}x} \left[\cos\left(\frac{\sqrt{3}}{2}x \right) + \frac{1}{\sqrt{3}} \left(\sin\frac{\sqrt{3}}{2}x \right) \right]$$

when $x \to \infty$, all the terms having exponential functions will go to zero and hence $y(x) \to 1$ in the limit. One can look into this system as a spring-mass system with damping. Observe that jump in the response function causes a jump in the function.

Example (23) : Solve the Volterra integral equation :

$$y(x) = f(x) + \int_0^x g(x - x')\, y(x')\, dx'$$

by Laplace transformation technique and then solve in particular

$$y(x) = e^{-x} - 4 \int_0^x \cos 2(x - x')y(x')dx'$$

In general $f(\cdot)$ and $g(\cdot)$ appearing in the given integral equation are known functions while $y(\cdot)$ is unknown function. The above convolution integral appearing in the RHS suggests us to apply Laplace transform on the given integral equation so that by convolution theorem we get :

$$Y(s) = F(s) + G(s)Y(s),$$

so that

$$Y(s) = \frac{F(s)}{1 - G(s)} \qquad (a)$$

By uniqueness of Laplace Inversion, one can get back $y(x)$ from (a). In the second part, $f(x) = e^{-x}$; $g(x) = 4\cos 2x$ so that

$$Y(s) = \mathcal{L}\left(e^{-x}\right) + 4\mathcal{L}\left(\int_0^x \cos 2(x - x')\, y(x')\, dx' \right)$$
$$= \frac{1}{s+1} - 4\frac{sY(s)}{s^2 + 4}$$
$$\therefore \quad Y(s) = \frac{s^2 + 4}{(s+1)(s+2)^2} = \frac{5}{s+1} - \frac{4}{s+2} + \frac{8}{(s+2)^2}$$

Applying inverse Laplace transform on both sides of the above equation and taking help of basic results, we get :

$$y(x) = 5e^{-x} - 4e^{-2x} - 8xe^{-2x}$$

Observe that when $x \to \infty$, $y(x) \to 0$.

Exercise 7

1. Find the Laplace transform of the following functions:

   (i) $e^{\lambda x} \sin^2 \mu x$ (ii) $x^2 \cos bx$ (iii) $x^2 e^{ax}$, (iv) $\sin(ax) \cos^2(ax)$

   (v) $\dfrac{\sin(ax) - (ax)\cos(ax)}{2a^3}$ (vi) $\cosh(bx)$ (vii) $a\cos(bx) + b\cos(ax)$

2. Find the Laplace transforms of the following functions:

   (a) $\quad f(x) = \begin{cases} 0 & , \ 0 < x < 2 \\ 3 & , \ 2 < x < 5 \\ 0 & , \ x > 5 \end{cases}$

   (b) $\quad f(x) = \begin{cases} 2 & , \ 0 < x < 3 \\ 0 & , \ 3 < x < 6 \\ 2 & , \ x > 6 \end{cases}$

   (c) $\quad f(x) = \begin{cases} 0 & , \ 0 < x < 2 \\ 2e^{-x} & , \ x > 2 \end{cases}$

   (d) $\quad f(x) = \begin{cases} 0 & , \ 0 < x < 4 \\ (x-4) & , \ 4 < x < 7 \\ 3 & , \ x > 7 \end{cases}$

3. Determine the function $f(x)$ whose Laplace transform $F(s)$ is :

   (i) $\dfrac{1}{s(s^2 + 4s + 13)}$ (ii) $\dfrac{1}{(s-1)(s^2+4)}$ (iii) $\dfrac{1}{(s^2+1)^2}$ (iv) $ln\left(\dfrac{s^2+1}{s}\right)$

   Hence in each case verify the convolution theorem.

4. Show that $f(x) = e^{-x^2}$ is not of exponential order but still $\mathcal{L}(e^{-x^2})$ converges absolutely by for all x.(This illustrates the conditions usually imposed on functions for ensuring existence of Laplace transforms, are sufficient and not necessary.

5. Use the formula for the derivative of Laplace transforms to calculate inverse Laplace transforms of the following functions :

(i) $F(s) = ln\left(\dfrac{s+3}{s-4}\right)$ (ii) $F(s) = ln\left(\dfrac{s-2}{s-1}\right)$

(iii) $F(s) = ln\left(\dfrac{s^2+4}{s^2+2}\right)$ (iv) $F(s) = \dfrac{s^2-b^2}{(s^2+b^2)^2}$ (v) $F(s) = \dfrac{s+1}{s^3}$

6. Show that the operation of convolution defined in the Laplace transforms is commutative, associative and distributive i.e.,

(i) $f*g = g*f$ (ii) $f*(g*h) = (f*g)*h$ (iii) $f*(g+h) = f*g+f*h$

7. Solve the following integral equations by means of Laplace transformation technique :

(a) $y(x) = 1 - \displaystyle\int_0^x (x-t)y(t)dt$

(b) $5\cos 2x = y(x) + \displaystyle\int_0^x (x-t)y(t)dt$

(c) $x = \displaystyle\int_0^x e^{(x-t)}g(t)dt$

(d) $g(x) = f(x) + \displaystyle\int_0^x (x-t)g(t)dt$

(e) $f(x) = \displaystyle\int_0^x g(t)(x-t)^{-\alpha}dt (0 < \infty < 1)$

8. Assuming that $f(x)$ is a function that admits of Laplace transform apply convolution theorem to the integral equation

$$\sin x = \int_0^x J_0(x-y)f(y)dy$$

to prove that

$$\int_0^x J_0(x-t)J_0(t)dt = \sin x$$

9. Apply convolution theorem of Laplace transforms to the integral $\int_0^x y^{m-1}(x-y)^{n-1}dy$ $(m, n > 0)$ to establish the relation

$$B(m,n) = \frac{\Gamma(m)\Gamma(n)}{\Gamma(m+n)}$$

10. Applying Laplace transforms for periodic functions, prove that

$$\mathcal{L}(\,\sin x + |\sin x|\,) = \frac{2}{(s^2 + 1)(1 - e^{-\pi s})}\,, \quad s > 0$$

and hence show that

$$\mathcal{L}(\,|\sin x|\,) = \frac{1}{s^2 + 1} \cdot \frac{1 + e^{-\pi s}}{1 - e^{-\pi s}}$$

11. It is known that if $F(s)$ be Laplace transform of $f(x)$, then

$$\mathcal{L}\left(\frac{f(x)}{x}\right) = \int_s^\infty F(s)ds, \quad s > 0$$

Use this result to find the Laplace transforms of

(i) $f(x) = \dfrac{1 - \cos x}{x}$ (ii) $\dfrac{1 - \cosh ax}{x}$ (iii) $\dfrac{\sin kx}{x}$ (iv) $\dfrac{e^{2x} - \sin x}{x}$

12. Find $\mathcal{L}(y)$ and y in each of the following cases when $y(x)$ satisfies the differential equation:

(a) $\dfrac{d^2y}{dx^2} + a^2 y = 0;\ y(0) = c;\ y'(0) = 0$

(b) $\dfrac{d^2y}{dx^2} + 2a\dfrac{dy}{dx} + b^2 y = 0;\ y(0) = c;\ y'(0) = 0$

(c) $\dfrac{d^2y}{dx^2} - 2k\dfrac{dy}{dx} + k^2 y = F(x),\ \text{provided}\ \mathcal{L}(f(x))\ \text{exists.}$

(d) $2\dfrac{d^2y}{dx^2} - 3\dfrac{dy}{dx} - 2y(x) = F(x),\quad \text{if}\quad y(0) = y'(0) = 0$

13. Show that $y(x) = 1$ satisfies the non-linear integral equation

$$2y(x) + \int_0^x y(t)y(x - t)dt = x + 2$$

14. Assumming $y(x)$ and its derivatives admit of Laplace transforms, solve the odes:

(i) $x^2\dfrac{d^2y}{dx^2} - 2y(x) = 2x;\ \text{if}\ y(2) = 2;\ y'(0) = 0$

(ii) $\dfrac{d^2y}{dx^2} - y = 1 + e^{3x}$

(iii) $\dfrac{d^2y}{dx^2} + y = 2\sin x\ ;\ \text{if}\ y(0) = 0\ \text{and}\ y'(0) = -1$

(iv) $\dfrac{d^2y}{dx^2} - 3\dfrac{dy}{dx} + 2y = h(x) = \begin{cases} 2, & \text{if } 0 < x < 4 \\ 0, & \text{if } x > 4 \end{cases}$; $y(0) = y'(0) = 0$

(v) $\dfrac{d^2y}{dx^2} + 7dydx + 10y = xe^{-3x}$; $y(0) = 0$; $y'(0) = -1$

(vi) $\dfrac{d^2y}{dx^2} + y = \begin{cases} x, & \text{if } 0 < x < \pi \\ \pi, & \text{if } x > \pi \end{cases}$; $y(0) = 2$; $y'(0) = 3$

15. Solve the following linear system using Laplace transform technique :

(a) $\left.\begin{array}{l} \dfrac{dy}{dt} + z = 3e^{2x} \\ \dfrac{dz}{dx} + y = 0 \end{array}\right\}$ subject to $\left.\begin{array}{l} y(0) = 2 \\ z(0) = 0 \end{array}\right\}$

(b) $\left.\begin{array}{l} 2\dfrac{dy}{dx} + \dfrac{dz}{dx} - y - z = e^{-x} \\ \dfrac{dy}{dx} + \dfrac{dz}{dx} + 2y + z = e^{x} \end{array}\right\}$ subject to $\left.\begin{array}{l} y(0) = 2 \\ z(0) = 1 \end{array}\right\}$

Chapter 8

Series Solutions of Linear Differential Equations

8.1 Introduction

In the preceding chapters we observed that only a handful class of linear differential equations can be solved explicitly in the closed form, i.e., in terms of some known elementary functions. This class includes linear differential equations of first order and linear differential equation of order two or more that are either constant co-efficient type or of Cauchy-Euler type. In chapter 6, where we worked with D' Alembert's order reduction method for solving differential equations of variable co-efficient type, we needed the intimation of one particular solution of the corresponding reduced ode. However, if no such intimation is available to us for a linear differential equation of higher order with variable co-efficients,then we have to seek alternative means. Solutions of differential equations in terms of convergent series provides us one such opening. Thus there develops a natural motivation to try a solution in the form of a power series but one should keep in mind that there exist differential equations where the solution is not given even in the form of a power series. Moreover, validity of such power series is a major issue since without getting enough assurance about the existence of the solution in the form of a convergent power series, hunting for it add to our woes.

By the time an attentive reader should have marked that for the class of differential equations solvable in terms of elementary functions, we tacitly presumed continuity of the co-efficient functions involved in the ode and concluded the differentiability of the solutions. However, to have a solution of differential equations in terms of a convergent power series, we must choose the co-efficient functions real analytic in some region G. Recall that a function f is analytic at a point c of its domain provided its Taylor series about c ; viz, $\sum_{n=0}^{\infty} \frac{f^{(n)}(c)}{n!}(x - c)^n$ exists and converges to $f(x)$ for all x in some neighborhood of the point c.The function f is said to be analytic in some region G iff it is analytic at every point of G.

A point c is said to be an **ordinary point** of the normalised second order differential equation

$$\frac{d^2y}{dx^2} + p(x)\frac{dy}{dx} + q(x)y = 0$$

provided $p(x)$ and $q(x)$ are both real analytic at c. If at least one of the functions $p(x)$ and $q(x)$ fails to be real analytic, the point c is referred to as a **singular point** of the second order ode. Thus it is clear that at ordinary points one can only think of solutions in the form od a convergent power series. In course of our discussion we shall discover that a convergent series solution of an ode can exist even if the co-efficient functions of some odes possess certain restricted type of singularities.

8.2 Review of Power Series

Power series constitute a very important class of series of functions. It is of the generic form $\sum_{n=0}^{\infty} a_n(x - x_0)^n$, or more simply, $\sum_{n=0}^{\infty} a_n x^n$. The former known as power series about the point x_0 while the letter as power series about the point 0. Without loss of generality we shall continue our discussion with $\sum_{n=0}^{\infty} a_n x^n$. Obviously any power series of the type $\sum_{n=0}^{\infty} a_n x^n$ converges at $x = 0$. However, we are interested whether it converges at some point other than $x = 0$. For instance, let's consider the three power series, viz,

$$(i) \sum_{n=0}^{\infty} n!x^n \quad (ii) \sum_{n=0}^{\infty} \frac{(n!)^2}{(2n)!} x^n \quad (iii) \sum_{n=0}^{\infty} \frac{x^n}{n!}$$

We now apply Ratio test on the corresponding positive term series:

$$(i)' \sum_{n=0}^{\infty} n! \mid x \mid^n \quad (ii)' \sum_{n=0}^{\infty} \frac{(n!)^2}{(2n)!} \mid x \mid^n \quad (iii)' \sum_{n=0}^{\infty} \frac{\mid x \mid^n}{n!}$$

In $(i)'$, $\lim_{n\to\infty} \frac{(n+1)!|x|^{n+1}}{n!|x|^n} = \lim_{n\to\infty}(n+1)|x| = 0$, only if $x = 0$ and so the series $\sum_{n=0}^{\infty} n!x^n$ is convergent only at $x = 0$

In $(ii)'$, $\displaystyle\lim_{n\to\infty} \frac{((n+1)!)^2}{(2n+2)!} \cdot \frac{(2n)!}{(n!)^2} \mid x \mid$

$$= \lim_{n\to\infty} \frac{(n+1)^2 \mid x \mid}{(2n+2)(2n+1)} = \lim_{n\to\infty} \frac{(1+\frac{1}{n}) \mid x \mid}{4(1+\frac{1}{2n})} = \frac{\mid x \mid}{4}$$

and so $\displaystyle\sum_{n=0}^{\infty} \frac{(n!)^2}{(2n)!} \mid x \mid^n$ is absolutely convergent provided $\mid x \mid < 4$

In $(iii)'$, $\displaystyle\lim_{n\to\infty} \frac{n! \mid x \mid}{(n+1)!} = \lim_{n\to\infty} \frac{\mid x \mid}{(n+1)} = 0$ and so the series $\displaystyle\sum_{n=0}^{\infty} \frac{x^n}{n!}$ is

absolutely convergent for all x.

The power series (i) is said to be nowhere convergent while that appearing in (iii) is everywhere convergent. The series (ii) is most interesting as it converges absolutely only in a subinterval (-4,4) of $\mathbb{R}$.

In context of convergence of a power series let's cite the following theoretical result without proof:

If a power series $\displaystyle\sum_{n=0}^{\infty} a_n x^n$ converges for $x = c$ but diverges for $x = d$, then $\displaystyle\sum_{n=0}^{\infty} a_n x^n$ converges absolutely for all x satisfying $\mid x \mid < \mid c \mid$ and diverges for all x satisfying $\mid x \mid > \mid d \mid$. If $\mid c \mid < \mid x \mid < \mid d \mid$, then $\displaystyle\sum_{n=0}^{\infty} a_n x^n$ may converge or may diverge and so this set is known as **annulus of doubt**.

This theoretical result eventually leads up to the concept of radius of convergence of a given power series. Observe that if $\mid c \mid < \mid x' \mid < d$, then the series $\sum a_n x'^n$ either converges or diverges.

In case $\displaystyle\sum_{n=0}^{\infty} a_n x'^n < \infty$, then due to the above result, $\displaystyle\sum_{n=0}^{\infty} a_n x^n$ converges absolutely for all x satisfying $\mid x \mid < \mid x' \mid$, forcing the annulus of doubt to dwindle. On the otherhand, if $\displaystyle\sum_{n=0}^{\infty} a_n x'^n = \infty$, then $\displaystyle\sum_{n=0}^{\infty} a_n x^n$ diverges for all x satisfying $\mid x \mid > \mid x' \mid$, again leading to slendering of annulus of doubt. If we iterate this process, the annulus of doubt ultimately shrinks to a set consisting of two equidistant points from $x = 0$. Infact, this inference is based on Nested Interval Property of $\mathbb{R}$. Labelling these two points as R and $-R$, we are led to the desired result that the power series $\displaystyle\sum_{n=0}^{\infty} a_n x^n$ converges for all x satisfying $\mid x \mid < R$ and diverges for all x satisfying $\mid x \mid > R$. We call $(-R, R)$ as the interval of convergence and R as the radius of convergence. So the series $\sum_{n=0}^{\infty} \frac{(n!)^2}{(2n)!} x^n$

has radius of convergence 4 while $\sum_{n=0}^{\infty} \frac{x^n}{n!}$ has radius of convergence ∞.

Since the radius of convergence of $\sum_{n=0}^{\infty} n! x^n$ is zero, this power series is

divergent. The working formula for finding the radius of convergence of a power series is given by either Cauchy Hadamard test or by Ratio test described in the following.

Cauchy-Hadamard Test:

If for the power series $\sum_{n=0}^{\infty} a_n x^n$, $R = \dfrac{1}{\overline{lim} \, | \, a_n \, |^{\frac{1}{n}}}$, then it

(i) converges absolutely for all x satisfying $| \, x \, | < R$.

(ii) diverges if $R=0$.

(iii) converges everywhere if $R = \infty$.

Ratio Test:

If for the power series $\sum_{n=0}^{\infty} a_n x^n$, $R = \lim_{n \to \infty} | \, \dfrac{a_n}{a_{n+1}} \, |$, then it

(i) converges absolutely for all x satisfying $| \, x \, | < R$.

(ii) diverges if $R = 0$.

(iii) converges everywhere if $R = \infty$.

Remark:

(a) In both the cases, the power series $\sum_{n=0}^{\infty} a_n x^n$ diverges for all x satisfying $| \, x \, | > R$ provided $0 < R < \infty$.

(b) The convergence of the power series at the endpoints of the interval of convergence is case-specific, i.e., it depends on whether the series $\sum_{n=0}^{\infty} a_n R^n$ or $\sum_{n=0}^{\infty} (-1)^n a_n R^n$ converges or not.

We now brief out some properties of power series in what follows. These will be very useful in the context of solving ordinary differential equations by power series technique.

(a) If R $(0 < R < \infty)$ be the radius of convergence of a power series $\sum_{n=0}^{\infty} a_n x^n$, then the series is uniformly convergent in any closed interval $[a, b]$ contained in the interval of convergence $(-R, R)$. Keeping the arbitrariness of choice of the closed interval in mind, we may consider $[-R + \epsilon, R - \epsilon]$ as the interval of uniform convergence, no matter how small $\epsilon > 0$ we opt for. So if a power series $\sum\limits_{n=0}^{\infty} a_n x^n$ converges absolutely in $(-R, R)$ $(0 < R < \infty)$ and uniformly in $[-R + \epsilon, R - \epsilon] \subseteq (-R, R)$ and $\exists$ a function $f : (-R, R) \longrightarrow \mathbb{R}$ such that $f(x) = \sum\limits_{n=0}^{\infty} a_n x^n$ for any $x \in (-R, R)$, then f is called the **sum-function** of the power-series.Moreover, this sum function f is continuous on $(-R, R)$.

(b) If $R > 0$ $(0 < R < \infty)$ be the radius of convergence of a power series $\sum\limits_{n=0}^{\infty} a_n x^n$ and $f(x)$ be its sum, then the power series $\sum\limits_{n=0}^{\infty} (n+1)a_{n+1} x^n$ (obtained from $\sum\limits_{n=0}^{\infty} a_n x^n$ by formal term-by-term differentiation) has the same radius of convergence R (since $\lim\limits_{n \to \infty} n^{\frac{1}{n}} = 1$).If $g(x)$ be the sum function of the new power series $\sum\limits_{n=0}^{\infty} (n+1)a_{n+1} x^n$, then it can be shown that $f(x)$, the sum function of the power series $\sum\limits_{n=0}^{\infty} a_n x^n$ is differentiable over any closed subinterval $[-R + \epsilon, R + \epsilon]$ $(\epsilon > 0)$ of $(-R, R)$ and moreover, $f'(x) = g(x)$.Since the choice of ϵ is upto us, one may conclude that the sum fuction $f(x)$ of the power series is differentiable everywhere in the interval of convergence $(-R, R)$ and its derivative $f'(x)$ can be calculated by using term-by-term differentiation of $\sum\limits_{n=0}^{\infty} a_n x^n$ representing $f(x)$. Clearly $a_0 = f(0)$ and $a_1 = f'(0)$.If we follow the same process then eventually we can have the power series $\sum\limits_{n=0}^{\infty} (n+1)(n+2)a_{n+2} x^n$ having the sum function $f''(x)$. Clearly $f''(0) = 2!a_2$, i.e., $a_2 = \frac{1}{2!}f''(0)$.One may formally continue this process upto any finite order and through successive differentiation get the formula $a_k = \frac{f^{(k)}(0)}{k!}$ $\forall k \geqslant 1$. This provides us an explicit way to determine the co-efficient of the power series expansion of a function.It also trivially follows from the above discussions that a power series $\sum\limits_{n=0}^{\infty} a_n x^n$ is identically equal to zero if each of its co-efficients is 0. In this context we should make the reader alert that although the sequence of functions $\{1, x, x^2, \ldots\ldots\ldots\infty\}$ forms a basis of the infinite dimensional vector space $P(\mathbb{R})$ and a power series $\sum\limits_{n=0}^{\infty} a_n x^n$ can be deemed of as some linear combination of these basis vectors,one must not be over-enthusiastic to draw the conclusion $a_n = 0$ for each n therefrom since the concept of linear independence of an infinite set is quite different from that of a finite set.

(c) One can consider linear combination of two power series $\sum\limits_{n=0}^{\infty} a_n x^n$

and $\sum\limits_{n=0}^{\infty} b_n x^n$. If $f(x)$ denotes the sum of $\sum\limits_{n=0}^{\infty} a_n x^n$ having radius of convergence $R_1(> 0)$ and $g(x)$ denotes the sum of $\sum\limits_{n=0}^{\infty} b_n x^n$ having radius of convergence $R_2(> 0)$,then the power series $\sum\limits_{n=0}^{\infty} (a_n + \lambda b_n) x^n$ has radius of convergence $R = Min(R_1, R_2)$, no matter what λ we choose. Moreover, the sum function of $\sum\limits_{n=0}^{\infty} (a_n + \lambda b_n) x^n$ is $f(x) + \lambda g(x)$, $x \epsilon(-R, R)$. In particular, if two different power series converge to the same function, i.e., have identical sum functions, then the two power series have the same set of co-efficients. This follows if we choose $\sum\limits_{n=0}^{\infty} a_n x^n = f(x) = \sum\limits_{n=0}^{\infty} b_n x^n$,and so for $\lambda = -1$, the power series $\sum\limits_{n=0}^{\infty} (a_n - b_n) x^n$ has sum function 0 . From discussions in (b) it follows that $a_n = b_n \ \forall n$.

(d) If $f(x)$ and $g(x)$ be the sum functions of the power series $\sum\limits_{n=0}^{\infty} a_n x^n$ and $\sum\limits_{n=0}^{\infty} b_n x^n$ respectively and R_1, R_2 be their respective radii of convergence, then their cauchy product $\sum\limits_{n=0}^{\infty} c_n x^n$ with $c_n = \sum\limits_{k=0}^{n} a_k b_{n-k}$ has the radius of convergence $R = Min\{R_1, R_2\}$ and moreover, it has the sum function $f(x).g(x)$ for all x satisfying $\mid x \mid < R$. We should recall that a power series is absolutely convergent within its interval of convergence and this absolute convergence plays a pivotal role in formulating the cauchy product as the cauchy product of two conditionally convergent series is not necessarily convergent.

(e) Suppose we have a function f that is infinitely differentiable in the interval $-R < x < R$. One may compute formally the co-efficient $a_n = \frac{f^{(n)}(0)}{n!}$ and then write down a formal power series $\sum\limits_{n=0}^{\infty} a_n x^n$. It is very interesting to observe that in most of the cases, this power series either diverges, or converges to a function $g(x)$ different from the original function $f(x)$. The function

$$f(x) = \begin{cases} e^{-\frac{1}{x^2}} & \text{, if } x \neq 0 \\ 0 & \text{, if } x = 0 \end{cases}$$

is infinitely differentiable and $f^{(n)}(0) = 0 \ \forall n$ but f has no power series expansion about $x = 0$.Thus there exists only a special class of functions where the power series converges back to $f(x)$ for all $x \epsilon(-R, R)$. These functions are known as **real analytic functions** that enjoy remarkable properties. If the power series $\sum\limits_{n=0}^{\infty} a_n x^n$ with $a_n = \frac{f^{(n)}(0)}{n!}$ $\forall n$ converges to $f(x)$, then we call $f(x)$ to be real analytic at $x = 0$. On the otherhand, if one computes the co-efficient $a_n = \frac{f^{(n)}(c)}{n!}$ $\forall n$ and the formal power series $\sum\limits_{n=0}^{\infty} a_n x^n$ converges to $f(x)$ $\forall x \epsilon (-R, R)$, then f is said to be analytic at $x = c$. The polynomials, the exponentials, sinusoidals and logarithmic functions are some of the distinguished examples of analytic functions. The linear combination of any two functions real analytic at some point $x = c$ is also real analytic at the same point. Finally we quote the very useful result that the sum function $f(x)$ of any power series $\sum\limits_{n=0}^{\infty} a_n x^n$ is real analytic at every point within its interval of convergence. Once the implication of real analyticity of a function f at a given point is understood, we can drop the word 'real' and carry on simply with analyticity at that point. The analyticity or singularity of the co-efficient functions of a given differential equation plays a key role in determining the line of attack in the series solution method. In the upcoming three articles we shall separately deal with the three different cases, viz, solution of ordinary differential equations at ordinary points, at regular singular points and also at irregular singular points (i.e., essential singularities). But even before the detailing, we shall give characterization of 'ordinary point', 'regular singular points' and 'irregular singular points' of an ode with illustrations.

8.3   Solutions about Ordinary Points in the Domain

Consider the second order ode in the normal form, viz,

$$y'' + P(x)y' + Q(x)y = 0$$

If $P(x)$ and $Q(x)$ are analytic at $x = c$, then $x = c$ is said to be an **ordinary point** of the ode chosen above. This implies that $P(x)$ and $Q(x)$ have convergent power series expansion about $x = c$. If R_1 be the radius of convergence of the power series representing $P(x)$ and R_2 be the radius of convergence of the power series representing $Q(x)$, then $\exists$ two linearly independent solutions of the form $y(x) = \sum\limits_{n=0}^{\infty} a_n(x-c)^n$, the radii of convergence of which will be less than or equal to R. However, only in handful cases, one is able to find sum function of such power series solutions in terms of elementary functions and give a compact solution form (The solutions of odes having constant co-efficient provide examples of this privileged class of odes. It is very easy to check that $x = 0$ is an ordinary point of the SHM governing ode, viz, $\frac{d^2y}{dx^2} + w^2 y = 0$. Similarly it is easy to show that $x = 0$ is an ordinary point of the ode $\frac{d^2y}{dx^2} + \frac{x}{1+x^2}\frac{dy}{dx} + \frac{1}{1+x^2}y = 0$ because the co-efficient functions $x(1 + x^2)^{-1}$ and $(1 + x^2)^{-1}$ are both analytic at $x = 0$ having convergent series expansions for $| x | < 1$. So by the above mentioned theory we are ensured that this ode has a power series solution of the form $y = \sum_{n=0}^{\infty} a_n x^n$, convergent for $| x | < 1$.

A useful result: Fuch's Theorem

If the differential equation $y'' + p(x)y' + q(x)y = 0$ has an ordinary point at $x = 0$, i.e., the co-efficient functions $p(x)$ and $q(x)$ have power series representations both converging for $| x | < R$, then there exist two fundamental solutions $y_1(x)$ and $y_2(x)$ satisfying $y_1(0) = 1; y_1'(0) = 0$ and $y_2(0) = 0; y_2'(0) = 1$ and each of $y_1(x)$ and $y_2(x)$ has a power series expansion converging for $| x | < R$. This result ensures that if α, β are two preassigned real constants, then $\exists$ a unique function $y = y(x)$ that satisfies the ode together with the initial conditions $y(0) = \alpha; y'(0) = \beta$ and admits of power series expansion for all $x \epsilon (-R, R)$. [In particular if $p(x)$ and $q(x)$ are polynomials, then as per Fuch's theorem, both the series solutions $y_1(x)$ and $y_2(x)$ converge for all $x \in \mathbb{R}$]

Observe that if $y_1(x)$ and $y_2(x)$ be two fundamental solutions of the ode, then the given ode has its general solution in the form

$$y(x) = c_1 y_1(x) + c_2 y_2(x)$$

Now $\quad y(0) = c_1 y_1(0) + c_2 y_2(0) \quad$ and $\quad y'(0) = c_1 y_1'(0) + c_2 y_2'(0)$

so that

$$\begin{pmatrix} y(0) \\ y'(0) \end{pmatrix} = \begin{pmatrix} y_1(0) & y_2(0) \\ y_1'(0) & y_2'(0) \end{pmatrix} \begin{pmatrix} c_1 \\ c_2 \end{pmatrix} = \begin{pmatrix} 1 & 0 \\ 0 & 1 \end{pmatrix} \begin{pmatrix} c_1 \\ c_2 \end{pmatrix} = \begin{pmatrix} c_1 \\ c_2 \end{pmatrix}$$

Therefore, $c_1 = y(0) = \alpha$ and $c_2 = y'(0) = \beta$ provided we assume that

$$y_1(0) = 1; y_1'(0) = 1; y_2(0) = 0; y_2'(0) = 1$$

Let's now describe the series solution method that is to be literally implemented to solve an ode about an ordinary point. If $x = c$ be an ordinary point of the ode $y''(x) + P(x)y + Q(x)y = 0$, then the solution

$y(x)$ is also analytic at $x = c$ and as already told before, one can write $y(x) = \sum_{n=0}^{\infty} a_n(x-c)^n$. However, one can apply a shifting transformation so as to make zero an ordinary point of the ode. Consequently one can write without loss of generality

$$y(x) = \sum_{n=0}^{\infty} a_n x^n \qquad (8.1)$$

From the earlier discussion it follows that in this case it is always possible to obtain two fundamental solutions of the form (8.1).

Applying term-by-term differentiation to (8.1) we have:

$$y' = \sum_{n=1}^{\infty} n a_n x^{n-1} = \sum_{n=0}^{\infty} (n+1)\, a_{n+1}\, x^n \qquad (8.2)$$

$$y'' = \sum_{n=2}^{\infty} n(n-1) a_n x^{n-2} = \sum_{n=0}^{\infty} (n+2)(n+1) a_{n+2}\, x^n \qquad (8.3)$$

By substituting equations (8.1), (8.2) and (8.3) into the given ode and demanding that the co-efficients of each power of x sum to zero, one obtains a recurrence relation expressing a_n in terms of $a_0, a_1, \ldots\ldots, a_{n-1}$. It is needless to say that the complexity of the actual recurrence relationship varies from one problem to the other.

Example (1) : Solve the ode $\frac{d^2 y}{dx^2} - y = 0$ by series method.

Observe that the co-efficients of the derivatives and that of y are constants and so are analytic at any point in $\mathbb{R}$. We therefore guess the solution in the form:

$$y = \sum_{n=0}^{\infty} c_n x^n \qquad (c_n's \text{ are yet to be determined })$$

Assuming validity of term-by-term differentiation, we get:

$$\frac{dy}{dx} = \sum_{n=1}^{\infty} n c_n x^{n-1} = \sum_{n=0}^{\infty} (n+1) c_{n+1} x^n$$

$$\frac{d^2 y}{dx^2} = \sum_{n=2}^{\infty} n(n-1) c_n x^{n-2} = \sum_{n=0}^{\infty} (n+2)(n+1) c_{n+2} x^n$$

Substituting in the original ode we get:

$$\sum_{n=0}^{\infty} \{ c_{n+2}(n+1)(n+2) - c_n \} x^n = 0$$

Each of the co-efficients appearing in above relation leads to the conclusion that it is zero. Thus we get the recurrence relation:

$$c_{n+2} = \frac{c_n}{(n+2)(n+1)} \forall n \geqslant 0$$

Therefore $c_0 = \dfrac{c_0}{2.1} = \dfrac{c_0}{2!}$

$$c_3 = \frac{c_1}{3.2} = \frac{c_1}{3!}$$

$$c_4 = \frac{c_2}{4.3} = \frac{c_0}{4.3.2.1} = \frac{c_0}{4!}$$

$$c_5 = \frac{c_3}{5.4} = \frac{c_1}{5.4.3!} = \frac{c_1}{5!} \quad \cdots$$

Hence we observe that the power series solution is of the form:

$$y = c_0(1 + \frac{x^2}{2!} + \frac{x^4}{4!} + \cdots \infty) + c_1(x + \frac{x^3}{3!} + \frac{x^5}{5!} + \cdots \infty)$$
$$= c_0 \cosh x + c_1 \sinh x,$$

where c_0 and c_1 are arbitrary constants.

Remarks : In this particular example, we are fortunate enough to get a compact form of a function corresponding to the series solution. However, we cannot expect the same for other problems.

Example (2) : Solve the ode: $y'' + xy' + 2y = 0$

The co-efficients of y' and y are x and 2 respectively and they are analytic everywhere in $\mathbb{R}$. So we can try a series solution in the form $\sum\limits_{n=0}^{\infty} c_n x^n$ for this ode. Using term-by-term differentiation we get:

$$y' = \sum_{n=1}^{\infty} n c_n x^{n-1}$$

$$y'' = \sum_{n=2}^{\infty} n(n-1) c_n x^{n-2} = \sum_{n=0}^{\infty} (n+2)(n+1) c_{n+2} x^n$$

Hence the given ode yields:

$$\sum_{n=0}^{\infty} (n+2)(n+1) c_{n+2} x^n + \sum_{n=1}^{\infty} x c_n x^n + 2 \sum_{n=0}^{\infty} c_n x^n = 0$$

Observe that in all the three summations, exponents of x are same, viz, n but ranges of summation are different. Since the minimum range is 1 to ∞, we shall regroup all the three summations over this minimum range, and keeping separate the isolated terms of the sums. With this mission we rewrite the above equation as:

$$\sum_{n=1}^{\infty}\{(n+2)(n+1)c_{n+2} + nc_n + 2c_n\}x^n + 2(c_2 + c_0) = 0$$

The only way the sum of above above power series can be zero is if each of the co-efficients is zero. Thus we have the relations:

$$c_2 = -c_0$$
$$(n+2)\{(n+1)c_{n+2} + c_n\} = 0 \; \forall n \geqslant 1$$

From the last one we get: $c_{n+2} = \frac{-c_n}{(n+1)\forall n \geqslant 1}$

If $n = 1$, $c_3 = \frac{-c_1}{2}$

If $n = 2$, $c_4 = -\frac{c_2}{3} = (-1)^2 \frac{c_0}{1.3}$

If $n = 3$, $c_5 = -\frac{c_3}{4} = \frac{(-1)^2 c_1}{2.4}$

If $n = 4$, $c_6 = -\frac{c_4}{5} = \frac{(-1)^3 c_0}{1.3.5}$ and so on.

Hence the series solution is:

$$y = c_0[1 - x^2 + \frac{x^4}{1.3} - \cdots \infty] + c_1[x - \frac{x^3}{2} + \frac{x^5}{2.4} - \cdots \infty]$$
$$\therefore \; y = c_0 \sum_{n=0}^{\infty} \frac{(-1)^n (x^2)^n}{1.3.5.....2n-1} + c_1 \, xe^{-\frac{x^2}{2}},$$

where c_0 and c_1 are arbitrary constants.

This shows that two linearly independent solutions of ode are:

$$y_1(x) \equiv 1 - x^2 + \frac{x^4}{1.3} - \frac{x^6}{1.3.5} + ...\infty = \sum_{n=0}^{\infty} \frac{(-1)^n (x^2)^n}{1.3.5......2n-1}$$

$$y_2(x) \equiv x - \frac{x^3}{2} + \frac{x^5}{2.4} - \frac{x^7}{2.4.6} + ...\infty = xe^{-\frac{x^2}{2}}$$

Example (3) : Solve by series method : $(1-x)y'' - y' + xy = 0$
In the normal form one can cast the ode as :

$$y'' - \frac{1}{1-x}y' + \frac{x}{1-x}y = 0$$

Comparing with $y'' + p(x)y' + q(x)y = 0$ it follows that $p(x) = \frac{-1}{(1-x)}$ and $q(x) = \frac{x}{1-x}$. Clearly $p(x)$ and $q(x)$ are not defined here at $x = 1$ and so $x = 1$ is a singular point. Observe that both $p(x)$ and $q(x)$ admit of power series expansion valid in $|x| < 1$. However, $x = 0$ is an ordinary point of the ode and so we seek a solution in the form $y(x) = \sum\limits_{n=0}^{\infty} c_n x^n$. Using term-by-term differentiation,

$$y'(x) = \sum_{n=1}^{\infty} n c_n x^{n-1} \; ; \; y''(x) = \sum_{n=2}^{\infty} n(n-1) c_n x^{n-2}$$

Putting all these in the given ode we have :

$$(1-x) \sum_{n=2}^{\infty} n(n-1) c_n x^{n-2} - \sum_{n=1}^{\infty} n c_n x^{n-1} + x \sum_{n=0}^{\infty} c_n x^n = 0$$

or, $$\sum_{n=2}^{\infty} n(n-1) c_n x^{n-2} - \sum_{n=2}^{\infty} n(n-1) c_n x^{n-1} - \sum_{n=1}^{\infty} n c_n x^{n-1}$$
$$+ \sum_{n=0}^{\infty} c_n x^{n+1} = 0$$

First we have to make the exponents of x same in all the four summations. To fulfil this goal, let's make the replacement $(n-2)$ by n in the first sum, $(n-1)$ by n in the second and third sum and n by $(n-1)$ in the fourth and last sum.

$$\therefore \sum_{n=0}^{\infty} (n+2)(n+1) c_{n+2} x^n - \sum_{n=1}^{\infty} (n+1)n \; c_{n+1} \; x^n$$
$$- \sum_{n=0}^{\infty} (n+1) c_{n+1} x^n + \sum_{n=1}^{\infty} c_{n-1} \; x^n = 0$$

Now the ranges are different - for first and third sum, the range is 0 to ∞ while for second and fourth sum, the range is 1 to ∞, showing that 1 to ∞ is the minimal range common to all the four sums. We therefore collate the four sums over this minimal range 1 to ∞ and keep aside the isolated spare terms. Hence we have :

$$\sum_{n=1}^{\infty} \{(n+2)(n+1)c_{n+2} - (n+1)n \; c_{n+1} - (n+1)c_{n+1} + c_{n-1}\} x^n + (2c_2 - c_1) = 0$$

The only way this can be realised is by making each of the co-efficients of x equal to zero. So we get :

$$2c_2 - c_1 = 0$$

and $(n+2)(n+1)c_{n+2} - (n+1)^2 c_{n+1} + c_{n-1} = 0 \ \forall \ n \geqslant 1.$

i, e, $c_{n+2} = \dfrac{(n+1)^2 c_{n+1} - c_{n-1}}{(n+2)(n+1)} \ \forall \ n \geqslant 1.$

From the first relation, viz, $2c_2 - c_1 = 0$ it follows that $c_2 = \frac{c_1}{2}$
The above recursion relation yields :

for $n = 1,$ $c_3 = \dfrac{2^2 c_2 - c_0}{3.2} = \dfrac{2c_1 - c_0}{3!}$

for $n = 2,$ $c_4 = \dfrac{9 c_3 - c_1}{4.3} = \dfrac{9 \left(\frac{2c_1 - c_0}{3!} \right) - c_1}{4.3} = \dfrac{4c_1 - 3c_0}{4!}$

for $n = 3,$ $c_5 = \dfrac{16 c_4 - c_2}{5.4} = \dfrac{16 \left(\frac{4c_1 - 3c_0}{4!} \right) - \frac{c_1}{2}}{5.4} = \dfrac{13c_1 - 6c_0}{5!}$

and so on. Hence the general form of the series solution reads :

$$
\begin{aligned}
y(x) &= \sum_{n=0}^{\infty} c_n x^n = c_0 + c_1\, x + c_2\, x^2 + c_3\, x^3 + \cdots \infty \\
&= c_0 + c_1\, x + \frac{c_1}{2!} x^2 + \left(\frac{2c_1 - c_0}{3!} \right) x^3 + \left(\frac{4c_1 - 3c_0}{4!} \right) x^4 + \cdots \infty \\
&= c_0 \left[1 - \frac{x^3}{6} - \frac{x^4}{8} + \cdots \right] + c_1 \left[x + \frac{x^2}{2} + \frac{x^3}{3} + \cdots \infty \right]
\end{aligned}
$$

Thus the two linearly independent series solutions are given by

$$
\left.
\begin{aligned}
y_1(x) &= 1 - \frac{x^3}{6} - \frac{x^4}{8} + \cdots \infty \\
\text{and} \quad y_2(x) &= x + \frac{x^2}{2} + \frac{x^3}{3} + \frac{x^4}{6} + \cdots \infty
\end{aligned}
\right\}
$$

It is a very simple observation that in particular if $c_0 = c_1 = 1$, then we get a particular integral of the ode as :

$$
\begin{aligned}
y &= y_1(x) + y_2(x) \\
&= 1 + x + \frac{x^2}{2} + x^3 \left(\frac{1}{3} - \frac{1}{6} \right) + x^4 \left(\frac{1}{6} - \frac{1}{8} \right) + \cdots \infty \\
&= 1 + \frac{x}{1!} + \frac{x^2}{2!} + \frac{x^3}{3!} + \frac{x^4}{4!} + \cdots \infty = e^x.
\end{aligned}
$$

Incidentally in Chapter 6 we found a trial method claiming that if the sum of co-efficients of a given ode of order 2 is zero, then $y_1(x) = e^x$ is one solution. To find another solution of the ode we have to use D'Alembert's order reduction method. The general solution is of the form $y = v(x)e^x$, where $v \in C^2(\mathbb{R})$.

$$\therefore \quad y' = e^x(v + v') \; ; \; y'' = e^x(v'' + 2v' + v)$$

i, e, $\quad (1 - x)(v'' + 2v' + v) - (v + v') + xv = 0$

or, $\quad v''(1 - x) + (1 - 2x)v' = 0$

Putting $v' = u$ and proceeding a few steps, we get $v(x) = \int \frac{e^{-2x}}{1-x} dx$ and so another linearly independent solution of the ode is : $y_2(x) = e^x \int \frac{e^{-2x} dx}{(1-x)}$. However, in the series solution method we have got two linearly independent solutions in one stroke!

Example (4) : Solve in series the Legendre's equation about $x = 0$

$$(1 - x^2)y'' - 2xy' + m(m + 1)y = 0, \quad m \text{ being a real parameter.}$$

In the normal form, the above differential equation reads:

$$y'' - \frac{2x}{1 - x^2}y' + \frac{m(m + 1)}{(1 - x^2)}y = 0$$

so that co-efficients of y' and that of y are respectively $\frac{-2x}{1-x^2}$ and $\frac{m(m+1)}{1-x^2}$, both of which being analytic at $x = 0$. Later on we shall see that $x = \pm 1$ are regular singular points of the Legendre equation (c.f §8.4). Since $x = 0$ is an ordinary point, we can try a power series solution of the Legendre equation in the form $y = \sum\limits_{n=0}^{\infty} c_n x^n$.

Therefore , $y' = \sum\limits_{n=1}^{\infty} n c_n x^{n-1}$ and $y'' = \sum\limits_{n=2}^{\infty} n(n - 1)c_n x^{n-2}$

On substituting these in Legendre's equation we get:

$$(1 - x^2) \sum\limits_{n=2}^{\infty} n(n - 1)c_n x^{n-2} - 2x \sum\limits_{n=1}^{\infty} n c_n x^{n-1} + m(m + 1) \sum\limits_{n=0}^{\infty} c_n x^n = 0$$

which on regrouping yields :

$$\sum\limits_{n=0}^{\infty} (n + 2)(n + 1)c_{n+2} x^n - \sum\limits_{n=2}^{\infty} n(n - 1)c_n x^n - 2 \sum\limits_{n=1}^{\infty} n c_n x^n +$$

$$m(m + 1) \sum\limits_{n=0}^{\infty} c_n x^n = 0$$

We now observe that the exponent of x is same for each of the four summations but only their ranges are different. Since the minimum range is 2 to ∞, we shall collate all these sums accordingly and keep the other spare terms isolated. Thus

$$\sum_{n=2}^{\infty}\{(n+2)(n+1)c_{n+2} - n(n-1)c_n - 2nc_n + m(m+1)c_n\}x^n + 2c_2$$

$$+(6c_3 - 2c_1)x + m(m+1)(c_0 + c_1 x) = 0$$

or, $$\sum_{n=2}^{\infty}\{(n+2)(n+1)c_{n+2} + (m(m+1) - n(n+1))c_n\}x^n + 2c_2$$

$$+m(m+1)c_0 + x[m(m+1)c_1 - 2c_1 + 6c_3] = 0$$

or, $$\sum_{n=2}^{\infty}\{(n+2)(n+1)c_{n+2} + c_n(m-n)(m+n+1)\}x^n + m(m+1)c_0$$

$$+2c_2 + x((m-1)(m+2)c_1 - 6c_3) = 0$$

This implies that the co-efficients of each power of x is zero.

$$\therefore \quad m(m+1)c_0 + 2c_2 = 0 \Rightarrow c_2 = \frac{-m(m+1)c_0}{2!}$$

$$(m-1)(m+2)c_1 = 6c_3 \Rightarrow c_3 = \frac{-(m-1)(m+2)c_1}{3!}$$

and $$c_{n+2} = \frac{-(m-n)(m+n+1)}{(n+2)(n+1)}c_n \quad \forall n \geqslant 2$$

$$\therefore \quad c_4 = \frac{-(m-2)(m+3)c_2}{4.3} = \frac{m(m+1)(m-2)(m+3)c_0}{4!}$$

$$c_5 = \frac{-(m-3)(m+4)}{5.4}c_3 = \frac{(m-1)(m+2)(m-3)(m+4)}{5!}c_1$$

and so on.

Hence the solution of Legendre equation is:

$$y = \sum_{n=0}^{\infty} c_n x^n = c_0 + c_1 x - \frac{m(m+1)c_0}{2!}x^2 - \frac{(m-1)(m+2)c_1 x^3}{3!} +$$

$$\frac{m(m+1)(m-2)(m+3)c_0 x^4}{4!} + \frac{(m-1)(m+2)(m-3)(m+4)c_1 x^5}{5!} + \cdots$$

$$= c_0\left[1 - \frac{m(m+1)}{2!}x^2 + \frac{m(m+1)(m-2)(m+3)x^4}{4!} - \cdots\infty\right] +$$

$$c_1\left[x - \frac{(m-1)(m+2)x^3}{3!} + \frac{(m-1)(m+2)(m-3)(m+4)x^5}{5!} - \cdots\infty\right]$$

Thus the two linearly independent solution are given by

$$y_1(x) = 1 - \frac{m(m+1)}{2!}x^2 + \frac{m(m+1)(m-2)(m+3)x^4}{4!} - \cdots \infty$$

$$y_2(x) = x - \frac{(m-1)(m+2)x^3}{3!} + \frac{(m-1)(m+2)(m-3)(m+4)x^5}{5!}$$
$$- \cdots \infty$$

Remark : (a) If m be a positive integer, then either m is an odd integer or an even integer. In case m is an odd integer, the second solution $y_2(x)$ terminates after a finite number of terms, yielding $y_2(x)$ a polynomial. In case m is an even positive integer, the first solution $y_1(x)$ terminates after a finite number of terms, making $y_1(x)$ a polynomial. These polynomials are known as **Legendre polynomials** which play crucial roles in Interpolation theory in numerical analysis and also in several areas of Mathematical Physics.

(b) The functions $y_1(x)$ and $y_2(x)$ are in general known as Legendre functions and are used in three-dimensional harmonic oscillators.

(c) By Cauchy-Hadamard formula for determining radius of convergence it is easy to check that the two power series representing $y_1(x)$ and $y_2(x)$ have the same radius of convergence, viz unity. So for all x satisfying $\mid x \mid < 1$, both the Legendre functions, viz, $y_1(x)$ and $y_2(x)$ exist.

Example (5) : Solve by series method Hermite equation:

$$y'' - 2xy' + 2py = 0 \; ; \; p \text{ being a constant.}$$

Observe that $x = 0$ is an ordinary point of this ode. (In fact the co-efficient of y being a polynomial, all points of $\mathbb{R}$ are ordinary points of this ode). So one can seek a power series solution in the form $y = \sum_{n=0}^{\infty} c_n x^n$. Using term-by-term differentiation we get:

$$y' = \sum_{n=1}^{\infty} nc_n x^{n-1} \quad \text{and} \quad y'' = \sum_{n=2}^{\infty} n(n-1)c_n x^{n-2}$$

Substituting these in given ode we have:

$$\sum_{n=2}^{\infty} n(n-1)c_n x^{n-2} - 2x \sum_{n=1}^{\infty} nc_n x^{n-1} + 2p \sum_{n=0}^{\infty} c_n x^n = 0$$

or, $\displaystyle\sum_{n=0}^{\infty}(n+2)(n+1)c_{n+2}x^n - 2\sum_{n=1}^{\infty}nc_nx^n + 2p\sum_{n=0}^{\infty}c_nx^n = 0$

We find that the exponent of x in each of the three summations is same but the ranges in the summation are different. Since the minimal range is 1 to ∞, we must collate all the terms having like powers and keep aside the spare terms. In this way we get:

$$\sum_{n=1}^{\infty}\{(n+2)(n+1)c_{n+2} + 2(p-n)c_n\}\, x^n + 2!c_2 + 2pc_0 = 0$$

The only way this can be realised is by setting co-efficients of each power of x equal to zero.

$\therefore \qquad\qquad c_2 + pc_0 = 0 \Leftrightarrow c_2 = \dfrac{-2pc_0}{2!}$

Also $\qquad (n+2)(n+1)c_{n+2} + 2(p-n)c_n = 0 \ \forall\ n \geqslant 1$

i.e., $\qquad\qquad c_{n+2} = \dfrac{2(n-p)c_n}{(n+2)(n+1)} \forall\ n \geqslant 1$

For $\ n = 1$, $\ c_3 = \dfrac{-2(p-1)c_1}{3.2}$

For $\ n = 2$, $\ c_4 = \dfrac{2(2-p)c_2}{4.3} = \dfrac{2^2p(p-2)c_0}{4!}$

For $\ n = 3$, $\ c_5 = \dfrac{2(3-p)c_3}{5.4} = \dfrac{2^2(p-1)(p-3)c_1}{5!}$

For $\ n = 4$, $\ c_6 = \dfrac{2(4-p)c_4}{6.5} = \dfrac{-2^3p(p-2)(p-4)c_0}{6!}$ and so on.

Thus the general solution of Hermite's differential equation will be:

$$y = \sum_{n=0}^{\infty}c_nx^n$$

$$= c_0[1 - \frac{2p}{2!}x^2 + \frac{2^2p(p-2)x^4}{4!} - \frac{2^3p(p-2)(p-4)x^6}{6!} + \cdots\infty]$$

$$+c_1[x - \frac{2(p-1)x^3}{3!} + \frac{2^2(p-1)(p-3)x^5}{5!} - \cdots\infty],$$

where c_0 and c_1 are arbitrary constants. So the two linearly independent solutions of Hermite's equation are:

$$y_1(x) = 1 - \tfrac{2p}{2!}x^2 + \tfrac{2^2p(p-2)x^4}{4!} - \tfrac{2^3p(p-2)(p-4)x^6}{6!} + \cdots\cdots\infty$$

$$y_2(x) = x - \frac{2(p-1)x^3}{3!} + \frac{2^2(p-1)(p-3)x^5}{5!} - \cdots\cdots \infty$$

Making n run through the even positive integers from the recursion relation (a) it directly follows from ratio test that the series representing $y_1(x)$ is convergent $\forall\ x\epsilon\mathbb{R}$ since for $k \to \infty$

$$\left| \frac{c_{2k+2}\ x^{2k+2}}{c_{2k} x^{2k}} \right| = \left| \frac{(2k-p)x^2}{(k+1)(2k+1)} \right| = \left| \frac{2k-p}{(k+1)(2k+1)} \right| |x^2| \to 0$$

Similarly if we make n run through the odd positive integers, then from the recursion relation (a) it also follows that the series representing $y_2(x)$ is convergent for all $x\epsilon\mathbb{R}$ since

$$\left| \frac{c_{2k+1}x^{2k+1}}{c_{2k-1}x^{2k-1}} \right| = \left| \frac{(2k-1-p)}{(2k+1)k} \right| |x|^2 \to 0 \ \ as \ \ k \to \infty$$

Remark : Similar to Legendre equation, there is a particular interesting case for Hermite equation when the constant p happens to be a positive integer. In particular if p is an odd positive integer of the form $(2m+1)$, the linearly independent solution $y_2(x)$ reduces to a polynomial of degree $(2m+1)$. On the otherhand, if p be an even positive integer of the form $2m$, the linearly independent solution $y_1(x)$ reduces to a polynomial of degree $2m$. All these polynomials are commonly known as **Hermite polynomials**. Hermite equation evolves in discussing hydrogen problem in elementary quantum mechanics while Hermite polynomials play vital role in Interpolation theory of numerical analysis.

Example (6) : Solve $y'' - 2x^2y' + 4xy = x^2 + 2x + 2$. Observe that $x = 0$ is an ordinary point of the ode given as the co-efficients of y' and y both being polynomials are analytic at $x = 0$. We choose $y = \sum\limits_{n=0}^{\infty} c_n x^n$ so that term-by-term differentiation gives :

$$y' = \sum_{n=1}^{\infty} nc_n x^{n-1} \ ; \ y'' = \sum_{n=2}^{\infty} n(n-1)c_n x^{n-2}$$

Putting back in the given inhomogeneous ode we have :

$$\sum_{n=2}^{\infty}(n+2)(n+1)c_{n+2}x^n - 2\sum_{n=1}^{\infty} nc_n x^{n+1} + 4\sum_{n=0}^{\infty} c_n x^{n+1} - (x^2+2x+2) = 0$$

Since the exponents of x in the three summations are different, we have to make necessary manipulations so as to bring these exponents at par.

To achieve this, let us replace n by $\overline{n-1}$ in the second and third sum so that

$$\sum_{n=1}^{\infty} nc_n x^{n+1} = \sum_{n=2}^{\infty}(n-1)c_{n-1}x^n \quad \text{and} \quad \sum_{n=0}^{\infty} c_n x^{n+1} = \sum_{n=1}^{\infty} c_{n-1}x^n$$

$$\therefore \sum_{n=0}^{\infty}(n+2)(n+1)c_{n+2}\,x^n - 2\sum_{n=2}^{\infty}(n-1)c_{n-1}x^n + 4\sum_{n=1}^{\infty} c_{n-1}(n-1)x^n$$
$$-(x^2 + 2x + 2) = 0$$

See that the ranges of summation are different viz, 0 to ∞ for the first 2 to ∞ for the second and 1 to ∞ for the third. Since 2 to ∞ is the minimum range, we have to collate the terms of all three series over this minimum range, viz, 2 to ∞ and keep aside the spare terms.

$$\therefore \quad 2c_2 + \sum_{n=2}^{\infty}\{(n+2)(n+1)c_{n+2} - 2(n-1)c_{n-1} + 4c_{n-1}\}x^n$$
$$+(6c_3 x + 4c_0)x - (x^2 + 2x + 2) = 0$$

$$\text{or,} \quad 2(c_2 - 1) + x(6c_3 + 4c_0 - 2) + (12c_4 + 2c_1 - 1)x^2$$
$$+ \sum_{n=3}^{\infty}\{(n+2)(n+1)c_{n+2} - (2n-6)c_{n-1}\}x^n = 0$$

The only way this can be realised is by setting the co-efficients of each power of x equal to zero.

Therefore $\quad c_2 = 1 \; ; \; 6c_3 + 4c_0 - 2 = 0 \; ; \; 12c_4 - 2c_1 - 1 = 0$

$$(n+2)(n+1)c_{n+2} - (2n-6)c_{n-1} = 0 \; \forall \; n \geqslant 3$$

i,e, $\quad c_{n+2} = \dfrac{2(n-3)}{(n+2)(n+1)}c_{n-1} \; \forall \; n \geqslant 3$.

Hence $\quad c_3 = \dfrac{1-2c_0}{3} \; ; \; c_4 = \dfrac{1-2c_1}{12} \; ; \; c_5 = 0 \; ;$

$$c_6 = \dfrac{1-2c_0}{45} \; ; \; c_7 = \dfrac{1-2c_1}{126} \; ; \; c_8 = 0 \text{ and so on.}$$

Thus the general series solution of the given ode reads :

$$y = \sum_{n=0}^{\infty} c_n \, x^n$$

$$= c_0 + c_1\,x + x^2 + \left(\frac{1-2c_0}{3}\right)x^3 + \left(\frac{1-2c_1}{12}\right)x^4 + \left(\frac{1-2c_0}{45}\right)x^6$$

$$+ \left(\frac{1-2c_1}{126}\right)x^7 + \cdots \infty$$

$$= c_0 \left[1 - \frac{2}{3}x^3 - \frac{2}{45}x^6 + \cdots\right] + c_1 \left[x - \frac{x^4}{6} - \frac{x^7}{63} + \cdots \infty\right]$$

$$+ \left[x^2 - \frac{x^3}{3} + \frac{x^4}{12} + \frac{x^6}{45} - \frac{x^7}{126} + \cdots \infty\right]$$

Remark (a) One can verify that each of these power series converges for all x. Theoretically it follows from Fuch's theory as the co-efficients of y and y' are polynomials.

(b) One can observe that in this example we have been able to determine a particular integral directly by series method. This has been feasible simply because the inhomogeneous term itself is a polynomial that keeps parity with the series form of solution sought for in the problem.

Example (7) : Solve by series method Airy equation : $y'' - xy = 0$
Since the equation is in normal form with the co-efficient of y being a polynomial, viz, $-x$, it follows $x = 0$ is an ordinary point of the form $y = \sum_{n=0}^{\infty} c_n x^n$ for the ode. Through term-by-term differentiation it follows that

$$y' = \sum_{n=1}^{\infty} n c_n x^{n-1} \; ; \; y'' = \sum_{n=2}^{\infty} n(n-1) c_n x^{n-2}$$

Substituting back in the ode given we get :

$$\sum_{n=2}^{\infty} n(n-1) c_n x^{n-2} - \sum_{n=0}^{\infty} c_n x^{n+1} = 0$$

or, $$\sum_{n=0}^{\infty} (n+2)(n+1) c_{n+2} x^n - \sum_{n=1}^{\infty} c_{n-1}\, x^n = 0$$

or, $$\sum_{n=1}^{\infty} \{(n+2)(n+1) c_{n+2} - c_{n-1}\} x^n + 2 c_2 = 0$$

This in turn implies that

$$c_2 = 0 \text{ and } c_{n+2} = \frac{c_{n-1}}{(n+2)(n+1)} \; \forall\, n \geqslant 1.$$

For $\quad n = 1, \quad c_3 = \dfrac{c_0}{3.2}$ $\quad ;\quad$ $n = 4, \quad c_6 = \dfrac{c_3}{6.5} = \dfrac{c_0}{6.5.3.2}$

$\quad n = 2, \quad c_4 = \dfrac{c_1}{4.3}$ $\quad ;\quad$ $n = 5, \quad c_7 = \dfrac{c_4}{7.6} = \dfrac{c_1}{7.6.4.3}$

$\quad n = 3, \quad c_5 = \dfrac{c_2}{5.4} = 0$ $\quad ;\quad$ $n = 6, \quad c_8 = 0 \quad$ and so on.

It is clear that

$$c_2 = c_5 = c_8 = \cdots = 0 \quad \text{i, e, } \quad c_{3k+2} = 0 \ \forall \ k = 0, 1, 2, \cdots$$

Further $c_3, c_6, \cdots$ are all multiples of c_0. To be more explicit,

$$c_{3k} = \frac{1}{(2.3)(5.6)\cdots((3k-1)(3k))}c_0 \quad \forall \ k = 1, 2, \cdots$$

Similarly all terms $c_4, c_7, \cdots$ are multiples of c_1. More elaborately,

$$c_{3k+1} = \frac{1}{(3.4)(6.7)\cdots((3k)(3k+1))}c_1 \quad \forall \ k = 1, 2, \cdots$$

Thus the general form of solution to Airy's equation reads :

$$y(x) = c_0 \left(1 + \frac{x^3}{6} + \frac{x^6}{180} + \cdots \infty \right) + c_1 \left(x + \frac{x^4}{12} + \frac{x^7}{504} + \cdots \infty \right)$$

where c_0 and c_1 are arbitrary constants.

So the linearly independent solutions are :

$$y_1(x) = \left(c_0 + \sum_{k=1}^{\infty} c_{3k}\, x^{3k} \right) \quad \text{and} \quad y_2(x) = \left(c_1 x + \sum_{k=1}^{\infty} c_{3k+1} x^{3k+1} \right)$$

It is clear that the two series representing $y_1(x)$ and $y_2(x)$ converge absolutely for all $x \in \mathbb{R}$ (c.f Fuch's theorem). These solutions are referred to as **Airy functions**.

Remark : (a) It is an interesting feature that for negative x, both $y_1(x)$ and $y_2(x)$ exhibit a wavy nature like sinusoidal functions while for positive x, both $y_1(x)$ and $y_2(x)$ show strictly monotone increasing behavior like exponential functions. Curve $y_1(x)$ passes through the point $(0, 1)$ while the curve $y_2(x)$ passes through the origin. It is clearly seen in the graph plotted below.

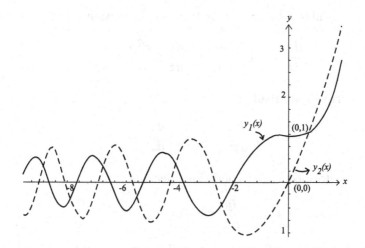

Fig 8.1 : Graph of linearly independent Airy functions $y_1(x)$ and $y_2(x)$ (appearing as Fuch's solutions) showing their zeros.

(b) From the graph it follows that both the Airy functions can have infinitely many zeros on negative x-axis and atmost one non-negative zero. But can one assert this truth analytically? The answer is in the affirmative. We first put $x = -t$ in Airy's ode so that it reduces to

$$\frac{d^2y}{dt^2} + ty = 0.$$

Comparing this with the normal form, viz, $\frac{d^2y}{dt^2} + Q(t)y = 0$, we get $Q(t) = t > 0 \ \forall \ t > 0$ and for any preassigned positive number a, $\int_a^\infty Q(t)dt = \int_a^\infty t\,dt = \infty$. Hence by Theorem 5 given in § 6.6 it follows that every non-trivial solution $y(t)$ of the equation $\frac{d^2y}{dt^2} + ty = 0$ has infinitely many zeros in $\mathbb{R}^+$. This is equivalent to state that the originally given Airy equation has infinitely many zeros in $\mathbb{R}^-$, i,e, the negative real axis.

(c) Airy equation evolves in the theory for diffraction in optics. Airy functions give the diffraction patterns.

Example (8) : Solve the IVP : $x^2y'' - xy' + ay = 0$; $y(1) = y'(1) = 1$

Since the information is given at the point $x = 1$, we must observe that $x = 1$ is an ordinary point of the ode. We apply the change of independent variable from x to z by the transformation $z = 1 - x$ so that

$$\frac{dz}{dx} = -1 \; ; \; \frac{dy}{dx} = -\frac{dy}{dz} \; ; \; \frac{d^2y}{dx^2} = \frac{d^2y}{dz^2}$$

The transformed ode reads :

$$(1 - z)^2 \frac{d^2y}{dz^2} + (1 - z)\frac{dy}{dz} + ay = 0$$

and the transformed initial conditions become :

$$y(z = 0) = 1 \; ; \; \frac{dy}{dz}(z = 0) = 1$$

Clearly $z = 0$ is an ordinary point of the new ode and so about $z = 0$ we may seek the power series solution in the form

$$y = \sum_{n=0}^{\infty} c_n z^n$$

Using term-by-term differentiation w.r to z we get.

$$\frac{dy}{dz} = \sum_{n=1}^{\infty} n c_n z^{n-1} \; ; \; \frac{d^2y}{dz^2} = \sum_{n=2}^{\infty} n(n-1)c_n z^{n-2}$$

Putting these in the recast ode we have :

$$(1 - z)^2 \left(\sum_{n=2}^{\infty} n(n-1)c_n z^{n-2} \right) + (1 - z)\left(\sum_{n=1}^{\infty} n c_n z^{n-1} \right) + a \sum_{n=0}^{\infty} c_n z^n = 0$$

i, e, $$\sum_{n=2}^{\infty} n(n-1)c_n z^{n-2} - 2\sum_{n=2}^{\infty} n(n-1)c_n z^{n-1} + \sum_{n=2}^{\infty} n(n-1)c_n z^n$$

$$+ \sum_{n=1}^{\infty} n c_n z^{n-1} - \sum_{n=1}^{\infty} n c_n z^n + a \sum_{n=0}^{\infty} c_n z^n = 0 \qquad ...(a)$$

Replacing n by $n + 2$, the sum $\sum_{n=2}^{\infty} n(n-1)c_n z^{n-2}$ becomes $\sum_{n=0}^{\infty} (n + 2)(n + 1)c_n + 2z^n$; replacing n by $(n + 1)$, the sum $\sum_{n=2}^{\infty} n(n-1)c_n z^{n-1}$ becomes $\sum_{n=1}^{\infty} (n + 1)n \, c_{n+1} z^n$ while replacing n by $(n+1)$ the sum $\sum_{n=1}^{\infty} n \, c_n z^{n-1}$ becomes $\sum_{n=0}^{\infty} (n+1)c_{n+1} z^n$. With all these manipulative works, (a) turns out to be

$$\sum_{n=0}^{\infty}(n+2)(n+1)c_{n+2}z^n - 2\sum_{n=1}^{\infty}(n+1)nc_{n+1}z^n + \sum_{n=0}^{\infty}(n+1)c_{n+1}z^n$$

$$+\sum_{n=2}^{\infty}n(n-1)c_nz^n - \sum_{n=1}^{\infty}nc_nz^n + a\sum_{n=0}^{\infty}c_nz^n = 0$$

i.e, $\sum_{n=2}^{\infty}\left[(n+2)(n+1)c_{n+2} + (1-2n)(n+1)c_{n+1} + \{n(n-2)+a\}c_n\right]z^n$

$$+(c_1 + 2c_2 + ac_0) + z(6c_3 - 2c_2 + (a-1)c_1) = 0$$

Equating co-efficients like power terms to zero we get :

$$\left.\begin{array}{l} c_1 + 2c_2 + ac_0 = 0 \\ 6c_3 - 2c_2 + (a-1)c_1 = 0 \end{array}\right\}$$

together with the general recurrence relation valid for $n \geqslant 2$:

$$(n+2)(n+1)c_{n+2} + (n+1)(1-2n)c_{n+1} + (n^2 - 2n + a)c_n = 0$$

Hence $c_2 = \dfrac{-(c_1 + ac_0)}{2!}$; $c_3 = \dfrac{-a(c_0 + c_1)}{3!}$

$$c_4 = \frac{a}{4!}\left[-2c_1 + (a-3)c_0\right], \quad \text{and so on.}$$

$\therefore \quad y(z) = c_0\left[1 - \dfrac{a}{2!}z^2 - \dfrac{a}{3!}z^3 + \dfrac{a-3}{4!}z^4 - ....\infty\right]$

$$+ c_1\left[z - \dfrac{z^2}{2!} - \dfrac{az^3}{3!} - \dfrac{z^4}{12}....\infty\right] \qquad ...(b)$$

From the initial conditions it follows that $c_0 = c_1 = 1$ and so the resultant solution will be expressible as (b) with c_0 and c_1 both unity. To convert the solution about $x = 1$, put back $z = 1 - x$ in answer.

8.4   Solution about Regular Singular Points

In §8.3 we have discussed at length the solution techniques of the second order odes about the ordinary points. We recall that the second order ode in the normal form, viz,

$$y''(x) + P(x)y' + Q(x)y = 0$$

is said to have $x = x_0$ an ordinary point provided $P(x)$ and $Q(x)$ are both analytic at x_0. This is equivalent to the statement that the ode

$$a_0(x)y'' + a_1(x)y' + a_2(x)y = 0 \qquad (8.4)$$

has $x = x_0$ as an ordinary point provided $a_0(x), a_1(x), a_2(x)$ are all analytic at x_0 and moreover $a_0(x_0) \neq 0$. If now $a_0(x_0) = 0$, then we say that $x = x_0$ is a singular point of ther ode (8.4). This is equivalent to state that the corresponding normal form of the ode, viz,

$$y''(x) + P(x)y' + Q(x)y = 0$$

has a singular point at $x = x_0$ if at least one of the two co-efficient functions $P(x)$ and $Q(x)$ fails to be analytic at x_0. So it is easy to identify the zeros of $a_0(x)$ as the root of all mischiefs! The singular point x_0 of the ode (8.4) is said to be a **regular singular point** provided we can recast the ode in the form

$$b_0(x)(x - x_0)^2 y'' + b_1(x)(x - x_0)y' + b_2(x)y = 0 \qquad (8.5)$$

with $b_0(x_0) \neq 0$ and $b_0(x), b_1(x), b_2(x)$ all analytic at x_0.

In case of the equivalent normal form, $y''(x) + P(x)y' + Q(x)y = 0$ the above definition of regular singular point x_0 shapes up in the form of following :

    (i) $(x - x_0)P(x)$ is analytic at $x = x_0$

    (ii) $(x - x_0)^2 Q(x)$ is analytic at $x = x_0$

In other words, this amounts to saying that x_0 is a regular singular point of the ode

$$y''(x) + P(x)y' + Q(x)y = 0$$

iff the co-efficient function $P(x)$ has a **pole of order atmost 1** while the co-efficient function $Q(x)$ has a **pole of order atmost 2**.

Summarily, the singular point x_0 of the ode (8.4) is said to be regular if one can cast the ode in the form :

$$(x - x_0)^2 y'' + (x - x_0)p_1(x)y' + q_1(x)y = 0 \qquad (8.6)$$

with $p_1(x)$ and $q_1(x)$ being analytic at x_0, i,e, both admitting of power series expansions about x_0 $(c.f.(8.6)$ with $(8.5))$.

However, for checking the regularity of a singular point, we shall use the following ansatz.

Compute $\lim_{x \to x_0} (x - x_0) p(x)$ and $\lim_{x \to x_0} (x - x_0)^2 q(x)$

and see whether they are finitely existent or not. If both the limits exist, then $x = x_0$ is a **regular singular point**. If either of the two fails to exist, $x = x_0$ is said to be an **irregular singular point**.

Formally we declare a singular point of the ode (8.4) to be irregular if it cannot be cast into either of the forms (8.5) or (8.6). This amounts to state that the co-efficient functions $P(x)$ and $Q(x)$ appearing in the normal form $y'' + P(x)y' + Q(x)y = 0$ has pole of order greater than 1 and greater than 2 respectively at $x = x_0$.

In the following we give a number of examples identifying the regular and irregular singular points of the odes.

Example (9) : Consider the ode : $c_0 x^2 y'' + c_1 x y' + c_2 y = 0$, where c_0, c_1, c_2 are constants.

This Cauchy-Euler ode has $x = 0$ as a regular singular point because it can be put in the normal form :

$$y'' + \frac{c_1}{c_0} \frac{1}{x} y' + \frac{c_2}{c_0} \frac{1}{x^2} y = 0$$

and moreover the limits

$$\lim_{x \to 0} x \left(\frac{c_1}{c_0 x} \right) = \frac{c_1}{c_0} \quad \text{and} \quad \lim_{x \to 0} x^2 \left(\frac{c^2}{c_0} \cdot \frac{1}{x^2} \right) = \frac{c^2}{c_0}$$

exist finitely.

Example (10) : Consider the ode : $x^3 y'' + (1 - \cos x)y' + 3xy = 0$. This ode with $x = 0$ as a singular point, can be put in the normal form:

$$y'' + \frac{1 - \cos x}{x^3} y' + \frac{3}{x^2} y = 0$$

Moreover, $\lim_{x \to 0} \frac{x(1 - \cos x)}{x^3} = \lim_{x \to 0} \frac{1}{2} \left(\frac{\sin \frac{x}{2}}{\frac{x}{2}} \right)^2 = \frac{1}{2}$

and $\lim_{x \to 0} x^2 \cdot \frac{3}{x^2} = 3$

This ensures that $x = 0$ is a regular singular point.

Example (11) : Consider the ode : $x(1 - x)y'' + (2 - 3x)y' - y = 0$ clearly $x = 0$ and $x = 1$ are the singular points. In the normal form this ode can be expressed as :

$$y'' + \frac{2 - 3x}{x(1 - x)}y' - \frac{y}{x(1 - x)} = 0$$

Observe that $x = 0$ and $x = 1$ are the singular points. To explore their nature we compute the following limits.

$$\lim_{x \to 0} \frac{x(2 - 3x)}{x(1 - x)} = 2 \; ; \; \lim_{x \to 0} x^2 \left(\frac{-1}{x(1 - x)} \right) = 0$$

$$\lim_{x \to 1} \frac{(x - 1)(2 - 3x)}{x(1 - x)} = \lim_{x \to 1} \frac{(3x - 2)}{x} = 1$$

$$\lim_{x \to 1} (x - 1)^2 \frac{1}{(x - 1)} = \lim_{x \to 1} (x - 1) = 0$$

Since all the limits are finite, we conclude that both $x = 0$ and $x = 1$ are regular singular points.

Example (12) : Consider $x^2(x - 1)^2 y'' + 3(\sin x)y' - (x - 1)y = 0$.

This ode has $x = 0$ and $x = 1$ as the only singular points. We now explore the nature of these singular points by putting the ode in the normal form :

$$y'' + \frac{3 \sin x}{x^2(x - 1)^2}y' - \frac{1}{x^2(x - 1)}y = 0$$

We find that $\quad \lim_{x \to 0} x \left(\frac{3 \sin x}{x^2(x - 1)^2} \right) = \lim_{x \to 0} 3 \left(\frac{\sin x}{x} \right) \frac{1}{(x - 1)^2} = 3$

$$\lim_{x \to 0} -\frac{1}{(x - 1)^2 x^2}x^2 = 1$$

$$\lim_{x \to 1} (x - 1)\frac{3 \sin x}{x^2(x - 1)^2} = \lim_{x \to 1} \frac{3 \sin x}{x^2(x - 1)} = \infty$$

$$\lim_{x \to 1} -(x - 1)^2 \frac{1}{x^2(x - 1)} = \lim_{x \to 1} \frac{-(x - 1)}{x^2} = 0$$

Hence $x = 0$ is a regular singular point while $x = 1$ is an irregular singular point.

Example (13) : Consider the ode: $(1 - x)y'' - y' + xy = 0$. In the normal form this ode can be written as :

$$y'' - \frac{1}{1-x}y' + \frac{x}{1-x}y = 0$$

This shows that $x = 1$ is a singular point of the ode. Moreover,

$$\lim_{x \to 1}(x - 1)\left(\frac{-1}{1-x}\right) = 1$$

$$\lim_{x \to 1}(x - 1)^2 \frac{x}{1-x} = Lt_{x \to 1}\, x(1-x) = 0$$

This confirms that $x = 1$ is a regular singular point of the ode.

Example (14) : Consider : $x^2y'' + xy' + \left(x^2 - \frac{1}{4}\right)y = 0$ for $x > 0$.

Clearly $x = 0$ is a regular singular point because the above ode can be rewritten as :

$$y'' + \frac{1}{x}y' + \left(1 - \frac{1}{4x^2}\right)y = 0$$

Moreover $\quad \lim_{x \to 0} x.\frac{1}{x} = 1 \; ; \; \lim_{x \to 0} x^2 \left(1 - \frac{1}{4x^2}\right) = -\frac{1}{4}$

Remark : This ode is known as **Bessel equation of order** $\frac{1}{2}$. It is a special case of

$$x^2y'' + xy' + (x^2 - p^2)y = 0 \; ; \; x > 0$$

known as **Bessel equation of order** p. $x = 0$ is a regular singular point of this more generalised version.

If we put $x = \frac{1}{z}$ in Bessel equation of order p, then it becomes : :

$$z^4 \frac{d^2y}{dz^2} + z^3 \frac{dy}{dz} + (1 - p^2z^2)y = 0$$

which in the normal form reads :

$$\frac{d^2y}{dz^2} + \frac{1}{z}\frac{dy}{dz} + \left(\frac{1 - p^2z^2}{z^4}\right)y = 0$$

Observe that $z = 0$ is an irregular singular point of the above ode as

$$\lim_{z \to 0} z.\frac{1}{z} = 1 \text{ but } \lim_{z \to 0} \frac{z^2(1 - p^2z^2)}{z^4} = \infty$$

So 'point at infinity' is an irregular singular point of the Bessel equation of order p.

Example (15) : In example (4) we dealt with Legendre equation $(1 - x^2)y'' - 2xy' + m(m + 1)y = 0$, m being a real parameter.

We claim that this ode has two singular points $x = 1$ and $x = -1$. We can rewrite viz, this ode in the normal form as :

$$y'' - \frac{2x}{1 - x^2}y' + \frac{m(m + 1)}{1 - x^2}y = 0$$

We observe that $\lim_{x \to 1} (x - 1)\left(\frac{-2x}{1 - x^2}\right) = \lim_{x \to 1} \frac{2x}{x + 1} = 1$

$$\lim_{x \to 1} (x - 1)^2 \frac{m(m + 1)}{(1 - x^2)} = \lim_{x \to 1} m(m + 1)\left(\frac{1 - x}{1 + x}\right) = 0$$

$$\lim_{x \to -1} (x + 1)\left(\frac{-2x}{-x^2 + 1}\right) = \lim_{x \to -1} \frac{-2x}{(1 - x)} = 1$$

$$\lim_{x \to -1} (x + 1)^2 \frac{m(m + 1)}{(1 - x^2)} = \lim_{x \to -1} \left(\frac{1 + x}{1 - x}\right)(m)(m + 1) = 0$$

Hence both the singular points, viz, $x = 1$ and $x = -1$ are regular.

Example (16) : Consider the ode : $\frac{d^2y}{dx^2} + y = 0$

First use the transformation : $x = \frac{1}{z}$ so that by chain rule :

$$\frac{dy}{dx} = -z^2 \frac{dy}{dz} \quad \text{and} \quad \frac{d^2y}{dx^2} = z^4 \frac{d^2y}{dz^2} + 2z^3 \frac{dy}{dz}$$

Thus the given ode is transformed to :

$$z^4 \frac{d^2y}{dz^2} + 2z^3 \frac{dy}{dz} + y = 0$$

i, e, $$\frac{d^2y}{dz^2} + \frac{2}{z}\frac{dy}{dz} + \frac{1}{z^4}y = 0$$

However $\lim_{z \to 0} z.\frac{2}{z} = 2$ but $\lim_{z \to 0} z^2\left(\frac{1}{z^4}\right) = \infty$.

It is clear that $z = 0$ is an irregular singular point of the new ode. This is equivalent to state that the given ode has an irregular singular point at infinity.

Example (17) : Consider the ode :

$$(x^4 - 2x^3 + x^2)y'' + 2(x - 1)y'' + x^2y = 0$$

The given ode can be written as :

$$x^2(x - 1)^2y'' + 2(x - 1)y' + x^2y = 0$$

Obviously $x = 0$ and $x = 1$ are singular points of the ode. We now explore the nature of these singular points. For this purpose the equation is first put into the normal form :

$$y'' + \frac{2y'}{x^2(x - 1)} + \frac{1}{(x - 1)^2}y = 0$$

We compute $\quad \lim_{x \to 0} \left(x \cdot \frac{2}{x^2(x - 1)} \right) = \lim_{x \to 0} \frac{2}{x(x - 1)} = \infty$

$$\lim_{x \to 0} x^2 \frac{1}{(x - 1)^2} = 0$$

$$\lim_{x \to 1} (x - 1)\frac{2}{x^2(x - 1)} = \lim_{x \to 1} \frac{2(x - 1)}{x^2} = 0$$

$$\lim_{x \to 1} (x - 1)^2 \frac{1}{(x - 1)^2} = 1$$

and observe that $x = 0$ is an irregular singular point while $x = 1$ is a regular singular point.

Alternatively one may proceed as follows :

The co-efficient function of y is $\frac{2}{x^2(x-1)}$, which can be written as

$$\frac{-2}{x^2}(1 - x)^{-1} = -\frac{2}{x^2}\sum_{k=0}^{\infty} x^k.$$

This shows that $x = 0$ is a pole of order 2 for this co-efficient function of y'. However, in our earlier discussion we found that if $x = 0$ is to be a regular singular point of the ode, then the co-efficient function of y' should have pole of order atmost 1. This led to the conclusion that $x = 0$ is an irregular singular point. To find nature of the singular point $x = 1$ it suffices to transform the ode by substituting $z = x - 1$ so that

$$\frac{2}{x^2(x - 1)} = \frac{2}{z(z + 1)^2} = \frac{2}{z}(1 + z)^{-2} = \frac{2}{z}(1 - 2z + 3z^2 - \cdots \infty)$$

This ensures that $z = 0$ or equivalently $x = 1$ is a pole of order 1 of the co-efficient function $\frac{2}{x^2(x-1)}$ of y'. Similarly the co-efficient function of y, viz, $\frac{1}{(x-1)^2}$ is transformed to $\frac{1}{z^2}$ by the same substitution and this shows that $z = 0$ or equivalently $x = 1$ is a pole of order 2 of $\frac{1}{(x-1)^2}$. Thus $x = 1$ is a regular singular point of the given ode.

Remark : From a comparative study of the two approaches it follows that the first approach was more compact and economical.

8.5 Frobenius Method

Once the nature of a singular point is explored, it is our binding to solve the ode about that point. In this article we shall discuss how to solve any linear ode about a regular singular point. The method of finding the solution of an ode about an irregular singular point is an advanced topic which we shall address in the last article. Meanwhile, for sake of simplicity we consider the case where $x = 0$ is a regular singular point. Because of the linearity of ode it is clear that by just a suitable linear transformation we can, in principle, shift any singular point x_0 to zero.

About regular singular points the solution technique is provided by Frobenius. He suggested a series solution of the form :

$$y(x) = x^\lambda \sum_{n=0}^{\infty} c_n x^n,$$

where the exponent λ is either an integer (positive/negative/zero) or even a fraction. But how can we justify such an ansatz ? The answer is hidden in Cauchy - Euler ode which has the general form :

$$a_0 x^2 \frac{d^2 y}{dx^2} + a_1 x \frac{dy}{dx} + a_2 y = 0,$$

where a_0, a_1, a_2 are constants. Since any fundamental solution of Cauchy-Euler ode is of the form $y = x^\lambda (\lambda \in \mathbb{R})$, and constant functions being analytic everywhere admit of power series expansion, one can propose this Frobenius form for the series solution of the ode about regular singular point $x = 0$. (Observe that we make use of the fact that constant functions are analytic but analytic functions are not necessarily constant functions. This argument also reminds us of the strategy adopted in generating the general solution of an inhomogeneous ode from its complementary function through the method of variation of parameters).

In the following we provide an outline of the Frobenius technique near a given regular singular point of a linear second order ode.In analogy with(8.6) we recall that $x = 0$ is a regular singular point of the ode (8.4) if it can be cast in the form :

$$x^2 y'' + x p_1(x) y' + q_1(x) y = 0 \qquad (8.4(a))$$

where $p_1(x)$ and $q_1(x)$ are analytic at $x = 0$. This amounts to state that $p_1(x)$ and $q_1(x)$ have Maclaurin's series expansion in some neighborhood of $x = 0$, say, $|x| < \rho$. Suppose

$$p_1(x) = \sum_{n=0}^{\infty} a_n x^n \; ; \; q_1(x) = \sum_{n=0}^{\infty} b_n x^n, \text{ where } |x| < \rho \qquad (8.7)$$

where ρ stands for the minimum of the radii of convergence of the two power series representing $p_1(x)$ and $q_1(x)$. Hence $\exists$ a real number $M > 0$ such that

$$|a_n| \leqslant \frac{M}{\rho^n} \quad \text{and} \quad |b_n| \leqslant \frac{M}{\rho^n} \; \forall \, n \geqslant 0$$

Let's propose $y = x^\lambda \sum_{n=0}^{\infty} c_n x^n = \sum_{n=0}^{\infty} c_n x^{n+\lambda}$ as solution of the ode (8.4) $[c_0 \neq 0]$

$$\left. \begin{array}{l} \text{Therefore, } y' = \sum_{n=0}^{\infty} c_n (n + \lambda) x^{n+\lambda-1} \\[3mm] y'' = \sum_{n=0}^{\infty} (n+\lambda)(n+\lambda-1) c_n \, x^{n+\lambda-2} \end{array} \right\} \qquad (8.8)$$

Substituting (8.8) into (8.4) we have :

$$\sum_{n=0}^{\infty} (n+\lambda)(n+\lambda-1) c_n x^{n+\lambda} + \left(\sum_{n=0}^{\infty} a_n x^n \right) \left(\sum_{n=0}^{\infty} c_n (n+\lambda) x^{n+\lambda} \right)$$

$$+ \left(\sum_{n=0}^{\infty} b_n \, x^n \right) \left(\sum_{n=0}^{\infty} c_n \, x^{n+\lambda} \right) = 0$$

Eliminating x^λ and using the Cauchy product of power series we can rewrite the above as :

$$\sum_{n=0}^{\infty} \left[(n+\lambda)(n+\lambda-1) c_n + \sum_{k=0}^{n} (a_k \, c_{n-k}(n-k+\lambda) + b_k \, c_{n-k}) \right] x^n = 0$$

Hence equating the co-efficients of individual powers of x we get:

$$c_0[\lambda(\lambda - 1) + a_0\lambda + b_0] = 0 \qquad (8.9)$$

$$\text{and} \quad (n + \lambda)(n + \lambda - 1)c_n + \sum_{k=0}^{n} \{a_k(n - k + \lambda) + b_k\}c_{n-k} = 0 \; \forall \; n \geqslant 1$$

$$(8.10)$$

In second term-set of (8.9)'s, replace $(n - k)$ by m so that $n = k + m$; $k = 0 \Leftrightarrow m = n$ and $k = n \Leftrightarrow m = 0$. Hence

$$(n + \lambda)(n + \lambda - 1)c_n + \sum_{m=0}^{n} \{a_{n-m}(m + \lambda) + b_{n-m}\}c_m = 0 \quad \forall \; n \geqslant 1$$

$$\text{Thus} \quad (n + \lambda)(n + \lambda - 1)c_n + \sum_{m=0}^{n-1} \{a_{n-m}(m + \lambda) + b_{n-m}\}c_m$$
$$+ (a_0(n + \lambda) + b_0)c_n = 0 \quad \forall \; n \geqslant 1$$

i, e, $\quad \{(n + \lambda)(n + \lambda - 1) + a_0(n + \lambda) + b_0\}c_n$
$$+ \sum_{m=0}^{n-1} \{a_{n-m}(m + \lambda) + b_{n-m}\}c_m = 0 \quad \forall \; n \geqslant 1$$

Since $c_0 \neq 0$ by our assumption, we get the quadratic equation in λ, viz,

$$f(\lambda) \equiv \lambda(\lambda - 1) + a_0\,\lambda + b_0 = 0 \qquad (8.11)$$

(8.11) is known as the **indicial equation** of the given ode. Theoretically speaking, there are two roots of this quadratic equation. Without justification we claim that both the roots of (8.11) are real. However to a fastidious reader this seems to be queer. We therefore follow the ode

$$x^2 y'' + x p_1(x)y' + q_1(x)y = 0 \qquad ...(8.4(a))$$

and use the substitution $z = xy'$ so that $z' = (xy'' + y')$ Hence we get the transformed version of (8.4(a)) as :

$$xz' = (1 - p_1(x))z - q_1(x)y$$

This gives us a pair of first order ode :

$$\left.\begin{array}{l} xy' = z \\ xz' = (1 - p_1(x))z - q_1(x)y \end{array}\right\} \qquad (8.12)$$

which can be put in the matrix form :

$$x\frac{d}{dx}\begin{pmatrix} y \\ z \end{pmatrix} = \begin{pmatrix} 0 & 1 \\ -q_1(x) & 1 - p_1(x) \end{pmatrix}\begin{pmatrix} y \\ z \end{pmatrix}$$

If we agree to treat $x\frac{d}{dx}$ as some linear operator T, then the above system can be deemed as :

$$T\begin{pmatrix} y \\ z \end{pmatrix} = \mathbf{A}\begin{pmatrix} y \\ z \end{pmatrix}$$

where $A(x)$ $\equiv \begin{pmatrix} 0 & 1 \\ -q_1(x) & 1 - p_1(x) \end{pmatrix}$

$$= \begin{pmatrix} 0 & 1 \\ (-b_0 - b_1 x - b_2 x^2 - \ldots \infty) & (1 - a_0 - a_1 x - a_2 x^2 - \ldots \infty) \end{pmatrix}$$

$$= \begin{pmatrix} 0 & 1 \\ -b_0 & 1 - a_0 \end{pmatrix} - x\begin{pmatrix} 0 & 1 \\ b_1 & a_1 \end{pmatrix} - x^2\begin{pmatrix} 0 & 1 \\ b_2 & a_2 \end{pmatrix} - \ldots \infty$$

Hence $A(0) = \begin{pmatrix} 0 & 1 \\ -b_0 & 1 - a_0 \end{pmatrix}$

It is a nice observation for us that the characteristic equation of this matrix $A(0)$ is the indicial equation (8.11) obtained before for the ode given. So without going into the trouble of series form one can be more tactical to determine the indicial equation of the ode. Still we have to address the question when the roots of the indicial equation (8.11) are real. Now we think the answer is readily available to us : if we demand y to be real, then z is real and the eigen vectors of $A(0)$ comprises of y and $A(0)$ is a real matrix, the eigenvalues of $A(0)$ must be all real. However, possibility of complex roots can't be ruled out (S.L. Ross : Differential Equations, Chapter 12).

Once the roots of the indicial equation are found, we have to deal with three subcases, viz,

(a) The roots of the indicial equation are distinct but they differ from each other by a non-integral quantity.

(b) The roots of the indicial equation are equal.

(c) The roots of the indicial equation differ by an integer

Subcase (a) : Here the two roots λ_1, λ_2 of the indicial equation differ by some non-integral quantity, say p. It can be shown that

$$\left.\begin{array}{l} y_1 \equiv y(x \; ; \; \lambda_1) = x^{\lambda_1} \sum\limits_{n=0}^{\infty} c_n(\lambda_1)x^n \; ; \\[4mm] y_2 \equiv y(x \; ; \; \lambda_2) = x^{\lambda_2} \sum\limits_{n=0}^{\infty} c_n(\lambda_2)x^n \end{array}\right\} \qquad (8.13)$$

are two linearly independent solutions of the ode and moreover both these power series converge for $|x| < \rho$. Note that $c_n(\lambda_1)$'s appearing in the first Frobenius series of (8.13) stands for c_n's in the recurrence relation (8.10) when $\lambda = \lambda_1$ is chosen. Similarly $c_n(\lambda_2)$ appearing in the second Frobenius series of (8.13) stands for c_n's in the recurrence relation (8.10) when $\lambda = \lambda_2$ is chosen.

If $\lambda_1 > \lambda_2$ then in this subcase, $f(\lambda_2 + n) \neq 0$ for any $n = 1, 2, \ldots\ldots$ and hence $\exists$ no apprehension of indeterminacy of the co-efficients $c_n(\lambda_2)$ corresponding to the smaller root λ_2 of the indicial equation.

Subcases (b) and (c) : Here we shall as usual first determine the Frobenius series solution corresponding to the larger root λ_1 and then apply the method of order reduction to determine another solution $y_2(x)$ of the ode that is linearly independent w.r.to $y_1(x)$.

In the same tune as subcase (a), if we agree to write

$$y_1(x) \equiv y(x; \lambda_1) = x^{\lambda_1} \sum_{n=0}^{\infty} c_n(\lambda_1)x^n$$

and assume the second linearly independent solution of the ode in the form $y_2(x) = v(x)y_1(x)$, then

$$y_2' = v(x)y_1'(x) + v'(x)y_1(x)$$
$$y_2'' = v(x)y_1''(x) + 2v'(x)y_1'(x) + v''(x)y_1(x)$$

and so $\left.\begin{array}{l} x^2 y_1'' + x p_1(x)y_1' + q_1(x)y_1 = 0 \\[2mm] x^2 y_2'' + x p_1(x)y_2' + q_1(x)y_2 = 0 \end{array}\right\}$

This leads to the relation :

$$x(2v'y_1' + v''y_1) + p_1(x)v'y_1 = 0$$

i, e, $x(2uy_1' + u'y_1) + p_1(x)uy_1 = 0$ [where $u \equiv v'$]

i, e, $\dfrac{u'}{u} + \dfrac{p_1(x)}{x} + \dfrac{2y_1'}{y_1} = 0$

Integrating this w.to x we get :

$$ln\ u + ln\ y_1^2 + \int \frac{p_1(x)}{x} dx = \text{constant}$$

i, e, $uy_1^2\ e^{\int \frac{p_1(x)}{x} dx} = \text{constant} = A, \text{say}.$

Hence $v'(x) = u = \dfrac{A}{y_1^2}\ e^{-\int \frac{p_1(x)}{x} dx} = \dfrac{A}{x^{2\lambda_1}} \cdot \dfrac{e^{-\int \frac{1}{x}\left(\sum\limits_{n=0}^{\infty} a_n x^n\right) dx}}{\left(\sum\limits_{n=0}^{\infty} c_n(\lambda_1)x^n\right)^2}$

$\therefore \quad v'(x) = \dfrac{A}{x^{2\lambda_1}\left(\sum\limits_{n=0}^{\infty} c_n(\lambda_1)x^n\right)^2} \cdot x^{-a_0}\cdot e^{-(a_1\ x + \frac{a_2}{2!}\ x^2 + \dots \infty)}$

$\qquad = \dfrac{A}{x^{2\lambda_1 + a_0}\left(\sum\limits_{n=0}^{\infty} c_n(\lambda_1)x^n\right)^2}\ e^{-(a_1 x + \frac{a_2}{2!} x^2 + \dots \infty)}$

$\qquad \equiv \dfrac{A}{x^{2\lambda_1 + a_0}}\ g(x), \text{ say},$

where $g(x)$ is clearly analytic at $x = 0$ and $g(0) = \dfrac{1}{c_0(\lambda_1)^2} \equiv g_0$ (say).

Hence $g(x) = \sum\limits_{n=0}^{\infty} g_n x^n$, g_n's being constants.

Now if λ_1, λ_2 are roots of the indicial equation (8.11), we must have:

$$\lambda_1 + \lambda_2 = 1 - a_0; \quad \lambda_1 \cdot \lambda_2 = b_0$$

If $\lambda_1 - \lambda_2 = k$, a positive integer, then $2\lambda_1 + a_0 = k + 1$ and so

$$v'(x) = \frac{A}{x^{k+1}} \sum_{n=0}^{\infty} g_n x^n = \frac{A}{x^{k+1}}(g_0 + g_1\ x + g_2\ x^2 + \dots \infty)$$

$$= A\left(\frac{g_0}{x^{k+1}} + \frac{g_1}{x^k} + \frac{g_2}{x^{k-1}} + \dots + \frac{g_k}{x} + g_{k+1} + \dots \infty\right) \quad (8.14)$$

Using term-by-term integration we have :

$$v(x) = A\left[\frac{g_0 x^{-k}}{-k} + \frac{g_1 x^{-(k-1)}}{-(k-1)} + \frac{g_2 x^{-(k-2)}}{-(k-2)} + \dots\right.$$

$$\left. + \frac{g_{k-1} x^{-1}}{(-1)} + g_k\ ln\ x + g_{k+1}\ x + \dots \infty\right]$$

Hence $y_2(x) = y_1(x)v(x)$

$$= Ag_k(ln\ x)y_1(x) + A[\frac{g_0x^{-k}}{-k} + \frac{g_1x^{-(k-1)}}{-(k-1)} + \ldots + \frac{g_{k-1}x^{-1}}{(-1)}$$

$$+ g_{k+1}\ x + \ldots \infty]\ y_1(x)$$

$$= B(ln\ x)y_1(x) + Ax^{\lambda_2}\left(\frac{g_0}{-k} + \frac{g_1x}{-(k-1)} + \frac{g_2x^2}{-(k-2)} + \right.$$

$$\left. \ldots + \frac{g_{k-1}}{(-1)}x^{k-1} + g_{k+1}\ x^{k+1} + \ldots\right)\left(\sum_{n=0}^{\infty}c_n(\lambda_1)x^n\right)$$

$$\equiv\ B(ln\ x)y_1(x) + Ax^{\lambda_2}\left(\sum_{n=0}^{\infty}h_n\ x^n\right)\left(\sum_{n=0}^{\infty}c_n(\lambda_1)x^n\right)$$

$$\equiv\ B\,(ln\ x)\,y_1(x) + Ax^{\lambda_2}\sum_{n=0}^{\infty}d_n\ x^n \qquad\qquad (8.15)$$

where $B \equiv Ag_k$ is a constant and

$$\sum_{n=0}^{\infty}h_nx^n \equiv \left(\frac{g_0}{-k} + \frac{g_1}{-(k-1)}x + \frac{g_2}{-(k-2)}x^2 + \cdots\right)$$

(we have obtained the last step through cauchy product of power series)

For subcase (b), we get a second linearly independent solution $y_2(x)$ of the ode. If B be zero, then $y_2(x)$ turns out to be another Frobenius series.

For subcase (c), it follows from (8.14) that as $\lambda_1 = \lambda_2$, $k = 0$ and so

$$v'(x) = A\left[\frac{g_0}{x} + g_1 + g_2x + \ldots \infty\right]$$

$$\therefore v(x) = A\left[g_0 ln\ x + g_1x + \frac{g_2x^2}{2} + \ldots \infty\right] \qquad (8.16)$$

On integration w.r.to x (8.16) yields :

$$v(x) = A\left[g_0\ ln\ x + g_1\ x + g_2\frac{x^2}{2} + \ldots \infty\right]$$

$$\therefore\quad y_2(x) = v(x)y_1(x) = A\left[g_0\ ln\ x + g_1\ x + g_2\frac{x^2}{2} + \ldots \infty\right]y_1(x)$$

$$(8.17)$$

In (8.17), $lnx-$ term is definitely present in $y_2(x)$ unless $g_0 = 0$. Only if $g_0 = 0$, the lnx term drops out and the second linearly independent solution $y_2(x)$ becomes a Frobenius series again.

Well, before wrapping up our theoretical discussions let's revisit equation (8.12) as it enables us to formulate an alternative way to determine a pair of independent series solutions to the ode (8.4) when the indicial equation (8.11) has two equal roots.

Since $L[y(x\ ;\ \lambda)] = c_0\ f(\lambda)x^{\lambda}$,

$$\therefore\quad \frac{\partial L}{\partial \lambda}\ [y(x\ ;\ \lambda)] = c_0\ [f'(\lambda) + f(\lambda)ln\ x]\ x^{\lambda}$$

where prime denotes derivative w.r. to λ

It is easy to verify that $\frac{\partial L}{\partial \lambda}[y(x\ ;\ \lambda)] = L\left[\frac{\partial}{\partial \lambda}y(x\ ;\ \lambda)\right]$ and so we conclude that when the indicial equation $f(\lambda) = 0$ has a double root, say, $\lambda = \lambda_1$, then $f(\lambda_1) = 0$ and $f'(\lambda_1) = 0$ and consequently

$$L\left[\frac{\partial y}{\partial \lambda}(x\ ;\ \lambda)\right]\Bigg|_{\lambda=\lambda_1} = \frac{\partial L}{\partial \lambda}\ \left[y(x,\lambda)\right]\Bigg|_{\lambda=\lambda_1} = 0,$$

implying that $\frac{\partial y}{\partial \lambda}(x;\lambda)\Big|_{\lambda=\lambda_1}$ is also a solution of the ode $L[y] = 0$. It is a routine exercise to check that under this circumstances, $y(x\ ;\ \lambda_1)$ and $\frac{\partial y}{\partial \lambda}(x;\lambda)\Big|_{\lambda=\lambda_1}$ are the fundamental solutions of the given ode (8.4).

Summarily, this completes the outline of Frobenius method. We hope the upcoming illustrations will give us better the insight to technique.

Example (18) : Solve the ode $(1 - x)y'' - y' + xy = 0$ about $x = 1$.

In example (13) we have already shown that $x = 1$ is a regular singular point of the given ode and so we apply the transformation $z = 1 - x$ so that $z = 0$ becomes a regular singular point of the transformed ode, viz,

$$zy'' + y' + (1 - z)y = 0,$$

where now y' and y'' denote differentiations w.r.to z instead of x. We propose Frobenius series solution of this transformed ode in the form :

$$y = z^{\lambda}\sum_{n=0}^{\infty} c_n z^n = \sum_{n=0}^{\infty} c_n\ z^{n+\lambda}$$

Therefore $\quad y' = \sum_{n=0}^{\infty}(n + \lambda)c_n\ z^{n+\lambda-1}$

$$y'' = \sum_{n=0}^{\infty}(n + \lambda)(n + \lambda - 1)c_n z^{n+\lambda-2}$$

Putting these in the new ode we get :

$$\sum_{n=0}^{\infty}(n+\lambda)(n+\lambda-1)c_n z^{n+\lambda-1} + \sum_{n=0}^{\infty}(n+\lambda)c_n\ z^{n+\lambda-1}$$

$$+(1-z)\sum_{n=0}^{\infty}c_n\ z^{n+\lambda} = 0$$

i,e, $\displaystyle\sum_{n=0}^{\infty}\{(n+\lambda)(n+\lambda-1)c_n + (n+\lambda)c_n\}z^{n+\lambda-1} + \sum_{n=0}^{\infty}c_n\ z^{n+\lambda}$

$$-\sum_{n=0}^{\infty}c_n\ z^{n+\lambda+1} = 0 \qquad \ldots(a)$$

In the second summation we replace n by $\overline{n-1}$ so that

$$\sum_{n=0}^{\infty}c_n z^{n+\lambda} = \sum_{n=1}^{\infty}c_{n-1}z^{n+\lambda-1}$$

while in the last summation we replace n by $\overline{n-2}$ so that

$$\sum_{n=0}^{\infty}c_n z^{n+\lambda+1} = \sum_{n=2}^{\infty}c_{n-2}z^{n+\lambda-1}$$

Thus the relation (a) eventually becomes :

$$\sum_{n=0}^{\infty}\{(n+\lambda)(n+\lambda-1) + (n+\lambda)\}c_n z^{n+\lambda-1} + \sum_{n=1}^{\infty}c_{n-1}z^{n+\lambda-1}$$

$$-\sum_{n=2}^{\infty}c_{n-2}\ z^{n+\lambda-1} = 0$$

Observe that after this exercise all the exponents of z are tailored to be the same and only differences being the respective ranges of sums.

$$\therefore \sum_{n=2}^{\infty}\left\{(n+\lambda)^2 c_n + c_{n-1} - c_{n-2}\right\}z^{n+\lambda-1} + \lambda^2 c_0 z^{\lambda-1}$$

$$+\left\{(\lambda+1)^2 c_1 + c_0\right\}z^{\lambda} = 0$$

Hence equating co-efficients of z^λ to zero we have :

$$\left.\begin{array}{c}\lambda^2 c_0 = 0\ ;\ (\lambda+1)^2 c_1 + c_0 = 0 \\ (n+\lambda)^2 c_n + c_{n-1} - c_{n-2} = 0\ \forall\ n \geqslant 2\end{array}\right\}$$

As $c_0 \neq 0$, $\lambda = 0$ is a double root of the indicial equation. Moreover,

$$c_1 = -c_0 \; ; \; c_n = \frac{c_{n-1} + c_{n-2}}{n^2} \quad \forall \, n \geqslant 2$$

$$\therefore \; c_2 = \frac{-c_1 + c_0}{2^2} = \frac{c_0}{2!}; \quad c_3 = \frac{-c_2 + c_1}{3^2} = \frac{-c_0}{3!}; \quad c_4 = \frac{c_0}{4!} \; \text{etc.}$$

Proceeding this way we shall get one Frobenius series solution, viz,

$$y_1(x) = c_0\left(1 - z + \frac{z^2}{2!} + \ldots \infty\right) = c_0 e^{-z} = (c_0 e^{-1})e^x$$

Setting $c_0 = e$, we get one linearly independent Frobenius series solution but luckily it was expressible in a compact analytic form.

Another linearly independent solution of this ode can be sought for by using D'Alembert's order reduction technique as follows :

The general solution of this second order ode is of the form :

$$y_2(x) = v(x)e^x, \text{where } v \in C^2(\mathbb{R})$$

$$\therefore \qquad y' = e^x(v + v'); y'' = e^x(v'' + 2v' + v)$$

i.e, $\qquad (1 - x)(v'' + 2v' + v) - (v + v^1) + xv = 0$

i.e, $\qquad v'' + v' - x(v'' + 2v') = 0 \qquad \qquad \ldots \text{(b)}$

Putting $v = u$ we get from (b), $\frac{u'}{u} + \frac{1-2x}{1-x} = 0$

Integrating both sides w.r. to x and using exponential form,

$$v = \int \frac{e^{-2x}}{(1 - x)} \, dx + c$$

Hence $\quad y(x) = v(x)e^x = e^x \left(\int \frac{e^{-2x}}{(1 - x)} \, dx + c\right)$

This general solution is a combination of two linearly independent solutions. On setting $c = 0$, we retrieve the second linearly independent solution.

Example (19) : Show that $x = 0$ is a regular singular point of the ode

$$2x^2 y'' + x(1 - x)y' - y = 0$$

and hence find two linearly independent series solutions. Verify that one of these solutions is expressible in terms of elementary functions.

Clearly $x = 0$ is a singular point of the ode given. In the normal form this can be written as :

$$y'' + \frac{1-x}{2x}y' - \frac{1}{2x^2}y = 0$$

Since $\quad \lim_{x \to 0} \frac{x(1-x)}{2x} = \frac{1}{2} \quad$ and $\quad \lim_{x \to 0} \; x^2\left(-\frac{1}{2x^2}\right) = -\frac{1}{2},$

the singular point $x = 0$ is regular. We therefore can apply the method of Frobenius and propose the solution to the ode in the form:

$$y = \sum_{n=0}^{\infty} c_n \, x^{n+\lambda}$$

On term-by-term differentiation this yields :

$$y' = \sum_{n=0}^{\infty}(n+\lambda)c_n x^{n+\lambda-1} \; ; \; y'' = \sum_{n=0}^{\infty}(n+\lambda)(n+\lambda-1)c_n x^{n+\lambda-2}$$

Putting these in the given ode we get :

$$2\sum_{n=0}^{\infty}(n+\lambda)(n+\lambda-1)c_n x^{n+\lambda} + (1-x)\sum_{n=0}^{\infty}(n+\lambda)c_n x^{n+\lambda}$$

$$-\sum_{n=0}^{\infty}c_n x^{n+\lambda} = 0$$

i, e, $\quad \sum_{n=0}^{\infty}\left\{ 2(n+\lambda)(n+\lambda-1) + (n+\lambda-1)\right\} c_n x^{n+\lambda}$

$$-\sum_{n=1}^{\infty}(n+\lambda-1)c_n x^{n+\lambda} = 0$$

(Replacing n by $\overline{n-1}$ in the second sum)

$\therefore \quad \sum_{n=1}^{\infty}\left\{ (2n+2\lambda+1)(n+\lambda-1)c_n - (n+\lambda-1)c_n\right\} x^n$

$$+(\lambda-1)(2\lambda+1)c_0 = 0$$

Equating the co-efficients of each power of x to zero we have :

$$c_0(\lambda)(2\lambda+1) = 0$$

and $\quad (n+\lambda-1)\{(2n+2\lambda+1)c_n - c_{n-1}\} = 0 \quad \forall \, n \geqslant 1$

If $c_0 \neq 0$, then the indicial equation is : $(\lambda - 1)(2\lambda + 1) = 0$, which has roots $\lambda = 1$ and $\lambda = -\frac{1}{2}$. The recurrence relation will be

$$(2n + 2\lambda - 1)c_n = c_{n-1} \ \forall \ n \geqslant 1 \tag{a}$$

Observe that the roots of the indicial equation are unequal and their difference is not an integer. The problem therefore falls under the jurisdiction of subcase (a) discussed in the Frobenius theory and two linearly independent Frobenius series solutions to this ode exist. We first find Frobenius series corresponding to the larger root, viz, $\lambda = 1$. In this case the recurrence relation(a) becomes :

$$c_n = \frac{c_{n-1}}{2n + 3} \ \forall \ n \geqslant 1$$

Therefore $\qquad c_1 = \frac{c_0}{5} \ ; \quad c_2 = \frac{c_1}{7} = \frac{c_0}{5.7} \ ; \quad c_3 = \frac{c_2}{9} = \frac{c_0}{5.7.9}$

$$\therefore \qquad y_1(x) \ \equiv \ y(x \ ; \ 1) = c_0 \ x \left(1 + \frac{x}{5} + \frac{x^2}{5.7} + \frac{x^3}{5.7.9} + \ldots \infty \right)$$

For the smaller root of the indicial equation, viz, $\lambda = -\frac{1}{2}$ the recurrence relation (a) becomes :

$$c_n = \frac{c_{n-1}}{2n} \ \forall \ n \geqslant 1$$

$$\therefore \qquad c_1 = \frac{c_0}{2} \ ; \quad c_2 = \frac{c_1}{4} = \frac{c_0}{8} \ ; \quad c_3 = \frac{c_2}{6} = \frac{c_0}{48} \qquad \text{and so on.}$$

$$\therefore \qquad y_2(x) \ \equiv \ y\left(x, -\frac{1}{2} \right) = c_0 \ x^{-\frac{1}{2}} \left(1 + \frac{x}{2} + \frac{x^2}{8} + \frac{x^3}{48} + \ldots \infty \right)$$

$$= \ c_0 x^{-\frac{1}{2}} \left(1 + \frac{\left(\frac{x}{2} \right)}{1!} + \frac{\left(\frac{x}{2} \right)^2}{2!} + \frac{\left(\frac{x}{2} \right)^3}{3!} + \ldots \infty \right)$$

$$= \ c_0 \ x^{-\frac{1}{2}} \ . \ e^{\frac{x}{2}}$$

Hence the second Frobenius series solution is expressible in terms of elementary functions.

As far as this problem is concerned, we are done. But now let's do something beyond the schedule.

(i) From the recurrence relation guiding the Frobenius series solution corresponding to $\lambda = 1$, it follows that $\lim\limits_{n \to \infty} \left| \frac{c_n}{c_{n-1}} \right| |x| = \infty$ and hence the series converges for all x.

Similarly the second Frobenius series also converges for all x. This outcome is of no surprise as it follows from the nature of the co-efficients of the given ode.

(ii) For this particular ode, the indicial equation pertaining to the series solution can be readily obtained once we formulate the matrix $A(x)$ associated in the simultaneous system equivalent to this ode. [See the theoretical discussion on Frobenius method]

$$\text{As} \quad A(x) = \begin{pmatrix} 0 & 1 \\ \frac{1}{2} & \frac{1}{2}(1+x) \end{pmatrix}, \quad A(0) = \begin{pmatrix} 0 & 1 \\ \frac{1}{2} & \frac{1}{2} \end{pmatrix}$$

Consequently the characteristic equation of $A(0)$ is given by

$$0 = |A(0) - \lambda I_2| = \begin{vmatrix} -\lambda & 1 \\ \frac{1}{2} & \frac{1}{2} - \lambda \end{vmatrix} = (\lambda - 1)\left(\lambda + \frac{1}{2}\right),$$

which is nothing but the desired indical equation. This saves a lot of labor and very easily predicts the nature of the series solutions. Obviously this articulation applies to all similar odes for which series solutions are being sought.

Example (20) : Find the general solution in the neighborhood of the singular point $x = 0$ of the ode

$$4x^2 y'' - 8x^2 y' + (4x^2 + 1)y = 0$$

The singular point $x = 0$ of this ode is regular because in the normal form the ode can be written as :

$$y'' - 2y' + \left(1 + \frac{1}{4x^2}\right) y = 0$$

and moreover, $\quad \lim_{x \to 0} (-2x) = 0 \; ; \; \lim_{x \to 0} x^2 \left(1 + \frac{1}{4x^2}\right) = \frac{1}{4}$

We seek the solution of this ode in the form of a Frobenius series having the generic form :

$$y = \sum_{n=0}^{\infty} c_n x^{n+\lambda} \quad \text{with} \quad c_0 \neq 0$$

$$\therefore y' = \sum_{n=0}^{\infty} (n+\lambda) c_n x^{n+\lambda-1} \; ; \; y'' = \sum_{n=0}^{\infty} (n+\lambda)(n+\lambda-1) c_n x^{n+\lambda-2}$$

Plugging these in the ode we get :

$$4 \sum_{n=0}^{\infty} (n + \lambda)(n + \lambda - 1) c_n x^{n+\lambda} - 8 \sum_{n=0}^{\infty} (n + \lambda) c_n x^{n+\lambda+1}$$

$$+ (4x^2 + 1) \sum_{n=0}^{\infty} c_n x^{n+\lambda} = 0$$

or, $\sum_{n=0}^{\infty} \left\{ 4(n + \lambda)(n + \lambda - 1) + 1 \right\} c_n x^{n+\lambda} - 8 \sum_{n=0}^{\infty} (n + \lambda)$

$$\times c_n x^{n+\lambda+1} + 4 \sum_{n=0}^{\infty} c_n x^{n+\lambda+2} = 0 \qquad \dots (a)$$

In the second sum we replace n by $\overline{n-1}$ so that

$$\sum_{n=0}^{\infty} (n + \lambda) c_n x^{n+\lambda+1} = \sum_{n=1}^{\infty} (n + \lambda - 1) c_{n-1} x^{n+\lambda}$$

In the last sum we replace n by $\overline{n-2}$ so that

$$\sum_{n=0}^{\infty} c_n x^{n+\lambda+2} = \sum_{n=2}^{\infty} c_{n-2} x^{n+\lambda}$$

Thus (a) turns out to be :

$$\sum_{n=0}^{\infty} \{4(n + \lambda)(n + \lambda - 1) + 1)\} c_n x^{n+\lambda} - 8 \sum_{n=1}^{\infty} (n + \lambda - 1) c_{n-1} x^{n+\lambda}$$

$$+ 4 \sum_{n=2}^{\infty} c_{n-2} x^{n+\lambda} = 0$$

Observe that exponents of x are made same but in doing so, range of the sums have got different. Collating the three sums and pruning the common factor x^{λ} we get :

$$\sum_{n=2}^{\infty} [4(n + \lambda)(n + \lambda - 1) + 1) c_n - 8(n + \lambda - 1) c_{n-1} + 4c_{n-2}] x^n$$

$$+ (2\lambda - 1)^2 c_0 + \{(2\lambda + 1)^2 c_1 - 8\lambda c_0\} x = 0 \qquad \dots.(b)$$

Equating the co-efficients of individual powers of x to zero we have :

(i) $(2\lambda - 1)^2 c_0 \neq 0 \iff (2\lambda - 1)^2 = 0 \iff \lambda = \frac{1}{2}, \frac{1}{2}$

(ii) $(2\lambda + 1)^2 c_1 - 8\lambda c_0 = 0$

(iii) $\{4(n+\lambda)(n+\lambda-1) + 1\}c_n - 8(n+\lambda-1)c_{n-1} + 4c_{n-2} = 0 \ \forall \ n \geqslant 2$

Since the roots of the indicial equation are equal, only one Frobenius series solution can be determined. As $\lambda = \frac{1}{2}$, (ii) gives $c_1 = c_0$ and the general recurrence relation (iii) gives:

$$\left\{ 4 \left(n + \frac{1}{2} \right) \left(n - \frac{1}{2} \right) + 1 \right\} c_n - 8 \left(n - \frac{1}{2} \right) c_{n-1} + 4c_{n-2} = 0$$

i,e,
$$c_n = \frac{(2n - 1)c_{n-1} - c_{n-2}}{n^2} \ \forall \ n \geqslant 2 \qquad \dots (c)$$

Now
$$n = 2 \Rightarrow c_2 = \frac{3c_1 - c_0}{2^2} = \frac{c_0}{2!}$$

$$n = 3 \Rightarrow c_3 = \frac{5c_2 - c_1}{3^2} = \frac{c_0}{3!}$$

$$n = 4 \Rightarrow c_4 = \frac{7c_3 - c_2}{4^2} = \frac{c_0}{4!} \qquad \text{and so on.}$$

By method of induction it follows that $c_n = \frac{c_0}{n!} \ \forall \ n \geqslant 1$ On setting $c_0 = 1$, the Frobenius series solution of the given ode therefore reads :

$$y_1(x) \equiv y \left(x \ ; \ \frac{1}{2} \right) = x^{\frac{1}{2}} \sum_{n=0}^{\infty} \frac{x^n}{n!} = x^{\frac{1}{2}} e^x$$

For finding a second solution $y_2(x)$ that stands linearly independent w.r.to $y_1(x)$ we apply D'Alembert's method.

$$\therefore \quad y_2(x) = v(x)y_1(x), \quad \text{where} \quad v(x) \in C^2(\mathbb{R}) \text{ is given by}$$

$$v(x) = \int \frac{1}{y_1^2(x)} e^{\int 2dx} dx = \int \frac{1}{x} dx = \ln x$$

Hence $\quad y_2(x) = (\ln x) \, x^{\frac{1}{2}} \, e^x$

The general solution of the ode is therefore :

$$y(x) = c_1 \, y_1(x) + c_2 \, y_2(x)$$

[**Remark**: The order reduction method used in this problem actually yields the general solution. However, in the present case we got $y_2(x)$ as we suppressed the integrating constant while finding expression of v]

Example (21) : Solve the ode $x(1-x)y'' - (1+3x)y' - y = 0$ about all its singular points.

From the ode itself, it becomes clear that $x = 0$ and $x = 1$ are both singular points in the finite part of the xy-plane. To test the nature of the singularity, we cast the ode in normal form :

$$y'' - \frac{(1+3x)}{x(1-x)} y' - \frac{1}{x(1-x)} y = 0$$

We now compute the following limits :

$$\lim_{x\to 0} x\left(\frac{-(1+3x)}{x(1-x)}\right) = 1 \; ; \; \lim_{x\to 0} x^2\left(\frac{-1}{x(1-x)}\right) = 0$$

$$\lim_{x\to 1}(1-x)\left(\frac{1+3x}{1-x}\right) = 4 \; ; \; \lim_{x\to 1}(x-1)^2\frac{1}{x(x-1)} = 0$$

Thus $x = 0$ and $x = 1$ are both regular singular points of the ode. We now apply the transformation $x = \frac{1}{z}$ to the ode to have:

$$\frac{1}{z}\left(1 - \frac{1}{z}\right)\left(2z^3\frac{dy}{dz} + z^4\frac{d^2y}{dz^2}\right) + \left(1 + \frac{3}{z}\right)\left(z^2\frac{dy}{dz}\right) - y = 0$$

or, $\quad (z-1)z^2\frac{d^2y}{dz^2} + z(3z+1)\frac{dy}{dz} - y = 0 \quad\quad\quad \dots(a)$

$$\left[\text{since } z = \frac{1}{x}, \; \frac{dy}{dx} = -\frac{1}{x^2}\frac{dy}{dz} \; ; \; \frac{d^2y}{dx^2} = 2z^3\frac{dy}{dz} + z^4\frac{d^2y}{dz^2}\right]$$

In the normal form (a) can be put as :

$$\frac{d^2y}{dz^2} + \frac{3z+1}{z(z-1)}\frac{dy}{dz} - \frac{1}{z^2(z-1)}y = 0,$$

Hence $\quad \lim_{z\to 0}\frac{z(3z+1)}{z(z-1)} = -1 \; ; \; \lim_{z\to 0} z^2\left(-\frac{1}{z^2(z-1)}\right) = 1$

This indicates that $z = 0$ is a regular singular point of the transformed ode and so 'point at infinity' is a regular singular point of the given ode. We now aim to find the series solutions about these singular points. Incidentally in all cases, Frobenius method comes into play.

We first find the series solution about $x = 0$ and propose the solution of the ode in the form :

$$y = x^\lambda \sum_{n=0}^{\infty} c_n x^n \quad \text{with} \quad c_0 \neq 0.$$

Since $y = \sum_{n=0}^{\infty} c_n x^{n+\lambda}$, on term-by-term differentiation we get :

$$y' = \sum_{n=0}^{\infty} (n + \lambda) c_n x^{n+\lambda-1} \; ; \; y'' = \sum_{n=0}^{\infty} (n + \lambda)(n + \lambda - 1) c_n \, x^{n+\lambda-2}$$

Plugging back into the ode we get :

$$x(1 - x) \sum_{n=0}^{\infty} (n + \lambda)(n + \lambda - 1) c_n x^{n+\lambda-2}$$

$$-(1 + 3x) \sum_{n=0}^{\infty} (n + \lambda) c_n \, x^{n+\lambda-1} - \sum_{n=0}^{\infty} c_n \, x^{n+\lambda} = 0$$

or, $$\sum_{n=0}^{\infty} (n + \lambda)(n + \lambda - 1) c_n \, x^{n+\lambda-1} - \sum_{n=0}^{\infty} (n + \lambda)(n + \lambda - 1) c_n \, x^{n+\lambda}$$

$$- \sum_{n=0}^{\infty} (n + \lambda) c_n \, x^{n+\lambda-1} - 3 \sum_{n=0}^{\infty} (n + \lambda) c_n \, x^{n+\lambda} - \sum_{n=0}^{\infty} c_n \, x^{n+\lambda} = 0$$

or, $$\sum_{n=0}^{\infty} (n + \lambda) (n + \lambda - 2) c_n \, x^{n+\lambda-1} - \sum_{n=0}^{\infty} (n + \lambda + 1)^2 c_n \, x^{n+\lambda} = 0$$

We now replace n by $\overline{n-1}$ in the second sum to get

$$\sum_{n=0}^{\infty} (n + \lambda)(n + \lambda - 2) c_n \, x^{n+\lambda-1} - \sum_{n=1}^{\infty} (n + \lambda)^2 c_{n-1} \, x^{n+\lambda-1} = 0$$

The exponents of x are now balanced but ranges of sums differ.

$$\therefore \quad \lambda(\lambda - 2) c_0 + \sum_{n=1}^{\infty} \{(n + \lambda)(n + \lambda - 2) c_n - (n + \lambda)^2 c_{n-1}\} x^n = 0$$

Proceeding before we get the indicial equation $\lambda(\lambda - 2) = 0$ together with the general recurrence relation

$$(n+\lambda)(n+\lambda-2)c_n - (n+\lambda)^2 c_{n-1} = 0 \; \forall \; n \geqslant 1 \qquad \ldots (a)$$

The indicial equation has two roots with the difference being an integer. We first find the Frobenius series corresponding to the larger root $\lambda = 2$. In this case the recurrence relation comes down to :

$$(n + 2)n \, c_n - (n + 2)^2 c_{n-1} = 0 \; \forall \; n \geqslant 1$$

i, e, $$c_n = \left(\frac{n + 2}{n} \right) c_{n-1} \; \forall \; n \geqslant 1$$

Putting
$$n = 1, \quad c_1 = 3c_0$$
$$n = 2, \quad c_2 = 2\, c_1 = 2.3\ c_0$$
$$n = 3, \quad c_3 = \frac{5}{3}\, c_2 = 10\ c_0$$
$$n = 4, \quad c_4 = \frac{6}{4}\, c_3 == 15\ c_0$$
$$n = 5, c_5 = \frac{7}{5}\, c_4 = 21\ c_0 \quad \text{and so on.}$$

This gives us a complete form of the Frobenius series :

$$
\begin{aligned}
y_1(x) = y(x\ ;2) &= (c_0 + c_1\, x + c_2\, x^2 + \ldots \infty)x^2 \\
&= c_0[1 + 3x + 6x^2 + 10x^3 + \ldots \infty]x^2
\end{aligned}
$$

Setting $c_0 = 1$, $y_1(x) = x^2(1-x)^{-3}$ (using binomial theorem)

The second linearly independent solution to this ode can be obtained by D'Alembert's order reduction method.

$\therefore \quad y_2(x) = v(x)y_1(x)$, where $v(x)$ is given by the relation

$$
\begin{aligned}
v(x) &= \int \frac{1}{y_1^2}\, e^{\int \frac{(1+3x)}{x(1-x)}dx}\, dx \\
&= \int \frac{1}{x^4(1-x)^{-6}}\, e^{(\int \frac{dx}{x} + 4\int \frac{dx}{1-x})}dx \\
&= \int \frac{(1-x)^6}{x^4}\, e^{(\ln x - 4\ln(1-x))}dx \\
&= \int \frac{(1-x)^6}{x^4}\, x(1-x)^{-4}dx \\
&= \left(\ln x + \frac{2}{x} - \frac{1}{2x^2}\right)
\end{aligned}
$$

Hence the expression for $y_2(x)$ is :

$$y_2(x) = \left(\ln x + \frac{2}{x} - \frac{1}{2x^2}\right)y_1(x) = x^2(1-x)^{-3}\left(\ln x + \frac{2}{x} - \frac{1}{2x^2}\right)$$

We now look for series solution about $x = 1$. We first apply linear transformation $z = (1-x)$ so that $\frac{dy}{dx} = -\frac{dy}{dz}$ and $\frac{d^2y}{dx^2} = \frac{d^2y}{dz^2}$.

Using the same notation y' for $\frac{dy}{dz}$ and y'' for $\frac{d^2y}{dz^2}$, the ode reads :

$$z(1-z)\, y'' + (4-3z)\, y' - y = 0 \tag{b}$$

As a routine exercise we propose the series solution in the form

$$y = \sum_{n=0}^{\infty} c_n z^{n+\lambda} \quad \text{so that}$$

$$y' = \sum_{n=0}^{\infty}(n+\lambda)c_n z^{n+\lambda-1} \; ; \; y'' = \sum_{n=0}^{\infty}(n+\lambda)(n+\lambda-1)c_n z^{n+\lambda-2}$$

On putting these into (b) we have :

$$z(1-z)\sum_{n=0}^{\infty}(n+\lambda)(n+\lambda-1)c_n z^{n+\lambda-2}$$

$$+ (4-3z)\sum_{n=0}^{\infty}(n+\lambda)\,c_n\,z^{n+\lambda-1} - \sum_{n=0}^{\infty}c_n z^n = 0$$

$$\text{i, e } = \sum_{n=0}^{\infty}(n+\lambda)(n+\lambda-1)c_n z^{n+\lambda-1} - \sum_{n=0}^{\infty}(n+\lambda)(n+\lambda-1)c_n z^{n+\lambda}$$

$$+4\sum_{n=0}^{\infty}(n+\lambda)x_n \, z^{n+\lambda-1} - 3\sum_{n=0}^{\infty}(n+\lambda)c_n z^{n+\lambda} - \sum_{n=0}^{\infty}c_n z^{n+\lambda} = 0$$

$$\therefore \quad \sum_{n=0}^{\infty}(n+\lambda)\,(n+\lambda+3)\,c_n z^{n+\lambda-1} - \sum_{n=0}^{\infty}\Big\{(n+\lambda)\,(n+\lambda-1)$$

$$+ 3(n+\lambda)+1\Big\}c_n\,z^{n+\lambda} = 0$$

$$\text{i, e } \sum_{n=0}^{\infty}(n+\lambda)(n+\lambda+3)c_n\,z^{n+\lambda-1} - \sum_{n=0}^{\infty}(n+\lambda+1)^2 c_n\,z^{n+\lambda} = 0$$

$$\sum_{n=1}^{\infty}(n+\lambda)\{(n+\lambda+3)c_n - (n+\lambda)c_{n-1}\}z^{n+\lambda-1} + \lambda(\lambda+3)c_0 z^{\lambda-1} = 0$$

(replacing n by $\overline{n-1}$ in the second sum)

Dropping $z^{\lambda-1}$ from both sides and equating co-efficients of different powers of z to zero we get :

$$\lambda(\lambda+3)c_0 = 0 \quad \text{so that} \quad \lambda = 0 \quad \text{or} \quad \lambda = -3 \; (\text{ as } c_0 \neq 0)$$

and the general recurrence relation :

$$(n+\lambda+3)c_n = (n+\lambda)c_{n-1} \quad \forall \, n \geqslant 1$$

For $\lambda = 0$, the above relation reduces to $(n+3)c_n = n\,c_{n-1}$

Now herefrom follows that

$$c_1 = \frac{1}{4} c_0 \ ; \ c_2 = \frac{1.2}{4.5} c_0 \ ; \ c_3 = \frac{1.2.3}{4.5.6} c_0 \quad \text{and so on.}$$

Hence the desired Frobenius series solution reads :

$$y_1(z) = c_0 \left[1 + \frac{1}{4}z + \frac{1.2}{4.5}z^2 + \frac{1.2.3}{4.5.6}z^3 + \ldots \infty \right] \ ; \ c_0 \neq 0.$$

One can find the second linearly independent solution to the ode by order reduction method. We shall alternatively use the Frobenius technique to find the same, but this time trying with the smaller root $\lambda = -3$ of the indicial equation. This makes the recurrence relation :

$$(n-3)\{nc_n - (n-3)c_{n-1}\} = 0 \quad \forall \, n \geqslant 1$$

i, e, $\quad n \, c_n = (n-3)c_{n-1}, \quad$ where n is any natural number except 3.

For $\quad n = 1$ and $n = 2$, we have $c_1 = -2c_0$ and $c_2 = c_0$

$$n = 4 \Rightarrow c_4 = \frac{1}{4} \, c_3;$$

$$n = 5 \Rightarrow c_5 = \frac{2}{5} \, c_4 = \frac{1}{4} \cdot \frac{2}{5} \, c_3;$$

$$n = 6 \Rightarrow c_6 = \frac{1}{4} \cdot \frac{2}{5} \cdot \frac{3}{6} \, c_3 \quad \text{and so on.}$$

Hence $\quad y(z) = z^{-3} \left[c_0(1 - 2z + z^2) + c_3 \left(1 + \frac{1}{4}z + \frac{1.2}{4.5}z^2 + \ldots \infty \right) x \right]$

$$= c_0 z^{-3}(1-z)^2 + c_3 \left(1 + \frac{1}{4}z + \frac{1.2}{4.5}z^2 + \ldots \infty \right)$$

So incidentally we get the general solution of the ode !

Replacing z by $(1-x)$ in $y(z)$ and putting $c_0 = 1$; $c_3 = 0$ or $c_0 = 0$; $c_3 = 1$ the two linearly independent solutions are found about the regular singular point $x = 1$.

We now find the series solution corresponding to the regular singularity 'at infinity'. Previously we obtained the transformed ode:

$$\frac{d^2y}{dz^2} + \frac{3z+1}{z(z-1)} \frac{dy}{dz} - \frac{1}{z^2(z-1)} \, y = 0 \ (z = 0 \text{ corrresponds to } x = \infty)$$

We again try the solution in the form : $y(z) = \sum_{n=0}^{\infty} c_n \, z^{n+\lambda}$

Hence $\quad y'(z) \equiv \dfrac{dy}{dz} = \sum_{n=0}^{\infty} (n+\lambda) \, c_n \, z^{n+\lambda-1}$

$$y''(z) \equiv \dfrac{d^2 y}{dz^2} = \sum_{n=0}^{\infty} (n+\lambda)(n+\lambda-1) \, c_n \, z^{n+\lambda-2}$$

Putting these in ode we get :

$$\sum_{n=0}^{\infty} (z-1)(n+\lambda)(n+\lambda-1)c_n z^{n+\lambda} + \sum_{n=0}^{\infty} (3z+1)(n+\lambda)c_n \, z^{n+\lambda}$$

$$- \sum_{n=0}^{\infty} c_n z^{n+\lambda} = 0$$

or, $\sum_{n=0}^{\infty} \{(n+\lambda)(n+\lambda-1) + 3(n+\lambda)\}c_n z^{n+\lambda+1}$

$$+ \sum_{n=0}^{\infty} \{-(n+\lambda)(n+\lambda-1) - 1 + (n+\lambda)\}c_n \, z^{n+\lambda} = 0$$

or, $\sum_{n=0}^{\infty} (n+\lambda)(n+\lambda+z)c_n z^{n+\lambda+1} - \sum_{n=0}^{\infty} (n+\lambda-1)^2 c_n z^{n+\lambda} = 0$

Replacing n by $\overline{n-1}$ in the first sum we get :

$$\sum_{n=1}^{\infty} (n+\lambda-1)(n+\lambda+1)c_{n-1} \, z^{n+\lambda} - \sum_{n=0}^{\infty} (n+\lambda-1)^2 c_n \, z^{n+\lambda} = 0$$

i, e, $(\lambda-1)^2 c_0 z^\lambda + \sum_{n=1}^{\infty} (n+\lambda-1)\{(n+\lambda+1)c_{n-1}$

$$-(n+\lambda-1)c_n\}z^{n+\lambda} = 0$$

Eliminating z^λ from both sides we get :

$$\left.\begin{array}{l} (\lambda-1)^2 c_0 = 0 \; ; \; c_0 \neq 0 \\ (n+\lambda+1)c_{n-1} = (n+\lambda-1)c_n \; \forall \, n > 1 \end{array}\right\}$$

Thus $\lambda = 1$ is a double root of the indicial equation. So we have only one Frobenius series solution here.

$$c_n = \dfrac{(n+\lambda+1)}{(n+\lambda-1)} \, c_{n-1} \forall \, n > 1$$

For $\lambda = 1$ this g.r.r. becomes : $c_n = \left(\dfrac{n+2}{n}\right) c_{n-1} \; \forall \; n \geqslant 1$

For $n = 1$, $c_1 = 3c_0$

For $n = 2$, $c_2 = 6c_0$

For $n = 3$, $c_3 = 10c_0$ and so on.

The Frobenius series becomes : $y_1(z) = \left(\displaystyle\sum_{n=0}^{\infty} c_n \, z^n\right) z^{-1}$

$$
\begin{aligned}
y_1(z) &= z^{-1}[c_0 + c_1 \, z + c_2 \, z^2 + c_3 \, z^3 + \ldots \infty] \\
&= c_0 \, z^{-1}[1 + 3z + 6z^2 + 10z^3 + \ldots \infty] = c_0 \, z^{-1}(1-z)^{-3}
\end{aligned}
$$

The second linearly independent solution is $y_2(z) = y_1(z)v(z)$, where $v(z)$ is given by :

$$
\begin{aligned}
v(z) &= \int \frac{1}{y_1^2(z)} \exp \left[-\int \frac{3z+1}{z(z-1)} dz\right] dz \\
&= \int c_0^{-2} \cdot z^2 (1-z)^6 \exp \left[-\int \frac{4dz}{z-1} + \int \frac{dz}{z}\right] dz \\
&= c_0^{-2} \int z^2 (1-z)^6 \left\{\frac{1}{(1-z)^4} + \frac{1}{z}\right\} dz \\
&= c_0^{-2} \left\{\int z^2 (1-z)^2 dz + \int z(1-z)^6 dz\right\} \\
&= -c_0^{-2} \left[\frac{1}{3}(1-z)^3 - \frac{1}{2}(1-z)^4 + \frac{1}{5}(1-z)^5 - \frac{1}{7}(1-z)^7 + \frac{1}{8}(1-z)^8\right] \\
\therefore \; y_2(z) &= -\frac{1}{zc_0} \left[\frac{1}{3} - \frac{1}{2}(1-z) + \frac{1}{5}(1-z)^2 - \frac{1}{7}(1-z)^4 + \frac{1}{8}(1-z)^5\right]
\end{aligned}
$$

This completes our workout.

Remark:

(a) In the case of the regular singular point $x = 0$, if we try to find the Frobenius series solution corresponding to the smaller root $\lambda = 0$ of the indicial equation then we experience the following :

Recurrence relation reads : $(n-2)c_n = n \, c_{n-1} \; \forall \; n \geqslant 1$

Here $n = 1$ gives $c_1 = -c_0$

$n = 2$ gives $0.c_2 = 2c_1$ so that $c_1 = c_0 = 0$,

which may be regarded as a contradiction to the assumption $c_0 \neq 0$. So this trial fails to deliver.

(b) In case of the regular singular point $x = 1$, we got two linearly independent solutions by Frobenius technique and infact discovered that the smaller root of the indicial equation could generate the general solution of the ode in one stroke. The following problem echoes the samething. This approach of using the smaller root of the indicial equation to find the general solution is known as **Pochhamer's technique.**

Example (22) : Solve the ode $x(1 - x)\frac{d^2y}{dx^2} - 3\frac{dy}{dx} + 2y = 0$ about the regular singular point $x = 1$.

In conformity with the approach adopted in the previous problem, we apply the transformation $x = (1 + t)$ so that the transformed ode

$$t(t+1)\frac{d^2y}{dt^2} + 3\frac{dy}{dt} - 2y = 0 \qquad \qquad \ldots (a)$$

has a regular singular point at $t = 0$. We seek a Frobenius series solution to this new ode in the form

$$y = \sum_{k=0}^{\infty} c_k \, t^{k+\lambda}$$

$$\therefore \ y' = \sum_{k=0}^{\infty}(k + \lambda)c_k \, t^{k+\lambda-1} \ ; \quad y'' = \sum_{k=0}^{\infty}(k + \lambda)(k + \lambda - 1)c_k \, t^{k+\lambda-2}$$

On plugging these in the given ode we get :

$$(t + 1)\sum_{k=0}^{\infty}(k + \lambda)(k + \lambda - 1)c_k t^{k+\lambda-1} + 3\sum_{k=0}^{\infty}(k + \lambda)c_k \, t^{k+\lambda-1}$$

$$-2\sum_{k=0}^{\infty} c_k \, t^{k+\lambda} = 0$$

$$\text{i, e,} \ \sum_{k=0}^{\infty}\{(k + \lambda)(k + \lambda - 1) + 3(k + \lambda)\}c_k \, t^{k+\lambda-1}$$

$$+ \sum_{k=0}^{\infty}\{(k + \lambda)(k + \lambda - 1) - 2\}c_k \, t^{k+\lambda} = 0$$

$$\text{i, e,} \ \sum_{k=1}^{\infty}[\{(k + \lambda)(k + \lambda - 1) + 3(k + \lambda)\}c_k$$

$$+ \{(k + \lambda - 1)(k + \lambda - 2) - 2\}c_{k-1}]t^{k+\lambda-1} + \{\lambda(\lambda - 1) + 3\lambda\}c_0 \, t^{\lambda-1} = 0,$$

where in the last sum we substituted k by $(k-1)$ and regrouped. Dropping the factor $t^{\lambda-1}$ from both sides and equating each of these co-efficients to zero we get :

$$\lambda(\lambda+2)c_0 = 0, \quad \text{i, e,} \quad \lambda(\lambda+2) = 0 \quad (\because c_0 \neq 0)$$

together with the general recurrence relation :

$$(k+\lambda)\{(k+\lambda+2)c_k + (k+\lambda-3)c_{k-1}\} \; \forall \; k \geqslant 1 \qquad \ldots \text{(b)}$$

The indicial equation has two roots, viz, 0 and -2. For the smaller root of the indicial equation, recurrence relation (b) reduces to :

$$(k-2)\{kc_k + (k-5)c_{k-1}\} \; \forall \; k \geqslant 1$$

i, e, $\qquad kc_k = (5-k)c_{k-1} \quad \text{for} \quad k \in \mathbb{N} - \{2\}$

Clearly for $\quad k = 1, \quad c_1 = 4c_0$

$\qquad$ for $\quad k = 3, \quad 3c_3 = 2c_2 \Leftrightarrow c_3 = \dfrac{2}{3}c_2$

$\qquad$ for $\quad k = 4, \quad 4c_4 = c_3 = \dfrac{2}{3}c_2 \Leftrightarrow c_4 = \dfrac{1}{6}c_2$

$\qquad$ for $\quad k = 5, \quad 5c_5 = 0 \Leftrightarrow c_5 = 0$

Consequently all subsequent co-efficients are zeros. Hence the solution boils down to

$$\begin{aligned} y(t) &= t^{-2}\left[c_0(1+4t) + c_2\left(1 + \frac{2}{3}t + \frac{1}{6}t^2\right)\right] \\ &= c_0\left(\frac{1}{t^2} + \frac{4}{t}\right) + c_2\left(1 + \frac{2}{3}t + \frac{1}{6}t^2\right), \end{aligned}$$

where c_0 and c_2 are arbitrary.

It is easy to verify that $y_1(t) = \left(\frac{1}{t^2} + \frac{4}{t}\right)$ and $y_2(t) = \left(1 + \frac{2}{3}t + \frac{1}{6}t^2\right)$ both satisfy the new ode (a) and moreover they are linearly independent because their Wronskian is non-zero. Thus the smaller root of the indicial equation gives the general solution of (a) in one stroke.

Remark : If we work with the larger root $\lambda = 0$, then the recurrence relation (b) boils down to

$$(k+2)c_k + (k-3)c_{k-1} = 0 \quad \forall \; k \geqslant 1$$

i, e, $\qquad c_k = \dfrac{(3-k)}{(k+2)}c_{k-1} \quad \forall \; k \geqslant 1$

For $\quad k = 1, \; c_1 = \dfrac{2}{3}c_0 \; ; \; k = 2 \Rightarrow c_2 = \dfrac{1}{4}c_1 = \dfrac{1}{6}c_0$

For $k = 3$, $c_3 = 0$ and all subsequent co-efficients are zeros.

Thus the Frobenius series solution reduces to the polynomial :

$$y_1(t) = \left(1 + \frac{2}{3}t + \frac{1}{6}t^2\right) \quad \text{(setting} \quad c_0 = 1)$$

Theoretically one Frobenius series solution is available. For finding the second linearly independent solution we may use the ansatz given in (8.15). However this seems to be rather clumsy for the present problem. In the next problem we make use of this ansatz quoted in (8.15) as we are compelled to since the indicial equation will have a double root.

Example (23) : Solve $x^2 y'' + x(x - 1)y' + (1 - x)y = 0$ about $x = 0$.

It is a routine exercise to verify that $x = 0$ is a regular singular point of the ode. We propose the Frobenius series solution in the form

$$y = \sum_{n=0}^{\infty} c_n \, x^{n+\lambda}.$$

Therefore, $y' = \sum_{n=0}^{\infty}(n+\lambda)c_n \, x^{n+\lambda-1}$; $y'' = \sum_{n=0}^{\infty}(n+\lambda)(n+\lambda-1)c_n \, x^{n+\lambda-2}$

On using these in the given ode we get :

$$\sum_{n=0}^{\infty}(n + \lambda)(n + \lambda - 1)c_n x^{n+\lambda} + \sum_{n=0}^{\infty}(n + \lambda)c_n(x - 1)x^{n+\lambda}$$

$$+(1 - x)\sum_{n=0}^{\infty} c_n \, x^{n+\lambda} = 0$$

i,e, $\sum_{n=0}^{\infty}(n + \lambda)(n + \lambda - 1)c_n \, x^{n+\lambda} - \sum_{n=0}^{\infty}(n + \lambda - 1)c_n \, x^{n+\lambda}$

$$+ \sum_{n=0}^{\infty}(n + \lambda - 1)x^{n+\lambda+1} = 0$$

Replacing n by $\overline{n - 1}$ in the last sum we get from above:

$$\sum_{n=0}^{\infty}(n + \lambda - 1)^2 c_n \, x^{n+\lambda} + \sum_{n=1}^{\infty}(n + \lambda - 2)c_{n-1} \, x^{n+\lambda} = 0 \qquad (*)$$

or, $\sum_{n=1}^{\infty}\{(n + \lambda - 1)^2 c_n + (n + \lambda - 2)c_{n-1}\}x^{n+\lambda} + (\lambda - 1)^2 c_0 \, x^{\lambda} = 0$

Dropping x^λ from both sides and proceeding as usual we conclude :

$$(\lambda - 1)^2 c_0 = 0 \quad \Leftrightarrow \quad \lambda = 1,1 \quad (\text{since} \quad c_0 \neq 0)$$

The general recurrence relation corresponding to $\lambda = 1$ will be :

$$n^2 c_n + (n-1)c_{n-1} = 0 \; \forall \, n \geqslant 1$$

$$\text{i, e,} \quad c_n = \frac{-(n-1)}{n^2} c_{n-1} \quad \forall \, n \geqslant 1$$

Clearly it follows from the recurrence relation that c_1 and all the following co-efficients are zeros.

Hence $y_1(x) = c_0 x$ is one solution. Setting $c_0 = 1$, we get one linearly independent solution $y_1(x) = x$.

Following the ansatz given in (8.15) we propose the second linearly independent solution of the ode in the form :

$$y_2(x) = A \sum_{n=0}^{\infty} d_n x^{n+1} + B y_1(x) \, ln \, x,$$

where A and B are arbitrary constants, that can, without loss of generality, be set to unity.

$$\therefore \; y_2(x) = \sum_{n=0}^{\infty} d_n x^{n+1} + x ln \, x$$

Through term-by-term differentiation we have :

$$y_2'(x) = \sum_{n=0}^{\infty} (n+1)d_n \, x^n + 1 + ln \, x$$

$$y_2''(x) = \sum_{n=0}^{\infty} (n+1)n d_n \, x^{n-1} + \frac{1}{x}$$

Putting these in the given ode we get :

$$\sum_{n=0}^{\infty} (n+1)n d_n x^{n+1} + x(x-1)\left\{ \sum_{n=0}^{\infty}(n+1)d_n x^n + (1+ln\,x) \right\}$$

$$+(1-x)\left\{ \sum_{n=0}^{\infty} d_n \, x^{n+1} + x \, ln \, x \right\} + x = 0$$

or, $\displaystyle\sum_{n=1}^{\infty}(n+1)nd_n\ x^{n+1} + x(x-1)\sum_{n=1}^{\infty}nd_nx^n + x^2 = 0$

or, $\displaystyle\sum_{n=1}^{\infty}n^2d_n\ x^{n+1} + \sum_{n=1}^{\infty}nd_n\ x^{n+2} = 0$

i, e, $\displaystyle\sum_{n=1}^{\infty}n^2d_nx^{n+1} + \sum_{n=2}^{\infty}(n-1)d_{n-1}x^{n+1} + x^2 = 0$

(where we have replaced n by $\overline{n-1}$ in the second sum)

i, e, $(d_1+1)x^2 + \displaystyle\sum_{n=2}^{\infty}\left\{(n-1)d_{n-1} + n^2d_n\right\}x^{n+1} = 0$ $\dots$ (a)

Equating co-efficients of each integral power of x to zero we get :

$$\left.\begin{array}{l} d_1 + 1 = 0 \ \Leftrightarrow\ d_1 = -1 \\ (n-1)d_{n-1} + n^2d_n = 0 \quad \forall\ n \geqslant 2 \end{array}\right\}$$

From the above recurrence relation it follows that

$$d_2 = -\frac{1}{2^2}d_1 = \frac{1}{2^2}$$

$$d_3 = -\frac{2}{3^2}d_2 = -\frac{1}{2.3^2}$$

and so on.

$$\begin{aligned} \therefore\ y_2(x) &= x\ ln\ x + \sum_{n=0}^{\infty}d_n\ x^{n+1} \\ &= \left(x\ ln\ x - x^2 + \frac{1}{1.2^2}x^3 - \frac{1}{2.3^2}x^4 - \dots \infty\right) + d_0x \end{aligned}$$

where d_0 is an arbitrary constant. Setting $d_0 = 0$, we get from $y_2(x)$ the linearly independent second solution to the ode.

From $(*)$ we get the generalised recurrence relation involving λ :

$$(n+\lambda-1)^2c_n + (n+\lambda-2)c_{n-1} = 0 \quad \forall\ n \geqslant 1$$

so that $c_n = \dfrac{-(n+\lambda-2)c_{n-1}}{(n+\lambda-1)^2} \quad \forall\ n \geqslant 1$

Thus $c_1 = \dfrac{-(\lambda-1)}{\lambda^2}\ c_0\ ;$

$$c_2 = \frac{(\lambda - 1)\lambda}{\lambda^2(\lambda + 1)^2} \, c_0$$

$$c_3 = \frac{-(\lambda - 1)\lambda(\lambda + 1)}{\lambda^2(\lambda + 1)^2(\lambda + 2)^2} \, c_0 \quad \text{and so on.}$$

Therefore,

$$y(x \; ; \; \lambda) = x^\lambda \sum_{n=0}^{\infty} c_n x^n$$

$$= c_0 x^\lambda \left[1 - \frac{(\lambda - 1)}{\lambda^2} x + \frac{(\lambda - 1)\lambda}{\lambda^2(\lambda + 1)^2} x^2 - \frac{(\lambda - 1)\lambda(\lambda + 1)}{\lambda^2(\lambda + 1)^2(\lambda + 2)^2} x^3 + \cdots \infty \right]$$

$$= c_0 x^\lambda \left[1 + \sum_{n=1}^{\infty} \frac{(-1)^n(\lambda - 1)\lambda \cdots (\lambda + \overline{n - 2})}{\lambda^2(\lambda + 1)^2 \cdots (\lambda + \overline{n - 1})^2} x^n \right]$$

$$\therefore \quad \frac{\partial y(x \; ; \; \lambda)}{\partial \lambda} = c_0 x^\lambda \left[x^n \left\{ \sum_{n=1}^{\infty} (-1)^n(\lambda - 1)\frac{d}{d\lambda}(\lambda(\lambda + 1) \cdots (\lambda + n - 2) \right. \right.$$

$$(\lambda + \overline{n - 1})^2) + \sum_{n=1}^{\infty} (-1)^n \frac{1}{\lambda(\lambda + 1) \cdots (\lambda + n - 2)(\lambda + n - 1)^2}$$

$$\left. \left. + \ln x \left(1 + \sum_{n=1}^{\infty} \frac{(-1)^n(\lambda - 1)}{\lambda(\lambda + 1) \cdots (\lambda + n - 2)(\lambda + n - 1)^2} \right) \right\} \right]$$

so that

$$\frac{\partial y(x \; ; \; \lambda)}{\partial \lambda} \bigg|_{\lambda=1} = c_0 \, x \left[x^n \left\{ \sum_{n=1}^{\infty} \frac{(-1)^n}{(n - 1)! \, n^2} \right\} + \ln x \right]$$

$$= c_0 \left[x \, \ln x + \sum_{n=1}^{\infty} \frac{(-1)^n x^{n+1}}{(n - 1)! \, n^2} \right]$$

$$= c_0 \left[x \, \ln x - x^2 + \frac{x^3}{1.2^2} - \frac{x^4}{2.3^2} + \cdots \infty \right]$$

As told before $\dfrac{\partial y(x \; ; \; \lambda)}{\partial \lambda}\bigg|_{\lambda=1}$ serves as the second linearly independent solution as told earlier. This gives us series alternative means to construct two linearly independent series solutions of the ode when exponents are equal.

Remark : This second linearly independent solution can be also found by order reduction method in the form $y_2(x) = v(x)y_1(x)$ where

$$v(x) = \int \frac{1}{y_1^2(x)} \exp \left[-\int \left(\frac{x - 1}{x} \right) dx \right] dx$$

$$= \int \frac{1}{x^2} \exp\,[-x + \ln\,x]\,dx = \int \frac{1}{x^2} \cdot x e^{-x} dx$$

$$= \int \frac{1}{x} \left(1 - x + \frac{x^2}{2!} - \frac{x^3}{3!} - \ldots \infty\right) dx$$

$$= \left(\ln\,x - x + \frac{x^2}{4} - \frac{x^3}{18} + \ldots \infty\right)$$

$$\text{Therefore}\quad y_2(x) = x\left[\ln\,x - x + \frac{x^2}{4} - \frac{x^3}{18} + \ldots \infty\right]$$

$$= x\,\ln\,x - x^2 + \frac{x^3}{1.2^2} - \frac{x^4}{2.3^2} + \ldots \infty$$

(Incidentally this is same as that obtained by using (8.15). We have intentionally suppressed constants of integration here)

Example (24) : Solve Bessel equation of order p, viz, $x^2 y'' + xy' + (x^2 - p^2)y = 0$ about the regular singular point $x = 0$, the parameter p being positive or zero.

Regarding this Bessel equation of order p we first make an interesting observation that the ode is invariant if p is replaced by $-p$. We propose the solution to this ode in the form :

$$y = \sum_{n=0}^{\infty} c_n x^{n+\lambda} \;;\; c_0 \neq 0.$$

Through term-by-term differentiation we have :

$$y' = \sum_{n=0}^{\infty} (n+\lambda)c_n\,x^{n+\lambda-1} \;;\; y'' = \sum_{n=0}^{\infty} (n+\lambda)(n+\lambda-1)c_n\,x^{n+\lambda-2}$$

On plugging these in Bessel equation we get :

$$\sum_{n=0}^{\infty} (n+\lambda)(n+\lambda-1)c_n x^{n+\lambda} + \sum_{n=0}^{\infty} (n+\lambda)c_n x^{n+\lambda}$$

$$+ (x^2 - p^2) \sum_{n=0}^{\infty} c_n x^{n+\lambda} = 0$$

$$\text{or,}\quad \sum_{n=0}^{\infty} \{(n+\lambda)^2 - p^2\}c_n\,x^{n+\lambda} + \sum_{n=0}^{\infty} c_n\,x^{n+\lambda+2} = 0$$

$$\text{or,}\quad \sum_{n=0}^{\infty} \{(n+\lambda)^2 - p^2\}c_n\,x^{n+\lambda} + \sum_{n=2}^{\infty} c_{n-2}\,x^{n+\lambda} = 0$$

where in the last sum we have replaced n by $\overline{n-2}$ to bring the exponents of x at par for both the sums.

$$\therefore \quad (\lambda^2 - p^2)c_0\ x^\lambda + ((\lambda+1)^2 - p^2)c_1 x^{\lambda+1}$$

$$+ \sum_{n=2}^{\infty}[\{(n+\lambda)^2 - p^2\}c_n + c_{n-2}]x^{n+\lambda} = 0$$

Dropping the common factor x^λ from both sides and equating the co-efficients of each power of x to zeros we get :

$(\lambda^2 - p^2)c_0 = 0 \quad \Rightarrow \lambda = \pm p \quad$ since $c_0 \neq 0$ by assumption.

$((\lambda+1)^2 - p^2)\, c_1 = 0 \qquad\qquad\qquad\qquad\qquad \dots \text{(a)}$

$\left\{ (\lambda+n)^2 - p^2 \right\} c_n + c_{n-2} = 0\ \forall\ n \geqslant 2 \qquad\qquad \dots \text{(b)}$

Equivalently one can write the g.r.r as : $c_n = \frac{-c_{n-2}}{(n+\lambda)^2 - p^2}\ \forall\ n \geqslant 2$

We start working with the special instance where $\lambda = p$. From (a) it follows that as $p \neq 0$, $c_1 = 0$. From recurrence relation (b) it follows that in all cases, the co-efficients $c_3, c_5, c_7, \dots$ bearing odd suffix are zeros. Hence the g.r.r reduces to $c_n = \frac{-c_{n-2}}{n(n+p)}\ \forall\ n \geqslant 2$ and in particular when n is odd, c_n's are zeros. If we put $n = 2k$ so that the above, recurrence relation becomes

$$c_{2k} = \frac{-c_{2k-2}}{2^2 k(k+p)}\ \forall\ k \geqslant 1,$$

$$\therefore\quad c_2 = \frac{-c_0}{2^2(1+p)} \quad \text{(for } k=1\text{)}$$

$$c_4 = \frac{-c_2}{2^2 \cdot 2 \cdot(2+p)} = \frac{(-1)^2 c_0}{2^{2.2} \cdot 2! \cdot(2+p)(1+p)} \quad \text{(for } k=2\text{)}$$

$$c_6 = \frac{-c_4}{2^2 \cdot 3 \cdot(3+p)} = \frac{(-1)^3 c_0}{2^6 \cdot 3! \cdot(3+p)(2+p)(1+p)} \quad \text{(for } k=3\text{)}$$

Inductively, $\quad c_{2k} = \dfrac{(-1)^k c_0}{2^{2k} k!(1+p)(2+p)\dots(k+p)}\ \forall\ k = 1,2,\dots$

To put the expression for c_{2k} in a more compact form we choose $c_0 = \frac{1}{2^p \Gamma(1+p)}$ so that through the successive use of the recurrence relation $m\Gamma(m) = \Gamma(m+1)\ \forall\ m > 0$ we have :

$$c_{2k} = \frac{(-1)^k}{2^{2k+p}\ k!(k+p)(k-1+p)\dots(2+p)(1+p)\Gamma(1+p)}$$

$$= \frac{(-1)^k}{2^{2k+p} \, k! \, \Gamma(k+p+1)} \quad \forall \, k = 1, 2, \dots \infty.$$

It is a nice little observation that our c_0 itself is in conformity with the above expression of c_{2k}. This extends the spectrum of the recurrence relation to $\mathbb{N} \bigcup \{0\}$ instead of $\mathbb{N}$. So we can have one Frobenius series solution, viz,

$$y_1(x) = x^p \sum_{n=0}^{\infty} c_{2n} \, x^{2n} = x^p \sum_{n=0}^{\infty} \frac{(-1)^n x^{2n}}{2^{2n+p} \, n! \, \Gamma(n+p+1)}$$

Applying Ratio test the reader may verify that this Frobenius series converges everywhere because

$$R = \lim_{n\to\infty} \left| \frac{c_{2n}}{c_{2n+2}} \right| = \lim_{n\to\infty} \frac{2^2(n+1)(n+p+1)}{x^2} = \infty$$

The sum of this series is conventionally denoted by $J_p(x)$ and known as Bessel function of the first kind of order p. Any reader will now aim to find the Frobenius series solution corresponding to the smaller root, viz, $\lambda = -p$ of the indicial equation. However if p is a non-integer, we can avoid this lengthy trot and simply make use of the symmetry inherent in the Bessel equation to predict a second linearly independent solution as $J_{-p}(x)$, where

$$J_{-p}(x) = \sum_{n=0}^{\infty} \frac{(-1)^n \cdot x^{2n-p}}{2^{2n-p} \, n! \, \Gamma(n-p+1)}$$

Thus $y_1(x) \equiv J_p(x)$ and $y_2(x) = J_{-p}(x)$ are two linearly independent Frobenius series solution provided p is not an integer.

In particular if $p = \frac{1}{2}$, we have Bessel equation of order $\frac{1}{2}$ and the corresponding linearly independent solutions are $J_{\frac{1}{2}}(x)$ and $J_{-\frac{1}{2}}(x)$, where

$$\begin{aligned} J_{\frac{1}{2}}(x) &= x^{\frac{1}{2}} \sum_{n=0}^{\infty} \frac{(-1)^n \cdot x^{2n}}{2^{2n+\frac{1}{2}} \cdot n! \, \Gamma\left(n+\frac{3}{2}\right)} \\ &= x^{\frac{1}{2}} \sum_{n=0}^{\infty} \frac{(-1)^n \cdot x^{2n}}{2^n . 2^{n+\frac{1}{2}} n! \left(n+\frac{1}{2}\right)\left(\left(n-\frac{1}{2}\right) \dots \frac{1}{2}\Gamma\left(\frac{1}{2}\right)\right)} \end{aligned}$$

$$= x^{\frac{1}{2}} \sum_{n=0}^{\infty} \frac{(-1)^n \cdot x^{2n}}{(2n+1)!\sqrt{\frac{\pi}{2}}}$$

$$= x^{-\frac{1}{2}} \sqrt{\frac{2}{\pi}} \sum_{n=0}^{\infty} \frac{(-1)^n \cdot x^{2n+1}}{(2n+1)!} = \sqrt{\frac{2}{\pi}} \cdot x^{-\frac{1}{2}} \sin x$$

$$J_{-\frac{1}{2}}(x) = x^{-\frac{1}{2}} \sum_{n=0}^{\infty} \frac{(-1)^n \cdot x^{2n}}{2^{n-\frac{1}{2}} \, n! \, \Gamma\left(\frac{1}{2}\right)}$$

$$= x^{-\frac{1}{2}} \sum_{n=0}^{\infty} \frac{(-1)^n \cdot x^{2n}}{(2n)!\sqrt{\frac{\pi}{2}}} = \sqrt{\frac{2}{\pi}} \, x^{-\frac{1}{2}} \cos x$$

It is a straightforward exercise to verify that $J_{\frac{1}{2}}(x)$ and $J_{-\frac{1}{2}}(x)$ are linearly independent as their Wronskian is non-zero.

In case the parameter p is zero, the $J_p(x)$ and $J_{-p}(x)$ coincide and we can have only one Frobenius series solution, viz,

$$y_1(x) = \sum_{n=0}^{\infty} \frac{(-1)^n x^{2n}}{2^{2n} \, n! \, \Gamma(n+1)} = \sum_{n=0}^{\infty} \frac{(-1)^n x^{2n}}{2^{2n}(n!)^2}$$

This solution $y_1(x)$ is known as Bessel function of order zero of the first kind and is denoted by $J_0(x)$.

For determining another series solution $y_2(x)$ that is linearly independent w.r to $y_1(x)$, we shall have to use the ansatz of (8.15) and predict the solution in the form :

$$y_2(x) = y_1(x)lnx + \sum_{n=0}^{\infty} d_n x^n \qquad \ldots(c)$$

where the co-efficients d_n's are yet to be determined. If we are very agile in this execution we shall discover that the anstaz (8.15) literally gives the general solution of the Bessel equation of order zero and by triggering the two linearly independent solutions.

Since $y_1(x)$ and $y_2(x)$ satisfy Bessel equation of order 0, it follows that

$$\left. \begin{array}{l} xy_1''(x) + y_1'(x) + xy_1(x) = 0 \\ xy_2''(x) + y_2'(x) + xy_2(x) = 0 \end{array} \right\} \qquad \ldots(d)$$

$$\therefore \quad y_2(x) = y_1(x)lnx + \sum_{n=0}^{\infty} d_n x^n$$

$$y_2'(x) = y_1'(x)lnx + \frac{1}{x}y_1(x) + \sum_{n=1}^{\infty} nd_n x^{n-1}$$

$$y_2''(x) = y_1''(x)lnx + \frac{2}{x}y_1'(x) - \frac{1}{x^2}y_1(x) + \sum_{n=2}^{\infty} n(n-1)d_n x^{n-2}$$

Plugging the expressions of $y_2(x), y_2'(x)$ and $y_2''(x)$ into the Bessel equation of order zero,

$$x^2 \left[y_1''(x)ln\ x + \frac{2}{x}y_1'(x) - \frac{1}{x^2}y_1(x) + \sum_{n=2}^{\infty} n(n-1)d_n x^{n-2} \right] +$$

$$x \left[y_1'(x)ln\ x + \frac{1}{x}y_1(x) + \sum_{n=1}^{\infty} nd_n x^{n-1} \right] + x^2 \left(y_1(x)ln\ x + \sum_{n=0}^{\infty} d_n x^n \right) = 0$$

i, e, $\quad 2xy_1'(x) + \sum_{n=1}^{\infty} nd_n x^n + \sum_{n=2}^{\infty} n(n-1)d_n x^n + \sum_{n=2}^{\infty} d_{n-2}x^n = 0$

(Replacing n by $\overline{n-2}$ in the last sum and using (d))

$\therefore \quad\quad 2xy_1'(x) + \sum_{n=2}^{\infty}(n^2 d_n + d_{n-2})x^n + d_1\ x = 0$

or, $\quad 2\sum_{n=1}^{\infty} \frac{(-1)^n 2n\ .\ x^{2n}}{2^{2n}(n!)^2} + \sum_{n=2}^{\infty}(n^2 d_n + d_{n-2})x^n + d_1 x = 0$

or, $\quad 2\sum_{n=1}^{\infty} \frac{(-1)^n\ .\ 2nx^{2n}}{2^{2n}(n!)^2} + \sum_{n=1}^{\infty}\{(2n)^2 d_{2n} + d_{2n-2}\}x^{2n-1}$

$\quad\quad + \sum_{n=1}^{\infty}\left((2n+1)^2 d_{2n+1} + d_{2n-1}\right)x^{2n+1} + d_1\ x = 0$

or, $\quad \sum_{n=1}^{\infty}\left[\frac{(-1)^n n}{2^{2n-2}(n!)^2} + (4n^2 d_{2n} + d_{2n-2})\right]x^{2n}$

$\quad\quad + \sum_{n=1}^{\infty}\{(2n+1)^2 d_{2n+1} + d_{2n-1}\}x^{2n+1} + d_1 x = 0$

Thus we get : $\quad\quad d_1 = 0$

$$d_{2n-1} + (2n+1)^2 d_{2n+1} = 0\ \forall\ n \geqslant 1 \quad \Leftrightarrow \quad d_{2n-1} = 0\ \forall\ n \geqslant 1$$

and $\quad \dfrac{(-1)^{n+1}}{2^{2n}\ .\ (n!)^2 n} - \dfrac{d_{2n-2}}{4n^2} = d_{2n}\ \forall\ n \geqslant 1 \quad\quad \dots (e)$

Hence only the terms having the odd co-efficients are zeros.

$$\text{In (e)}, \quad n = 1 \quad \text{gives} : d_2 = \frac{1 - d_0}{2^2}$$

$$n = 2 \quad \text{gives} : d_4 = \frac{2d_0 - 3}{27} \quad \text{and so on.}$$

$$\text{If} \quad d_0 = 0, \quad n = 3 \quad \text{gives} : d_6 = \frac{(-1)^{3+1}}{2^6 \cdot (3!)^2 \cdot 3} - \frac{d_4}{2^2 \cdot 3^2}$$

$$= \frac{(-1)^{3+1}}{2^6 \cdot (3!)^2} \left[\frac{1}{3} + \frac{3}{2} \right] = \frac{(-1)^{3+1}}{2^6 \cdot (3!)^2} \left(1 + \frac{1}{2} + \frac{1}{3} \right)$$

and so on.

$$\therefore \; y_2(x) = y_1(x) ln \; x + \sum_{k=0}^{\infty} \frac{(-1)^{k+1} x^{2k}}{2^{2k} (k!)^2} \left(1 + \frac{1}{2} + \frac{1}{3} + \ldots + \frac{1}{k} \right)$$

The function $y_2(x)$ is known as 'Bessel function of order zero of the second kind' and is conventionally denoted by $Y_0(x)$. The graphs of $J_0(x)$ and $Y_0(x)$ are jointly given in the following figure 8.2

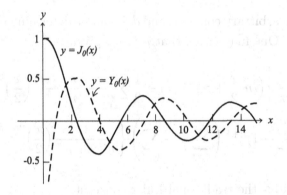

Fig 8.2 : $J_0(x)$: Bessel functions of order zero of first kind

$Y_0(x)$: Bessel functions of order zero of second kind

We again comeback to the case where p is a positive integer.

We recall $J_p(x) = \sum_{n=0}^{\infty} \frac{(-1)^n x^{2n+p}}{2^{2n+p} \; n! \; \Gamma(n + p + 1)}$ is one solution of

the Bessel equation of order p $(p \in \mathbb{N})$. Again we observe that

$$J_{-p}(x) = \sum_{n=0}^{\infty} \frac{(-1)^n x^{2n-p}}{2^{2n-p} n! \; \Gamma(n - p + 1)} \quad \text{is not linearly independent of } J_p(x)$$

in this case as $J_{-p}(x) = (-1)^p J_p(x)$. This claim is formally verified as follows:

In the expression of $J_{-p}(x)$, there appears the term $\Gamma(n - p + 1)$ in the denominator but it is defined only if $n \geqslant p$ if we keep an eye to convergence criterion of gamma functions.So effectively the sum range of $J_{-p}(x)$ boils down $n = p$ to $n = \infty$. Thus

$$J_{-p}(x) = \sum_{n=p}^{\infty} \frac{(-1)^n \cdot x^{2n-p}}{2^{2n-p} \, n! \, \Gamma(n - p + 1)} = \sum_{k=0}^{\infty} \frac{(-1)^{k+p} \cdot x^{2k+p}}{2^{2k+p} \cdot (k + p)! \, \Gamma(k + 1)}$$

(Put $n = k + p$ on R.H.S to get the last form above)
Hence $J_{-p}(x) = (-1)^p J_p(x)$, as claimed.

Keeping this situation in mind we can use ansatz (8.15) to propose the form of a second linearly independent solution in the form :

$$y_2(x) = \mu J_p(x) lnx + x^{-p} \sum_{n=0}^{\infty} d_n x^n,$$

where μ is an arbitrary constant and d_n's can be determined from routine calculations. One may check that $y_2(x)$ is given by :

$$y_2(x) = \frac{2}{\pi} \left[\left(ln\frac{x}{2} + \gamma \right) J_p(x) - \frac{1}{2} \sum_{n=0}^{p-1} \frac{(p - n - 1)!}{n!} \left(\frac{x}{2}\right)^{2n-p} \right.$$

$$\left. + \frac{1}{2} \sum_{n=0}^{\infty} (-1)^{n+1} \left(\sum_{k=1}^{n} \frac{1}{k} + \sum_{k=1}^{n+p} \right) \left\{ \frac{1}{n!(n + p)!} \left(\frac{x}{2}\right)^{2n+p} \right\} \right],$$

where γ denotes the traditional Euler constant.

This function $y_2(x)$ is known as Bessel's function of the 2nd kind of order p and conventionally denoted by $Y_p(x)$. So when p is a positive integer, the general solution of Bessel equation is of the form :

$$y(x) = a J_p(x) + b Y_p(x)$$

with a and b being arbitrary constants.

Example (25) : Solve for large values of x, Legendre's equation, viz, $(1 - x^2)y'' - 2xy' + m(m + 1)y = 0$; m being a positive constant.

We substitute $x = \frac{1}{z}$ in the ode given so that $\frac{dz}{dx} = -z^2$

$$\therefore \quad \frac{dy}{dx} = -z^2 \frac{dy}{dz} \ ; \quad \frac{d^2y}{dx^2} = z^2 \left(2z \frac{dy}{dz} + z^2 \frac{d^2y}{dz^2} \right)$$

Hence $\quad z^2 \left(1 - \frac{1}{z^2} \right) \left(2z \frac{dy}{dz} + z^2 \frac{d^2y}{dz^2} \right) + 2z \frac{dy}{dz} + m(m+1)y = 0$

or, $\quad z^2(z^2 - 1) \frac{d^2y}{dz^2} + 2z^3 \frac{dy}{dz} + m(m+1)y = 0 \qquad \cdots (a)$

Clearly $z = 0$ is a point of regular singularity of the ode (a) since

$$\lim_{z \to 0} \frac{2z^2}{z^2 - 1} = 0 \quad \text{and} \quad \lim_{z \to 0} \frac{z^2 m(m+1)}{z^2(z^2 - 1)} = m(m+1) < \infty$$

We therefore try a Frobenius series of the form : $y = \sum\limits_{n=0}^{\infty} c_n z^{n+\lambda}$ as series solution of (a). On term-by-term differentiation we get :

$$\frac{dy}{dz} = \sum_{n=0}^{\infty} c_n z^{n+\lambda} \quad \text{and} \quad \frac{d^2y}{dz^2} = \sum_{n=0}^{\infty} c_n (n+\lambda)(n+\lambda-1) z^{n+\lambda-2}$$

Plugging these into (a) we have :

$$(z^2 - 1) \sum_{n=0}^{\infty} c_n (n+\lambda)(n+\lambda-1) z^{n+\lambda} + 2z^2 \sum_{n=0}^{\infty} c_n (n+\lambda) z^{n+\lambda}$$

$$+ m(m+1) \sum_{n=0}^{\infty} c_n z^{n+\lambda} = 0$$

or, $\quad \sum\limits_{n=0}^{\infty} c_n \{(n+\lambda)(n+\lambda-1) + 2(n+\lambda)\} z^{n+\lambda+2}$

$$+ \sum_{n=0}^{\infty} \{m(m+1) - (n+\lambda)(n+\lambda-1)\} c_n \, z^{n+\lambda} = 0$$

or, $\quad \sum\limits_{n=0}^{\infty} c_n (n+\lambda)(n+\lambda+1) z^{n+\lambda+2} +$

$$\sum_{n=0}^{\infty} \{m(m+1) - (n+\lambda)(n+\lambda-1)\} c_n \, z^{n+\lambda} = 0$$

Replacing n by $\overline{n-2}$ in the first sum we get the above in the form

$$\sum_{n=2}^{\infty} c_{n-2} (n+\lambda-2)(n+\lambda-1) z^{n+\lambda} +$$

$$\sum_{n=0}^{\infty}\{m(m+1)-(n+\lambda)(n+\lambda-1)\}c_n\ z^{n+\lambda}=0$$

$$\therefore \sum_{n=2}^{\infty}[(n+\lambda-2)(n+\lambda-1)c_{n-2}+\{m(m+1)-(n+\lambda)(n+\lambda-1)\}c_n]z^{n+\lambda}$$

$$+\{m(m+1)-\lambda(\lambda-1)\}c_0\ z^{\lambda}+\{m(m+1)-(\lambda+1)\lambda\}c_1\ z^{1+\lambda}=0$$

Dropping the factor z^{λ} from both sides and equating the co-efficients of different powers of z to zero we have :

$$\{m(m+1)-\lambda(\lambda-1)\}c_0=0 \qquad\qquad\qquad \ldots\text{(i)}$$

$$\{m(m+1)-\lambda(\lambda+1)\}c_1=0 \qquad\qquad\qquad \ldots\text{(ii)}$$

$$(n+\lambda-2)(n+\lambda-1)c_{n-2}$$
$$+\{m(m+1)-(n+\lambda)(n+\lambda-1)\}c_n=0\ \forall\ n\geqslant 2 \qquad \ldots\text{(iii)}$$

(i) gives the indicial equation : $(m+\lambda)(m-\lambda+1)=0$

    i,e, $\lambda=-m\ ;\ \lambda=(m+1)$.   (since $c_0\neq 0$)

The recurrence relation (iii) ensures that $c_1=c_3=c_5=\ldots=0$ because from (ii) it follows that $c_1=0$. So only terms involving even powers of x survive in the proposed series solution. We first try with the smaller root of the indicial equation, i,e, $\lambda=-m$. Hence (iii) yields:

$$c_n=\frac{-(n-m-2)(n-m-1)}{n[2m-(n-1)]}\ c_{n-2}\ \forall\ n\geqslant 2$$

If $n=2$, $c_2=\dfrac{-m(m-1)}{2(2m-1)}\ c_0$

If $n=4$, $c_4=\dfrac{-(2-m)(3-m)}{4(2m-3)}\ c_2=\dfrac{\dot{m}(m-1)(m-2)(m-3)}{(2m-1)(2m-3)2.4}c_0$

and so on. Hence one Frobenius series solution reads :

$$y=c_0\ z^{-m}\left[1-\frac{m(m-1)}{(2m-1).2}z^2+\frac{m(m-1)(m-2)(m-3)}{(2m-1)(2m-3).2.4}z^4-\ldots\right]$$

$$=c_0\ x^m\left[1-\frac{m(m-1)}{(2m-1).2}x^{-2}+\frac{m(m-1)(m-2)(m-3)}{(2m-1)(2m-3).2.4}x^{-4}-\ldots\right]$$

$$=c_0\left[x^m-\frac{m(m-1)}{(2m-1).2}x^{m-2}+\frac{m(m-1)(m-2)(m-3)}{(2m-1)(2m-3).\ 2.4}x^{m-4}-\ldots\right]$$

On choosing the arbitrary constant c_0 to be $\dfrac{1.3.5\ldots(2n-1)}{n!}$ it follows that $y_1(x)$ becomes :

$$y_1\,(x) = \frac{1.3.5\ldots(2m-1)}{m!}\left[x^m - \frac{m(m-1)}{(2m-1)2}x^{m-2}\right.$$
$$\left.+\frac{m(m-1)(m-2)(m-3)}{(2m-1)(2m-3).2.4}x^{m-4} - \ldots\infty\right]$$

and is said to represent **Legendre function of first kind**. $y_1(x)$ is visibly a descending power-series in x and as per conventions, denoted by $P_m(x)$. If the parameter p happens to be a positive integer, $P_m(x)$ reduces to a polynomial, the degree of which depending on the choice of the integer value assumed by the parameter m.

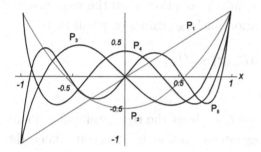

$P_1(x) =$	x
$P_2(x) =$	$\frac{1}{2}[3x^2 - 1]$
$P_3(x) =$	$\frac{1}{2}[5x^3 - 3x]$
$P_4(x) =$	$\frac{1}{8}[35x^4 - 30x^2 + 3]$
$P_5(x) =$	$\frac{1}{8}[35x^4\ 70x^3 + 15x]$

$P_n(x), n = 1(1)5$

Fig 8.3 : Legendre polynomials $P_n(x)$

Now choose $\lambda = m + 1$ so that recurrence relation (iii) gives :

$$c_n = \frac{(n+m)(n+m-1)}{n[2m+(n+1)]}c_{n-2} \quad \forall\ n \geqslant 2$$

$$\therefore \quad c_2 = \frac{(m+2)(m+1)}{2.(2m+3)}c_0$$

$$\therefore \quad c_4 = \frac{(m+4)(m+3)}{(2m+5).4}c_2 = \frac{(m+1)(m+2)(m+3)(m+4)}{(2m+3)(2m+5).4.2}c_0$$

and so on. Hence another Frobenius series solution is found as :

$$y\ =\ c_0\,z^{(m+1)}\left[1 + \frac{(m+1)(m+2)}{(2m+3).2}z^2+\right.$$
$$\left.\frac{(m+1)(m+2)(m+3)(m+4)}{(2m+3)(2m+5).2.4}z^4 + \ldots\infty\right]$$

$$\therefore \quad y_2(x)\ =\ c_0\left[x^{-(m+1)} + \frac{(m+1)(m+2)}{(2m+3).2}x^{-(m+3)}+\right.$$

$$\frac{(m+1)(m+2)(m+3)(m+4)}{(2m+3)(2m+5).2.4}x^{-(m+5)}+\dots\infty\Bigg]$$

On choosing the arbitrary constant c_0 as $\frac{m!}{1.3.5\dots(2m+1)}$ it follows that

$$y_2(x)=\frac{m!}{1.3.5\dots(2m+1)}\Bigg[x^{-(m+1)}+\frac{(m+1)(m+2)}{(2m+3).2}x^{-(m+3)}$$
$$+\frac{(m+1)(m+2)(m+3)(m+4)}{(2m+3)(2m+5).2.4}x^{-(m+5)}+\dots\infty\Bigg]$$

and it is said to represent Legendre function of the second kind, conventionally denoted by $Q_m(x)$. If $P_m(x)$ and $Q_m(x)$ are two linearly independent solution of the Legendre's equation, then the most general solution of the Legendre's equation for large values of x will be :

$$y(x)=aP_m(x)+bQ_m(x),$$

a and b being arbitrary constants.

Let's now solve Legendre equation about the regular singular points $x=\pm1$. Observe that the Legendre equation is invariant under the change of independent variable $x\to-x$. This is why the nature of the solution about $x=1$ is sufficient for discussion.

On using the linear transformation $x=1+z$, the given ode becomes:

$$z(2+z)y''+2(1+z)y'-m(m+1)y=0\ ,$$

where y' and y'' now denote $\frac{dy}{dz}$ and $\frac{d^2y}{dz^2}$ respectively. Clearly $x=1$ corresponds to $z=0$. We propose the Frobenius series solution in the form $y=\sum_{n=0}^{\infty}c_nz^{n+\lambda}$ so that on term-by-term differentiation we get :

$$y'=\sum_{n=0}^{\infty}c_n(n+\lambda)z^{n+\lambda-1}$$

$$y''=\sum_{n=0}^{\infty}c_n(n+\lambda)(n+\lambda-1)z^{n+\lambda-2}$$

Using these in the ode we get:

$$(2+z)\sum_{n=0}^{\infty}c_n(n+\lambda)(n+\lambda-1)z^{n+\lambda-1}+2(1+z)\sum_{n=0}^{\infty}c_n(n+\lambda)z^{n+\lambda-1}$$

$$-m(m+1)\sum_{n=0}^{\infty} c_n z^{n+\lambda} = 0$$

or, $$2\sum_{n=0}^{\infty} c_n(n+\lambda)(n+\lambda-1)z^{n+\lambda-1} + \sum_{n=0}^{\infty} c_n(n+\lambda)(n+\lambda-1)z^{n+\lambda}$$

$$+2\sum_{n=0}^{\infty} c_n(n+\lambda)z^{n+\lambda-1} + 2\sum_{n=0}^{\infty} c_n(n+\lambda)z^{n+\lambda}$$

$$-m(m+1)\sum_{n=0}^{\infty} c_n z^{n+\lambda} = 0$$

or, $$2\sum_{n=0}^{\infty} c_n(n+\lambda)^2 z^{n+\lambda-1} + \sum_{n=0}^{\infty}\{(n+\lambda)(n+\lambda+1)$$

$$-m(m+1)\}c_n z^{n+\lambda} = 0$$

or, $$2\sum_{n=0}^{\infty}(n+\lambda)^2 c_n z^{n+\lambda-1} + \sum_{n=1}^{\infty}\{(n+\lambda-1)(n+\lambda)$$

$$-m(m+1)\}c_{n-1} z^{n+\lambda-1} = 0$$

Dropping $z^{\lambda-1}$ on both sides and setting co-efficients of each individual powers of z to 0 we get :

$$2c_0\lambda^2 = 0 \Rightarrow \lambda = 0,0 \quad (\because c_0 \neq 0)$$
and $2c_n(n+\lambda)^2 + \{(n+\lambda-1)(n+\lambda) - m(m+1)\}c_{n-1} = 0 \; \forall \, n \geqslant 1$

Setting $\lambda = 0$ we get from the recurrence relation :

$$2c_n \, n^2 + \{(n-1)(n) - m(m+1)\}c_{n-1} = 0$$

i, e, $$c_n = \frac{(m-n+1)(m+n)}{2n^2}c_{n-1} \quad \forall \, n \geqslant 1$$

$$n = 1 \Rightarrow c_1 = \frac{(m+1)m}{2.1^2}c_0 = \frac{(m+1)m}{(1!)^2.2}c_0$$

$$n = 2 \Rightarrow c_2 = \frac{(m+2)(m+1)}{2.2^2}c_1 = \frac{(m+2)(m+1)^2 m}{2^2(2!)^2}c_0$$

Proceeding this way we ultimately get one Frobenius series solution

$$y_1(z) = c_0 \left[\frac{1 + (m+1)m}{(1!)^2}\left(\frac{z}{2}\right) + \frac{(m+2)(m+1)^2 \, m}{2!}\left(\frac{z}{2}\right)^2 + \cdots\right]$$

By Ratio test we can verify that this series solution converges if $|z| < 2$ i,e, if $-1 < x < 3$. By symmetry, the Frobenius series about $x = -1$

will be convergent for $-3 < x < 1$. The second solution about $z = 0$ (corresponding to $x = 1$) will be of the form :

$$y(z) = c \, ln \, z + g_0 + g_1 z + ...\infty, \quad \text{where} \quad z = (x - 1)$$

Remark : We have shown that Legendre equation has three regular singular points, viz, $-1, 1$ and ∞ and moreover, we can find two linearly independent series solutions about each of these three points by using Frobenius method. We observed (c.f Example 15) that about the ordinary point $x = 0$, there exist linearly independent power series solutions, both having the interval of convergence $(-1, 1)$. We also found that one of the series solutions about the point $x = 1$ has interval of convergence $(-1, 3)$ while one of the series solutions about the point $x = -1$ has interval of convergence $(-3, 1)$ (c.f Example 25). So the series solutions about the point $x = 0$ fail at $x = \pm 1$, the two nearby singular points. Similarly the series solution at $x = 1$ fail at $x = -1$, the nearest singular point w.r.to $x = 1$ while the series solution at $x = -1$ fail at $x = 1$, the nearest singular point w.r.to $x = -1$. However, the process of 'analytic continuation' can be used to find a series for a wider range of the independent variable x.

Example (26) : Use $x = 2z - 1$ to transform Legendre equation :

$$(1 - x^2)y'' - 2xy' + m(m + 1)y = 0.$$

Since $x = 2z - 1$, $dx = 2dz$ and so the transformed ode will be :

$$z(1 - z)\frac{d^2y}{dz^2} + (1 - 2z)\frac{dy}{dz} + m(m + 1)\, y = 0$$

$$\left[\text{keep in mind } y' \equiv \frac{dy}{dx} = \frac{1}{2}\frac{dy}{dz} \text{ and } y'' = \frac{1}{4}\frac{d^2y}{dz^2}\right]$$

The newly transformed ode resembles the form of a Hypergeometric equation as it can be written as

$$z(1 - z)\frac{d^2y}{dz^2} + [1 - \{(m + 1) + (-m) + 1\}z]\frac{dy}{dz} - (-m)(m + 1)y = 0$$

i,e, of the form

$$z(1 - z)\frac{d^2y}{dz^2} + [c - (a + b + 1)z]\frac{dy}{dz} - aby = 0$$

provided we agree to write $a = (m+1)$; $b = -m$; $c = 1$

Remark :

(i) This example is only a special case of the general result that every second order ode having atmost three regular singular points can be cast to the hypergeometric differential equation by means of a suitable transformation. Legendre equation, Chebyshev equation, Bessel equation, Laguerre equation all can be transformed to the hypergeometric equation. So it is best to study hypergeometric equation separately as it gives rise to hypergeometric series and hypergeometric functions whereform emerge the most interesting special functions of Mathematical Physics.

(ii) The hypergeometric equation

$$z(1-z)\frac{d^2y}{dz^2} + [c - (a+b+1)z]\frac{dy}{dz} - aby = 0$$

has three regular singularities, viz, $z = 0$, $z = 1$ and $z = \infty$. From previous example it follows that $z = 0$ corresponds to $x = -1$ of the Legendre equation while $z = 1$ corresponds to $x = 1$ and $z = \infty$ corresponds to $x = \infty$. It was Kummer who first derived a set of six distinct series solutions of the hypergeometric differential equation — two apiece for the three regular singularities talked of. So it will be worthwhile a decision to dedicate a separate section to the hypergeometric equation of Euler and its solutions in terms of hypergeometric functions of Gauss.

8.6 Hypergeometric Equation

Euler's hypergeometric equation reads :

$$x(1-x)y'' + [c - (a+b+1)x]y' - aby = 0 \qquad (8.18)$$

where a, b, c are real constants. These parameters or constants play a pivotal role in determining the nature of the solutions. Earlier we pointed out that in $\mathbb{R}$, $x = 0$ and $x = 1$ are its only regular singular points. In what follows, let's test this in the following after rewriting the hypergeometric equation in the normal form :

$$y'' + \frac{\{c - (a+b+1)c\}}{x(1-x)}y' - \frac{ab}{x(1-x)}y = 0$$

Observe that $\displaystyle\lim_{x \to 0} \frac{x\{c - (a + b + 1)x\}}{x(1 - x)} = c \;;\; \lim_{x \to 0} \frac{x^2 ab}{x(1 - x)} = 0$

$$\lim_{x \to 1} \frac{(1 - x)\{c - (a + b + 1)x\}}{x(1 - x)} = c - (a + b + 1)$$

$$\lim_{x \to 1} (1 - x)^2 \frac{ab}{x(1 - x)} = \lim_{x \to 1} \frac{(1 - x)ab}{x} = 0$$

Alternatively, one could have shown the analyticity of the co-efficient funtions $p(x) = \frac{c - (a+b+1)x}{x(1-x)}$ and $q(x) = \frac{-ab}{x(1-x)}$ in the neighborhoods of $x = 0$ and $x = 1$ and thereby draw one and the same conclusion.

We start working for the Frobenius series solution about $x = 0$. We therefore propose the solution in the form

$$y(x) = \sum_{n=0}^{\infty} d_n x^{n+\lambda}, \text{ where } d_0 \neq 0$$

Using term-by-term differentiation we get :

$$y'(x) = \sum_{n=0}^{\infty} (n + \lambda) d_n x^{n+\lambda-1} \;;\; y''(x) = \sum_{n=0}^{\infty} (n + \lambda)(n + \lambda - 1) d_n x^{n+\lambda-2}$$

Plugging these into the hypergeometric equation we get :

$$(1 - x) \sum_{n=0}^{\infty} (n + \lambda)(n + \lambda - 1) d_n x^{n+\lambda-1} +$$

$$\left[c - (a + b + 1)x \right] \sum_{n=0}^{\infty} (n + \lambda) d_n \, x^{n+\lambda-1} - ab \sum_{n=0}^{\infty} d_n \, x^{n+\lambda} = 0$$

or, $\displaystyle\sum_{n=0}^{\infty} \{(n + \lambda)(n + \lambda - 1) + c(n + \lambda)\} d_n \, x^{n+\lambda-1} -$

$$\sum_{n=0}^{\infty} \left\{ (n + \lambda)(n + \lambda - 1) + (a + b + 1)(n + \lambda) + ab \right\} d_n \, x^{n+\lambda} = 0$$

or, $\displaystyle\sum_{n=0}^{\infty} (n + \lambda)(n + \lambda - 1 + c) d_n \, x^{n+\lambda-1}$

$$- \sum_{n=0}^{\infty} \{(n + \lambda)(n + \lambda + a + b) + ab\} d_n x^{n+\lambda} = 0$$

Substitute n by $\overline{n - 1}$ in the second sum so as to get :

$$\sum_{n=1}^{\infty} \Big[(n + \lambda)(n + \lambda - 1 + c) d_n - \{(n + \lambda - 1)(n + \lambda + a + b - 1) + ab\} \times$$

$$d_{n-1}\Big] x^{n+\lambda-1} + \lambda(\lambda - 1 + c)d_0\, x^{\lambda-1} = 0$$

Eliminating the common factor $x^{\lambda-1}$ from both sides and equating to zero the co-efficients of each individual power of x we have :

$$\lambda(\lambda - 1 + c)d_0 = 0 \Rightarrow \text{ either } \lambda = 0 \text{ or, } \lambda = 1 - c \ (\because d_0 \neq 0)$$

$$(n + \lambda)(n + \lambda - 1 + c)d_n - \Big\{ (n + \lambda - 1)(n + \lambda + a + b - 1) + ab\Big\} d_{n-1}$$
$$= 0 \ \forall\, n \geqslant 1$$

If c be not an integer, then the two roots of the indicial equation donot differ by an integer and so theoretically we are ensured of the existence of two linearly independent Frobenius series solutions, one corresponding to $\lambda = 0$ and another corresponding to $\lambda = 1 - c$. Corresponding to $\lambda = 0$, the recurrence relation becomes :

$$n(n - 1 + c)d_n = \{(n - 1)(n + a + b - 1) + ab\}d_{n-1} \quad \forall\, n \geqslant 1$$

or, $\quad n(n + c - 1)d_n = (n + a - 1)(n + b - 1)d_{n-1}$

$\therefore \quad d_n = \dfrac{(n + a - 1)(n + b - 1)}{n(n + c - 1)}\, d_{n-1} \ \forall\, n \geqslant 1$

$\therefore \quad n = 1 \Rightarrow d_1 = \dfrac{ab}{1!c}d_0$

$n = 2 \Rightarrow d_2 = \dfrac{ab(a + 1)(b + 1)}{2!c(c + 1)}\, d_0$

$n = 3 \Rightarrow d_3 = \dfrac{a(a + 1)(a + 2)b(b + 1)(b + 2)}{3!c(c + 1)(c + 2)}\, d_0$

Inductively, $\ d_n = \dfrac{a(a + 1)\ldots(a + \overline{n - 1})b(b + 1)\ldots(b + \overline{n - 1})}{n!c(c + 1)\ldots(c + \overline{n - 1})}\, d_0$

Hence series solution corresponding to $\lambda = 0$ will be (setting $d_0 = 1$):

$$y(x) \ = \ x^0 \sum_{n=0}^{\infty} d_n x^n$$

$$= \ \sum_{n=0}^{\infty} \frac{a(a + 1)\ldots(a + \overline{n - 1})b(b + 1)\ldots(b + \overline{n - 1})}{n!\, c(c + 1)\ldots(c + \overline{n - 1})}\, x^n$$

The radius of convergence of this power series is unity because

$$R = \lim_{n \to \infty} \left| \frac{d_n}{d_{n+1}} \right| = \lim_{n \to \infty} \left| \frac{(n + 1)(c + n - 1)}{(a + n - 1)(b + n - 1)} \right|$$

$$= \lim_{n \to \infty} \left| \frac{\left(1 + \frac{1}{n}\right)\left(1 + \frac{c-1}{n}\right)}{\left(1 + \frac{a-1}{n}\right)\left(1 + \frac{b-1}{n}\right)} \right| = 1$$

The sum function of the hypergeometric series obtained above is known as hypergeometric function and commonly denoted by $F(a, b, c\ ; x)$.

Corresponding to exponent $\lambda = (1 - c)$ the recurrence relation will be :

$$d_n = \frac{(n + a - c)(n + b - c)}{n(n + 1 - c)}\, d_{n-1} \,\forall\, n \geqslant 1$$

$$n = 1 \Rightarrow d_1 = \frac{(a + 1 - c)(b + 1 - c)}{1(2 - c)}\, d_0$$

$$n = 2 \Rightarrow d_2 = \frac{(a + 1 - c)(a + 2 - c)(b + 1 - c)(b + 2 - c)d_0}{2!(2 - c)(3 - c)}$$

and inductively,

$$d_n = \frac{(a + 1 - c)\ldots(a + n - c)(b + 1 - c)\ldots(b + n - c)}{n!(2 - c)(3 - c)\ldots(n + 1 - c)}\, d_0$$

Consequently the second Frobenius series solution will be :

$$y(x) = x^{1-c}\, F(a - c + 1\ ,\ b - c + 1, 2 - c\ ;\ x)\ .$$

Let's now digress a little and apply tricky transformation of the dependent variable y appearing in the hypergeometric equation. It is a simple observation that if we replace y by $x^{1-c}u$, then the given hypergeometric equation will be transformed again to a hypergeometric equation of the form :

$$x(1-x)u'' + [(2-c) - \{(a-c+1) + (b-c+1) + 1\}x]u' - (a-c+1)(b-c+1)u = 0$$

This transformed hypergeometric equation can be rewritten as :

$$x(1 - x)u'' + [c' - (a' + b' + 1)x]u' - a'b'u = 0 \quad (8.19)$$

with $\quad a' = a - c + 1\ ;\ b' = b - c + 1\ ;\ c' = 2 - c$

Using the result of the series solution corresponding to $\lambda = 0$ of the original hypergeometric equation (8.18) we can directly predict one Frobenius series solution of the newlook hypergeometric equation (8.19) in the form : $u = F(a', b', c'; x)$.

$$\therefore \quad y(x) = x^{1-c}F(a - c + 1,\ b - c + 1,\ 2 - c\ ;\ x)$$

turns out to be the second linearly independent solution of the original hypergeometric equation (8.18). Hence form invariance of the hypergeometric equation gives us a shortcut to find the general solution of the equation (8.18) in the generic form :

$$y(x) = AF(a, b, c \; ; \; x) + Bx^{1-c}F(a - c + 1, \; b - c + 1, \; 2 - c; \; x) \quad (8.20)$$

where A and B are arbitrary constants and c is not an integer.

In case c is not an integer, we are not sure of the existence two linearly independent series solution of the hypergeometric equation (8.18).

If we are interested for series solution of (8.18) about the point $x = 1$, then we shall apply the linear transformation $z = 1 - x$ of the independent variable so that $y' = -\frac{dy}{dz}$ and $y'' = \frac{d^2y}{dz^2}$

Therefore, equation (8.18) will be transformed to :

$$z(1 - z)\frac{d^2y}{dz^2} + \{(a + b - c + 1) - (a + b + 1)z\}\frac{dy}{dz} - aby = 0 \quad (8.21)$$

This new ode (8.21) is in conformity with (8.18) with the parameter c replaced by $(a + b - c + 1)$ and other parameters a and b remaining unaltered. Clearly $z = 0$ corresponds to $x = 1$ and hence the solution - finding about $x = 1$ is in letter and spirit same as the deal done with the original hypergeometric equation (8.17).

Remark :

(a) In example (26) we showed that Legendre equation is transformed to hypergeometric equation. However $c = 1$ in this case and so $\lambda = 0$ is a double root of indicial equation corresponding to the transformed hypergeometric equation. This at once leads us to one power series solution about $z = 0$:

$$y(z) = F(m + 1, -m, 1 \; ; \; z)$$

i,e, one power series solution about the point $x = -1$ is :

$$\bar{y}(x) \equiv y\left(\frac{1}{2}(1 + x)\right) = F\left(m + 1, -m, 1 \; ; \; \frac{1}{2}(1 + x)\right)$$

The second independent solution about $x = -1$ will be obtained by the order reduction technique if we propose it in the form :

$$y(x) = v(x)F\left(m + 1, -m, 1 \; ; \; \frac{1}{2}(1 + x)\right), \quad \text{where} \quad v \in C^2(\mathbb{R})$$

(b) The transcendental functions like $e^x, ln(1 + x)$, $\sin^{-1} x$, $\tan^{-1} x$, $\sin x$, $\cos x$ and many others can be shown as special instances of the hypergeometric functions.

8.7   Irregular singular points

Discussion on irregular singular points of any linear ode can be started with the simplest deal of a first order linear homogeneous ode having the form $\frac{dy}{dx} + P(x)y = 0$ the general solution of which is of the form $y(x) = Ce^{-\int P(x)dx}$, C being a constant. If $P(x)$ has a pole of order $(k+1)$ at $x = 0$ then $P(x)$ will have a Laurent expansion with $(k+1)$ terms having negative powers of x.

$$\therefore \quad P(x) = \sum_{n=-(k+1)}^{\infty} a_n x^n = \sum_{n=-(k+1)}^{n=-2} a_n x^n + \frac{a_{-1}}{x} + \sum_{n=0}^{\infty} a_n x^n$$

Using term-by term integration we get :

$$\int P(x)dx = \sum_{n=-(k+1)}^{n=-2} \left(\frac{a_n}{n+1}\right) x^{n+1} + a_{-1} \, ln \, x + \sum_{n=0}^{\infty} \left(\frac{a_n}{n+1}\right) x^{n+1}$$

In the first term, viz, $\sum_{n=-(k+1)}^{n=-2} \left(\frac{a_n}{n+1}\right) x^n$ if we put $n = -(m+1)$, then m ranges from 1 to k and the term looks like $\sum_{m=1}^{k} \frac{a_{-(m+1)}}{-m} x^{-m}$

$$\therefore \quad \int P(x)dx = \sum_{m=1}^{k} \frac{a_{-(m+1)}}{-mx^m} + a_{-1}ln \, x + \sum_{n=0}^{\infty} \frac{a_n}{n+1}x^{n+1}$$

This gives the solution of the chosen first order linear ode as :

$$y(x) = C \, \exp\left[\sum_{m=1}^{k} \frac{a_{-(m+1)}}{-mx^m} - a_{-1} \, ln \, x - \sum_{n=0}^{\infty} \frac{a_n}{(n+1)} x^{n+1}\right]$$

$$= C \, \exp\left[\sum_{m=1}^{k} \frac{a_{-(m+1)}}{mx^m}\right] x^{-a_{-1}} . \exp\left[-\sum_{n=0}^{\infty} \frac{a_n}{n+1} x^{n+1}\right]$$

Observe that the sum function of the power series $\sum_{n=0}^{\infty} \frac{a_n}{n+1} x^{n+1}$ is an analytic function of x and so also the function $\exp\left[-\sum_{n=0}^{\infty} \frac{a_n}{n+1} x^{n+1}\right]$ is analytic at $x = 0$. This is why one can express the latter function in the form of a convergent Taylor series of the form $\sum_{n=0}^{\infty} b_n x^n$.

i, e, $y(x) = C \exp\left[\sum_{m=1}^{k} \frac{a_{-(m+1)}}{mx^m}\right] x^{-a_1} \left(\sum_{n=0}^{\infty} b_n \, x^n\right),$

showing that the solution of the first order linear ode appears in the form of an exponential function times a Frobenius series. This solution pattern of the first order liners ode enables us to propose the pattern of the solutions of a second order linear homogeneous ode having the normal form :

$$y''(x) + p(x)y' + q(x)y = 0 \qquad \text{c.f (6.24)}$$

with $x = 0$ being an irregular singular point. In fact, we christen $x = 0$ to be an irregular singular point if it fails to be a regular singular point. To be more precise, we define $x = 0$ to be an **irregular singular point of rank k** $(k \in \mathbb{N})$ provided $p(x)$ has a pole of order $(k + 1)$ while $q(x)$ has a pole of order $2(k + 1)$ at $x = 0$ and moreover, at least one of the limits $\lim_{x \to 0} x^{k+1} p(x)$ and $\lim_{x \to 0} x^{2(k+1)} q(x)$ is non-zero. If $p(x)$ has a pole of order $(r + 1)$ and $q(x)$ has a pole of order $2(r' + 1)$ at $x = 0$, then the rank of the irregular singular point $x = 0$ is $Max\{r, r'\} = k$, say. It follows naturally form the definition, that when $x = 0$ is an irregular singular point of rank k of a second order ode, then we may consider its solution in the general form :

$$y(x) = \exp\,[g(x)] \sum_{n=0}^{\infty} c_n x^{n+\lambda}, \qquad (8.22)$$

where $g(x)$ is a polynomial of degree k in $\frac{1}{x}$,

i,e, $\quad g(x) = \sum_{m=1}^{k} \frac{A_m}{x^m}, \quad A_m$'s being constants.

If for some second order linear ode, $x = 0$ turns out to be an irregular singular point of rank 1, then $g(x) = \frac{A_1}{x}$ and so $y(x)$ assumes the form :

$$y(x) = e^{\frac{A_1}{x}} \sum_{n=0}^{\infty} c_n \, x^{n+\lambda}, \qquad (8.23)$$

where $A_1 \neq 0$ and λ being some constants not yet determined.

However, it is known to us that $x = 0$ is an essential singularity of the function $exp(\frac{A_1}{x})$. From the above discussion one may draw a generalised conclusion that if $x = 0$ be an irregular singular point of a linear ode, then $x = 0$ turns out to be a point of essential singularity of its solution. It is obvious that if instead of $x = 0$, $x = x_0$ be an irregular singular point, then by the simple linear transformation $z = x - x_0$ of the independent variable x, the whole problem can be translated in terms of $z = 0$. Thus it is a protocol that whenever $x = x_0$ is an irregular singular point of the second order ode having the general form

$$a_0(x)y'' + a_1(x)y' + a_2(x)y = 0 \qquad \text{(c.f (6.9))}$$

then $x = x_0$ will appear as an 'essential singularity' of the solution of that ode.

We now focus on some illustrative examples that will clarify how to solve an ode about an irregular singular point. Cognate to the upcoming example, let us do a brief discussion relating to irregular singular points in the following.

(a) Let's begin with the Bessel equation of order $\frac{1}{2}$, viz,

$$x^2 \frac{d^2 y}{dx^2} + x \frac{dy}{dx} + \left(x^2 - \frac{1}{4} \right) y = 0$$

We apply the transformation $y = x^{-\frac{1}{2}} z$ to the above equation. The reader can easily check that the transformed ode is

$$\frac{d^2 z}{dx^2} + z = 0,$$

the standard SHM equation. Since the general solution of the latter is readily available in the form

$$z = c_1 \sin x + c_2 \cos x,$$

the general solution of the Bessel equation of order $\frac{1}{2}$ is :

$$y(x) = x^{-\frac{1}{2}} z = c_1 \left(x^{-\frac{1}{2}} \sin x \right) + c_2 \left(x^{-\frac{1}{2}} \cos x \right)$$
$$= A J_{\frac{1}{2}}(x) + B J_{-\frac{1}{2}}(x),$$

where $\quad A \equiv C_1 \sqrt{\dfrac{\pi}{2}} \; ; \; B \equiv C_2 \sqrt{\dfrac{\pi}{2}}$

Thus we could retrieve the solutions of Bessel equation of order half from those of the SHM equation. This happens near the regular singular point $x = 0$.

(b) Had we applied $y = x^{\frac{1}{2}} z$ to the Bessel equation of order $\frac{1}{2}$, then it will be reduced to the form

$$\frac{d^2 z}{dx^2} + \frac{2}{x} \frac{dz}{dx} + z = 0 \qquad (8.23)$$

On applying the transformation $x = \frac{1}{t}$, keeping z unchanged, the equation (8.23) reduces to

$$\frac{d^2 z}{dt^2} + \frac{1}{t^4} z = 0$$

For this final ode, $t = 0$ is an irregular singular point. So solving this ode at $t = 0$ is equivalent to solving (8.23) at $x = \infty$, i,e, more accurately for very large values of x. Intuitively one may predict that this in turn is equivalent to exploring the solution of Bessel equation of order $\frac{1}{2}$ at $x = \infty$, the only irregular singular point of Bessel equation. However, this intuition works! Let's see how the situation grows.

Example (27) : Find series solution of $\frac{d^2 y}{dx^2} + \frac{1}{x^4} y = 0$ for small of x.

The point $x = 0$ is obviously an irregular singular point with rank 1. Hence by the ansatz(8.22) the form of the solution about $x = 0$ can be proposed as :

$$y(x) = e^{\frac{A_1}{x}} \sum_{n=0}^{\infty} c_n x^{n+\lambda} \equiv e^{\frac{A_1}{x}} Y(x) \qquad \cdots (a)$$

where $Y(x) \equiv \sum_{n=0}^{\infty} c_n x^{n+\lambda}$ is a Frobenius series.

Differentiating (a) we get :

$$\frac{dy}{dx} = e^{\frac{A_1}{x}} \left[\frac{dy}{dx} - \frac{A_1}{x^2} Y \right]$$

$$\frac{d^2 y}{dx^2} = e^{\frac{A_1}{x}} \left[\frac{d^2 y}{dx^2} - \frac{A_1}{x^2} \frac{dy}{dx} + \frac{2A_1}{x^3} Y - \frac{A_1}{x^2} \left(\frac{dy}{dx} - \frac{A_1}{x^2} Y \right) \right]$$

$$= e^{\frac{A_1}{x}} \left[\frac{d^2 y}{dx^2} - \frac{2A_1}{x^2} \frac{dy}{dx} + \left(\frac{2A_1}{x^3} + \frac{A_1^2}{x^4} \right) Y \right]$$

Hence putting these expressions for $\frac{dy}{dx}$ and $\frac{d^2 y}{dx^2}$ in the ode and dropping the common factor $e^{\frac{A_1}{x}}$ we have :

$$\frac{d^2 y}{dx^2} - \frac{2A_1}{x^2} \frac{dy}{dx} + \left(\frac{2A_1}{x^3} + \frac{A_1^2 + 1}{x^4} \right) Y = 0 \qquad \cdots (b)$$

The value of A_1 is determined by demanding the most divergent term in the co-efficient of Y in (b) to vanish. So we get :

$$A_1^2 + 1 = 0 \quad \Leftrightarrow \quad A_1 = \pm\, i$$

Plugging these values of A_1 in (b) we get :

$$\frac{d^2y}{dx^2} - \frac{2i}{x^2}\frac{dy}{dx} + \left(\frac{2i}{x^3}\right)Y = 0$$

and $\qquad \dfrac{d^2y}{dx^2} + \dfrac{2i}{x^2}\dfrac{dy}{dx} - \dfrac{2i}{x^3}Y = 0$

or, equivalently, $\quad x^3\dfrac{d^2y}{dx^2} - 2ix\dfrac{dy}{dx} + 2iY = 0$

and $\qquad x^3\dfrac{d^2y}{dx^2} + 2ix\dfrac{dy}{dx} - 2iY = 0$

However, we need not work with both of them because one is the complex conjugate of the other. For definiteness, let's work former.

Since $\quad Y(x) = \displaystyle\sum_{n=0}^{\infty} c_n x^{n+\lambda}$, $(c_0 \neq 0)$

$$\frac{dY}{dx} = \sum_{n=0}^{\infty} c_n(n+\lambda)x^{n+\lambda-1};$$

$$\frac{d^2Y}{dx^2} = \sum_{n=0}^{\infty} c_n(n+\lambda)(n+\lambda-1)x^{n+\lambda-2}$$

$$\therefore \quad \sum_{n=0}^{\infty} c_n(n+\lambda)(n+\lambda-1)x^{n+\lambda+1} - 2i\sum_{n=0}^{\infty} c_n(n+\lambda-1)x^{n+\lambda} = 0$$

Replacing n by $\overline{n-1}$ in the first term we get after regrouping :

$$\sum_{n=1}^{\infty}(n+\lambda-1)\{(n+\lambda-2)c_{n-1} - 2ic_n\}x^{n+\lambda} - 2ic_0(\lambda-1)x^\lambda = 0$$

Dropping x^λ from both sides and setting co-efficients of each power of x to zero we get :

$$2(\lambda-1)c_0 = 0 \quad \Leftrightarrow \quad \lambda = 1$$

and $\quad (n+\lambda-2)c_{n-1} - 2ic_n = 0 \;\forall\, n \geqslant 1$

For $\lambda = 1$, the above recurrence relation reduces to :

$$(n-1)c_{n-1} = 2ic_n \;\forall\, n \geqslant 1$$

Thus $c_n = 0 \;\forall\, n \geqslant 1$ and the Frobenius series terminates after one term.

This gives us one linearly independent solution of the given ode :

$$y_1(x) = x \, \exp\left(\frac{i}{x}\right) = x \left[\cos\left(\frac{1}{x}\right) + i \, \sin\left(\frac{1}{x}\right)\right] \qquad [\because Y(x) = c_0 x]$$

By complex conjugation one can have the second linearly independent solution, viz,

$$y_2(x) = x \exp\left(\frac{-i}{x}\right) = x \left[\cos\left(\frac{1}{x}\right) - i \, \sin\left(\frac{1}{x}\right)\right]$$

It is a routine exercise to show that $y = x \sin\left(\frac{1}{x}\right)$ and $y = x \cos\left(\frac{1}{x}\right)$ are two linearly independent solutions of the ode, originally given. Obviously these solutions are valid for small values of x, i,e, in the neighborhood of the origin. From the discussion (b) immediately preceding this example it follows that

$$z_1(x) = \frac{1}{x} \sin x \quad \text{and} \quad z_2(x) = \frac{1}{x} \cos x$$

are solutions of ode (8.23) near ∞, i,e, for large values of x. Therefore

$$\left. \begin{aligned} \bar{y}_1(x) &= x^{\frac{1}{2}} \, z_1(x) = x^{-\frac{1}{2}} \sin x \\ \bar{y}_2(x) &= x^{\frac{1}{2}} \, z_2(x) = x^{-\frac{1}{2}} \cos x \end{aligned} \right\}$$

are two linearly independent solutions of Bessel equation of order $\frac{1}{2}$ for large values of x. Hence without loss of generality, one can conclude that for large x (i,e, $x \gg 1$),

$$J_{\frac{1}{2}}(x) \equiv \sqrt{\frac{2}{\pi}} \, \bar{y}_1(x) = \sqrt{\frac{2}{\pi}} x^{-\frac{1}{2}} \sin x$$

and $\qquad J_{-\frac{1}{2}}(x) \equiv \sqrt{\frac{2}{\pi}} \, \bar{y}_2(x) = \sqrt{\frac{2}{\pi}} x^{-\frac{1}{2}} \cos x$

are again linearly independent solutions of Bessel equation of order $\frac{1}{2}$.

Remark : By an approach similar to that shown in the previous example, the reader can verify that for $x \gg 1$,

$$J_p(x) = \sqrt{\frac{2}{\pi x}} \cos\left(x - \frac{\pi}{4} - \frac{p\pi}{2}\right)$$

and $\qquad J_{-p}(x) \equiv \sqrt{\frac{2}{\pi x}} \cos\left(x - \frac{\pi}{4} + \frac{p\pi}{2}\right)$

are two linearly independent solutions of the Bessel equation of order p.

Example (28) : Solve the ode $x^2(x-2)^2y'' + 2(x-2)y' + (x+1)y = 0$
about the singular point $x = 0$

The singular point $x = 0$ is an irregular singular point of rank unity and so following the ansatz given in (8.22) the solution of this ode can be taken in the form :

$$y(x) = e^{\frac{A}{x}}Y(x) \; ; \; A \neq 0$$

where $Y(x)$ denotes the sum of the Frobenius series $\sum\limits_{n=0}^{\infty} c_n x^n$.

Term by term differentiation yields from (a)

$$y'(x) = e^{\frac{A}{x}}\left[\frac{dY}{dx} + \frac{A}{x^2}Y(x)\right]$$

$$y''(x) = e^{\frac{A}{x}}\left[\frac{d^2Y}{dx^2} - \frac{2A}{x}\frac{dY}{dx} + \left(\frac{2A}{x^3} + \frac{A^2}{x^4}\right)Y\right]$$

Plugging these in the given ode and dropping out factor $e^{\frac{A}{x}}$, we get :

$$x^2(x-2)^2\frac{d^2Y}{dx^2} + 2(x-2)\{1 - Ax(x-2)\}\frac{dY}{dx} +$$

$$Y\left[\frac{4}{x^2}(A^2 - A) + \frac{1}{x}(-4A^2 + 10A) + (1 - 8A + A^2) + x(1+2A)\right] = 0$$

$$\dots(a)$$

We now choose A so that the most divergent term in the co-efficient of Y vanishes. This leads us to

$$A^2 - A = 0 \iff A = 1 \text{ and } A = 0$$

Substituting $A = 1$ in (a) we get after some simplification :

$$x^3(x-2)^2\frac{d^2Y}{dx^2} + 2(x-2)x(1 + 2x - x^2)\frac{dY}{dx} + 3(x^2 - 2x + 2)Y = 0 \dots(b)$$

Now since $Y(x) = \sum\limits_{n=0}^{\infty} c_n x^{n+\lambda} \; ; \; c_0 \neq 0,$

$$\frac{dY}{dx} = \sum\limits_{n=0}^{\infty} c_n(n+\lambda)x^{n+\lambda-1} \; ; \; \frac{d^2Y}{dx^2} = \sum\limits_{n=0}^{\infty} c_n(n+\lambda)(n+\lambda-1)x^{n+\lambda-2}$$

Putting these in ode (b) we get :

$$(x-2)^2 \sum_{n=0}^{\infty} c_n(n+\lambda)(n+\lambda-1)x^{n+\lambda+1} + 2(x-2)(1+2x-x^2) \times$$

$$\sum_{n=0}^{\infty} c_n(n+\lambda)x^{n+\lambda} + 3(x^2-2x+2)\sum_{n=0}^{\infty} c_n x^{n+\lambda} = 0$$

i, e,
$$\sum_{n=0}^{\infty}\{c_n(n+\lambda)(n+\lambda-1) - 2c_n(n+\lambda)\}x^{n+\lambda+3}$$

$$+ \sum_{n=0}^{\infty}\{-4c_n(n+\lambda)(n+\lambda-1) + 8c_n(n+\lambda) + c_n\}x^{n+\lambda+2}$$

$$+ \sum_{n=0}^{\infty}\{4c_n(n+\lambda)(n+\lambda-1) - 6c_n(n+\lambda) - 2c_n\}x^{n+\lambda+1}$$

$$+ \sum_{n=0}^{\infty}\{-4c_n(n+\lambda) + 2c_n\}x^{n+\lambda} = 0$$

Dropping x^λ factor and writing the expressions compactly,

$$\sum_{n=0}^{\infty}(n+\lambda)(n+\lambda-3)c_n \, x^{n+3} +$$

$$\sum_{n=0}^{\infty}\{-4(n+\lambda)(n+\lambda-3) + 1\}c_n \, x^{n+2} +$$

$$\sum_{n=0}^{\infty}\{2(n+\lambda)(2n+2\lambda-5) - 2\}c_n \, x^{n+1} +$$

$$\sum_{n=0}^{\infty}\{-4(n+\lambda) + 2\}c_n \, x^n = 0$$

Replace n by $\overline{n-3}$ in the first sum, by $\overline{n-2}$ in the second sum and by $\overline{n-1}$ in the third sum to get :

$$\sum_{n=3}^{\infty}(n+\lambda-3)(n+\lambda-6)c_{n-3}x^n +$$

$$\sum_{n=2}^{\infty}\{1 - 4(n+\lambda-2)(n+\lambda-5)\}c_{n-2}x^n +$$

$$\sum_{n=1}^{\infty}\{2(n+\lambda-1)(2n+2\lambda-7) - 2\}c_{n-1}x^n +$$

$$\sum_{n=0}^{\infty}\{-4(n+\lambda)+2\}c_n x^n = 0$$

i, e, $(2-4\lambda)c_0 + \{(2-4(\lambda+1))c_1 + (4\lambda^2 - 10\lambda - 2)c_0\}x$

$+\{(2-4(\lambda+2))c_2 + (2(\lambda+1)(2\lambda-3)-2)c_1 + (1-4\lambda(\lambda-3))c_0\}x^2$

$+\sum_{n=3}^{\infty}[(n+\lambda-3)(n+\lambda-6)c_{n-3} + \{1-4(n+\lambda-2)(n+\lambda-5)\}c_{n-2}$

$+\{2(n+\lambda-1)(2n+2\lambda-7)-2\}c_{n-1} + \{-4(n+\lambda)+2\}c_n]x^n = 0$

Equating the co-efficients of each power of x to zero we get :

$$(2-4\lambda)c_0 = 0 \quad \Leftrightarrow \quad \lambda = \frac{1}{2} \quad (\because \ c_0 \neq 0)$$

$$-(2+4\lambda)c_1 + (4\lambda^2 - 10\lambda - 2)c_0 = 0 \qquad \dots (c)$$

$$-(4\lambda+6)c_2 + (4\lambda^2 - 2\lambda - 8)c_1 + (1 - 4\lambda^2 + 12\lambda)c_0 = 0 \quad \dots (d)$$

and the recurrence relation :

$$(n+\lambda-3)(n+\lambda-6)c_{n-3} + \{1-4(n+\lambda-2)(n+\lambda-5)\}c_{n-2}$$

$$+\{2(n+\lambda-1)(2n+2\lambda-7)-2\}c_{n-1} + \{-4(n+\lambda)+2\}c_n = 0 \ \dots (e)$$

Putting $\lambda = \frac{1}{2}$ in (c), (d) we have :

$$-(4c_1 + 6c_0) = 0 \quad \Rightarrow \quad c_1 = -\frac{3}{2}c_0$$

$$-8c_2 - 8c_1 + 6c_0 = 0 \quad \Rightarrow \quad c_2 = \frac{9}{4}c_0$$

From (e), we have, on substituting $\lambda = \frac{1}{2}$, the recurrence relation :

$$\frac{1}{4}(2n-5)(2n-11)c_{n-3} + \{1 - (2n-3)(2n-9)\}c_{n-2}$$

$$+(2n-1)(2n-6)c_{n-1} - 4n\, c_n = 0 \ \ \forall \, n \geqslant 3 \qquad \dots (f)$$

For $n = 3$ we get from (f),

$$-\frac{5}{4}c_0 + 10c_1 - 12c_3 = 0 \quad \Rightarrow \quad c_3 = \frac{-65}{48}c_0$$

For $n = 4$ we get from (f),

$$-\frac{9}{4}c_1 + 6c_2 + 14c_3 - 16c_4 = 0$$

or, $\left(\frac{9}{4}\right)\left(\frac{3}{2}\right)c_0 + \left(\frac{27}{2}\right)c_0 + 14\left(\frac{-65}{48}\right)c_0 - 16c_4 = 0$

$$\therefore \quad c_4 = \frac{-25}{12}c_0 \ \text{ and so on.}$$

Hence the Frobenius term reads :

$$Y(x) = c_0 x^{\frac{1}{2}} \left[1 - \frac{3}{2}x + \frac{9}{4}x^2 - \frac{65}{48}x^3 - \frac{25}{12}x^4 + \ldots \infty \right]$$

So one solution of the ode originally given will read :

$$y(x) = c_0 e^{\frac{1}{x}} x^{\frac{1}{2}} \left[1 - \frac{3}{2}x + \frac{9}{4}x^2 - \frac{65}{48}x^3 - \frac{25}{12}x^4 + \ldots \infty \right],$$

c_0 being a non-zero constant. The second linearly independent solution can now be obtained by D'Alembert's order reduction method.

Remark : The solution of the ode $y'' + p(x)y' + q(x)y = 0$ can be taken in the form $y(x) = e^{\phi(x)} Y(x)$ provided we can choose $\phi(x)$ in such a way that the transformed ode

$$Y''(x) + (p(x) + 2\phi'(x))Y'(x) + \{q(x) + p(x)\phi'(x) + \phi''(x) + \phi'^2(x)\}Y(x) = 0$$

has a regular solution, i,e, a solution having the form of either a Frobenius series or the product of a Frobenius series with a logarithmic multiplier. It can be shown that to achieve this target $\phi(x)$ must be a polynomial in $\frac{1}{x}$. In general these integrals about an irregular singular point are known as **normal integrals**. The normal integrals, in general, are divergent but they are asymptotic expansions of the solution to the original ode. So despite their divergent character, they carry an immense importance in the study of solutions of odes.

Exercise 8

1. Locate the singular point, if any, of the following differential equations and classify them as regular and irregular. (You are supposed to restrict your search in the finite part of the complex plane).

   (a) $x(1-x)y'' - 3y' + 2y = 0$
   (b) $x(1-x)y'' - 3xy' - y = 0$
   (c) $x(1-x)y'' + (2-3x)y' - y = 0$
   (d) $x^2(1+2x)y'' + 2x(1+6x)y' - 2y = 0$
   (e) $2x^3 y'' - x(2-5x)y' + y = 0$
   (f) $x(1-x)y'' + (\gamma - (1+\alpha+\beta)x)y' - \alpha\beta y = 0,$
       $(\alpha, \beta, \gamma$ being constants$)$

(g) $2x^2(1-x)y'' - 5x(1+x)y' + (5-x)y = 0$
(h) $x^3(x-2)(x+3)y'' + x^2y' + (x-2)y = 0$

2. For each of the problems given in the previous set check whether $x = \infty$ is a singular point or not. If yes, determine its nature.

3. Prove that the singular points of the ode

$$(x\sin x)y'' + (\cos x)y' + e^x y = 0$$

are in A.P. and hence identify the regular and irregular ones.

4. Solve the following differential equations about the indicated ordinary points by Power-series method and also determine the radius of convergence of each solution :

(a) $y'' + xy = 0$ $[x = 0]$
(b) $y'' + (1+x)y = 0$ $[x = 0]$
(c) $y'' + (x^2-1)y = x^2$ $[x = 0]$
(d) $y'' + x^2y = 0$ $[x = 0]$
(e) $(x-2)y'' + y = 0$ $[x = 1]$
(f) $(x^2-1)y'' - 4y = 0$ $[x = 0]$

5. Solve the following initial-value problems by Power-series method about the origin $x = 0$.

(a) $y'' + (1-x)y = 0$; $y(0) = 1,\ y'(0) = 0$
(b) $y'' - x^2y' + (\sin x)y = 0$; $y(0) = 0,\ y'(0) = 1$
(c) $(1-x^2)y'' + 2y = 0$; $y(0) = 4,\ y'(0) = 5$

6. For each of the following differential equations $x = 0$ is a regular singular point. Use Frobenius method to find a general solution of each of these equations in the neighborhood of $x = 0$ and determine the region of convergence of the series solutions in each case.

(a) $2x^2y'' + x^2y' - y = 0$
(b) $2x^2y'' + x(4x-1)y' + 2(3x-1)y = 0$
(c) $x^2y'' - xy' + (x^2-3)y = 0$
(d) $xy'' - xy' + \left(2x^2 + \frac{5}{9}\right)y = 0$
(e) $xy'' - (x^2+2)y' + xy = 0$

7. Find the general solution of each of the following differential equations about each of the regular singular points by using Frobenius technique. Without computing literally the radius of convergence for these Frobenius series, can you predict the region of convergence of each of the series solutions.

   (a) $2x(1-x)y'' + y' - y = 0$
   (b) $4x^2(1-x)y'' - xy' + (1-x)y = 0$
   (c) $2x^2(1-x)y'' - x(1+7x)y' + y = 0$
   (d) $x(x-2)y'' + 2(x-1)y' - 2y = 0$
   (e) $(2x^2 - x)y'' + 2(x-1)y' - (2x^2 - 3x + 2)y = 0$
   (f) $xy'' + (1-x)y' + \lambda y = 0$, λ is a constant [Laguerre equation]
   (g) $(1-x^2)y'' - xy' + \alpha^2 y = 0$, α is a constant [Chebyshev equation]
   (h) $4x^2(x^2-1)y'' + 8x^3 y' - y = 0$

8. Using matrix method can you provide the indicial equation of each of the differential equation in problem no. 6 ?

9. Identify the ode $x(1-x)y'' + 2(1-x)y' + 2y = 0$ as a hypergeometric equation and hence write down its linearly independent solutions about $x = 0$ and $x = 1$ in terms of hypergeometric functions.

10. Describe the behavior of the integral curves near the singular points for each of the following differential equations.

    (a) $(1-x^2)y'' - 2xy' + m(m+1)y = 0$ [Legendre]
    (b) $xy'' + (1-x)y' + \lambda y = 0$ [Laguerre's]
    (c) $x^2 y'' + xy' + (x^2 - p^2)y = 0$ [Bessel]
    (d) $x(1-x)y'' + \{c - (1+a+b)x\}y' - aby = 0$ [Hypergeometric].

    [Clue : Negative exponent for the ode, i,e, negative root of the indicial equation indicates existence of a pole in the solution. Existence of positive fractional exponent i,e, positive fractional root of the indicial equation indicates existence of a branch point in the solution.

    These behaviors occur at a regular singular point.]

11. Show that $x = 0$ is an irregular singular point of rank 1 for the differential equation

$$x^2 y'' - (1-3x)y' + y = 0$$

and that $y = \frac{1}{x}e^{-\frac{1}{x}}$ is one of its solutions. Can you find another linearly independent solution ?

12. (a) Solve the following differential equations about $x = 0$.

   (i) $x^4 y'' + 2x^3 y' - y = 0$
   (ii) $x^4 y'' + x(1 + 2x^2)y' + 5y = 0$

   For show that both have $x = 0$ as irregular singular point. If one tries Frobenius series solution for these equation, why should the approach fail ? Try the ansatz prescribed in (8.23). Also find the solution for large values of x.

   (b) Show that $x = 0$ is an irregular singular point of rank 1 of the differential equation

   $$x^2 y'' + (3x - 1)y' + y = 0$$

   Explain why should we not be successful if we dare to use Frobenius method for its solution ? Now apply the ansatz of (8.23) for finding one solution of this equation. Can you venture to solve this ode by Laplace transform ?

   (c) Check that $y = c_1 \cos\left(\frac{1}{\sqrt{x}}\right) + c_2 \sin\left(\frac{1}{\sqrt{x}}\right)$ is the general solution of the ode

   $$4x^3 y'' + 6x^2 y' + y = 0$$

   about the irregular singular point $x = 0$

13. Solve the ode $y'' - k^2 xy = 0$ and check that its general solution is:

$$y(x) = x^{\frac{1}{2}}\left[c_1 J_{\frac{1}{3}}\left(\frac{2}{3}kx^{\frac{3}{2}}\right) + c_2 J_{-\frac{1}{3}}\left(\frac{2}{3}kx^{\frac{3}{2}}\right)\right],$$

where $J(\cdot)$ is a Bessel function of the first kind.

14. (a) Show that one Frobenius series solution exists for the ode $x^2 y'' + xy' + (x - 1)y = 0$ and that is given by

$$y_1(x) = x\left[1 + \sum_{n=1}^{\infty} \frac{(-1)^n x^n}{n!(n + 2)!}\right]$$

(b) show that the general solution of the ode (about $x = 0$)

$$xy'' + (x - 1)y' - y = 0$$

will be $y(x) = c_1(x - 1 + e^{-x}) + c_2 e^{-x}$.

(c) Show that the power series $(1 - 3x + 2x^2 - \frac{2}{3}x^3 + \cdots \infty)$ is a solution of the ode

$$2xy'' + (x + 1)y' + 3y = 0.$$

Can you provide another linearly independent solution?

Chapter 9

Solving Linear Systems by Matrix Methods

9.1 Introduction

Solving a system of simultaneous linear differential equations by the matrix methods is the objective of our final chapter. We shall see here how the techniques of linear algebra provide us ways to achieve this target. We shall henceforth look into a system of simultaneous linear differential equations as a linear vector differential equation.

We first begin with a linear system having the form :

$$y_1' = a_{11}(x)y_1 + a_{12}(x)y_2 + \cdots\cdots + a_{1n}(x)y_n + f_1(x)$$
$$y_2' = a_{21}(x)y_1 + a_{22}(x)y_2 + \cdots\cdots + a_{2n}(x)y_n + f_2(x)$$

$$\cdots\cdots\cdots\cdots\cdots\cdots\cdots\cdots\cdots\cdots\cdots\cdots\cdots\cdots\cdots\cdots\cdots\cdots$$

$$\cdots\cdots\cdots\cdots\cdots\cdots\cdots\cdots\cdots\cdots\cdots\cdots\cdots\cdots\cdots\cdots\cdots\cdots$$

$$y_n' = a_{n1}(x)y_1 + a_{n2}(x)y_2 + \cdots\cdots + a_{nn}(x)y_n + f_n(x)$$

where we presume $a_{ij}(x)$; $i, j = 1, 2, .., n$ and $f_i(x)(i = 1, 2, .., n)$ to be continuous real-valued functions of independent variable x over some interval I. We say that the set of n functions $\{y_1, y_2, ..., y_n\}$ is a solution of this linear system over I if these n functions are continuously differentiable over I and satisfies the above system of odes, Technically speaking, $y_r \in C'(I)$ for $r = 1, 2, .., n$ and $a_{ij} \in C(I)$ for $i = 1, 2, .., n$.

If one introduces the vector functions

$$\underset{\sim}{y} = \begin{pmatrix} y_1 \\ y_2 \\ \vdots \\ y_n \end{pmatrix} \; ; \; \underset{\sim}{y}' = \begin{pmatrix} y_1' \\ y_2' \\ \vdots \\ y_n' \end{pmatrix} \; ; \; \underset{\sim}{f}(x) = \begin{pmatrix} f_1(x) \\ f_2(x) \\ \vdots \\ f_n(x) \end{pmatrix}$$

and the matrix function

$$A(x) = [a_{ij}(x)]_{n \times n} \; ; \; x \in I,$$

then the above system can be expressed in equivalent vector equation
form as :

$$y' = A(x)y + f(x) \tag{9.1}$$

where $A(x)$ is a $n \times n$ matrix function continuous over I and $f(x)$ is
a $n \times 1$ matrix function continuous over I while y is a $n \times 1$ matrix
function continuously differentiable over I. In the light of vector calculus
it follows that the above matrix functions owe their properties from the
common properties shared by their components.

Let's now explain why we refer to the system (9.1) as a linear vector
differential equation. Recalling that the set $C(I)$ of all real-valued continuous functions defined over I forms a real vector space w.r.to pointwise addition and scalar multiplication, it is easy to show that the set of
$n \times 1$ matrix functions continuous over I forms a vector space $\prod_{1 \leqslant i \leqslant n} C(I)$
under usual matrix addition and scalar multiplication. Mathematically
speaking, if u and v belong to $\prod_{1 \leqslant i \leqslant n} C(I)$, then for any $\alpha, \beta \in \mathbb{R}$, $(\alpha u + \beta v)$
is again a continuous vector function over I, i,e, $(\alpha u + \beta v) \in \prod_{1 \leqslant i \leqslant n} C(I)$.
Similarly it is easy to verify that the set of $n \times 1$ matrix functions contin-
uously differentiable over I forms a vector space $\prod_{1 \leqslant i \leqslant n} C'(I)$ under usual
matrix addition and scalar multiplication. Mathematically, if u, v belong
to $\prod_{1 \leqslant i \leqslant n} C'(I)$, then $\alpha u + \beta v \in \prod_{1 \leqslant i \leqslant n} C'(I)$ for any $\alpha, \beta \in \mathbb{R}$. For brevity
let's write

$$V \equiv \prod_{1 \leqslant i \leqslant n} C'(I) \; ; \; W \equiv \prod_{1 \leqslant i \leqslant n} C(I)$$

and define a transformation L from V to W by the following ansatz :

$$Ly(x) \equiv y'(x) - A(x)y(x) \; ,$$

where $x \in I$ and $A(x)$ is a given $n \times n$ continuous matrix function. L
defined above is a linear transformation as can be shown in the following.

Since V is a vector space, for any $\alpha, \beta \in \mathbb{R}$ and any $y(x), z(x) \in V$,
$\alpha y + \beta z \in V$ and so

$$\begin{aligned}
L[\alpha y + \beta z](x) &= (\alpha y' + \beta z')(x) - A(x)(\alpha y + \beta z)(x) \\
&= \alpha\{y'(x) - A(x)y(x)\} + \beta\{z'(x) - A(x)z(x)\} \\
&= \alpha Ly(x) + \beta Lz(x)
\end{aligned}$$

Hence the linear system can be written as :

$$Ly(x) = f(x) \tag{9.2}$$

Since L is a linear transformation from the vector space V to the vector space W, we can look into the linear system (9.1) as a linear vector differential equation (9.2). In particular if $f = 0$, (9.2) is called 'homogeneous linear vector differential equation' and if $f \neq 0$, (9.2) is referred to as 'inhomogeneous linear vector differential equation'.

Any reader can at once realise that any higher order linear ode as equivalent to a linear vector differential equation with the companion matrix playing the role of $A(x)$ in (9.1). We start with equation (5.3) which represents the normal form of the nth order ode :

$$y^{(n)} = -p_1(x)y^{(n-1)} - p_2(x)y^{(n-2)} - \cdots\cdots - p_{n-1}(x)y' - p_n(x)y + q(x)$$

where $x \in I$, some interval. If we introduce new functions $z_1, z_2, \cdots, z_n$ defined by

$$z_i(x) = y^{(i-1)}(x) \quad \text{for} \quad i = 1, 2, \cdots n, \quad \text{then}$$

$$z_i'(x) \equiv \frac{dz_i}{dx} = z_{i+1}(x) \quad \text{for} \quad i = 1, 2, \cdots, \overline{n-1} \ ;$$

$$z_n'(x) \equiv \frac{dz_n}{dx} = -p_1(x)z_n - p_2(x)z_{n-1} - \cdots - p_n(x)z_1 + q(x)$$

In the matrix form this reads :

$$\frac{d}{dx}\begin{pmatrix} z_1 \\ z_2 \\ \vdots \\ z_n \end{pmatrix} = \begin{bmatrix} 0 & 1 & 0 & \cdots & 0 \\ 0 & 0 & 1 & \cdots & 0 \\ \vdots & \vdots & & & \\ -p_n & -p_{n-1} & -p_{n-2} & \cdots & -p_1 \end{bmatrix}\begin{bmatrix} z_1 \\ z_2 \\ \vdots \\ z_n \end{bmatrix} + \begin{bmatrix} 0 \\ 0 \\ \vdots \\ q(x) \end{bmatrix}$$

i,e, $\quad \dfrac{d}{dx}(z(x)) = \begin{bmatrix} 0 & 1 & 0 & \cdots & 0 \\ 0 & 0 & 1 & \cdots & 0 \\ \vdots & \vdots & & & \\ -p_n & -p_{n-1} & -p_{n-2} & \cdots & -p_1 \end{bmatrix} z(x) + f(x)$

where we define $\quad z(x) \equiv \begin{pmatrix} z_1 \\ z_2 \\ \vdots \\ z_n \end{pmatrix} \ ; \ f(x) \equiv \begin{pmatrix} 0 \\ 0 \\ \vdots \\ q(x) \end{pmatrix}$

Note that the companion matrix plays the role of matrix $A(x)$ here.

We first start working with homogeneous linear vector differential equation, viz,

$$Ly(x) = 0 \quad \text{i, e, } y'(x) = A(x)\, y(x) \tag{9.3}$$

If $\phi_1(x), \phi_2(x), \cdots, \phi_n(x)$ be any n solution of (9.3), that is if $L\phi_k(x) = 0 \ \forall \ k = 1, 2, \cdots, n$, then for any $c_1, c_2, \cdots, c_n$ belonging to $\mathbb{R}$, the linear combination $c_1 \underset{\sim}{\phi}_1(x) + c_2 \underset{\sim}{\phi}_2(x) + \cdots + c_n \underset{\sim}{\phi}_n(x)$ satisfies (9.3) because

$$L\left[c_1 \underset{\sim}{\phi}_1(x) + c_2 \underset{\sim}{\phi}_2(x) + \cdots + c_n \underset{\sim}{\phi}_n(x)\right] = \sum_{k=1}^{n} c_k \, L\underset{\sim}{\phi}_k(x) = \underset{\sim}{0} \tag{9.4}$$

This shows that the set S of all possible solutions of the equation (9.3) forms a vector space. If $\underset{\sim}{\psi}_1(x), \underset{\sim}{\psi}_2(x), \cdots, \underset{\sim}{\psi}_n(x)$ be n elements of this vector space such that they are linearly independent and they are capable of spanning this vector space S, then $\{\underset{\sim}{\psi}_1, \underset{\sim}{\psi}_2, \cdots, \underset{\sim}{\psi}_n\}$ will be a basis of S. In our workout the main objective is therefore to determine a basis $\{\underset{\sim}{\psi}_1, \underset{\sim}{\psi}_2, \cdots, \underset{\sim}{\psi}_n\}$ of the solution space S of (9.3). Once that is done, we can at once write down the general solution $\underset{\sim}{\psi}$ as a linear combination of them :

$$\underset{\sim}{\psi}(x) = c_1 \underset{\sim}{\psi}_1(x) + c_2 \underset{\sim}{\psi}_2(x) + \cdots + c_n \underset{\sim}{\psi}_n(x) \ ; \ c_i \in \mathbb{R} \quad \forall i = 1(1)n$$

We now enunciate two important results related to the theory of vector differential equations in the form of the following theorems.

Theorem 9.1. [Existence-Uniqueness Theorem]

If the $n \times n$ matrix function A and the n-column vector function b are continuous functions over some interval I, then the IVP

$$\underset{\sim}{y}'(x) = A(x)\underset{\sim}{y}(x) + \underset{\sim}{b}(x) \ ; \ \underset{\sim}{y}(x_0) = \underset{\sim}{\xi}_0 \,,$$

where $x_0 \in I$ and $\underset{\sim}{\xi}_0$ is a constant n-column vector, admits of a unique solution that exists over the whole interval.

Note (a): In the language of linear operator L introduced in our discussion, the IVP can alternatively be written as

$$L\underset{\sim}{y}(x) = \underset{\sim}{b}(x) \ ; \ \underset{\sim}{y}(x_0) = \underset{\sim}{\xi}_0,$$

with interpretations of x_0 and ξ_0 remaining the same.

(b) This result is a generalisation of the Existence-Uniqueness theorem for higher order linear odes stated earlier.

Theorem 9.2. The vector differential equation (9.3), viz,

$$Ly(x) \equiv y'(x) = A(x)y(x) = 0 \; ; \; x \in I$$

admits of n linearly independent solutions and if $\psi_1(x), \psi_2(x), \cdots, \psi_n(x)$ are n linearly independent solutions of (9.3) on I, then every solution of (9.3) is a linear combination of $\psi_1(x), \cdots, \psi_n(x)$.

Proof : We begin our proof with the following surrogate I.V.P :

$$y'(x) = A(x)y(x) \text{ subject to } y(x_0) = \xi_k; \; k = 1, 2, \cdots, n \; ; \; x_0 \in I \;,$$

where ξ_k's are linearly independent n-column vectors.

Suppose $\psi_1(x), \psi_2(x), \cdots, \psi_n(x)$ are n solutions of this I.V.P. taken into consideration.

Assume the relation

$$c_1 \psi_1(x) + c_2 \psi_2(x) + \cdots + c_n\psi_n(x) = 0 \; ; \quad x \in I.$$

In particular, if $x = x_0$, we get :

$$c_1 \psi_1(x_0) + c_2 \psi_2(x_0) + \cdots + c_n \psi_n(x_0) = 0$$

i,e, $\quad c_1 \xi_1 + c_2 \xi_2 + \cdots + c_n \xi_n = 0$

However, $\xi_1, \xi_2, \cdots, \xi_n$ being linearly independent, $c_1 = \cdots = c_n = 0$ and consequently $\psi_1(x), \psi_2(x), \cdots, \psi_n(x)$ are n linearly independent solutions of the vector differential equation (9.3). This completes the proof of the existence of n linearly independent solutions.

Due to the relation (9.4) stated earlier, one can directly conclude that any linear combination of ψ_k's is again a solution of (9.3), i,e, in other words, span $\{\psi_1(x), \psi_2(x), \cdots, \psi_n(x)\} \subseteq S$, the set of all possible solutions of (9.3). We are therefore left with the task of proving that every solution of (9.3) is a linear combination of ψ_k's ($k = 1, 2, \cdots, n$), i,e, in mathematical language, $S \subseteq$ Span $\{\psi_1(x), \psi_2(x), \cdots, \psi_n(x)\}$. To prove this, let's define n constant vectors $\underset{\sim}{\eta}_k \equiv \underset{\sim}{\psi}_k(x_0)$ with $x_0 \in I$ and assume the relation

$$c_1' \, \underset{\sim}{\eta}_1 + c_2' \, \underset{\sim}{\eta}_2 + \cdots + c_n' \, \underset{\sim}{\eta}_n = 0 \ , \ c_k' \text{'s being scalars.}$$

Define $\qquad \underset{\sim}{u}(x) \equiv \sum_{k=1}^{n} c_k' \, \underset{\sim}{\psi}_k(x) \ ; \ x \in I$

$$\therefore \quad \underset{\sim}{u}(x_0) = \sum_{k=1}^{n} c_k' \, \underset{\sim}{\psi}_k(x_0) = \sum_{k=1}^{n} c_k' \, \underset{\sim}{\eta}_k = \underset{\sim}{0} \ .$$

Using the Uniqueness theorem 9.1, it follows that $u(x)$ is the trivial zero solution and hence $\sum_{k=1}^{n} c_k' \underset{\sim}{\psi}_k(x) = \underset{\sim}{0}$. But since $\underset{\sim}{\psi}_k(x)$'s are linearly independent, c_k''s are all zeros. This in turn leads to the conclusion that $\underset{\sim}{\eta}_1 \equiv \underset{\sim}{\psi}_1(x_0)$, $\underset{\sim}{\eta}_2 \equiv \underset{\sim}{\psi}_2(x_0) \cdots, \underset{\sim}{\eta}_n \equiv \underset{\sim}{\psi}_n(x_0)$ are all linearly independent.

Let $\underset{\sim}{v}(x)$ be any arbitrary solution of (9.3). As $x_0 \in I$, the constant vector $\underset{\sim}{v}(x_0)$ is expressible uniquely as a linear combination of n constant vectors $\underset{\sim}{\eta}_1, \underset{\sim}{\eta}_2, \cdots, \underset{\sim}{\eta}_n$. So $\exists$ scalars $r_1, r_2, \cdots, r_n$ such that

$$r_1 \, \underset{\sim}{v}_1(x_0) + r_2 \, \underset{\sim}{v}_2(x_0) + \cdots + r_n \, \underset{\sim}{v}_n(x_0) = v(x_0)$$

By Theorem (9.1) it now follows that

$$\underset{\sim}{v}(x) = r_1 \, \underset{\sim}{v}_1(x) + r_2 \, \underset{\sim}{v}_2(x) + \cdots + r_n \, \underset{\sim}{v}_n(x) \quad \forall x \in I \ ,$$

i,e, $\quad \underset{\sim}{v} \in \text{Span} \{\underset{\sim}{v}_1(x), \underset{\sim}{v}_2(x), \cdots, \underset{\sim}{v}_n(x)\}$.

This completes the proof of our claim that $\{\underset{\sim}{v}_1(x), \underset{\sim}{v}_2(x), \cdots, \underset{\sim}{v}_n(x)\}$ serves as a basis of the solution space S of the linear vector differential equation (9.3).

Before we proceed further, we must be aware of one of the weakness in our development made so far. We have assumed that the elements of the matrix $A(x) \equiv [a_{ij}(x)]_{n \times n}$ are all real-valued continuous functions defined on I and the scalars appearing in the linear combinations like(9.4) are all real. However, one should replace $\mathbb{R}$ by $\mathbb{C}$ as otherwise we shall not be able to encompass all sorts of vector differential equations having the form (9.3).

Once the theorems (9.1) and (9.2) are in our repertoire, we may venture to find the solution of the vector differential equation

$$\underset{\sim}{y}'(x) = A\underset{\sim}{y}(x) \ , \tag{9.5}$$

with A being a constant $n \times n$ matrix. The treatment we are going to pursue is based on theory of eigenvalues and eigenvectors of A.

We recall that a non-null vector $\underset{\sim}{x} \in \mathbb{R}^n$ or $\mathbb{C}^n$ is an eigenvector of a constant matrix $A \equiv [a_{ij}]_{n \times n}$ (with $a_{ij} \in \mathbb{R}$ or $\mathbb{C}$) provided $\exists$ scalar $\lambda(\lambda \in \mathbb{R}$ or $\mathbb{C})$ such that $Ax = \lambda x$. The scalar λ is referred to as eigenvalue of A corresponding to the eigenvector $\underset{\sim}{x}$. The pair $(\lambda, \underset{\sim}{x}$) is known to be an eigenpair of matrix A. Here from follows the fact that the linear homogeneous vector equation $(A - \lambda I_n)\underset{\sim}{x} = \underset{\sim}{0}$ admits of non-null/non-trivial solution provided

$$\det (A - \lambda I_n) = 0 \qquad (9.6)$$

This equation is a nth degree polynomial equation in λ and so by Fundamental Theorem of Algebra, admits of exactly n roots. Thus the constant matrix A has n eigenvalues. Hence solving(9.5) is tantamount to solving the eigenvalue problem for the associated matrix A. One should note that this availability of n roots of the characteristic equation (9.6) depends on the assumption that the underlying field is $\mathbb{C}$.

Our next section is devoted to the various possible results and examples related to eigenvalue problems. Any advanced learner of linear algebra can safely escape it but we have a brief discussion of the same to make our treatment self-sufficient and complete.

9.2   Eigenvalue Problems of a square matrix : Diagonalisability

We begin the section with the following example.

Example 9.1. Find the eigenpairs of the 2×2 matrix $A = \begin{pmatrix} 0 & 1 \\ -2 & -3 \end{pmatrix}$.

The characteristic equation for A is given by

$$0 = \det (A - \lambda I_2) = \begin{vmatrix} -\lambda & 1 \\ -2 & -3-\lambda \end{vmatrix} = (\lambda+1)(\lambda+2)$$

so that the eigenvalues of A are -1 and -2.

The eigenvector corresponding to $\lambda = -1$ is $\begin{pmatrix} x_1 \\ x_2 \end{pmatrix}$, where

$$\begin{pmatrix} 0 \\ 0 \end{pmatrix} = \begin{pmatrix} 1 & 1 \\ -2 & -2 \end{pmatrix} \begin{pmatrix} x_1 \\ x_2 \end{pmatrix} \equiv \begin{pmatrix} x_1 + x_2 \\ -2x_1 - 2x_2 \end{pmatrix}$$

i,e, $x_1 = -x_2 = c \neq 0$, say.

Hence $c \begin{pmatrix} 1 \\ -1 \end{pmatrix}$; $c \neq 0$ are the eigenvectors of A corresponding to

$\lambda = -1$ and consequently $\left\{ -1, \begin{pmatrix} 1 \\ -1 \end{pmatrix} \right\}$ is an eigenpair of A.

The eigenvector corresponding to $\lambda = -2$ is $\begin{pmatrix} y_1 \\ y_2 \end{pmatrix}$, where

$$\begin{pmatrix} 0 \\ 0 \end{pmatrix} = \begin{pmatrix} 2 & 1 \\ -2 & -1 \end{pmatrix} \begin{pmatrix} y_1 \\ y_2 \end{pmatrix} \equiv \begin{pmatrix} 2y_1 + y_2 \\ -2y_1 - y_2 \end{pmatrix}$$

i,e, $y_2 = -2y_1 = -2c' \neq 0$

i,e, $c' \begin{pmatrix} 1 \\ -2 \end{pmatrix}$; $c' \neq 0$ are the eigenvectors of A corresponding to $\lambda = -2$

and so $\left\{ -2, \begin{pmatrix} 1 \\ -2 \end{pmatrix} \right\}$ is another eigenpair of A.

The eigenvectors $\begin{pmatrix} 1 \\ -1 \end{pmatrix}$ and $\begin{pmatrix} 1 \\ -2 \end{pmatrix}$ of A are linearly independent

since $\begin{vmatrix} 1 & 1 \\ -1 & -2 \end{vmatrix} = -1 \neq 0.$

This implies that the two one-dimensional subspaces, viz, V_1 spanned

by $\begin{pmatrix} 1 \\ -1 \end{pmatrix}$ and V_2 spanned by $\begin{pmatrix} 1 \\ -2 \end{pmatrix}$ are disjoint in the sense

$$V_1 \cap V_2 = \left\{ \begin{pmatrix} 0 \\ 0 \end{pmatrix} \right\} \text{ and moreover }, \quad \mathbb{R}^2 = V_1 \oplus V_2$$

So the constant matrix A has procured a basis of $\mathbb{R}^2$ through its eigenvectors.

The example discussed above illustrates the simple truth that the eigenvectors corresponding to distinct eigenvalues of a square matrix are linearly independent and moreover, when A is such that none of its eigenvalues are repeated, then the eigenvectors of A procure a basis of the vector space over which A acts as a linear operator. (A is a called **regular** matrix in this case).

Example 9.2. Find the eigenpairs of the matrix $A = \begin{pmatrix} 2 & 2 & 1 \\ 1 & 3 & 1 \\ 1 & 2 & 2 \end{pmatrix}$

The characteristic equation of A is $0 = (1 - \lambda)^2 (5 - \lambda)$,

so that the eigenvalues are 5,1,1.

If $\begin{pmatrix} x_1 \\ x_2 \\ x_3 \end{pmatrix}$ be an eigenvector corresponding to $\lambda = 5$, then

$$\begin{pmatrix} -3 & 2 & 1 \\ 1 & -2 & 1 \\ 1 & 2 & -3 \end{pmatrix} \begin{pmatrix} x_1 \\ x_2 \\ x_3 \end{pmatrix} = \begin{pmatrix} 0 \\ 0 \\ 0 \end{pmatrix} \qquad (a)$$

i,e, equivalently,

$$\left. \begin{array}{r} x_1 + 2x_2 - 3x_3 = 0 \\ x_2 - x_3 = 0 \end{array} \right\}$$

[This equivalent system is obtainable if we apply elementary row operations on the above system (a)]

$\therefore \ x_1 = x_2 = x_3 = c(\neq 0)$ is a non-trivial solution of (a) and hence

$c \begin{pmatrix} 1 \\ 1 \\ 1 \end{pmatrix}$; $c \neq 0$ are the eigenvectors corresponding to $\lambda = 5$.

Now let's determine eigenvector corresponding to $\lambda = 1$. It is also

of the generic form $\begin{pmatrix} x_1 \\ x_2 \\ x_3 \end{pmatrix}$, where $x_1 + 2x_2 + x_3 = 0$. If we choose

$x_2 = a, x_3 = b$, then $x_1 = -2a - b$

Therefore, $\begin{pmatrix} x_1 \\ x_2 \\ x_3 \end{pmatrix} = a \begin{pmatrix} -2 \\ 1 \\ 0 \end{pmatrix} + b \begin{pmatrix} -1 \\ 0 \\ 1 \end{pmatrix}$

This shows that corresponding to $\lambda = 1$ there exist two linearly independent eigenvectors, viz, $(-2, 1, 0)^T$ and $(-1, 0, 1)^T$.

The three eigenvectors of A determined above are linearly independent as can be verified by computing determinants. We observe that eigenvalue 5 of A is non-repeated while the eigenvalue 1 of A is 2-fold since its algebraic as well as geometric multiplicity is 2. Since in the present example all the eigenvalues have their geometric multiplicity equal to the algebraic multiplicity, A is regular. The eigenvectors of A provide a basis of $\mathbb{R}^3$ since one can express $\mathbb{R}^3$ as :

$$\mathbb{R}^3 = \text{Span} \left\{ \begin{pmatrix} -2 \\ 1 \\ 0 \end{pmatrix} \right\} \oplus \text{Span} \left\{ \begin{pmatrix} -1 \\ 0 \\ 1 \end{pmatrix} \right\} \oplus \text{Span} \left\{ \begin{pmatrix} 1 \\ 1 \\ 1 \end{pmatrix} \right\},$$

where $\oplus$ denotes direct sum of subspaces of $\mathbb{R}^3$.

It is a general observation that for regular matrices(no matter whether eigenvalues are distinct or not) one can construct a basis (made up of the eigenvectors of A) for the underlying vector space of linear transformation associated with matrix A.

Let's march on with the previous example [i,e, Example 9.2] and construct the non-singular modal matrix P given by

$$P = \begin{pmatrix} -2 & -1 & 1 \\ 1 & 0 & 1 \\ 0 & 1 & 1 \end{pmatrix}$$

Obviously, one can compute $det\ P = 4$ and P^{-1} to be

$$P^{-1} = \frac{1}{4} \begin{pmatrix} -1 & 2 & -1 \\ -1 & -2 & 3 \\ 1 & 2 & 1 \end{pmatrix}$$

Hence $P^{-1}\ AP = \frac{1}{4} \begin{pmatrix} -1 & 2 & -1 \\ -1 & -2 & 3 \\ 1 & 2 & 1 \end{pmatrix} \begin{pmatrix} 2 & 2 & 1 \\ 1 & 3 & 1 \\ 1 & 2 & 2 \end{pmatrix} \begin{pmatrix} -2 & -1 & 1 \\ 1 & 0 & 1 \\ 0 & 1 & 1 \end{pmatrix}$

$$= \begin{pmatrix} 1 & 0 & 0 \\ 0 & 1 & 0 \\ 0 & 0 & 5 \end{pmatrix}$$

Thus the similarity transformation of A by modal matrix has diagonalised A. Infact this feature of diagonalisation of matrix A owes its origin from the fact that one can have a basis of the underlying vector space from the knowledge of eigenvectors. This property is common to all simple matrices (i,e, matrices all of whose eigenvalues are regular).[One of the most remarkable theorems in matrix theory is that real symmetric matrices are simple].

We now prove two salient results related to eigenvalue problems.

Theorem 9.3. The eigenvectors corresponding to two different eigenvalues of a square matrix are linearly independent.

Proof. Let λ and μ be two distinct eigenvalues of square matrix A of order n and let $\underset{\sim}{x}_\lambda$ and $\underset{\sim}{x}_\mu$ be its corresponding eigenvectors. We want to prove that $\underset{\sim}{x}_\lambda$ and $\underset{\sim}{x}_\mu$ are linearly independent.

Consider the relation : $a\underset{\sim}{x}_\lambda + b\underset{\sim}{x}_\mu = \underset{\sim}{0}$

Apply A on both sides to yield :

$\underset{\sim}{0} = aA\underset{\sim}{x}_\lambda + bA\underset{\sim}{x}_\mu = a\lambda\underset{\sim}{x}_\lambda + b\mu\underset{\sim}{x}_\lambda$

Therefore, $a(\mu - \lambda)\,\underset{\sim}{x}_\lambda = \underset{\sim}{0}$ (why?)

But $\mu \neq \lambda$ and $\underset{\sim}{x}_\lambda \neq \underset{\sim}{0}$ implies $a = 0$ and so $b = 0$. Hence the claim.

Theorem 9.4. If A be simple (i,e, all its eigenvalues are regular) then A can be diagonalised by a similarity transformation carried out by its modal matrix. Conversely, if A be similar to a diagonal matrix, then A is simple.

Proof. Let A be a non-simple matrix of order n. So $\exists$ n linearly independent eigenvectors of A. If $\underset{\sim}{x}_1, \underset{\sim}{x}_2, \cdots, \underset{\sim}{x}_n$ be the eigenvectors of A with $\lambda_1, \lambda_2, \cdots, \lambda_n$ being the corresponding eigenvalues (might not all λ_k's be distinct), then

$$A\underset{\sim}{x}_k = \lambda_k\underset{\sim}{x}_k \ \forall \ k = 1, 2, \cdot, n$$

and hence $P = (\underset{\sim}{x}_1, \underset{\sim}{x}_2, \cdots, \underset{\sim}{x}_n)$ is the modal matrix. It goes without saying that P is non-singular.

$$\therefore \qquad AP \quad = \quad (A\underset{\sim}{x}_1, A\underset{\sim}{x}_2, \cdots, A\underset{\sim}{x}_n) = (\lambda_1 \underset{\sim}{x}_1, \lambda_2 \underset{\sim}{x}_2, \cdots, \lambda_n \underset{\sim}{x}_n)$$

$$= \quad (\underset{\sim}{x}_1 \ \underset{\sim}{x}_2 \cdots \underset{\sim}{x}_n) \begin{pmatrix} \lambda_1 & & & O \\ & \lambda_2 & & \\ & & \ddots & \\ O & & & \lambda_n \end{pmatrix}$$

$$= \quad P \begin{pmatrix} \lambda_1 & & & O \\ & \lambda_2 & & \\ & & \ddots & \\ O & & & \lambda_n \end{pmatrix}$$

Now P being non-singular, P^{-1} exists uniquely and so

$$P^{-1}AP = \begin{pmatrix} \lambda_1 & & & O \\ & \lambda_2 & & \\ & & \ddots & \\ O & & & \lambda_n \end{pmatrix} = D, \text{ a diagonal matrix.}$$

This completes proof of first part.

For the converse, let A be diagonalisable in the sense that there exists a non-singular matrix Q of order n such that

$$Q^{-1}AQ = D \quad \text{i, e,} \quad AQ = QD$$

Denote the columns of Q as $\underset{\sim}{v}_1, \underset{\sim}{v}_2, \cdots, \underset{\sim}{v}_n$ and write D as $D = [\alpha_i \ \delta_{ij}]_{n \times n}$; δ_{ij}'s being the Kronecker delta.

$$\text{Thus} \qquad A(\underset{\sim}{v}_1, \underset{\sim}{v}_2, \cdots, \underset{\sim}{v}_n) = (\underset{\sim}{v}_1, \underset{\sim}{v}_2, \cdots \underset{\sim}{v}_n) \begin{pmatrix} \alpha_1 & & & O \\ & \alpha_2 & & \\ & & \ddots & \\ O & & & \alpha_n \end{pmatrix}$$

$$\text{i.e,} \qquad A\underset{\sim}{v}_k = \alpha_k \, \underset{\sim}{v}_k \quad \forall \quad k = 1, 2, \cdots, n$$

i.e, $\underset{\sim}{v}_k$'s are eigenvectors of A with α_k's being the corresponding eigenvalues.

As Q is non-singular, the set $\{ \underset{\sim}{v}_1, \underset{\sim}{v}_2 \cdots, \underset{\sim}{v}_n\}$ is linearly independent and hence the eigenvalues α_k's of A are regular, (i,e, A is simple). This completes the proof of the second part.

Example 9.3. Consider the matrix $\begin{bmatrix} 1 & 2 \\ 0 & 1 \end{bmatrix}$ to find its eigenvalues and eigenvectors.

This matrix being upper triangular we can at once identify its eigenvalues. Here 1 is the eigenvalue that has algebraic multiplicity 2. Let's find eigenvector(s) corresponding to eigenvalue 1. If $(x_1, x_2)^T$ be an eigenvector, then $2x_2 = 0$ and hence $c\begin{pmatrix} 1 \\ 0 \end{pmatrix}$, $c \neq 0$ is an eigenvector. Unfortunately we cannot find two linearly independent eigenvectors of the given matrix corresponding to eigenvalue 1. Thus the geometric multiplicity of eigenvalue 1 falls short of the algebraic multiplicity of the same. In a formal language we refer to such an eigenvalue as an irregular one and the matrix given is christened to be **defective**.

In general a n-square matrix is said to be 'defective' if there exists at least one eigenvalue of A which is **irregular** i.e, for which geometric multiplicity is less than its algebraic multiplicity. For a defective matrix we cannot have a complete basis of linearly independent eigenvectors. Simple matrices are square matrices of 'full rank' while defective matrices are 'rank-deficient'. Simple matrices are therefore diagonalisable while defective matrices are not diagonalisable. Since for defective matrices a complete basis of eigenvectors is not available, we augment the set of available linearly independent eigenvectors with generalised eigenvectors, that are necessary for solving defective system of ordinary differential equations. The truth that for any square matrix each eigenvalue having algebraic multiplicity $r(r \geq 1)$ has r linearly independent generalised eigenvectors is the key to success in this context.

Any Jordan block of size $n \times n$ like the following is defective.

$$J = \begin{bmatrix} \alpha & 1 & 0 & \cdots & 0 \\ & \alpha & 1 & \cdots & 0 \\ & & \ddots & \ddots & \vdots \\ & \mathbf{O} & & \ddots & \ddots & 1 \\ & & & & \ddots & \alpha \end{bmatrix}_{n \times n}$$

J has eigenvalue α with algebraic multiplicity n, but possesses only one distinct eigenvector, viz, $\underset{\sim}{x} = (1\ 0\ \cdots\cdots 0)^T$ as it is the only non-trivial solution of $(J - \alpha I_n)\underset{\sim}{x} = 0$.

In what follows, we quote a powerful general result in linear algebra without proof. However, we shall illustrate it with examples.

Theorem 9.5 For every square matrix over the complex field C, $\exists$ a unitary transformation S such that $S^{-1}AS$ is an upper triangular matrix with the eigenvalues of A appearing as its diagonal elements.

Remark:

(i) As S is unitary, $S^{-1} = (\overline{S^T})$, the bar denoting complex conjugation.

(ii) If S is unitary matrix S be real, it becomes an orthogonal matrix and in this case S is constructed with the help of Gram-Schmidt orthogonalisation process provided $\mathbb{R}^n$ (n is the order of the square matrix A) is endowed with an inner product structure so as to become an Euclidean space.

(iii) If the matrix A itself be Hermitian (or real symmetric). then $\exists$ a unitary transformation S such that $S^{-1}AS$ is diagonal matrix with eigenvalues being the diagonal entries. As a particular case to it evolves the result that every real symmetric matrix is orthogonally diagonalisable.

We now pick up three examples, first one being of a defective matrix, second one of a simple but non-Hermitian matrix and the third one of a real symmetric matrix to illustrate the theoretical results given above.

Example 9.4 Find the eigenpairs of the matrix $A = \begin{pmatrix} 1 & 2 & 2 \\ 0 & 2 & 1 \\ -1 & 2 & 2 \end{pmatrix}$ and hence upper triangularise A.

The eigenvalues of A are given by

$$0 = \begin{vmatrix} 1-\lambda & 2 & 2 \\ 0 & 2-\lambda & 1 \\ -1 & 2 & 2-\lambda \end{vmatrix} = (1-\lambda)\,(\lambda-2)^2$$

The eigenvector corresponding to eigenvalue 2 is $c(2,1,0)^T$, $c \neq 0$. Since algebraic multiplicity of the eigenvalue $\lambda = 2$ is 2 while its geometric multiplicity is unity, the matrix A is defective. The eigenvector corresponding to $\lambda = 1$ is $c'(1\ 1\ -1)^T$, $c' \neq 0$. It is obvious that the eigenvectors $\begin{pmatrix} 2 \\ 1 \\ 0 \end{pmatrix}$ and $\begin{pmatrix} 1 \\ 1 \\ -1 \end{pmatrix}$ are linearly independent.

We now augment this linearly independent set into a basis of $\mathbb{R}^3$:

$$\left\{ \begin{pmatrix} 2 \\ 1 \\ 0 \end{pmatrix}, \begin{pmatrix} 0 \\ 0 \\ 1 \end{pmatrix}, \begin{pmatrix} 1 \\ 1 \\ -1 \end{pmatrix} \right\}$$

This basis is however not orthogonal. We make it orthogonal by means of Gram-Schmidt process. Write

$$\underset{\sim}{x_1} \equiv \begin{pmatrix} 2 \\ 1 \\ 0 \end{pmatrix} ; \; \underset{\sim}{x_2} \equiv \begin{pmatrix} 0 \\ 0 \\ 1 \end{pmatrix} \text{ and } \underset{\sim}{x_3} = \begin{pmatrix} 1 \\ 1 \\ -1 \end{pmatrix}$$

Define $\quad \underset{\sim}{y_1} \equiv \underset{\sim}{x_1}$

$$\underset{\sim}{y_2} = \underset{\sim}{x_2} - \frac{<\underset{\sim}{x_2}, \underset{\sim}{y_1}>}{\| \underset{\sim}{y_1} \|^2} \underset{\sim}{y_1} = \begin{pmatrix} 0 \\ 0 \\ 1 \end{pmatrix}$$

and $\quad \underset{\sim}{y_3} = \underset{\sim}{x_3} - \frac{<\underset{\sim}{x_3}, \underset{\sim}{y_1}>}{\| \underset{\sim}{y_1} \|^2} \underset{\sim}{y_1} - \frac{<\underset{\sim}{x_3}, \underset{\sim}{y_2}>}{\| \underset{\sim}{y_2} \|^2} \underset{\sim}{y_2}$

$$= \begin{pmatrix} 1 \\ 1 \\ -1 \end{pmatrix} - \frac{3}{5} \begin{pmatrix} 2 \\ 1 \\ 0 \end{pmatrix} + \begin{pmatrix} 0 \\ 0 \\ 1 \end{pmatrix} = \begin{pmatrix} -\frac{1}{5} \\ \frac{2}{5} \\ 0 \end{pmatrix}$$

and corresponding unit vectors are :

$$\underset{\sim}{z_1} \equiv \frac{\underset{\sim}{y_1}}{\| \underset{\sim}{y_1} \|} = \begin{pmatrix} \frac{2}{\sqrt{5}} \\ \frac{1}{\sqrt{5}} \\ 0 \end{pmatrix}$$

$$\underset{\sim}{z_2} \equiv \frac{\underset{\sim}{y_2}}{\| \underset{\sim}{y_2} \|} = \begin{pmatrix} 0 \\ 0 \\ 1 \end{pmatrix}$$

$$\underset{\sim}{z_3} \equiv \frac{\underset{\sim}{y_3}}{\| \underset{\sim}{y_3} \|} = \begin{pmatrix} -\frac{1}{\sqrt{5}} \\ \frac{2}{\sqrt{5}} \\ 0 \end{pmatrix}$$

We therefore construct the orthogonal matrix S given by

$$S = \begin{pmatrix} \frac{2}{\sqrt{5}} & 0 & -\frac{1}{\sqrt{5}} \\ \frac{1}{\sqrt{5}} & 0 & \frac{2}{\sqrt{5}} \\ 0 & 1 & 0 \end{pmatrix}$$

$$\therefore S^{-1}AS = \begin{pmatrix} \frac{2}{\sqrt{5}} & \frac{1}{\sqrt{5}} & 0 \\ 0 & 0 & 1 \\ -\frac{1}{\sqrt{5}} & \frac{2}{\sqrt{5}} & 0 \end{pmatrix} \begin{pmatrix} 1 & 2 & 2 \\ 0 & 2 & 1 \\ -1 & 2 & 2 \end{pmatrix} \begin{pmatrix} \frac{2}{\sqrt{5}} & 0 & -\frac{1}{\sqrt{5}} \\ \frac{1}{\sqrt{5}} & 0 & \frac{2}{\sqrt{5}} \\ 0 & 1 & 0 \end{pmatrix}$$

$$= \begin{pmatrix} 2 & \sqrt{5} & 2 \\ 0 & 2 & \sqrt{5} \\ 0 & 0 & 1 \end{pmatrix}$$

This shows that the matrix A, although defective, can be transformed to an upper triangular matrix with its diagonal elements being the eigenvalues of A.

Example 9.5. Find the eigenpairs of the matrix given by

$$A = \begin{pmatrix} 0 & 1 & 1 \\ -2 & 3 & 2 \\ -3 & 3 & 4 \end{pmatrix} \quad \text{and hence upper triangularise.}$$

The eigenvalues of A are 1,1,5. The eigenvectors of A corresponding to eigenvalue 1 are $\underset{\sim}{x}_1 \equiv \begin{pmatrix} 1 \\ 1 \\ 0 \end{pmatrix}$ and $\underset{\sim}{x}_2 \equiv \begin{pmatrix} 1 \\ 0 \\ 1 \end{pmatrix}$ and the eigenvector of A corresponding to eigenvalue 5 is $\underset{\sim}{x}_3 \equiv \begin{pmatrix} 1 \\ 2 \\ 3 \end{pmatrix}$

Though $\{\underset{\sim}{x}_1, \underset{\sim}{x}_2, \underset{\sim}{x}_3\}$ is linearly independent and hence form a basis of $\mathbb{R}^3$, this set is non-orthogonal as can be shown by direct computations. Define an orthogonal basis in $\mathbb{R}^3$ from the above linearly independent set $\{\underset{\sim}{x}_1, \underset{\sim}{x}_2, \underset{\sim}{x}_3\}$ on using Gram-Schmidt process. To achieve this goal we define :

$$\underset{\sim}{y}_1 \equiv \begin{pmatrix} 1 \\ 1 \\ 0 \end{pmatrix} = \underset{\sim}{x}_1$$

$$\underset{\sim}{y_2} \equiv \underset{\sim}{x_2} - \frac{<\underset{\sim}{x_2}, \underset{\sim}{y_2}>}{\|\underset{\sim}{y_2}\|^2}\underset{\sim}{y_1} = \begin{pmatrix} 1 \\ 0 \\ 1 \end{pmatrix} - \frac{1}{2}\begin{pmatrix} 1 \\ 1 \\ 0 \end{pmatrix} = \begin{pmatrix} \frac{1}{2} \\ -\frac{1}{2} \\ 1 \end{pmatrix}$$

$$\underset{\sim}{y_3} \equiv \underset{\sim}{x_3} - \frac{<\underset{\sim}{x_3}, \underset{\sim}{y_1}>}{\|\underset{\sim}{y_1}\|^2}\underset{\sim}{y_1} - \frac{<\underset{\sim}{x_3}, \underset{\sim}{y_2}>}{\|\underset{\sim}{y_2}\|^2}\underset{\sim}{y_2}$$

$$= \begin{pmatrix} 1 \\ 2 \\ 1 \end{pmatrix} - \frac{3}{2}\begin{pmatrix} 1 \\ 1 \\ 0 \end{pmatrix} - \frac{5}{3}\begin{pmatrix} \frac{1}{2} \\ -\frac{1}{2} \\ 1 \end{pmatrix} = \begin{pmatrix} -\frac{4}{3} \\ \frac{4}{3} \\ \frac{4}{3} \end{pmatrix}$$

We construct the orthogonal matrix S given by

$$S = \begin{pmatrix} \frac{1}{\sqrt{2}} & \frac{1}{\sqrt{6}} & -\frac{1}{\sqrt{3}} \\ \frac{1}{\sqrt{2}} & -\frac{1}{\sqrt{6}} & \frac{1}{\sqrt{3}} \\ 0 & \frac{2}{\sqrt{6}} & \frac{1}{\sqrt{3}} \end{pmatrix} \equiv (\underset{\sim}{z_1}\, \underset{\sim}{z_2}\, \underset{\sim}{z_3})$$

obtained by $\underset{\sim}{z_1} \equiv \dfrac{\underset{\sim}{y_1}}{\|\underset{\sim}{y_1}\|}$; $\underset{\sim}{z_2} \equiv \dfrac{\underset{\sim}{y_2}}{\|\underset{\sim}{y_2}\|}$; $\underset{\sim}{z_3} \equiv \dfrac{\underset{\sim}{y_3}}{\|\underset{\sim}{y_3}\|}$

Therefore, $S^{-1}AS$

$$= \begin{pmatrix} \frac{1}{\sqrt{2}} & \frac{1}{\sqrt{2}} & 0 \\ \frac{1}{\sqrt{6}} & -\frac{1}{\sqrt{6}} & \frac{2}{\sqrt{6}} \\ -\frac{1}{\sqrt{3}} & \frac{1}{\sqrt{3}} & \frac{1}{\sqrt{3}} \end{pmatrix}\begin{pmatrix} 0 & 1 & 1 \\ -2 & 3 & 2 \\ -3 & 3 & 4 \end{pmatrix}\begin{pmatrix} \frac{1}{\sqrt{2}} & \frac{1}{\sqrt{6}} & -\frac{1}{\sqrt{3}} \\ \frac{1}{\sqrt{2}} & -\frac{1}{\sqrt{6}} & \frac{1}{\sqrt{3}} \\ 0 & \frac{2}{\sqrt{6}} & \frac{1}{\sqrt{3}} \end{pmatrix}$$

$$= \begin{pmatrix} 1 & 0 & \sqrt{\frac{3}{2}} \\ 0 & 1 & \frac{5}{\sqrt{2}} \\ 0 & 0 & 5 \end{pmatrix}$$

This is a 3×3 upper triangular matrix to which A has been reduced. Note that the matrix being regular, we can find a non-singular matrix P (whose columns are eigenvectors of A) such that $P^{-1}AP$ is a diagonal matrix. To be specific,

$$P = \begin{pmatrix} 1 & 1 & 1 \\ 1 & 0 & 2 \\ 0 & 1 & 3 \end{pmatrix} \text{ so that}$$

$$P^{-1}AP = \frac{1}{4}\begin{pmatrix} 2 & 2 & -2 \\ 3 & -3 & 1 \\ -1 & 1 & 1 \end{pmatrix}\begin{pmatrix} 0 & 1 & 1 \\ -2 & 3 & 2 \\ -3 & 3 & 4 \end{pmatrix}\begin{pmatrix} 1 & 1 & 1 \\ 1 & 0 & 2 \\ 0 & 1 & 3 \end{pmatrix}$$

$$= \begin{pmatrix} 1 & 0 & 0 \\ 0 & 1 & 0 \\ 0 & 0 & 5 \end{pmatrix}$$

In this example we find that A is diagonalisable but not orthogonally diagonalisable as in constructing the orthogonal basis set $\{z_1, z_2, z_3\}$ of $\mathbb{R}^3$ from the eigenbasis $\{x_1, x_2, x_3\}$, we have to remain satisfied with upper-triangularised form of A in lieu of a completely diagonalised form. The third and final example verifies the truth that for real symmetric matrices orthogonal diagonalisation is feasible.

Example 9.6. Find the eigenpairs of the symmetric matrix A given by

$$A = \begin{pmatrix} 6 & -2 & 2 \\ -2 & 3 & -1 \\ 2 & -1 & 3 \end{pmatrix}$$

and hence find an orthogonal similarity transform to diagonalise A.

The eigenvalues of A are given by

$$0 = \begin{vmatrix} \lambda - 6 & 2 & -2 \\ 2 & \lambda - 3 & 1 \\ -2 & 1 & \lambda - 3 \end{vmatrix} = (\lambda - 2)^2 (\lambda - 8)$$

so that 8 is a simple eigenvalue and 2 is an eigenvalue having algebraic multiplicity 2. By standard procedure one can prove that an eigenvector corresponding to $\lambda = 8$ is $x_1 \equiv (2\ -1\ 1)^T$ while two linearly independent eigenvectors corresponding to $\lambda = 2$ are $x_2 \equiv (1\ 2\ 0)^T$ and $x_3 \equiv (-1\ 0\ 2)^T$. So for the given matrix A we have availed a complete set of eigenvectors. However, this set of vectors is not orthogonal. We therefore construct as before a complete set of orthogonal eigenvectors of A, using Gram-Schmidt process.

We define a new set of vectors $\{y_1, y_2, y_3,\}$ by

$$y_1 \equiv x_1 = (2\ -1\ 1)^T$$

$$y_2 \equiv x_2 - \frac{<x_2, y_1>}{\|y_1\|^2} y_1 = (1\ 2\ 0)^T = x_2$$

$$y_3 \equiv x_3 - \frac{<x_3, y_1>}{\|y_1\|^2} y_1 - \frac{<x_3, y_2>}{\|y_2\|^2} y_2$$

$$= (-1\ 0\ 2)^T + \left(\frac{1}{5}\ \frac{2}{5}\ 0\right)^T = \left(-\frac{4}{5}\ \frac{2}{5}\ 2\right)^T$$

Let us define an orthogonal matrix S as follows:

$$S = \begin{bmatrix} \dfrac{\underset{\sim}{y_1}}{\|\underset{\sim}{y_1}\|} & \dfrac{\underset{\sim}{y_2}}{\|\underset{\sim}{y_2}\|} & \dfrac{\underset{\sim}{y_3}}{\|\underset{\sim}{y_3}\|} \end{bmatrix} = \begin{bmatrix} \dfrac{2}{\sqrt{6}} & \dfrac{1}{\sqrt{5}} & -\dfrac{2}{\sqrt{30}} \\ -\dfrac{1}{\sqrt{6}} & \dfrac{2}{\sqrt{5}} & \dfrac{1}{\sqrt{30}} \\ \dfrac{1}{\sqrt{6}} & 0 & \dfrac{5}{\sqrt{30}} \end{bmatrix}$$

and find that $S^{-1}AS$ is given by

$$\begin{pmatrix} \dfrac{2}{\sqrt{6}} & -\dfrac{1}{\sqrt{6}} & \dfrac{1}{\sqrt{6}} \\ \dfrac{1}{\sqrt{5}} & \dfrac{2}{\sqrt{5}} & 0 \\ -\dfrac{2}{\sqrt{30}} & \dfrac{1}{\sqrt{30}} & \dfrac{5}{\sqrt{30}} \end{pmatrix} \begin{pmatrix} 6 & -2 & 2 \\ -2 & 3 & -1 \\ 2 & -1 & 3 \end{pmatrix} \begin{pmatrix} \dfrac{2}{\sqrt{6}} & \dfrac{1}{\sqrt{5}} & -\dfrac{2}{\sqrt{30}} \\ -\dfrac{1}{\sqrt{6}} & \dfrac{2}{\sqrt{5}} & \dfrac{1}{\sqrt{30}} \\ \dfrac{1}{\sqrt{6}} & 0 & \dfrac{5}{\sqrt{30}} \end{pmatrix}$$

$$\therefore \qquad S^{-1}AS = \begin{pmatrix} 8 & 0 & 0 \\ 0 & 2 & 0 \\ 0 & 0 & 2 \end{pmatrix}$$

Thus in this example, we have got something more than upper triangularisation via the orthogonal similarity transformation. This is nothing unexpected as the symmetry of the matrix A accounts for its orthogonal diagonalisability.

We pack up this optional article with a few more results quoted without proof as we shall use them in our subsequent development.

(a) From the characteristic equation $det(A - \lambda I_n) = 0$ of a n-square matrix A it follows that trace A, formally defined as $\sum_{i=1}^{n} a_{ii}$ [$\because A = (a_{ij})_{n \times n}$] is equal to the sum of eigenvalues of A (with multiplicity counted for).

(b) From the characteristic equation it also follows that determinant of A is the product of its eigenvalues. So if one of the eigenvalues be zero, then the matrix A is singular and conversely if A be singular, then at least one eigenvalue of A is zero. These results are useful in context of Liouville's theorem coming up in § 9.7

(c) Cayley-Hamilton Theorem : Every square matrix satisfies its own characteristic equation. This result will be used in 'Putzer Algorithm' discussed in § 9.8

9.3   Solution of Vector Differential Equation using Eigenvalues of Associated Matrix.

In the previous section we discussed about eigenpairs of square matrices and possible of their diagonalisation when matrices are simple. It is quite natural a question: 'Why are we so concerned with finding of eigenvalues and eigenvectors of square matrices and also about the diagonalisation while sobring vector differential equation?' The simple answer to the first part of the above query is provided by the following:

Theorem 9.6 If $\left(\lambda_k, \underset{\sim}{y_k}\right)$ be an eigenpair of a n-square constant matrix A, then the column vector $\underset{\sim}{y}(x) = \underset{\sim}{y_k} e^{\lambda_k x}, x \in \mathbb{R}$ is a solution of the vector differential equation $\underset{\sim}{y'} = A\underset{\sim}{y}$, where prime denotes differentiation w.r. to x.

Proof. Since $\underset{\sim}{y}(x) = \underset{\sim}{y_k} e^{\lambda_k x}$, $x \in \mathbb{R}$,

$$\underset{\sim}{y'}(x) = \frac{d}{dx}\left(\underset{\sim}{y_k}\ e^{\lambda_k x}\right) = \underset{\sim}{y_k}\left(\lambda_k\ e^{\lambda_k x}\right) = A\ \underset{\sim}{y_k}\ e^{\lambda_k x} = A\underset{\sim}{y}(x)$$

This result is a basic one as it ensures that from the knowledge of n linearly independent eigenvectors one can generate the general solution of the vector differential equation $\underset{\sim}{y'}(x) = A\underset{\sim}{y}(x)$. In answering the second query, theorem 9.4 stated and proved in the previous section gives the clue. According to that theorem any simple n-square constant matrix A can be diagonalised by a similarity transformation carried out by its modal matrix P, i,e, $P^{-1}AP = D$. To be specific, let the modal matrix P comprise of n linearly independent eigenvectors $\underset{\sim}{v_1}, \underset{\sim}{v_2}, \cdots, \underset{\sim}{v_n}$ of A and $D = [\lambda_j\ \delta_{ij}]_{n \times n}$.

First we define a linear transformation of co-ordinates as :

$$\underset{\sim}{z} = P^{-1}\underset{\sim}{y} \Leftrightarrow \underset{\sim}{y} = P\underset{\sim}{z}$$

$$\therefore \quad \frac{d\underset{\sim}{z}}{dx} = \frac{d}{dx}\left(P^{-1}\underset{\sim}{y}\right) = P^{-1}\frac{d\underset{\sim}{y}}{dx} = P^{-1}A\underset{\sim}{y} = \left(P^{-1}AP\right)\underset{\sim}{z} = D\underset{\sim}{z}$$

i.e,
$$\begin{pmatrix} \frac{dz_1}{dx} \\ \frac{dz_2}{dx} \\ \vdots \\ \frac{dz_n}{dx} \end{pmatrix} = \begin{pmatrix} \lambda_1 & & & O \\ & \lambda_2 & & \\ & & \ddots & \\ O & & & \lambda_n \end{pmatrix} \begin{pmatrix} z_1 \\ z_2 \\ \vdots \\ z_n \end{pmatrix}$$

Hence we have a system of n decoupled linear first order odes :

$$\frac{dz_k}{dx} = \lambda_k \, z_k \quad \text{so that} \quad z_k = c_k \, e^{\lambda_k x} \; (k = 1, 2, \cdots, n)$$

This means

$$\begin{pmatrix} z_1 \\ z_2 \\ \vdots \\ z_n \end{pmatrix} = \begin{pmatrix} c_1 \, e^{\lambda_1 x} \\ c_2 \, e^{\lambda_2 x} \\ \vdots \\ c_n \, e^{\lambda_n x} \end{pmatrix} \quad \text{i, e,} \quad P^{-1} \begin{pmatrix} y_1 \\ y_2 \\ \vdots \\ y_n \end{pmatrix} = \begin{pmatrix} c_1 \, e^{\lambda_1 x} \\ c_2 \, e^{\lambda_2 x} \\ \vdots \\ c_n \, e^{\lambda_n x} \end{pmatrix}$$

$$\therefore \; \underset{\sim}{y}(x) \equiv \begin{pmatrix} y_1 \\ y_2 \\ \vdots \\ y_n \end{pmatrix} = P \begin{pmatrix} c_1 \, e^{\lambda_1 x} \\ c_2 \, e^{\lambda_2 x} \\ \vdots \\ c_n \, e^{\lambda_n x} \end{pmatrix} = (\underset{\sim}{v}_1 \; \underset{\sim}{v}_2 \cdots \underset{\sim}{v}_n) \begin{pmatrix} c_1 \, e^{\lambda_1 x} \\ c_2 \, e^{\lambda_2 x} \\ \vdots \\ c_n \, e^{\lambda_n x} \end{pmatrix}$$

i, e, $\; \underset{\sim}{y}(x) = c_1 \, e^{\lambda_1 x} \, \underset{\sim}{v}_1 + c_2 \, e^{\lambda_2 x} \, \underset{\sim}{v}_2 + \cdots + c_n \, e^{\lambda_n x} \, \underset{\sim}{v}_n$

So for a vector differential equation $\underset{\sim}{y} = A\underset{\sim}{y}$ we can write the general solution as a linear combination of its n linearly independent solutions written in terms of the eigenpairs of the associated n-square matrix A whenever A is simple.

Example 9.7. Solve the linear system

$$\left. \begin{array}{l} \dot{y}_1 = y_2 \\ \dot{y}_2 = -2y_1 - 3y_2 \end{array} \right\}$$

In the matrix form this is written as :

$$\begin{pmatrix} \dot{y}_1 \\ \dot{y}_2 \end{pmatrix} = \begin{pmatrix} 0 & 1 \\ -2 & -3 \end{pmatrix} \begin{pmatrix} y_1 \\ y_2 \end{pmatrix}$$

As found in the example 9.1, the eigenpairs of this matrix are

$$\left\{ -1, \begin{pmatrix} 1 \\ -1 \end{pmatrix} \right\} \quad \text{and} \quad \left\{ -2, \begin{pmatrix} 1 \\ -2 \end{pmatrix} \right\}$$

.

Hence the general solution of the system is :

$$y(x) = c_1 \, e^{-x} \begin{pmatrix} 1 \\ -1 \end{pmatrix} + c_2 \, e^{-2x} \begin{pmatrix} 1 \\ -2 \end{pmatrix}$$

$$\left. \begin{array}{l} y_1(x) = c_1\ e^{-x} + c_2\ e^{-2x} \\[2mm] y_2(x) = -c_1\ e^{-x} - 2c_2\ e^{-2x} \end{array} \right\} \qquad (a)$$

where c_1, c_2 are arbitrary constants. (The phase potrait for the linear system of this example can be found by sketching the solution curves defined by (a)) .

Example 9.8. Solve the coupled system of linear odes :

$$\left. \begin{array}{l} \frac{dy_1}{dx} = 2y_1 + 2y_2 + y_3 \\[4mm] \frac{dy_2}{dx} = y_1 + 3y_2 + y_3 \\[4mm] \frac{dy_3}{dx} = y_1 + 2y_2 + 2y_3 \end{array} \right\}$$

where y_1, y_2, y_3 are functions of x.

In matrix form this system can be cast as : $\dfrac{dy}{dx} = Ay$,

where $\quad y = (y_1\ \ y_2\ \ y_3)^T \quad$ and $\quad A = \begin{pmatrix} 2 & 2 & 1 \\ 1 & 3 & 1 \\ 1 & 2 & 2 \end{pmatrix}$

The matrix A associated with this differential equation is simple as both its eigenvalues viz, 5 and 1 are regular [cf. Example 9.2].

In example 9.2 we found that $\begin{pmatrix} 1 \\ 1 \\ 1 \end{pmatrix}$ is an eigenvector of A correspond-

ing to eigenvalue 5 while $\begin{pmatrix} -2 \\ 1 \\ 0 \end{pmatrix}$ and $\begin{pmatrix} -1 \\ 0 \\ 1 \end{pmatrix}$ are linearly independent eigenvectors corresponding to eigenvalue 1. Hence the general solution of the system is :

$$y(x) = c_3 e^{5x} \begin{pmatrix} 1 \\ 1 \\ 1 \end{pmatrix} + c_1 e^{x} \begin{pmatrix} -2 \\ 1 \\ 0 \end{pmatrix} + c_2 e^{x} \begin{pmatrix} -1 \\ 0 \\ 1 \end{pmatrix}$$

i, e,
$$\left.\begin{array}{l} y_1(x) = c_3\, e^{5x} - 2c_1\, e^x - c_2 e^x \\ y_2(x) = c_3\, e^{5x} + c_1\, e^x \\ y_3(x) = c_3\, e^{5x} + c_2\, e^x \end{array}\right\}$$

i, e,
$$\underset{\sim}{y}(x) = \begin{pmatrix} -2e^x & -e^x & e^{5x} \\ e^x & 0 & e^{5x} \\ 0 & e^x & e^{5x} \end{pmatrix} \begin{pmatrix} c_1 \\ c_2 \\ c_3 \end{pmatrix}$$

$\therefore$
$$\underset{\sim}{y}(0) = \begin{pmatrix} -2 & -1 & 1 \\ 1 & 0 & 1 \\ 0 & 1 & 1 \end{pmatrix} \begin{pmatrix} c_1 \\ c_2 \\ c_3 \end{pmatrix} \equiv P \begin{pmatrix} c_1 \\ c_2 \\ c_3 \end{pmatrix}$$

i, e,
$$\begin{pmatrix} c_1 \\ c_2 \\ c_3 \end{pmatrix} = P^{-1}\, \underset{\sim}{y}(0) = \frac{1}{4}\begin{pmatrix} -1 & 2 & -1 \\ -1 & -2 & 3 \\ 1 & 2 & 1 \end{pmatrix} \underset{\sim}{y}(0)$$

Thus
$$\underset{\sim}{y}(x) = \frac{1}{4}\begin{pmatrix} -2e^x & -e^x & e^{5x} \\ e^x & 0 & e^{5x} \\ 0 & e^x & e^{5x} \end{pmatrix} \begin{pmatrix} -1 & 2 & -1 \\ -1 & -2 & 3 \\ 1 & 2 & 1 \end{pmatrix} \underset{\sim}{y}(0)$$

$$= \frac{1}{4}\begin{pmatrix} 3e^x + e^{5x} & -3e^x + 2e^{5x} & e^{5x} - e^x \\ -e^x + e^{5x} & 2e^{5x} + 2e^x & -e^x + e^{5x} \\ -e^x + e^{5x} & -2e^x + 2e^{5x} & 3e^x + e^{5x} \end{pmatrix} \underset{\sim}{y}_0, \text{say}$$

This can be proved as the unique solution of the IVP (for all $x \in \mathbb{R}$) $\underset{\sim}{y}'(x) = A\underset{\sim}{y}(x)$ subject to $\underset{\sim}{y}(0) = \underset{\sim}{y}_0$ where A is as quoted before.

Later we shall see that the unique solution of this IVP is presentable as $\underset{\sim}{y}(x) = e^{Ax}\, \underset{\sim}{y}_0$.

In the present case A being diagonalisable,

$$e^{Ax} = P \begin{pmatrix} e^x & 0 & 0 \\ 0 & e^x & 0 \\ 0 & 0 & e^{5x} \end{pmatrix} P^{-1}$$

$$= \frac{1}{4}\begin{pmatrix} -2 & -1 & 1 \\ 1 & 0 & 1 \\ 0 & 1 & 1 \end{pmatrix} \begin{pmatrix} e^x & 0 & 0 \\ 0 & e^x & 0 \\ 0 & 0 & e^{5x} \end{pmatrix} \begin{pmatrix} -1 & 2 & -1 \\ -1 & -2 & 3 \\ 1 & 2 & 1 \end{pmatrix}$$

$$= \frac{1}{4}\begin{pmatrix} 3e^x + e^{5x} & -3e^x + 2e^{5x} & e^{5x} - e^x \\ -e^x + e^{5x} & 2e^x + 2e^{5x} & e^{5x} - e^x \\ -e^x + e^{5x} & -2e^x + 2e^{5x} & e^{5x} + 3e^x \end{pmatrix}$$

We shall prove this result of uniqueness theoretically, no matter what the square matrix A is. Infact computation of e^{Ax} is our main headache when A is not diagonalisable.

We have not yet defined e^{Ax} when A is any arbitrary n-square real matrix, i,e, $A \in M_n(\mathbb{R})$. To define this exponential function of matrices we need some preliminary ideas of operator norm and its properties. The next section is devoted to that development followed by the **fundamental theorem for linear system** that spells out existence and uniqueness of solution of the IVP

$$\underset{\sim}{y}' = A\underset{\sim}{y} \quad \text{subject to} \quad \underset{\sim}{y}(0) = \underset{\sim}{y}_0 \ ; \ \underset{\sim}{y}_0 \in \mathbb{R}^n \quad \text{and} \quad A \in M_n(\mathbb{R})$$

9.4 Operator Norm and Its use in defining Exponentials of Matrices.

Recall from § **4.6**, the definition of norm of a bounded linear operator $T : \mathbb{R}^n \to \mathbb{R}^n$ ($\mathbb{R}^n$ being endowed with Euclidean norm $|\underset{\sim}{x}| = \left| \sum_{i=1}^n x_i^2 \right|^{\frac{1}{2}}$ for any $\underset{\sim}{x} = (x_1, x_2, \cdots, x_n)^T \in \mathbb{R}^n$). Our entire treatment is tailored to suit the needs of our main goal of solving odes, but it holds equally good generically with necessary modifications because these are core results of operators on normed linear spaces.

$$\| T \| = Sup \left\{ \frac{|T(\underset{\sim}{y})|}{|\underset{\sim}{y}|} \ ; \ \underset{\sim}{y} \in \mathbb{R}^n \right\}$$

$$\therefore \quad |T(\underset{\sim}{y})| \leqslant \| T \| \, |\underset{\sim}{y}| \quad \forall \, \underset{\sim}{y} \in \mathbb{R}^n \qquad \cdots (9.6)$$

Again if $\mathcal{L}(\mathbb{R}^n)$ denotes the set of all linear operators defined on $\mathbb{R}^n$, then for any $T, S \in \mathcal{L}(\mathbb{R}^n)$, $TS \in \mathcal{L}(\mathbb{R}^n)$ and moreover

$$\| TS \| \leqslant \| T \| \, \| S \| \quad \text{(submultiplicative property)} \qquad (9.7)$$

$$\left[\ \| TS \| = Sup \left\{ \frac{|(TS)(\underset{\sim}{y})|}{|\underset{\sim}{y}|} \ ; \ \underset{\sim}{y} \in \mathbb{R}^n \right\} \right.$$

Since S is a linear operator on $\mathbb{R}^n$, $|S(\underset{\sim}{y})| \leqslant \|S\| \, |\underset{\sim}{y}|$

$$|(TS)(\underset{\sim}{y})| \leqslant \|T\| |S(\underset{\sim}{y})| \leqslant \|T\| \, \|S\| \, |\underset{\sim}{y}|$$

$$\text{i,e,} \quad \frac{|(TS)(\underset{\sim}{y})|}{|\underset{\sim}{y}|} \leqslant \|T\| \, \|S\|$$

$$\text{i, e,} \qquad Sup\left\{\frac{|(TS)(\underset{\sim}{y})|}{|\underset{\sim}{y}|} \; : \; \underset{\sim}{y} \in \mathbb{R}^n\right\} \leqslant ||T|| \, ||S||$$

$$\text{i, e,} \qquad ||TS|| \leqslant ||T|| \, ||S|| \qquad \Big]$$

In particular if $S = T$, we get $||T^2|| \leqslant ||T||^2$

Inductively one can prove that for any finite $k \in N$,

$$||T^k|| \leqslant ||T||^k \tag{9.8}$$

We now want to prove that the operator series $\sum_{k=0}^{\infty} \frac{T^k x^k}{k!}$ is uniformly and absolutely convergent for $|x| \leqslant x_0$ for any preassigned $x_0 > 0$. Observe that for any $\underset{\sim}{y} \in \mathbb{R}^n$,

$$\left|\left(\frac{T^k x^k}{k!}\right)(\underset{\sim}{y})\right| = \left|\frac{x^k}{k!}\right| \left|T^k(\underset{\sim}{y})\right|$$

$$\leqslant \left|\frac{x^k}{k!}\right| \, ||T^k|| \, |\underset{\sim}{y}| \quad \text{(using (9.6))}$$

$$\leqslant \frac{x_0^k}{k!} ||T||^k \, |\underset{\sim}{y}| \quad \text{(using (9.8))} \tag{9.9}$$

$$\therefore \; Sup\left\{\frac{\left|\left(\frac{T^k x^k}{k!}\right)(\underset{\sim}{y})\right|}{|\underset{\sim}{y}|} \; ; \; \underset{\sim}{y} \in \mathbb{R}^n\right\} \leqslant \frac{x_0^k}{k!} ||T||^k \equiv \frac{x_0^k}{k!} c^k \; \forall \, k \in \mathbb{N} \cup \{0\}$$

where we agree to write $||T||$ as c. Hence $\left|\left|\frac{(Tx)^k}{k!}\right|\right|$ is well defined for all $k \in \mathbb{N} \cup \{0\}$. It follows from (9.9) that

$$\left|\left(\frac{T^k x^k}{k!}\right)\left(\frac{\underset{\sim}{y}}{|\underset{\sim}{y}|}\right)\right| \leqslant \frac{x_0^k \, c^k}{k!} \; \forall \, k \in N \; \cup \{0\}$$

and as the series $\sum_{k=0}^{\infty} \frac{x_0^k \, c^k}{k!}$ is convergent (to e^{cx_0}), by Weierstrass M-test it follows that the operator series $\sum_{k=0}^{\infty} \frac{T^k x^k}{k!}$ converges uniformly as well as absolutely for $|x| \leqslant x_0$, The sum of this uniformly and absolutely convergent series is defined to be the operator e^{Tx}.

$$\therefore \qquad e^{Tx} \equiv \sum_{k=0}^{\infty} \frac{T^k x^k}{k!}$$

It is a simple exercise to prove that e^{Tx} is a linear operator on $\mathbb{R}^n$.

Let $a, b \in \mathbb{R}$ and $\underset{\sim}{y}, \underset{\sim}{y'} \in \mathbb{R}^n$.

$$\because\ e^{Tx}\left(\underset{\sim}{y}\ \right) = \sum_{k=0}^{\infty} \frac{T^k x^k}{k!}(\underset{\sim}{y}) = \sum_{k=0}^{\infty} \frac{x^k}{k!}\, T^k \underset{\sim}{y}\ ,$$

$$e^{Tx}\left(a\underset{\sim}{y} + b\underset{\sim}{y'}\right) = \sum_{k=0}^{\infty} \frac{x^k}{k!} T^k (a\underset{\sim}{y} + b\underset{\sim}{y'})$$

$$= \sum_{k=0}^{\infty} \frac{x^k}{k!} \left[aT^k(\underset{\sim}{y}) + bT^k(\underset{\sim}{y'}) \right]$$

$$= a \sum_{k=0}^{\infty} \frac{x^k}{k!}\, T^k(\underset{\sim}{y}) + b \sum_{k=0}^{\infty} \frac{x^k}{k!}\, T^k\left(\underset{\sim}{y'}\right)$$

$$= a e^{Tx}(y) + b\, e^{Tx}(y'),$$

ensuring that e^{Tx} is linear and hence $||e^{Tx}||$ is well-defined.

$$\therefore \qquad ||e^{Tx}|| = \left\|\sum_{k=0}^{\infty} \frac{T^k x^k}{k!}\right\|$$

$$= \left\| \lim_{m \to \infty} \sum_{k=0}^{m} \frac{T^k x^k}{k!} \right\|$$

$$= \lim_{m \to \infty} \left\| \sum_{k=0}^{m} \frac{T^k x^k}{k!} \right\| \quad (\because \text{ norm is } ||\cdot|| \text{ continuous })$$

$$\leqslant \lim_{m \to \infty} \sum_{k=0}^{m} \left\| \frac{T^k x^k}{k!} \right\|$$

$$= \sum_{k=0}^{\infty} \frac{|x^k|}{k!}\, \left\| T^k \right\| = e^{||T||\, |x|}$$

where in the penultimate steps we used the homogeneity property and triangle inequality satisfied by linear operators.

[If $T, S \in \mathcal{L}\ (\mathbb{R}^n)$, then hold good the following relations :

(i) $||S + T|| \leqslant ||S|| + ||T||$ (Triangle inequality)

(ii) $||\lambda T|| = |\lambda|\ ||T||\ \forall\ \lambda \in \mathbb{R}$ (Homogeneity)

(iii) $||T|| \geqslant 0$ and $||T|| = 0$ iff $T = 0$ (Positive definiteness)

Proof. By definition,

$$||S + T|| = Sup \left\{ \frac{|(S + T)(\underset{\sim}{y})|}{|\underset{\sim}{y}|}\ ;\ \underset{\sim}{y} \in \mathbb{R}^n \right\}$$

Now $\dfrac{|(S+T)(\underset{\sim}{y})|}{|\underset{\sim}{y}|} \leqslant \dfrac{|S(\underset{\sim}{y})|}{|\underset{\sim}{y}|} + \dfrac{|T(\underset{\sim}{y})|}{|\underset{\sim}{y}|}$

$\leqslant Sup\left\{ \dfrac{|T(\underset{\sim}{y})|}{|\underset{\sim}{y}|} \; ; \; \underset{\sim}{y} \in \mathbb{R}^n \right\} + Sup\left\{ \dfrac{|S(\underset{\sim}{y})|}{|\underset{\sim}{y}|} \; ; \; \underset{\sim}{y} \in \mathbb{R}^n \right\}$

$\therefore \quad Sup\left\{ \dfrac{|(S+T)(\underset{\sim}{y})|}{|\underset{\sim}{y}|} \; ; \; \underset{\sim}{y} \in \mathbb{R}^n \right\} \leqslant Sup\left\{ \dfrac{|T(\underset{\sim}{y})|}{|\underset{\sim}{y}|} \; ; \; \underset{\sim}{y} \in \mathbb{R}^n \right\}$

$+ Sup\left\{ \dfrac{|S(\underset{\sim}{y})|}{|\underset{\sim}{y}|} \; ; \; \underset{\sim}{y} \in \mathbb{R}^n \right\}$

i,e, $||S+T|| \leqslant ||S|| + ||T||$

(ii) $||\lambda T|| = Sup\left\{ \dfrac{|(\lambda T)(\underset{\sim}{y})|}{|\underset{\sim}{y}|} \; ; \; \underset{\sim}{y} \in \mathbb{R}^n \right\} = |\lambda| \; ||T||$

Again $||T|| \geqslant 0$ is obvious from definition itself.

If for $T \in \mathcal{L}\ (\mathbb{R}^n)$ $||T|| = 0$, then

$$\dfrac{|T(\underset{\sim}{y})|}{|\underset{\sim}{y}|} = 0 \;\; \forall \; \underset{\sim}{y} \in \mathbb{R}^n \; \Leftrightarrow \; T(\underset{\sim}{y}) = 0 \; \forall \; \underset{\sim}{y} \in \mathbb{R}^n \; \Leftrightarrow \; T = 0$$

Once (i),(ii),(iii) are proved, we claim that $\mathcal{L}\ (\mathbb{R}^n)$ endowed with $|| \cdot ||$ is a normed linear space in its own right.]

As in the present chapter our main objective is to solve a linear system having the form

$$\underset{\sim}{y}'(x) = A\underset{\sim}{y}(x) \tag{9.5}$$

with A being a constant $n \times n$ matrix over field $\mathbb{R}$, we shall presume that the linear transformation T on $\mathbb{R}^n$ is represented as the $n \times n$ matrix A w.r.to the canonical basis $\{ \underset{\sim}{e}_1, \underset{\sim}{e}_2, \cdots, \underset{\sim}{e}_n \}$ of $\mathbb{R}^n$ and define for any $x \in \mathbb{R}$, the $n \times n$ matrix e^{Ax} by the rule :

$$e^{Ax} = \sum_{k=0}^{\infty} \dfrac{A^k x^k}{k!}$$

so that $||e^{Ax}|| \leqslant e^{||A||\,|x|}$, where $||A||$ denotes the norm of the matrix A. In other words, the norm of the linear operator T that has been represented as matrix A w.r.to the canonical basis $\{ \underset{\sim}{e}_1, \underset{\sim}{e}_2, \cdots, \underset{\sim}{e}_n \}$ of $\mathbb{R}^n$ is denoted by $||A||$.

A reader who goes through the account presented so far in this article may hurl the question to us : 'why are we so much focussed on a particular linear operator e^{Ax} with $A \in M_n(\mathbb{R})$ and $x \in \mathbb{R}$ in our discussion?' The answer to this question is provided by the "Fundamental Theorem for linear systems" stated as :

Theorem 9.7 : The IVP $\underset{\sim}{y}'(x) = A\underset{\sim}{y}(x)$; $\underset{\sim}{y}(0) = \underset{\sim}{y_0}$ with $\underset{\sim}{y_0} \in \mathbb{R}^n$ has a unique solution for all $x \in \mathbb{R}$ and is given by $\underset{\sim}{y}(x) = e^{Ax}\underset{\sim}{y_0}$

Proof : To prove this result we first compute the derivative of exponential matrix function e^{Ax}. Recall that derivative of a function is a limiting process $\left[f'(x) = \underset{h\to 0}{\mathrm{Lt}}\, \frac{f(x+h)-f(x)}{h} \right]$ and the exponential function e^{Ax} is the limit of the partial sum sequence $\left\{ \sum_{k=0}^{m} \frac{A^k x^k}{k!} \right\}$ of the uniformly and absolutely convergent series $\sum_{k=0}^{\infty} \frac{A^k x^k}{k!}$.

$$\therefore \quad \frac{d}{dx}\left(e^{Ax}\right) = \underset{h\to 0}{\mathrm{Lt}}\, \frac{e^{A(x+h)} - e^{Ax}}{h} = e^{Ax} \underset{h\to 0}{\mathrm{Lt}} \left(\frac{e^{Ah} - I_n}{h}\right)$$

$$= e^{Ax} \underset{h\to 0}{\mathrm{Lt}}\, \frac{1}{h} \left[\underset{m\to\infty}{\mathrm{Lt}} \left(\sum_{k=0}^{m} \frac{A^k\, h^k}{k!} - I_n \right) \right]$$

$$= e^{Ax} \underset{h\to 0}{\mathrm{Lt}}\, \underset{m\to\infty}{\mathrm{Lt}}\, \sum_{k=1}^{m} \frac{A^k\, h^{k-1}}{k!}$$

$$= e^{Ax} \underset{m\to\infty}{\mathrm{Lt}}\, \underset{h\to 0}{\mathrm{Lt}}\, \sum_{k=1}^{m} \frac{A^k\, h^{k-1}}{k!} = e^{Ax} \cdot A = Ae^{Ax}$$

where in the penultimate steps we have used Moore's theorem that ensures interchangability of the two convergent limit processes and also the fact that A commutes with e^{Ax}. Again

$$\frac{d}{dx}\left(\underset{\sim}{y}(x)\right) = \frac{d}{dx}\left(e^{Ax}\underset{\sim}{y_0}\right) = \frac{d}{dx}\left(e^{Ax}\right)\underset{\sim}{y_0} = Ae^{Ax}\underset{\sim}{y_0} = A\underset{\sim}{y}(x)$$

for all $x \in \mathbb{R}$. Also, $\underset{\sim}{y}(0) = \underset{\sim}{y_0}$. This ensures that $\underset{\sim}{y}(x) = e^{Ax}\underset{\sim}{y_0}$ is the solution of the IVP given

To prove the uniqueness of the solution of the IVP let's assume $\underset{\sim}{z}(x)$ to be any solution and define $\underset{\sim}{u}(x) \equiv e^{-Ax}\underset{\sim}{z}(x)$.

Therefore, $\quad \frac{d}{dx}(\underset{\sim}{u}(x)) = \frac{d}{dx}\left(e^{Ax}\underset{\sim}{z}(x)\right)$

$$= -Ae^{-Ax}\underset{\sim}{z}(x) + e^{Ax}\frac{d}{dx}\left(\underset{\sim}{z}(x)\right)$$

$$= -Ae^{-Ax}\underset{\sim}{z}(x) + e^{-Ax}A\underset{\sim}{z}(x) = 0 \ \forall \ x \in \mathbb{R}$$

$(\because e^{-Ax}$ commutes with $A)$

Thus $\underset{\sim}{u}(x)$ is a constant vector. By definition $\underset{\sim}{u}(x) \equiv e^{-Ax} \underset{\sim}{z}(x)$

$$\therefore \quad \underset{\sim}{u}(0) = \underset{\sim}{u}(x) = \underset{\sim}{z}(0) = \underset{\sim}{y}_0$$

Hence $\quad \underset{\sim}{y}_0 = e^{-Ax} \underset{\sim}{z}(x) \Leftrightarrow \underset{\sim}{z}(x) = e^{Ax}\underset{\sim}{y}_0$

This shows that $\underset{\sim}{y}(x) = e^{Ax}\underset{\sim}{y}_0$ is the unique solution of IVP given.

Once the fundamental theorem quoted above is established, our aim is to determine e^{Ax} when the solution of IVP quoted in the above theorem is being sought. We therefore explore a very simple property of matrix e^A in the following:

Theorem 9.8: If P is any non-singular n-square matrix and A is a n-square matrix, then the matrix $e^{PAP^{-1}}$ is also a n-square matrix and moreover,

$$e^{PAP^{-1}} = Pe^A \ P^{-1} \tag{9.10}$$

Proof. Write $PAP^{-1} = S.$ S is clearly non-singular.

$$\therefore \quad e^{PAP^{-1}} = e^S = \sum_{k=0}^{\infty} \frac{S^k}{k!} = \sum_{k=0}^{\infty} \frac{(PAP^{-1})^k}{k!}$$

$$= \lim_{m\to\infty} \sum_{k=0}^{m} \frac{(PAP^{-1})^k}{k!}$$

$$= \lim_{m\to\infty} \sum_{k=0}^{m} \frac{PA^kP^{-1}}{k!} = P\left(\sum_{k=0}^{\infty} \frac{A^k}{k!}\right) P^{-1} = Pe^A P^{-1}$$

Now in particular, if PAP^{-1} be a diagonal matrix of the form $[\lambda_j \delta_{ij}]_{n\times n}$, then we can write

$$e^{Sx} = e^{(PAP^{-1})x} = \sum_{k=0}^{\infty} \frac{(PAP^{-1})^k \ x^k}{k!} = \sum_{k=0}^{\infty} \left([\lambda_j\delta_{ij}]_{n\times n}\right)^k \frac{x^k}{k!}$$

$$= \sum_{k=0}^{\infty} \left[\frac{\lambda_j^k \ x^k}{k!} \delta_{ij}\right]_{n\times n} = \left[\sum_{k=0}^{\infty} \frac{(\lambda_j x)^k}{k!} \delta_{ij}\right]_{n\times n}$$

i, e, $\quad Pe^{Ax}P^{-1} = \left[\sum_{k=0}^{\infty} \frac{(\lambda_j \ x)^k}{k!} \delta_{ij}\right]_{n\times n} = \left[e^{\lambda_j x} \delta_{ij}\right]_{n\times n}$

i, e, $\quad e^{Ax} = P^{-1} \left[e^{\lambda_j x} \delta_{ij}\right]_{n\times n} P$

Thus in the matrix form one can write e^{Ax} as :

$$e^{Ax} = P^{-1} \begin{pmatrix} e^{\lambda_1 x} & & & \\ & e^{\lambda_2 x} & & \text{\LARGE O} \\ & & \ddots & \\ \text{\LARGE O} & & & e^{\lambda_n x} \end{pmatrix} P \qquad (9.11)$$

From the above discussion it transpires that for computation of e^{Ax} we need construct a non-singular matrix P that could force diagonalisation of A whenever A is simple. In the previous section we circumvented the problems involving defective matrix A. Now our stage is all set to pursue this problem provided we enrich our repertoire with the concept of generalised eigenvectors of matrix A so as to construct the desired P or the like meant for diagonalisation. We therefore pass onto a short discussion on generalised eigenvectors in the upcoming article followed by Jordanization in the next.

9.5 Generalised Eigenvectors : Solution to IVP

In computing e^{Ax} for a $n \times n$ defective matrix A (i,e, a matrix having at least one irregular eigenvalue) we need incorporate the concept of generalised eigenvectors and generalised eigenspaces of matrices together with the idea of nilpotent matrices. The formal definitions and basic results are cited without proof in the following.

If λ_r be an eigenvalue of $A \in M_n(\mathbb{R})$ having multiplicity $r \leqslant n$, then any non-trivial solution of the homogeneous matrix equations

$$(A - \lambda_r I_n)^k \, \underset{\sim}{v} = \underset{\sim}{0} \quad \text{for} \quad k = 1, 2, \cdots, r.$$

is called a **generalised eigenvector** of matrix A. The generalised eigenvectors corresponding to a given multiple eigenvalue λ_r form a subspace of $\mathbb{R}^n$ of course when the null vector $\underset{\sim}{0}$ in $\mathbb{R}^n$ is appended. Infact it may be shown that this subspace is nothing but $Ker\left((A - \lambda_r I_n)^k\right)$ and is known as the **generalised eigenspace** of that particular eigenvalue λ_r having multiplicity r. The dimension of this generalised eigenspace of λ_r is also r.

It is a very interesting result of linear algebra that every eigenvector of a square matrix is a generalised eigenvector but the converse is not true. We will not furnish formal proof to this claim but shall cite one example of a defective matrix whose generalised eigenvectors are not necessarily eigenvectors.

Consider the matrix $\quad A = \begin{pmatrix} 2 & 0 & 0 \\ 1 & 2 & 0 \\ 0 & 1 & 2 \end{pmatrix} \in M_3(\mathbb{R})$. Since this matrix

A is lower triangular, it is clear that 2 is the eigenvalue of A with algebraic multiplicity 3. It is a routine matter to check that $\exists$ only one

eigenvector, viz, $\begin{pmatrix} 0 \\ 0 \\ 1 \end{pmatrix}$ corresponding to 2. Hence A is defective.

To compute the generalised eigenvectors of A for eigenvalue 2 we begin solving

$$(A - 2I_3)^2 \begin{pmatrix} x_1 \\ x_2 \\ x_3 \end{pmatrix} = \begin{pmatrix} 0 \\ 0 \\ 0 \end{pmatrix}$$

i, e, $\quad \begin{pmatrix} 0 & 0 & 0 \\ 1 & 0 & 0 \\ 0 & 1 & 0 \end{pmatrix}^2 \begin{pmatrix} x_1 \\ x_2 \\ x_3 \end{pmatrix} = \begin{pmatrix} 0 \\ 0 \\ 0 \end{pmatrix}$

i, e, $\quad \begin{pmatrix} 0 & 0 & 0 \\ 0 & 0 & 0 \\ 1 & 0 & 0 \end{pmatrix} \begin{pmatrix} x_1 \\ x_2 \\ x_3 \end{pmatrix} = \begin{pmatrix} 0 \\ 0 \\ 0 \end{pmatrix}$

So $\begin{pmatrix} 0 \\ x_2 \\ x_3 \end{pmatrix}$ is a generalised eigenvector of A for arbitrary x_2, x_3. (not both zeros).

This implies $\begin{pmatrix} 0 \\ 1 \\ 0 \end{pmatrix}$ and $\begin{pmatrix} 0 \\ 0 \\ 1 \end{pmatrix}$ both are linearly independent gener-

alised eigenvectors of A. As claimed, $\begin{pmatrix} 0 \\ 0 \\ 1 \end{pmatrix}$ was an eigenvector of A

corresponding to eigenvalue 2 but $\begin{pmatrix} 0 \\ 1 \\ 0 \end{pmatrix}$ is not an eigenvector of A.

We further try $(A - 2I_3)^3 \begin{pmatrix} x_1 \\ x_2 \\ x_3 \end{pmatrix} = \begin{pmatrix} 0 \\ 0 \\ 0 \end{pmatrix}$

But it means $\begin{pmatrix} 0 & 0 & 0 \\ 0 & 0 & 0 \\ 0 & 0 & 0 \end{pmatrix} \begin{pmatrix} x_1 \\ x_2 \\ x_3 \end{pmatrix} = \begin{pmatrix} 0 \\ 0 \\ 0 \end{pmatrix}$, so that x_1, x_2, x_3

are all arbitrary. Thus $\begin{pmatrix} 1 \\ 0 \\ 0 \end{pmatrix}, \begin{pmatrix} 0 \\ 1 \\ 0 \end{pmatrix}, \begin{pmatrix} 0 \\ 0 \\ 1 \end{pmatrix}$ are all generalised eigenvectors of A.

It is observed that the matrix $(A - 2I_3)$ is nilpotent of index 3. See that here the three generalised eigenvectors of A are the canonical basis vectors of $\mathbb{R}^3$. This result is no fluke or special for a particular A. It can be theoretically established that the n linearly independent generalised eigenvectors of $A \in M_n(\mathbb{R})$ form a basis of the vector space $\mathbb{R}^n$ on which the n-square matrix A or its parent linear operator is defined.

Further there is a celebrated result in this connexion, known as **Decomposition Theorem**, which states that if the eigenvalues of A are all real, then one can express $\mathbb{R}^n$ as the direct sum of the generalised eigenspaces of the distinct eigenvalues of A. (Ref : Hirsh & Smale/Sheldon Axler).

To complete our discussion we introduce the idea of nilpotent matrices. A square matrix N of order n is said to be **nilpotent of index p** iff p is the least positive integer for which $N^k = 0$.

As an example, the 2×2 matrix N given by

$$N = \begin{pmatrix} ab & b^2 \\ -a^2 & -ab \end{pmatrix}$$ is nilpotent of index 2 as $N^2 = 0$.

Similarly the 4×4 matrix

$$N = \begin{pmatrix} -1 & -2 & -1 & -1 \\ 1 & 1 & 1 & 1 \\ 0 & 1 & 0 & 0 \\ 0 & 0 & 0 & 0 \end{pmatrix}$$

is nilpotent of index 3 as $N^3 = 0$ although

$$N^2 = \begin{pmatrix} -1 & -1 & -1 & -1 \\ 0 & 0 & 0 & 0 \\ 1 & 1 & 1 & 1 \\ 0 & 0 & 0 & 0 \end{pmatrix} \neq 0$$

Generally, if $N \in \mathcal{L}\ (\mathbb{R}^n)$ is nilpotent, then $N^{dim(\mathbb{R}^n)} = N^n = 0$. Contrapositively, if N be a n-square matrix but $N^n \neq 0$, then N is not nilpotent.

Now the stage is all set to introduce the following result : If A be a real n-square matrix having real eigenvalues $\lambda_1, \lambda_2, \cdots, \lambda_n$ (repeated according to their multiplicity) and if the generalised eigenvectors $\{\ \underset{\sim}{v}_1, \underset{\sim}{v}_2, \cdots, \underset{\sim}{v}_n\ \}$ of A forms a basis of $\mathbb{R}^n$, then A can be uniquely expressed as sum of two n-square matrices S and N, where S is diagonalisable and N is nilpotent. In its train following results come up:

(a) S can be diagonalised by a similarity transformation carried out by a non-singular n-square matrix P whose columns are the n generalised eigenvectors of A . In symbols this means that $\exists\ P \equiv [\ \underset{\sim}{v}_1, \underset{\sim}{v}_2, \cdots, \underset{\sim}{v}_n]$ such that $\det P \neq 0$ and moreover,

$$S = P\,[\ \lambda_j\ \delta_{ij}\ ]_{n \times n}\ P^{-1}$$

(b) $N \equiv A - S$ is nilpotent of index $p \leqslant n$ as $N^{dim(\mathbb{R}^n)} = 0$

(c) $[S, N] \equiv SN - NS = 0$ i,e, N and S commute with each other and so also with A.

$$\therefore \qquad \begin{aligned} e^{Ax} = e^{(S+N)x} &= e^{Sx+Nx} \\ &= e^{Sx} \cdot e^{Nx} \quad \text{[cf BCH formula (9.16)]} \\ &= P\left(e^{\lambda_j x}\ \delta_{ij}\right) P^{-1}\ e^{Nx} \end{aligned}$$

Baker Campbell – Hausdorff (BCH) formula is related to Lie Groups and Lie Algebras

$$\text{i,e,} \qquad e^{Ax} = P\left(e^{\lambda_j x}\ \delta_{ij}\right) P^{-1} \left[I_n + Nx + \cdots + \frac{N^{p-1}x^{p-1}}{(p-1)!}\right] \qquad (9.12)$$

Once (9.12) is derived, we can use it together with the Theorem 9.7 to solve the IVP :

$$\underset{\sim}{y}'(x) = A(x)\, \underset{\sim}{y}(x) \quad \text{subject to} \quad \underset{\sim}{y}(0) = \underset{\sim}{y}_0$$

The unique solution of the above IVP is given by

$$\underset{\sim}{y}(x) = e^{Ax}\, \underset{\sim}{y}_0 = P\left(e^{\lambda_j x}\, \delta_{ij}\right) P^{-1}\left(I_n + Nx + \cdots + \frac{N^{p-1}\, x^{p-1}}{(p-1)!}\right) \underset{\sim}{y}_0$$

$$(9.13)$$

If in particular A be itself diagonalisable, then $S = A$ and $N = 0$. Consequently, the unique solution of the IVP boils down to

$$\underset{\sim}{y}(x) = P\left(e^{\lambda_j x}\, \delta_{ij}\right) P^{-1}\, \underset{\sim}{y}_0 \qquad (9.14)$$

From (9.13) follows the fact that if all the eigenvalues of A are one and the same, i,e, $\lambda_j = \lambda \; \forall \; j = 1, 2, \cdots, n$, then the matrix $\left(e^{\lambda_j x}\, \delta_{ij}\right)$ boils down to a scalar matrix of order n and hence $P\left(e^{\lambda_j x}\, \delta_{ij}\right) P^{-1} = e^{\lambda x} I_n$.

Under this particular circumstances, (9.13) would reduce to

$$\underset{\sim}{y}(x) = e^{\lambda x}\left(I_n + Nx + \frac{N^2 x^2}{2!} + \cdots + \frac{N^{p-1}\, x^{p-1}}{(p-1)!}\right) \underset{\sim}{y}_0 \qquad (9.15)$$

So here we could successfully avoid construction of P, the non-singular n-matrix comprising of n linearly independent generalised eigen-vectors of the matrix A governing the system.

The lower-triangular matrix $A = \begin{pmatrix} 2 & 0 & 0 \\ 1 & 2 & 0 \\ 0 & 1 & 2 \end{pmatrix}$ cited at the beginning of this article is of this species, As $N \equiv (A - 2I_3)$ is nilpotent of index 2, i,e, $N^2 = 0$, it follows that the solution of the IVP

$$\underset{\sim}{y}'(x) = \begin{pmatrix} 2 & 0 & 0 \\ 1 & 2 & 0 \\ 0 & 1 & 2 \end{pmatrix} \underset{\sim}{y}(x) \quad \text{subject to} \quad \underset{\sim}{y}(0) = \begin{pmatrix} 1 \\ 2 \\ 3 \end{pmatrix} \qquad (*)$$

will be of the form :

$$\underset{\sim}{y}(x) = e^{2x}\left(I_3 + \frac{Nx}{1!}\right)\begin{pmatrix} 1 \\ 2 \\ 3 \end{pmatrix}$$

$$= e^{2x} \left\{ \begin{pmatrix} 1 & 0 & 0 \\ 0 & 1 & 0 \\ 0 & 0 & 1 \end{pmatrix} + \begin{pmatrix} 0 & 0 & 0 \\ x & 0 & 0 \\ 0 & x & 0 \end{pmatrix} \right\} \begin{pmatrix} 1 \\ 2 \\ 3 \end{pmatrix}$$

$$= e^{2x} \begin{pmatrix} 1 & 0 & 0 \\ x & 1 & 0 \\ 0 & x & 1 \end{pmatrix} \begin{pmatrix} 1 \\ 2 \\ 3 \end{pmatrix} = \begin{pmatrix} 1 \\ x+2 \\ 2x+3 \end{pmatrix} e^{2x}$$

The reader could at once recognise this solution of the above vector differential equation (9.17) as cognate to the solution of

$$(D-2)^3 y = 0 \quad \text{subject to} \quad y(0) = 3 \ ; \ y'(0) = 2 \ ; \ y''(0) = 1$$

Example 9.8. Solve $\underset{\sim}{y}'(x) = A\underset{\sim}{y}(x)$ subject to $\underset{\sim}{y}(0) = \underset{\sim}{y}_0$

$$\text{where} \quad A = \begin{pmatrix} 0 & 0 & 0 & 0 \\ 1 & 0 & 0 & 1 \\ 1 & 0 & 0 & 1 \\ 0 & -1 & 1 & 0 \end{pmatrix}$$

The characteristic equation of A is :

$$0 = |A - \lambda I| = \begin{vmatrix} -\lambda & 0 & 0 & 0 \\ 1 & -\lambda & 0 & 1 \\ 1 & 0 & -\lambda & 1 \\ 0 & -1 & 1 & -\lambda \end{vmatrix} = \lambda^4$$

so that $\lambda = 0$ has multiplicity 4.

By routine method one finds the only eigenvector of A corresponding to $\lambda = 0$ to be $(1 \ 1 \ 1 \ -1)^T$. So we have to seek three more linearly independent generalised eigenvectors of A so as to form a basis of $\mathbb{R}^4$ comprising of generalised eigenvectors of A. Now

$$A^2 = \begin{pmatrix} 0 & 0 & 0 & 0 \\ 0 & -1 & 1 & 0 \\ 0 & -1 & 1 & 0 \\ 0 & 0 & 0 & 0 \end{pmatrix} \quad \text{while} \quad A^3 = \begin{pmatrix} 0 & 0 & 0 & 0 \\ 0 & 0 & 0 & 0 \\ 0 & 0 & 0 & 0 \\ 0 & 0 & 0 & 0 \end{pmatrix}$$

We compute generalised eigenvectors of A by solving the equation

$$\begin{pmatrix} 0 \\ 0 \\ 0 \\ 0 \end{pmatrix} = A^2 \begin{pmatrix} x_1 \\ x_2 \\ x_3 \\ x_4 \end{pmatrix} = \begin{pmatrix} 0 & 0 & 0 & 0 \\ 0 & -1 & 1 & 0 \\ 0 & -1 & 1 & 0 \\ 0 & 0 & 0 & 0 \end{pmatrix} \begin{pmatrix} x_1 \\ x_2 \\ x_3 \\ x_4 \end{pmatrix} = \begin{pmatrix} 0 \\ x_3 - x_2 \\ x_3 - x_2 \\ 0 \end{pmatrix}$$

so that four linearly independent generalised eigenvectors of A are :

$$\begin{pmatrix} 1 \\ 0 \\ 0 \\ 0 \end{pmatrix}, \begin{pmatrix} 0 \\ 1 \\ 1 \\ 0 \end{pmatrix}, \begin{pmatrix} 0 \\ 0 \\ 0 \\ 1 \end{pmatrix} \text{ and } \begin{pmatrix} 1 \\ 1 \\ 1 \\ -1 \end{pmatrix}$$

Construction of P is needless here because the matrix $(\lambda_j \delta_{ij})_{4 \times 4} = 0$.

$$\therefore \qquad S = P(\lambda_j \delta_{ij})_{4 \times 4} \, P^{-1} = 0.$$

Thus $N \equiv A - S = A$ and consequently $N^3 = A^3 = 0$

The general solution will be given by

$$\underset{\sim}{y}(x) = \left(I_4 + Nx + \frac{N^2 x^2}{2!} \right) \underset{\sim}{y_0} = \begin{pmatrix} 1 & 0 & 0 & 0 \\ x & 1 - \frac{x^2}{2} & \frac{x^2}{2} & x \\ x & -\frac{x^2}{2} & 1 + \frac{x^2}{2} & x \\ 0 & -x & x & 1 \end{pmatrix} \underset{\sim}{y_0}$$

The four linearly independent vectors of the solution-space $\mathbb{R}^4$ are

$$\begin{pmatrix} 1 \\ x \\ x \\ 0 \end{pmatrix} ; \begin{pmatrix} 0 \\ 1 - \frac{x^2}{2} \\ -\frac{x^2}{2} \\ -x \end{pmatrix} ; \begin{pmatrix} 0 \\ \frac{x^2}{2} \\ 1 + \frac{x^2}{2} \\ x \end{pmatrix} \text{ and } \begin{pmatrix} 0 \\ x \\ x \\ 1 \end{pmatrix} \text{ with } x_0 \in \mathbb{R}$$

Example 9.9 Solve the vector differential equation

$$\underset{\sim}{y'}(x) = \begin{pmatrix} 2 & 0 & 0 \\ 1 & 2 & 0 \\ 1 & 0 & 3 \end{pmatrix} \underset{\sim}{y}(x) \text{ subject to } \underset{\sim}{y}(0) = \begin{pmatrix} c_1 \\ c_2 \\ c_3 \end{pmatrix}$$

Solution: The matrix A associated with the above problem is a lower triangular matrix and so its eigenvalues are readily recognised to be

2,2,3. Through stereotype approach of determining eigenvectors one can

find that $\begin{pmatrix} 0 \\ 0 \\ 1 \end{pmatrix}$ is an eigenvector corresponding to eigenvalue 3 while

for the eigenvalue 2 $\exists$ only one linearly independent eigenvector, viz,

$\begin{pmatrix} 0 \\ 1 \\ 0 \end{pmatrix}$. Thus A is defective and hence we have to find one generalised

eigenvector corresponding to 2 so that we can frame a basis of $\mathbb{R}^3$ made up of generalised eigenvectors of A. From the equation

$$(A - 2I_3)^2 \begin{pmatrix} x_1 \\ x_2 \\ x_3 \end{pmatrix} = \begin{pmatrix} 0 \\ 0 \\ 0 \end{pmatrix} \text{ we find that } \begin{pmatrix} 0 \\ 1 \\ 0 \end{pmatrix} \text{ and } \begin{pmatrix} 1 \\ 0 \\ -1 \end{pmatrix} \text{ are}$$

generalised eigenvectors, corresponding to eigenvalue 2.

Once this is done, one may construct matrix P given by

$$P = \begin{pmatrix} 1 & 0 & 0 \\ 0 & 1 & 0 \\ -1 & 0 & 1 \end{pmatrix} \; ; \; P^{-1} = \begin{pmatrix} 1 & 0 & 0 \\ 0 & 1 & 0 \\ 1 & 0 & 1 \end{pmatrix}$$

$$\therefore \quad S = \begin{pmatrix} 1 & 0 & 0 \\ 0 & 1 & 0 \\ -1 & 0 & 1 \end{pmatrix} \begin{pmatrix} 2 & 0 & 0 \\ 0 & 2 & 0 \\ 0 & 0 & 3 \end{pmatrix} \begin{pmatrix} 1 & 0 & 0 \\ 0 & 1 & 0 \\ 1 & 0 & 1 \end{pmatrix} = \begin{pmatrix} 2 & 0 & 0 \\ 0 & 2 & 0 \\ 1 & 0 & 3 \end{pmatrix}$$

$$\therefore \quad N \equiv A - S = \begin{pmatrix} 0 & 0 & 0 \\ 1 & 0 & 0 \\ 0 & 0 & 0 \end{pmatrix} \quad \text{[vide Decomposition Theorem]}$$

It is a simple task to verify that $N^2 = 0$ and moreover,

$$SN = NS = \begin{pmatrix} 0 & 0 & 0 \\ 2 & 0 & 0 \\ 0 & 0 & 0 \end{pmatrix}$$

Hence by (9.13) one can write down the unique solution of the IVP as :

$$\underset{\sim}{y}(x) = P \left(e^{\lambda_j x} \, \delta_{ij} \right) P^{-1} \left(I_3 + Nx \right) \begin{pmatrix} c_1 \\ c_2 \\ c_3 \end{pmatrix}$$

$$= \begin{pmatrix} 1 & 0 & 0 \\ 0 & 1 & 0 \\ -1 & 0 & 1 \end{pmatrix} \begin{pmatrix} e^{2x} & 0 & 0 \\ 0 & e^{2x} & 0 \\ 0 & 0 & e^{3x} \end{pmatrix} \begin{pmatrix} 1 & 0 & 0 \\ 0 & 1 & 0 \\ 1 & 0 & 1 \end{pmatrix} \times$$

$$\begin{pmatrix} 1 & 0 & 0 \\ x & 1 & 0 \\ 0 & 0 & 1 \end{pmatrix} \begin{pmatrix} c_1 \\ c_2 \\ c_3 \end{pmatrix}$$

$$= \begin{pmatrix} e^{2x} & 0 & 0 \\ xe^{2x} & e^{2x} & 0 \\ e^{3x} - e^{2x} & 0 & e^{3x} \end{pmatrix} \begin{pmatrix} c_1 \\ c_2 \\ c_3 \end{pmatrix} \qquad \text{(i)}$$

$$= c_1 \begin{pmatrix} e^{2x} \\ xe^{2x} \\ e^{3x} - e^{2x} \end{pmatrix} + c_2 \begin{pmatrix} 0 \\ e^{2x} \\ 0 \end{pmatrix} + c_3 \begin{pmatrix} 0 \\ 0 \\ e^{3x} \end{pmatrix}$$

Remark (a) The matrix appearing in (i) is again lower triangular.

(b) Since the initial condition $\underset{\sim}{y}(0) = \begin{pmatrix} c_1 \\ c_2 \\ c_3 \end{pmatrix}$ given in this problem is
at our disposal, we may look into it as the general solution of the vector
differential equation $\underset{\sim}{y}'(x) = A\underset{\sim}{y}(x)$ (same A as in present problem) as a
linear combination of three linearly independent solutions

$$\underset{\sim}{\phi_1}(x) \equiv \begin{pmatrix} e^{2x} \\ xe^{2x} \\ e^{3x} - e^{2x} \end{pmatrix} \; ; \; \underset{\sim}{\phi_2}(x) \equiv \begin{pmatrix} 0 \\ e^{2x} \\ 0 \end{pmatrix} \; ; \; \underset{\sim}{\phi_3}(x) \equiv \begin{pmatrix} 0 \\ 0 \\ e^{3x} \end{pmatrix}$$

9.6 Jordanization of Matrices and Solution of Ode

Computing the exponential function e^A of a non-diagonalisable, i,e,
defective matrix A in the closed form is in general not a simple task.
For this purpose we make use of the well-known result in linear algebra
that if A be any complex matrix of order n, i,e, $A \in M_n(\mathbb{C})$, then $\exists$ a
non-singular matrix $P \in M_n(\mathbb{C})$ such that $P^{-1}AP$ has Jordan form, viz,

$$J = \begin{pmatrix} J_1 & & & & O \\ & J_2 & & & \\ & & \ddots & & \\ & & & \ddots & \\ O & & & & J_n \end{pmatrix},$$

where each J_i $(i = 1, 2, \cdots, k)$ is an elementary Jordan block with pos-
sibly different eigenvalues λ of A. The Jordan form J is unique upto

the ordering of the J_i's, but the non-singular transforming matrix P involved in the process is not unique.

Two subcases may arise — (i) λ is real (ii) λ is complex. In the first subcase, J_i's are of the generic form

$$J_i = \begin{pmatrix} \lambda_i & 1 & 0 & \cdots & \cdots & 0 \\ 0 & \lambda_i & 1 & 0 & \cdots & 0 \\ \vdots & & & & & \vdots \\ 0 & 0 & \cdots & \cdots & \lambda_i & 1 \\ 0 & 0 & \cdots & \cdots & 0 & \lambda_i \end{pmatrix}$$

so that $(J_i)_{jl} = \lambda_i \delta_{jl} + \delta_{j,\overline{j+1}}$ for $j, l = 1, 2, \cdots, n$

In the second subcase, i,e, for complex eigenvalues of the form $(a + ib)$, the generic form of the J_i's are

$$J_i = \begin{pmatrix} D_2 & I_2 & O & \cdots & \cdots & O \\ O & D_2 & I_2 & \cdots & \cdots & O \\ \vdots & & & & & \\ O & O & \cdots & \cdots & D_2 & I_2 \\ O & O & \cdots & \cdots & O & D_2 \end{pmatrix}$$

with $D_2 = \begin{pmatrix} a & -b \\ b & a \end{pmatrix}$; $I_2 = \begin{pmatrix} 1 & 0 \\ 0 & 1 \end{pmatrix}$ and $O = \begin{pmatrix} 0 & 0 \\ 0 & 0 \end{pmatrix}$

We shall take up the case of real eigenvalues of A. For the complex eigenvalues, discussions will not be touched upon in this text. It's a trivial observation that in the first subcase, each of the elementary Jordan blocks J_i is a defective matrix since it has only one eigenvector. (In the second subcase, each of the elementary Jordan blocks J_i is a defective matrix as it has only two eigenvectors).

In what follows, we present an algorithm for Jordanisation of $A \in M_n(\mathbb{C})$:

Step-I: Determine the characteristic polynomial $f_A(\lambda) \equiv det(A - \lambda I)$ and because $A \in M_n(\mathbb{C}), f_A(\lambda)$ can be expressed as a product of linear polynomials in λ. This gives us the eigenvalues of A with respective algebraic multiplicities.

Step-II: For each fixed eigenvalues λ, define $T \equiv A - \lambda I_n$.

Step-III: If λ be an eigenvalue of A with algebraic multiplicity m_λ, then one can establish the theoretical result that $\exists$ some positive integer $p(\leqslant m_\lambda)$ such that $\dim(Ker(T^p)) = m_\lambda$. Further observe that

$$Ker(T^{k-1}) \subseteq Ker(T^k) \text{ since } \underset{\sim}{\xi} \in Ker(T^{k-1}) \Leftrightarrow T^{k-1}\underset{\sim}{\xi} = \underset{\sim}{0} \Rightarrow$$
$$T^k\underset{\sim}{\xi} = T(T^{k-1}\underset{\sim}{\xi}) = T(\underset{\sim}{0}) = \underset{\sim}{0} \Leftrightarrow \underset{\sim}{\xi} \in Ker(I^k).$$

Hence $\dim Ker(T^{k-1}) \leqslant m_\lambda$.

Letting $d_k \equiv \dim(Ker(T^k))$ it follows that $\{d_k\}$ is an increasing finite sequence bounded by m_λ. d_k's are known as deficiency indices and can be found by Gaussian reduction. Indeed d_k stands for the number of zero-rows in the row-reduced echelon form of T^k.

Again, $\quad d_k - d_{k-1} = \dim(Ker\ T^k) - \dim\left(Ker(T^{k-1})\right)$

$$= \# \text{ Jordan blocks of size at least } k$$

Parallelly, $d_{k+1} - d_k = \#$ Jordan blocks of size at least $(k+1)$.

So $(d_k - d_{k-1}) - (d_{k+1} - d_k)$ gives the exact number of Jordan blocks of size k. We must compute the sequence $\{d_k\}$ for each given λ until we get $d_p = m_\lambda$.

One may interpret $d_k's$ in another way:

d_1 denotes the number of chains of generalised eigenvectors corresponding to λ. As $d_p = m_\lambda$, one of these chains must be of length p. Compute the chains of generalised eigenvectors corresponding to given λ, starting from the chain with length p. Obviously $(2d_k - d_{k-1} - d_{k+1})$ gives us the number of Jordan chains of length k corresponding to that particular eigenvalue λ.

Step-IV: We form a chain of p generalised eigenvectors by choosing a vector $\underset{\sim}{v}_p$ in $Ker(T^p)$ and then define $\underset{\sim}{v}_{p-1} = T\underset{\sim}{v}_p$, $\underset{\sim}{v}_{p-2} = T\underset{\sim}{v}_{p-1} = T^2\underset{\sim}{v}_p$ and so on. To ensure that all these vectors $\underset{\sim}{v}_{p-1}, \cdots \underset{\sim}{v}_1$ are non-zero, we must have $\underset{\sim}{v}_p \in Ker(T^q)$ for $q = 1, 2, \cdots, p-1$, i,e, p must be the smallest positive integer for which $T^k\underset{\sim}{v}_p = \underset{\sim}{0}$

Let's try to explain the ideas through illustrative example.

Example 9.10 Consider the Jordanisation of the matrix A given by

$$A = \begin{pmatrix} 0 & -1 & -2 & -1 \\ 1 & 2 & 1 & 1 \\ 0 & 0 & 1 & 0 \\ 0 & 0 & 1 & 1 \end{pmatrix}$$

It is a trival exercise to check that $\lambda = 1$ is an eigenvalue of A with algebraic multiplicity 4.

The row-reduced echelon form of $T \equiv A - I_4$ is

$$\begin{pmatrix} 1 & 1 & 1 & 1 \\ 0 & 0 & 1 & 0 \\ 0 & 0 & 0 & 0 \\ 0 & 0 & 0 & 0 \end{pmatrix}$$

If $\begin{pmatrix} x_1 \\ x_2 \\ x_3 \\ x_4 \end{pmatrix}$ be an eigenvector corresponding to $\lambda = 1$, then

$$\underset{\sim}{0} = T \begin{pmatrix} x_1 \\ x_2 \\ x_3 \\ x_4 \end{pmatrix} \text{ and so } x_1 + x_2 + x_3 + x_4 = 0 \text{ and } x_3 = 0$$

Set $x_2 = c_1, x_4 = c_2$ and so $x_1 = -c_1 - c_2$

$(x_1 \ x_2 \ x_3 \ x_4)^T = c_1(-1 \ 1 \ 0 \ 0)^T + c_2(-1 \ 0 \ 0 \ 1)^T$

$KerT = Ker(A - I_4) = \text{Span} \left\{ (-1 \ 1 \ 0 \ 0)^T, (-1 \ 0 \ 0 \ 1)^T \right\}$

i,e, $\dim(KerT) = 2$, indicating the existence of 2 chains of generalised eigenvectors of A.

Further, $(A - I_4)^2 = T^2 = 0$, showing that each of the Jordan chains is of length 2.

To find the generalised eigenvector of one Jordan chain, let's start with the canonical basis $\left\{ \underset{\sim}{e}_1, \underset{\sim}{e}_2, \underset{\sim}{e}_3, \underset{\sim}{e}_4 \right\}$ of $\mathbb{R}^4$. Observe that $(1 \ 0 \ 0 \ 0)^T \notin$ $Ker(A - I_4)$ and therefore we set $\underset{\sim}{v}_2 \equiv 1 \ 0 \ 0 \ 0)^T$ as one generalised eigenvector $\underset{\sim}{v_2}$ and define $\underset{\sim}{v_1} = T\underset{\sim}{v_2} = (-1\ 1\ 0\ 0)^T$, which is an ordinary eigenvector. To find the second chain of generalised eigenvector we have to satisfy two conditions. So $\{\underset{\sim}{w_2}, \underset{\sim}{w_1}\}$ can be a legitimate second Jordan chain provided (i) $\underset{\sim}{w_2}$ is independent of $\underset{\sim}{v_2}$ and should $\underset{\sim}{w_2}$ not lie in the space $KerT$ (ii) $\underset{\sim}{w_1}$ must be independent of $\underset{\sim}{v_2}$. The canonical ordered basis $(\underset{\sim}{e_1}, \underset{\sim}{e_2}, \underset{\sim}{e_3}, \underset{\sim}{e_4},)$ of $\mathbb{R}^4$ shows us the path how to construct second chain of generalised eigenvectors. Since $\underset{\sim}{e_2} \equiv \underset{\sim}{v_2}$, we begin with this our trial and error method and set $\underset{\sim}{w_2} = \underset{\sim}{e_2}$.

However, as $T\underset{\sim}{w_2} = (-1\ 1\ 0\ 0)^T \in KerT$, $\underset{\sim}{e_2}$ is not an admissible choice of $\underset{\sim}{w_2}$. Next let's try $\underset{\sim}{e_3}$ for $\underset{\sim}{w_2}$ and see that

$$\underset{\sim}{w_1} = T\underset{\sim}{w_2} = T\underset{\sim}{e_3} = (-2\ 1\ 0\ 1)^T.$$

Further $\underset{\sim}{w_1}$ being not a multiple of $\underset{\sim}{v_1}$, we have a second Jordan chain.

We may construct the matrix P made of generalised eigenvectors as :

$$P = \left(\ \underset{\sim}{v_1}\ \underset{\sim}{v_2}\ \underset{\sim}{w_1}\ \underset{\sim}{w_2}\ \right) = \begin{pmatrix} -1 & 1 & -2 & 0 \\ 1 & 0 & 1 & 0 \\ 0 & 0 & 0 & 1 \\ 0 & 0 & 1 & 0 \end{pmatrix}$$

and apply on A the transformation $P^{-1}AP$ to get the desired J.

$$J = P^{-1}AP$$

$$= \begin{pmatrix} 0 & 1 & 0 & -1 \\ 1 & 1 & 0 & 1 \\ 0 & 0 & 0 & 1 \\ 0 & 0 & 1 & 0 \end{pmatrix} \begin{pmatrix} 0 & -1 & -2 & -1 \\ 1 & 2 & 1 & 1 \\ 0 & 0 & 1 & 0 \\ 0 & 0 & 1 & 1 \end{pmatrix} \begin{pmatrix} -1 & 1 & -2 & 0 \\ 1 & 0 & 1 & 0 \\ 0 & 0 & 0 & 1 \\ 0 & 0 & 1 & 0 \end{pmatrix}$$

$$= \begin{pmatrix} -1 & 1 & \vdots & 0 & 0 \\ 0 & 1 & \vdots & 0 & 0 \\ \cdots & \cdots & \vdots & \cdots & \cdots \\ 0 & 0 & \vdots & 1 & 1 \\ 0 & 0 & \vdots & 0 & 1 \end{pmatrix}$$

Observe that this Jordan form J is unique and it consists of two elementary Jordan blocks as expected, each of length 2. However, the matrix P

that transforms A to J is non-unique. In the present problem, another legitimate choice of P is :

$$P \equiv \begin{pmatrix} -1 & 1 & 1 & -1 \\ 1 & 0 & 0 & 0 \\ 0 & 0 & 0 & 1 \\ 0 & 0 & 1 & 0 \end{pmatrix}$$

Obviously these two choices of P's are row-equivalent, i,e, they have the same row-space. P therefore serving as the base-changing matrix.

Remark : As $KerT \subseteq KerT^2$, it follows that rank-1 generalised eigenvectors are necessarily within the span of rank-2 generalised eigenvectors.

Alternative Procedure

Step I and Step II same as previous procedure.

Step III : If λ be a particular eigenvalue of A with algebraic multiplicity m_λ, then find the subspaces $Im(T^k)$ for $k = 1, 2, \cdots, m_\lambda$ and also a basis for each of them. It's quite obvious that if $k > m_\lambda$, then $Im(T^k)$'s are all identical with $Im(T^{m_\lambda})$ since there exists m_λ linearly independent generalised eigenvectors with such an eigenvalue λ.

Step IV : Find the subspaces $W_k = \left(Im(T^{k-1})\backslash Im(T^k)\right) \bigcap KerT$ for each $k = 1, 2, \cdots, m_\lambda$. Each non-null vector ξ in this subspace W_k will be a first vector of Jordan chain of length k and so will correspond to one Jordan block of size k.

Step V : For each W_k, construct a basis. If $\mathcal{B}_\lambda$ denotes union of these bases, then for each $\underset{\sim}{v}_1 \neq \underset{\sim}{0}$ in $\mathbb{B}_\lambda$, solve $T\underset{\sim}{v}_2 = \underset{\sim}{v}_1$ to find $\underset{\sim}{v}_2$, then $T\underset{\sim}{v}_3 = \underset{\sim}{v}_2$ and so on until $\underset{\sim}{v}_i \notin Im\ T$. The vectors $\underset{\sim}{v}_1, \underset{\sim}{v}_2, \underset{\sim}{v}_3, \cdots$ obtained in this are ordered. Once this is done for all eigenvalues of A, we get a non-singular matrix P comprising of such $\underset{\sim}{v}$'s etc so that $P^{-1}AP = J$.

Remark : This P is non-unique while J is unique.

With the problem given in Example 9.10 let's illustrate the alternative procedure. In that example,

$$T = \begin{pmatrix} -1 & -1 & -2 & -1 \\ 1 & 1 & 1 & 1 \\ 0 & 0 & 0 & 0 \\ 0 & 0 & 1 & 0 \end{pmatrix} \text{ and } ImT = Sp\left\{ \begin{pmatrix} -1 \\ 1 \\ 0 \\ 0 \end{pmatrix}, \begin{pmatrix} 0 \\ -1 \\ 0 \\ 1 \end{pmatrix} \right\}$$

$$Ker\ T = Sp\left\{\begin{pmatrix} 1 \\ 1 \\ 1 \\ 1 \end{pmatrix}, \begin{pmatrix} 0 \\ 0 \\ 1 \\ 0 \end{pmatrix}\right\}; \quad Im\ T^2 = Sp\left\{\begin{pmatrix} 0 \\ 0 \\ 0 \\ 0 \end{pmatrix}\right\}$$

Hence $(Im\ T - Im\ T^2)\bigcap Ker\ T = (Im\ T \bigcap Ker\ T)\ Im(T^2)$

Let $\underset{\sim}{u} \in Im(T) \bigcap Ker(T)$

$$\therefore \quad u \in Sp\left\{\begin{pmatrix} -1 \\ 1 \\ 0 \\ 0 \end{pmatrix}, \begin{pmatrix} 0 \\ -1 \\ 0 \\ 0 \end{pmatrix}\right\}\bigcap Sp\left\{\begin{pmatrix} 1 \\ 1 \\ 1 \\ 1 \end{pmatrix}, \begin{pmatrix} 0 \\ 0 \\ 1 \\ 0 \end{pmatrix}\right\}$$

After little workout on linear algebra it follows that

$$\underset{\sim}{u} = t\begin{pmatrix} 1 \\ 0 \\ 0 \\ -1 \end{pmatrix} + s\begin{pmatrix} -1 \\ 1 \\ 0 \\ 0 \end{pmatrix}$$

where s and t are scalars, not both simultaneously zeros.

Hence $\begin{pmatrix} 1 \\ 0 \\ 0 \\ -1 \end{pmatrix}$ and $\begin{pmatrix} -1 \\ 1 \\ 0 \\ 0 \end{pmatrix}$ work as first vectors $\underset{\sim}{v_1}$ & $\underset{\sim}{w_1}$ of the two

Jordan chains of length 2. Obviously $\underset{\sim}{v_2}$ that satisfies $T\underset{\sim}{v_2} = \underset{\sim}{v_1}$ is given by $(0\ 0\ 0\ 1)^T$ and $\underset{\sim}{v_2} \notin Im\ T$. Similarly $\underset{\sim}{w_2}$ that satisfies $T\underset{\sim}{w_2} = \underset{\sim}{w_1}$ is given by $(0\ 0\ -1\ 1)^T$ and $\underset{\sim}{w_2} \notin Im\ T$

This constructs the base-changing matrix P, viz,

$$P = \begin{pmatrix} 1 & 0 & -1 & 0 \\ 0 & 0 & 1 & 0 \\ 0 & -1 & 0 & 0 \\ -1 & 1 & 0 & 1 \end{pmatrix}, \text{ so that } P^{-1} = \begin{pmatrix} 1 & 1 & 0 & 0 \\ 0 & 0 & -1 & 0 \\ 0 & 1 & 0 & 0 \\ 1 & 1 & 1 & 1 \end{pmatrix}$$

Hence

$$P^{-1}AP = \begin{pmatrix} 1 & 1 & 0 & 0 \\ 0 & 0 & -1 & 0 \\ 0 & 1 & 0 & 0 \\ 1 & 1 & 1 & 1 \end{pmatrix}\begin{pmatrix} 0 & -1 & -2 & -1 \\ 1 & 2 & 1 & 1 \\ 0 & 0 & 1 & 0 \\ 0 & 0 & 1 & 1 \end{pmatrix}\begin{pmatrix} 1 & 0 & -1 & 0 \\ 0 & 0 & 1 & 0 \\ 0 & -1 & 0 & 0 \\ -1 & 1 & 0 & 1 \end{pmatrix}$$

$$= \begin{pmatrix} 1 & 1 & \vdots & 0 & 0 \\ 0 & 1 & \vdots & 0 & 0 \\ \cdots & \cdots & \vdots & \cdots & \cdots \\ 0 & 0 & \vdots & 1 & 1 \\ 0 & 0 & \vdots & 0 & 1 \end{pmatrix} = J \equiv \begin{pmatrix} J_1(1) & O \\ O & J_2(1) \end{pmatrix}$$

J shows existence of 2 blocks, each of size 2.

Example 9.11 Consider the Jordanisation of the matrix A given by

$$A = \begin{bmatrix} -2 & -1 & -3 \\ 4 & 3 & 3 \\ -2 & 1 & -1 \end{bmatrix}$$

The characteristic polynomial of A is $(-\lambda + 4)(2 - \lambda)^2$.

Define $T \equiv \begin{pmatrix} -4 & -1 & -3 \\ 4 & 1 & 3 \\ -2 & 1 & -3 \end{pmatrix} \equiv (A - 2I)$;

Clearly $Ker T = Sp\left\{ \begin{pmatrix} 1 \\ -1 \\ -1 \end{pmatrix} \right\}$; $Im\, T = Sp\left\{ \begin{pmatrix} 1 \\ -1 \\ -1 \end{pmatrix}, \begin{pmatrix} 0 \\ 0 \\ 1 \end{pmatrix} \right\}$

and $Im\, T^2 = Sp\left\{ \begin{pmatrix} 1 \\ -1 \\ 1 \end{pmatrix} \right\}$

As per the second procedure,

$$\left(Im\, T \backslash Im\, T^2\right) \bigcap Ker\, T = \left(Im\, T \bigcap Ker\, T\right) \backslash Im(T^2)$$

$$= \left(Sp\left\{ \begin{pmatrix} 1 \\ -1 \\ -1 \end{pmatrix}, \begin{pmatrix} 0 \\ 0 \\ 1 \end{pmatrix} \right\} - Sp\left\{ \begin{pmatrix} 1 \\ -1 \\ 1 \end{pmatrix} \right\} \right) \bigcap Sp\left\{ \begin{pmatrix} 1 \\ -1 \\ -1 \end{pmatrix} \right\}$$

$$= Sp\left\{ \begin{pmatrix} 1 \\ -1 \\ -1 \end{pmatrix} \right\} - Sp\left\{ \begin{pmatrix} 1 \\ -1 \\ -1 \end{pmatrix} \right\}$$

which turns out to be set of all null vectors in $Sp\left\{\begin{pmatrix} 1 \\ -1 \\ -1 \end{pmatrix}\right\}$

Without loss of generality we choose $\underset{\sim}{v_2} = \begin{pmatrix} 1 \\ -1 \\ -1 \end{pmatrix}$ and find $\underset{\sim}{v_2}$ through

the relation $T\underset{\sim}{v_2} = \underset{\sim}{v_1}$. A bit of algebraic exercise shows that $\underset{\sim}{v_2} = (0 \ -1 \ 0)^T$ and moreover, $\underset{\sim}{v_2} \notin Im \ T$.

Again for $\lambda = -4$, the eigenvectors are scalar multiples of $(1 \ -1 \ 1)^T$. All these enable us to construct the base-changing matrix P :

$$P = \begin{pmatrix} 1 & 0 & 1 \\ -1 & -1 & -1 \\ -1 & 0 & 1 \end{pmatrix} \text{ so that } P^{-1} = \begin{pmatrix} \frac{1}{2} & 0 & \frac{1}{2} \\ -1 & -1 & 0 \\ \frac{1}{2} & 0 & \frac{1}{2} \end{pmatrix} \text{ and}$$

$$P^{-1}AP = \begin{pmatrix} \frac{1}{2} & 0 & -\frac{1}{2} \\ -1 & -1 & 0 \\ \frac{1}{2} & 0 & \frac{1}{2} \end{pmatrix} \begin{pmatrix} -2 & -1 & -3 \\ 4 & 3 & 3 \\ -2 & 1 & -1 \end{pmatrix} \begin{pmatrix} 1 & 0 & 1 \\ -1 & -1 & -1 \\ -1 & 0 & -1 \end{pmatrix}$$

$$= \begin{pmatrix} 2 & 1 & \vdots & 0 \\ 0 & 2 & \vdots & 0 \\ \cdots & \cdots & \cdots & \cdots \\ 0 & 0 & \vdots & -4 \end{pmatrix} = J \equiv \begin{pmatrix} J_1(2) & O \\ O & J_2(-4) \end{pmatrix}$$

Once we obtain a Jordan form J from A, we claim that each of the elementary Jordan block J_i can be uniquely expressed as sum of two square matrices, viz, D_i and N_i, where D_i is a diagonal matrix having the same diagonal entries as those of J_i, while N_i is a nilpotent matrix.

We shall skip the details of the proof of this claim but quote two lemmas that work as its basic ingredients.

Lemma-A : If λ, μ be two distinct eigenvalues of $A \in M_n(\mathbb{C})$ and K_λ, K_μ denote their respective generalised eigenspaces, then

$$K_\lambda \bigcap K_\mu = \{\underline{0}\} \ , \text{ the null space of } \mathbb{C}^n.$$

Lemma-B : Strictly upper triangular matrices are nilpotent.

Lemma-A can be proved by employing the definition of generalised eigenvectors while Lemma-B follows directly from Cayley-Hamilton theorem. It is a simple observation that $N = J - D$ is strictly upper triangular and hence by Lemma-B, N is nilpotent.

$$PJ \ P^{-1} = P(D+N)P^{-1} = PD \ P^{-1} + PN \ P^{-1},$$

where $PDP^{-1} \equiv S$ is a diagonalisable matrix and $N' = PNP^{-1}$ is nilpotent of the same index as that of N. If r be the degree of nilpotency of N, then r is the smallest positive integer such that $N^r = 0$, the null matrix. Further observe that

(i) $(N')^r = (PNP^{-1})^r = PN^rP^{-1} = 0 \ (\because N^r = 0),$

indicating that the degree of nilpotency is invariant under similarity transformation.

(ii) $SN' = (PDP^{-1})(PNP^{-1}) = P(DN)P^{-1} = P(ND)P^{-1}$
$$= (PNP^{-1})(PDP^{-1}) = N'S,$$

where we used the fact that N and D commute with each other.

Thus we get back the celebrated S-N decomposition theorem (succinctly referred to as "Decomposition Theorem" earlier in the present chapter) that claims that any $A \in M_n(\mathbb{C})$ is uniquely expressible as sum of two commuting matrices S and N', where S is diagonalisable and N' is nilpotent.

We can now apply either the S-N decomposition of A or the D-N decomposition of $J = P^{-1}AP$ together with Campbell-Baker-Hausdorff relation which states that for any two square matrices C_1 and C_2,

$$\exp (C_1) \cdot \exp (C_2) = \exp \left[C_1 + C_2 + \frac{1}{2} [C_1, C_2] + \cdots \infty \right]$$
$$= \exp [C_1 + C_2], \text{ provided } [C_1, C_2] = 0 \qquad (9.16)$$

In the present context,

$$e^A = e^{PJP^{-1}} = \sum_{n=0}^{\infty} \frac{(PJP^{-1})^n}{n!} = P \left(\sum_{n=0}^{\infty} \frac{J^n}{n!} \right) P^{-1} = Pe^JP^{-1}$$

Therefore, $e^A = Pe^JP^{-1} = P \left[e^{(D+N)} \right] P^{-1} = P(e^De^N)P^{-1}$

$(\because D$ and N commute, $e^D \cdot e^N = e^{D+N}$ by CBH) i, e $\quad e^A = P(e^{\xi_i}\,\delta_{ij})_{n\times n}\left(\displaystyle\sum_{k=0}^{\infty}\frac{N^k}{k!}\right)P^{-1}$ $\quad$ (assuming $D = [\xi_i\delta_{ij}]$)

Again N being nilpotent of order n, its index cannot exceed n. Without loss of generality, if we assume index of N to be q, then $N^k = 0 \;\forall\, k \geqslant q$ and $N^k \neq 0$ if $k < q$. So one may write down

$$e^A = P\left[e^{\xi_i}\,\delta_{ij}\right]_{n\times n}\left(\sum_{k=0}^{q-1}\frac{N^k}{k!}\right)P^{-1}.$$

Now the infinite matrix series $\sum_{k=0}^{\infty}\frac{N^k}{k!}$, standing for e^N, eventually decimates to a matrix-polynomial of degree atmost $q-1$ and $\left(e^\xi\delta_{ij}\right)_{n\times n}$ is a diagonal matrix of order n, whose diagonal entries are exponential function of the respective diagonal entries of D.

One can also write down the expression for e^A as :

$$\begin{aligned}
e^A &= \left(P\left[e^{\xi_i}\delta_{ij}\right]_{n\times n}P^{-1}\right)\left(P\sum_{k=0}^{q-1}\frac{N^k}{k!}P^{-1}\right)\\[2mm]
&= \left(P\left[e^{\xi_i}\delta_{ij}\right]_{n\times n}P^{-1}\right)\left(\sum_{k=0}^{q-1}\frac{(PNP^{-1})^k}{k!}\right) = e^S\sum_{k=0}^{q-1}\frac{(N')^k}{k!},
\end{aligned}$$

where $N' = PNP^{-1}$ is also nilpotent, i, e, $e^{S+N} = e^S\cdot\displaystyle\sum_{k=0}^{q-1}\frac{(N')^k}{k!}$.

Because e^S is non $-$ singular, $e^N = \displaystyle\sum_{k=0}^{q-1}\frac{(N')^k}{k!}$,

Remark: **(a)** Jordan decomposition theorem cannot work for a matrix whose characteristic polynomial does not split on the field wherefrom the matrix elements are chosen. This is why we had chosen the underlying field to be $\mathbb{C}$ that is algebraically closed.

(b) In the first procedure, construction of Jordan chain of generalised eigenvectors (g.v) is done from the tail, i,e, the hiearchy is :

$$\text{rank-4 g.v} \longrightarrow \text{rank-3 g.v} \longrightarrow \text{rank-2 g.v} \longrightarrow \text{rank-1 g.v.}$$

In the second procedure, construction of Jordan chain of generalised eigenvectors (g.v) is done from the beginning. i,e, the hiearchy is :

rank-1 g.v $\longrightarrow$ rank-2 g.v $\longrightarrow$ rank-3 g.v $\longrightarrow$ rank-4 g.v.

We now discuss the utility of Jordan canonical form in context of solving vector differential equations.

Suppose the vector differential equation to be solved reads :

$$\frac{dy}{dx} = A\underset{\sim}{y} \quad \text{with } \underset{\sim}{y}(0) = \underset{\sim}{y}_0 \text{ and } A \in M_n(\mathbb{C})$$

We know that every matrix $A \in M_n(\mathbb{C})$ can be cast into the Jordan form by means of a similarity transformation. If $P^{-1}AP = J$, a Jordan canonical form (JCF), then one substitutes $\underset{\sim}{y} = P\underset{\sim}{z}$ to recast the given I.V.P as :

$$\frac{d}{dx}(P\underset{\sim}{z}) = (PJP^{-1})(P\underset{\sim}{z})$$

$$\text{i, e,} \quad \frac{d\underset{\sim}{z}}{dx} = P^{-1}(PJ\underset{\sim}{z}) = J\underset{\sim}{z}$$

This system naturally breaks into a finite number of smaller systems, each involving one Jordan block. Thus it suffices to concentrate on one elementary Jordan block J_λ of size m, say, corresponding to a particular eigenvalue λ having algebraic multiplicity m_λ and find the general expression of the solution of $\frac{d\underset{\sim}{z}}{dx} = J_\lambda \underset{\sim}{z}$. If $(z_{r_1}, z_{r_2}, \cdots, z_{r_m})^T$ be part of the n-component column matrix $\underset{\sim}{z} \equiv (z_1, z_2, \cdots, z_n)^T$ that corresponds to the said elementary Jordan block J_λ. To make life simple, we shall deliberately write $(z_{r_1}, z_{r_2}, \cdots, z_{r_m})^T$ as $(z'_1, z'_2, \cdots, z'_m)^T = \underset{\sim}{z}'$.

Corresponding to the Jordan block J_λ, the given equation $\frac{d\underset{\sim}{z}}{dx} = J\underset{\sim}{z}$, has the fragment $\frac{d\underset{\sim}{z}'}{dx} = J_\lambda\underset{\sim}{z}'$ that can be explicitly written as :

$$\begin{pmatrix} \frac{dz'_1}{dx} \\[2mm] \frac{dz'_2}{dx} \\[2mm] \vdots \\[2mm] \frac{dz'_{m-1}}{dx} \\[2mm] \frac{dz'_m}{dx} \end{pmatrix} = \begin{pmatrix} \lambda z' + z'_2 \\[2mm] \lambda z'_2 + z'_3 \\[2mm] \vdots \\[2mm] \lambda z_{m-1} + z'_m \\[2mm] \lambda z'_m \end{pmatrix} \quad \cdots\cdots$$

From the last equation of the above, viz, $\dfrac{dz'_m}{dx} = \lambda z'_m$,

we get $z'_m = c_m\, e^{\lambda x}$. Hence the penultimate equation gives :

$$\frac{dz'_{m-1}}{dx} - \lambda z'_{m-1} = z'_m = c_m e^{\lambda x}$$

Standard method of first order linear ode gives the solution :

$$z'_{m-1} = (c_m x + c_{m-1})e^{\lambda x}.$$

Herefrom by principle of induction one can determine that

$$\left.\begin{aligned}
z'_1 &= \left(c_m \frac{x^{m-1}}{(m-1)!} + c_{m-1}\frac{x^{m-2}}{(m-2)!} + \cdots\cdots + c_2 x + c_1\right) e^{\lambda x} \\
z'_2 &= \left(c_m \frac{x^{m-2}}{(m-2)!} + c_{m-1}\frac{x^{m-3}}{(m-3)!} + \cdots\cdots + c_3 x + c_2\right) e^{\lambda x} \\
&\cdots\cdots\cdots\cdots\cdots\cdots\cdots\cdots\cdots\cdots\cdots\cdots\cdots\cdots\cdots \\
&\cdots\cdots\cdots\cdots\cdots\cdots\cdots\cdots\cdots\cdots\cdots\cdots\cdots\cdots\cdots \\
z'_{m-1} &= (c_m x + c_{m-1})e^{\lambda x} \\
z'_m &= c_m e^{\lambda x}
\end{aligned}\right\}$$

In matrix form this can be expressed as :

$$\begin{pmatrix} z'_1 \\ z'_2 \\ \vdots \\ \vdots \\ z'_{m-1} \\ z'_m \end{pmatrix} = K_\lambda(x) \begin{pmatrix} c_1 \\ c_2 \\ \vdots \\ \vdots \\ c_{m-1} \\ c_m \end{pmatrix},$$

where $K_\lambda(x) \equiv$

$$\begin{pmatrix}
1 & x & \frac{x^2}{2!} & \cdots & \cdots & \frac{x^{m-1}}{(m-1)!} \\
 & 1 & x & \cdots & \cdots & \frac{x^{m-2}}{(m-2)!} \\
 & & \ddots & & & \vdots \\
 & & & \ddots & & \vdots \\
 & & & & \ddots & \vdots \\
 & \mathbf{O} & & & 1 & x \\
 & & & & & 1
\end{pmatrix}_{m\times m} e^{\lambda x}$$

Therefore, $\underset{\sim}{z}'(0) = K_\lambda(0) \begin{pmatrix} c_1 \\ c_2 \\ \vdots \\ c_{m-1} \\ c_m \end{pmatrix}$ since $K_\lambda(0) = I_m$

Observe that this block $K_\lambda(x)$ corresponds in a one-to-one manner to the elementary Jordan block J_λ (having size $m \times m$) of the original matrix A. Hence we have

$$\underset{\sim}{z}'(x) = K_\lambda(x) \, \underset{\sim}{z}'(0),$$

corresponding to the elementary Jordan block J_λ.

Parallel treatment can be done for the other Jordan blocks. Hence the solution of the surrogate IVP, viz.

$$\frac{d\underset{\sim}{z}}{dx} = J\underset{\sim}{z} \ ; \ \underset{\sim}{z}(0) = P^{-1}\underset{\sim}{y_0}$$

is of the form $\underset{\sim}{z}(x) = K(x)P^{-1}\underset{\sim}{y_0}$, $K(x)$ being the block-diagonal matrix :

$$\begin{pmatrix} K_1(x) & & \\ & K_2(x) & \quad O \\ O & & K_r(x) \end{pmatrix}$$

having the constituent blocks of the generic form $K_\lambda(x)$ defined before.

Hence $\underset{\sim}{y}(x) = PK(x)P^{-1}\underset{\sim}{y_0}$ serves as the solution of the original IVP :

$$\frac{d\underset{\sim}{y}}{dx} = A\underset{\sim}{y} \ ; \ \underset{\sim}{y}(0) = \underset{\sim}{y_0} \text{ with } A \in M_n(\mathbb{C})$$

All these intrigues of theory cannot be digested without working out problems straightway on them. This is why we shall workout two practice problems in the following :

Example 9.12 Solve the vector differential equation $\frac{d\underset{\sim}{y}}{dx} = A\underset{\sim}{y}$, subject to the initial condition $\underset{\sim}{y}(0) \equiv \underset{\sim}{y_0} = (1 \ 0 \ 1)^T$, where A is given as in Example 9.11

In the example 9.11 we observed that A can be jordanized into J by means of a similarity transformation $P^{-1}AP$, where

$$J = \begin{pmatrix} 2 & 1 & \vdots & 0 \\ 0 & 2 & \vdots & 0 \\ \cdots & \cdots & \vdots & \cdots \\ 0 & 0 & \vdots & -4 \end{pmatrix} ; \quad P = \begin{pmatrix} 1 & 0 & 1 \\ -1 & -1 & -1 \\ -1 & 0 & 1 \end{pmatrix} ;$$

$$P^{-1} = \begin{pmatrix} \frac{1}{2} & 0 & -\frac{1}{2} \\ -1 & -1 & 0 \\ \frac{1}{2} & 0 & \frac{1}{2} \end{pmatrix}$$

As per our scheme discussed,

$$K_1(x) = e^{2x} \begin{pmatrix} 1 & x \\ 0 & 1 \end{pmatrix} ; \quad K_2(x) = e^{-4x}$$

and so $K(x) = \begin{pmatrix} K_1(x) & \vdots & O \\ \cdots & \cdots & \cdots \\ O & \vdots & K_2(x) \end{pmatrix} = \begin{pmatrix} e^{2x} & xe^{2x} & 0 \\ 0 & e^{2x} & 0 \\ 0 & 0 & e^{-4x} \end{pmatrix}$

$\therefore\ \underset{\sim}{y}(x) = PK(x)P^{-1}\,\underset{\sim}{y}_0$

$$= \begin{pmatrix} 1 & 0 & 1 \\ -1 & -1 & -1 \\ -1 & 0 & 1 \end{pmatrix} \begin{pmatrix} e^{2x} & xe^{2x} & 0 \\ 0 & e^{2x} & 0 \\ 0 & 0 & e^{-4x} \end{pmatrix} \begin{pmatrix} \frac{1}{2} & 0 & -\frac{1}{2} \\ -1 & -1 & 0 \\ \frac{1}{2} & 0 & \frac{1}{2} \end{pmatrix} \begin{pmatrix} 1 \\ 0 \\ 1 \end{pmatrix}$$

$$= \begin{pmatrix} 1 & 0 & 1 \\ -1 & -1 & -1 \\ -1 & 0 & 1 \end{pmatrix} \begin{pmatrix} -xe^{2x} \\ -e^{2x} \\ e^{-4x} \end{pmatrix} = \begin{pmatrix} -xe^{2x} & + & e^{-4x} \\ (x+1)e^{2x} & - & e^{-4x} \\ xe^{2x} & + & e^{-4x} \end{pmatrix}$$

The solution of the vector differential equation $\frac{dy}{dx} = A\underset{\sim}{y}$ subject to the condition $\underset{\sim}{y}(0) = \underset{\sim}{y}_0$ can also be put in terms of matrix exponentials.

$\because\ J = PAP^{-1}, \quad e^{Ax} = Pe^{xJ}\,P^{-1}$

Hence computing e^{xJ} is our main objective. As J consists of the elementary Jordan blocks $J_1, J_2, \cdots, J_k$, e^{xJ} consists of the blocks $e^{xJ_1}, e^{xJ_2}, \cdots, e^{xJ_k}$.

Let J_λ be an elementary Jordan block of size m pertaining to the eigenvalue λ of A. Since $J_\lambda = D_\lambda + N_\lambda$, it follows that

$$e^{xJ_\lambda} = e^{x(D_\lambda + N_\lambda)} = e^{xD_\lambda} \cdot e^{xN_\lambda} \, ,$$

where e^{xD_λ} is a scalar matrix whose diagonal entries are all $e^{\lambda x}$, i,e, $e^{xD_\lambda} = e^{\lambda x} I_m$. Further N_λ being a nilpotent matrix of order m, its index is $\not> m$ and hence e^{xN_λ} is of the form

$$\begin{pmatrix} 1 & x & \frac{x^2}{2!} & \cdots & \cdots & \frac{x^{m-1}}{(m-1)!} \\ & 1 & x & \cdots & \cdots & \frac{x^{m-2}}{(m-2)!} \\ & & \ddots & & & \vdots \\ & & & \ddots & & \vdots \\ & & & & \ddots & \vdots \\ & O & & & 1 & x \\ & & & & & 1 \end{pmatrix}$$

which is nothing but $K_\lambda(x)$ computed earlier. Hence $e^{xJ_\lambda} = e^{\lambda x} K_\lambda(x)$. This spells out that e^{xJ} is a block diagonal matrix whose constituents $e^{\lambda x} K_\lambda(x)$. We therefore have

$$e^{xA} = P e^{xJ} P^{-1} = P \begin{pmatrix} e^{\lambda_1 x} K_{\lambda_1}(x) & & & \\ & e^{\lambda_2 x} K_{\lambda_2}(x) & & O \\ O & & & \\ & & & e^{\lambda_n x} K_{\lambda_n}(x) \end{pmatrix}$$

Hence the solution of the IVP reads : $\underset{\sim}{y}(x) = e^{xA} \underset{\sim}{y_0}$

In the previous example, viz, example 9.12

$$e^{xJ} = e^{x(D+N)} = e^{xD} \cdot e^{xN} = \begin{pmatrix} e^{2x} & 0 & 0 \\ 0 & e^{2x} & 0 \\ 0 & 0 & e^{-4x} \end{pmatrix} (I_3 + xN)$$

$$= \begin{pmatrix} e^{2x} & 0 & 0 \\ 0 & e^{2x} & 0 \\ 0 & 0 & e^{-4x} \end{pmatrix} \begin{pmatrix} 1 & x & 0 \\ 0 & 1 & 0 \\ 0 & 0 & 1 \end{pmatrix} = \begin{pmatrix} e^{2x} & xe^{2x} & 0 \\ 0 & e^{2x} & 0 \\ 0 & 0 & e^{-4x} \end{pmatrix}$$

Hence the solution of the IVP is :

$$\underset{\sim}{y}(x) = e^{xA} \underset{\sim}{y_0} = \begin{pmatrix} e^{2x} & xe^{2x} & 0 \\ 0 & e^{2x} & 0 \\ 0 & 0 & e^{-4x} \end{pmatrix} \begin{pmatrix} 1 \\ 0 \\ 1 \end{pmatrix} = \begin{pmatrix} -xe^{2x} + e^{-4x} \\ (x+1)e^{2x} - e^{-4x} \\ xe^{2x} + e^{-4x} \end{pmatrix}$$

Remark : This form $y(x) = e^{xA}\, y_0$ of the IVP may be looked into as a generalization of the solution $y = y_0\, e^{ax}$ of the single variable IVP $\frac{dy}{dx} = ay$ subject to the initial condition $y(0) = y_0$.

In context of the last example, it is clear that the solution of the differential equation $\frac{d\underset{\sim}{z}}{dx} = J\underset{\sim}{z}$ [not IVP] is given by

$$\underset{\sim}{z}(x) = K_\lambda(x) \begin{pmatrix} c_1 \\ c_2 \\ c_3 \end{pmatrix}, \text{where } K_\lambda(x) \text{ is same as before.}$$

$$\therefore\ \underset{\sim}{z}(x) = C_1 e^{2x} \begin{pmatrix} 1 \\ 0 \\ 0 \end{pmatrix} + c_2 e^{2x} \begin{pmatrix} x \\ 1 \\ 0 \end{pmatrix} + c_3 e^{-4x} \begin{pmatrix} 0 \\ 0 \\ 1 \end{pmatrix}$$

$$\therefore\ \underset{\sim}{y}(x) = P\underset{\sim}{z}(x) = c_1 e^{2x} P \begin{pmatrix} 1 \\ 0 \\ 0 \end{pmatrix} + c_2 e^{2x} P \begin{pmatrix} x \\ 1 \\ 0 \end{pmatrix} + c_3 e^{-4x} P \begin{pmatrix} 0 \\ 0 \\ 1 \end{pmatrix}$$

$$= c_1 e^{2x} \begin{pmatrix} -1 \\ -1 \\ 1 \end{pmatrix} + c_2 e^{2x} \left[x \begin{pmatrix} 1 \\ -1 \\ -1 \end{pmatrix} + \begin{pmatrix} 0 \\ -1 \\ 0 \end{pmatrix} \right] + c_3 e^{-4x} \begin{pmatrix} 1 \\ -1 \\ -1 \end{pmatrix}$$

Observe that $\underset{\sim}{v_1} \equiv (-1\ -1\ 1)^T$ is a rank-1 generalised eigenvector of A, corresponding to $\lambda = 2$ while $\underset{\sim}{v_2} \equiv (0\ -1\ 0)^T$ is a rank-2 generalised eigenvector of A corresponding to the same eigenvalue.

Further, $\underset{\sim}{w} \equiv (1\ -1\ -1)^T$ is a rank-1 generalised eigenvector of A corresponding to $\lambda = -4$. It is a routine exercise to check that

$$\bar{y}_1(x) = e^{2x}\underset{\sim}{v_1}\ ;\ \bar{y}_2(x) = e^{2x}[x\underset{\sim}{v_1} + \underset{\sim}{v_2}]\ \text{ and }\ \bar{y}_3(x) = e^{-4x}\underset{\sim}{w}$$

are linearly independent solutions of the ode :

$$\frac{d\underset{\sim}{y}}{dx} = A\underset{\sim}{y}, \text{ where } A \equiv \begin{pmatrix} -2 & -1 & -3 \\ 4 & 3 & 3 \\ -2 & 1 & -1 \end{pmatrix}$$

Let's now solve the vector differential equation

$$\frac{d\underset{\sim}{y}}{dx} = A\underset{\sim}{y} \text{ where } A = \begin{pmatrix} 0 & -1 & -2 & -1 \\ 1 & 2 & 1 & 1 \\ 0 & 0 & 1 & 0 \\ 0 & 0 & 1 & 1 \end{pmatrix}$$

From our previous workout (c.f Example 9.10) it is found that 1 is an eigenvalue of A with algebraic multiplicity 4. Jordanised form of A is :

$$J = \begin{pmatrix} 1 & 1 & 0 & 0 \\ 0 & 1 & 0 & 0 \\ 0 & 0 & 1 & 1 \\ 0 & 0 & 0 & 1 \end{pmatrix} \text{ and } P = \left(\begin{array}{c|c|c|c} \underset{\sim}{v_1} & \underset{\sim}{v_2} & \underset{\sim}{w_1} & \underset{\sim}{w_2} \end{array} \right).$$

Using the transformation $\underset{\sim}{y} = P\underset{\sim}{z}$ the given vector differential equation reduces to $\frac{d\underset{\sim}{z}}{dx} = J\underset{\sim}{z}$, which on solving gives :

$$\underset{\sim}{z}(x) = K_1(x) \begin{pmatrix} c_1 \\ c_2 \\ c_3 \\ c_4 \end{pmatrix} \text{ with } K_1(x) = e^x \begin{pmatrix} 1 & x & x^2/2! & x^3/3! \\ 0 & 1 & x & x^2/2 \\ 0 & 0 & 1 & x \\ 0 & 0 & 0 & 1 \end{pmatrix}$$

$$\therefore \ \underset{\sim}{z}(x) = c_1 e^x \begin{pmatrix} 1 \\ 0 \\ 0 \\ 0 \end{pmatrix} + c_2 e^x \begin{pmatrix} x \\ 1 \\ 0 \\ 0 \end{pmatrix} + c_3 e^x \begin{pmatrix} x^2/2! \\ x \\ 1 \\ 0 \end{pmatrix} + c_4 e^x \begin{pmatrix} x^3/3! \\ x^2/2! \\ x \\ 1 \end{pmatrix}$$

i,e, $\underset{\sim}{y}(x) = P\underset{\sim}{z}(x) = c_1 e^x \underset{\sim}{v_1} + c_2 e^x (x\underset{\sim}{v_1} + \underset{\sim}{v_2}) + c_3 e^x (\frac{x^2}{2!} \underset{\sim}{v_1} + x\underset{\sim}{v_2} + \underset{\sim}{w_1})$

$$+ c_4 e^x \left(\frac{x^3}{3!} \underset{\sim}{v_1} + \frac{x^2}{2!} \underset{\sim}{v_2} + x\underset{\sim}{w_1} + \underset{\sim}{w_2} \right)$$

It's routine exercise to establish that

$$\underset{\sim}{\bar{y}}_1(x) = e^x \underset{\sim}{v_1} \qquad\qquad ; \quad \underset{\sim}{\bar{y}}_2(x) = e^x (x\underset{\sim}{v_1} + \underset{\sim}{v_2})$$

$$\underset{\sim}{\bar{y}}_3(x) = e^x \left(\frac{x^2}{2!} \underset{\sim}{v_1} + x\underset{\sim}{v_2} + \underset{\sim}{w_1} \right) ; \quad \underset{\sim}{\bar{y}}_4(x) = e^x \left(\frac{x^3}{3!} \underset{\sim}{v_1} + \frac{x^2}{2!} \underset{\sim}{v_2} + x\underset{\sim}{w_1} + \underset{\sim}{w_2} \right)$$

are four linearly independent solution of the vector ode $\frac{d\underset{\sim}{y}}{dx} = A\underset{\sim}{y}$.

Remark : If T is defined to be $(A - \lambda I)$ and $\underset{\sim}{v_1}, \underset{\sim}{v_2}, \cdots, \underset{\sim}{v_m}$ be a chain of generalised eigenvectors of A with eigenvalue λ, i,e,

$$\underset{\sim}{v_m} \xrightarrow{T} \underset{\sim}{v_{m-1}} \xrightarrow{T} \underset{\sim}{v_{m-2}} \xrightarrow{T} \cdots\cdots \xrightarrow{T} \underset{\sim}{v_2} \xrightarrow{T} \underset{\sim}{v_1} \xrightarrow{T} \underset{\sim}{0} ,$$

then

$$\underset{\sim}{\bar{y}}_1(x) = e^{\lambda x} \underset{\sim}{v_1}$$

$$\bar{y}_2(x) = e^{\lambda x}\left(x\underset{\sim}{v}_1 + \underset{\sim}{v}_2\right)$$

$$\bar{y}_3(x) = e^{\lambda x}\left(\frac{x^2}{2!}\underset{\sim}{v}_1 + x\underset{\sim}{v}_2 + \underset{\sim}{v}_3\right)$$

$$\cdots\cdots\cdots\cdots\cdots\cdots\cdots\cdots\cdots\cdots\cdots\cdots\cdots$$

$$\cdots\cdots\cdots\cdots\cdots\cdots\cdots\cdots\cdots\cdots\cdots$$

$$\bar{y}_m(x) = e^{\lambda x}\left(\frac{x^{m-1}}{(m-1)!}\underset{\sim}{v}_1 + \frac{x^{m-2}}{(m-2)!}\underset{\sim}{v}_2 + \cdots + x\underset{\sim}{v}_{m-1} + \underset{\sim}{v}_m\right)$$

form a set of m linearly independent solutions of the vector differential equation $\frac{dy}{dx} = A\underset{\sim}{y}$.

9.7 Fundamental Matrix and Liouville Theorem

The result we found in the example 9.9 was quite expected because corresponding to every vector differential equation $\underset{\sim}{y}'(x) = A\underset{\sim}{y}(x)$ with $A \in M_n(\mathbb{R})$, $\exists$ a set of n linearly independent n-column vectors $\{\underset{\sim}{\phi}_1(x), \underset{\sim}{\phi}_2(x), \cdots, \underset{\sim}{\phi}_n(x)\}$ which are solutions of the ode. If one constructs a n-square matrix $\Phi(x)$ whose columns are these solutions, then that matrix

$$\Phi(x) \equiv [\underset{\sim}{\phi}_1(x)\ \underset{\sim}{\phi}_2(x)\cdots\underset{\sim}{\phi}_n(x)]$$

is called **fundamental matrix** related to this vector differential equation. (We have assumed that $A(x)$ (or succinctly A) is a continuous matrix function over I.)

So general solution to the vector differential equation $\underset{\sim}{y}'(x) = A\underset{\sim}{y}(x)$ is given by $\underset{\sim}{y}(x) = \Phi(x)\underset{\sim}{c}$ where $\underset{\sim}{c}$ is an arbitrary vector in $\mathbb{R}^n$.

A set of n solutions of the vector differential equation $\underset{\sim}{y}'(x) = A(x)\underset{\sim}{y}(x)$ is linearly independent iff a fundamental matrix can be formed out of these n solutions.

Example 9.13 Check whether $\underset{\sim}{\phi}_1(x) \equiv \begin{pmatrix} e^{4x} \\ 2e^{4x} \\ e^{4x} \end{pmatrix}$; $\underset{\sim}{\phi}_2(x) \equiv \begin{pmatrix} 0 \\ 3e^x \\ e^x \end{pmatrix}$

and $\underset{\sim}{\phi_3}(x) \equiv \begin{pmatrix} 2e^{4x} \\ 4e^{4x} \\ 2e^{4x} \end{pmatrix}$ form a fundamental matrix of the ode.

$$\frac{d\underset{\sim}{y}(x)}{dx} = \begin{pmatrix} 1 & -3 & 9 \\ 0 & -5 & 18 \\ 0 & -3 & 10 \end{pmatrix} \underset{\sim}{y}(x) \equiv A\underset{\sim}{y}(x)$$

$$\therefore \quad A\underset{\sim}{\phi_1}(x) = \begin{pmatrix} 1 & -3 & 9 \\ 0 & -5 & 18 \\ 0 & -3 & 10 \end{pmatrix} \begin{pmatrix} e^{4x} \\ 2e^{4x} \\ e^{4x} \end{pmatrix} = 4 \begin{pmatrix} e^{4x} \\ 2e^{4x} \\ e^{4x} \end{pmatrix}$$

and $\quad \underset{\sim}{\phi_1'}(x) = \dfrac{d}{dx} \begin{pmatrix} e^{4x} \\ 2e^{4x} \\ e^{4x} \end{pmatrix} = \begin{pmatrix} 4e^{4x} \\ 8e^{4x} \\ 4e^{4x} \end{pmatrix} = 4 \begin{pmatrix} e^{4x} \\ 8e^{4x} \\ e^{4x} \end{pmatrix}$,

$$\underset{\sim}{\phi_1'}(x) = A\underset{\sim}{\phi_1}(x)$$

Again $\quad A\underset{\sim}{\phi_2}(x) = \begin{pmatrix} 1 & -3 & 9 \\ 0 & -5 & 18 \\ 0 & -3 & 10 \end{pmatrix} \begin{pmatrix} 0 \\ 3e^{x} \\ e^{x} \end{pmatrix} = \begin{pmatrix} 0 \\ 3e^{x} \\ e^{x} \end{pmatrix}$

and $\quad \underset{\sim}{\phi_2'}(x) = \begin{pmatrix} 0 \\ 3e^{x} \\ e^{x} \end{pmatrix} \Rightarrow \underset{\sim}{\phi_2'}(x) = A\underset{\sim}{\phi_2}(x)$

Finally $\quad A\underset{\sim}{\phi_3}(x) = \begin{pmatrix} 1 & -3 & 9 \\ 0 & -5 & 18 \\ 0 & -3 & 10 \end{pmatrix} \begin{pmatrix} 2e^{4x} \\ 4e^{4x} \\ 8e^{4x} \end{pmatrix} = 4 \begin{pmatrix} 2e^{4x} \\ 4e^{4x} \\ 2e^{4x} \end{pmatrix}$

and $\quad \underset{\sim}{\phi_3'}(x) = 4 \begin{pmatrix} 2e^{4x} \\ 4e^{4x} \\ 2e^{4x} \end{pmatrix} \Rightarrow \underset{\sim}{\phi_3'}(x) = A\underset{\sim}{\phi_3}(x)$

However, $\underset{\sim}{\phi_1}(x), \underset{\sim}{\phi_2}(x), \underset{\sim}{\phi_3}(x)$ are not linearly independent since

$$det\,[\underset{\sim}{\phi_1}(x) \ \underset{\sim}{\phi_2}(x) \ \underset{\sim}{\phi_3}(x)] = \begin{vmatrix} e^{4x} & 0 & 2e^{4x} \\ 2e^{4x} & 3e^{x} & 4e^{4x} \\ e^{4x} & e^{x} & 2e^{4x} \end{vmatrix} = 0$$

So $[\phi_1 \ \phi_2 \ \phi_3]$ is not a fundamental matrix.

Example 9.14 Check if $\begin{bmatrix} e^x & xe^x & e^{2x} \\ e^x & (x+1)e^x & 2e^{2x} \\ e^x & (x+2)e^x & 4e^{2x} \end{bmatrix}$ associated with the

vector differential equation $\dfrac{d\underset{\sim}{y}}{dx} = \begin{pmatrix} 0 & 1 & 0 \\ 0 & 0 & 1 \\ 2 & -5 & 4 \end{pmatrix} \underset{\sim}{y}(x)$ is a fundamental

matrix or not.

Observe that

$$\underset{\sim}{\phi}_1'(x) \equiv \begin{pmatrix} e^x \\ e^x \\ e^x \end{pmatrix} = \begin{pmatrix} 0 & 1 & 0 \\ 0 & 0 & 1 \\ 2 & -5 & 4 \end{pmatrix} \begin{pmatrix} e^x \\ e^x \\ e^x \end{pmatrix} \equiv A\underset{\sim}{\phi}_1(x)$$

$$\underset{\sim}{\phi}_2'(x) \equiv \begin{pmatrix} (x+1)e^x \\ (x+2)e^x \\ (x+3)e^x \end{pmatrix} = \begin{pmatrix} 0 & 1 & 0 \\ 0 & 0 & 1 \\ 2 & -5 & 4 \end{pmatrix} \begin{pmatrix} xe^x \\ (x+1)e^x \\ (x+2)e^x \end{pmatrix} \equiv A\underset{\sim}{\phi}_2(x)$$

$$\underset{\sim}{\phi}_3'(x) \equiv \begin{pmatrix} 2e^{2x} \\ 4e^{2x} \\ 8e^{2x} \end{pmatrix} = \begin{pmatrix} 0 & 1 & 0 \\ 0 & 0 & 1 \\ 2 & -5 & 4 \end{pmatrix} \begin{pmatrix} e^{2x} \\ 2e^{2x} \\ 4e^{2x} \end{pmatrix} \equiv A\underset{\sim}{\phi}_3(x)$$

Therefore $\underset{\sim}{\phi}_1(x), \underset{\sim}{\phi}_2(x), \underset{\sim}{\phi}_3(x)$ are solutions of the vector differential equation given. Now we observe that they are linearly independent since

$$\begin{vmatrix} e^x & xe^x & e^{2x} \\ e^x & (x+1)e^x & 2e^{2x} \\ e^x & (x+2)e^x & 4e^{2x} \end{vmatrix} = e^{4x} \begin{vmatrix} 1 & x & 1 \\ 1 & x+1 & 2 \\ 1 & x+2 & 4 \end{vmatrix} = e^{4x} \neq 0 \ \forall \ x \in \mathbb{R}$$

So $[\underset{\sim}{\phi}_1(x) \ \underset{\sim}{\phi}_2(x) \ \underset{\sim}{\phi}_3(x)]$ is a fundamental matrix of the given vector differential equation.

Remark : The computation of the determinants in these problems may be gruelling if the expressions of the available solutions are very complicated. One respite from this may be the Liouville's Theorem that ensures that if a system of n solutions associated with a vector differential equation $\underset{\sim}{y}'(x) = A(x)\underset{\sim}{y}(x)$ (A being continuous function of x) be linearly independent for some point $x_0 \in I$, then they are linearly independent for all $x \in I$. In the following we state and prove this theorem which might be thought of as generalisation of Ostrogradsky-Liouville formula proved in Chapter 5.

Liouville's Theorem : If $\phi_1(x), \phi_2(x), \cdots, \phi_n(x)$ be any n solutions of the vector differential equation $\underset{\sim}{y}'(x) = A(x)\underset{\sim}{y}(x)$ on an interval I and $\Phi(x)$ is the matrix function having its columns $\phi_1(x), \phi_2(x), \cdots, \phi_n(x)$, then for $x_0 \in I$,

$$det\ \Phi(x) = exp\left[\int_{x_0}^{x} tr(A(t))dt\right] det\ [\Phi(x_0)] \ \forall\ x \in I$$

Proof. By definition,

$$\Phi(x) = \begin{bmatrix} \phi_{11}(x) & \phi_{12}(x) & \cdots & \phi_{1n}(x) \\ \phi_{21}(x) & \phi_{22}(x) & \cdots & \phi_{2n}(x) \\ \vdots & \vdots & & \vdots \\ \phi_{n1}(x) & \phi_{n2}(x) & \cdots & \phi_{nn}(x) \end{bmatrix},$$

where $\quad \underset{\sim}{\phi_1}(x) = \begin{pmatrix} \phi_{11}(x) \\ \phi_{21}(x) \\ \vdots \\ \phi_{n1}(x) \end{pmatrix}, \cdots\cdots, \underset{\sim}{\phi_n}(x) = \begin{pmatrix} \phi_{1n}(x) \\ \phi_{2n}(x) \\ \vdots \\ \phi_{nn}(x) \end{pmatrix}$

and $\qquad \phi_{ij} \in C^1\ (I) \quad$ for $\quad \left.\begin{matrix} i = 1, 2, \cdots, n \\ j = 1, 2, \cdots, n \end{matrix}\right\}$

$$\therefore\ \frac{d}{dx}\{det[\Phi(x)]\} = \frac{d}{dx} \begin{bmatrix} \phi_{11}(x) & \phi_{12}(x) & \cdots & \phi_{1n}(x) \\ \phi_{21}(x) & \phi_{22}(x) & \cdots & \phi_{2n}(x) \\ \vdots & \vdots & & \vdots \\ \phi_{n1}(x) & \phi_{n2}(x) & \cdots & \phi_{nn}(x) \end{bmatrix}$$

$$= \begin{bmatrix} \phi'_{11}(x) & \cdots & \phi'_{1n}(x) \\ \phi_{21}(x) & \cdots & \phi_{2n}(x) \\ \vdots & & \vdots \\ \phi_{n1}(x) & \cdots & \phi_{nn}(x) \end{bmatrix} + \cdots + \begin{bmatrix} \phi_{11}(x) & \cdots & \phi_{1n}(x) \\ \phi_{21}(x) & \cdots & \phi_{2n}(x) \\ \vdots & & \vdots \\ \phi'_{n1}(x) & \cdots & \phi'_{nn}(x) \end{bmatrix}$$

$$= \begin{vmatrix} c\sum_{k=1}^{n} a_{1k}\phi_{k1}(x) & \sum_{k=1}^{n} a_{1k}\phi_{k2}(x) & \cdots & \sum_{k=1}^{n} a_{1k}\phi_{kn}(x) \\ \phi_{21}(x) & \phi_{22}(x) & \cdots & \phi_{2n}(x) \\ \vdots & \vdots & & \vdots \\ \phi_{n1}(x) & \phi_{n2}(x) & \cdots & \phi_{nn}(x) \end{vmatrix}$$

$$+ \begin{vmatrix} \phi_{11}(x) & \phi_{12}(x) & \cdots & \phi_{1n}(x) \\ \sum\limits_{k=1}^{n} a_{2k}\phi_{k1}(x) & \sum\limits_{k=1}^{n} a_{2k}\phi_{k2}(x) & \cdots & \sum\limits_{k=1}^{n} a_{2k}\phi_{2k}(x) \\ \phi_{31}(x) & \phi_{32}(x) & \cdots & \phi_{3n}(x) \\ \phi_{n1}(x) & \phi_{n2}(x) & \cdots & \phi_{nn}(x) \end{vmatrix} + \cdots$$

$$+ \begin{vmatrix} \phi_{11}(x) & \phi_{12}(x) & \cdots & \phi_{1n}(x) \\ \phi_{21}(x) & \phi_{22}(x) & \cdots & \phi_{2n}(x) \\ \vdots & \vdots & & \vdots \\ \sum\limits_{k=1}^{n} a_{nk}\phi_{k1}(x) & \sum\limits_{k=1}^{n} a_{nk}\phi_{k2}(x) & \cdots & \sum\limits_{k=1}^{n} a_{nk}\phi_{kn}(x) \end{vmatrix}$$

$\equiv \Delta_1 + \Delta_2 + \cdots + \Delta_n$ where Δ_k stands for k th determinant.

Δ_k's can be computed by using the elementary row operations :

$$R_1' = R_1 - a_{12}R_2 - a_{13}R_3 - \cdots - a_{1n}R_n$$
$$R_2' = R_2 - a_{21}R_1 - a_{23}R_3 - \cdots - a_{2n}R_n$$
$$\cdots\cdots\cdots\cdots\cdots\cdots\cdots\cdots\cdots\cdots\cdots\cdots$$
$$R_n' = R_n - a_{n1}R_1 - a_{n2}R_2 - \cdots - a_{n\,\overline{n-1}}\,R_{n-1}$$

$$\therefore \frac{d}{dx}\{det[\Phi(x)]\}$$

$$= (a_{11} + a_{22} + \cdots + a_{nn}) \begin{vmatrix} \phi_{11}(x) & \phi_{12}(x) & \cdots & \phi_{1n}(x) \\ \phi_{21}(x) & \phi_{22}(x) & \cdots & \phi_{2n}(x) \\ \vdots & \vdots & & \vdots \\ \phi_{n1}(x) & \phi_{n2}(x) & \cdots & \phi_{nn}(x) \end{vmatrix}$$

$$= (tr\ A)\ det[\Phi(x)]$$

This gives us on integration between x_0 and x :

$$det[\Phi(x)] = exp\left[\int_{x_0}^{x} tr(A(t))dt\right] det\ [\Phi(x_0)] \qquad \forall\ x \in I$$

Few simple observations :

(i) If $det[\Phi(x_0)] = 0$, then $det[\Phi(x)] = 0\ \forall\ x \in I$ and if $det[\Phi(x_0)] \neq 0$ then $det[\Phi(x)] = 0\ \forall\ x \in I$. So if the linear independence of the solution vectors $\{\underset{\sim}{\phi}_1(x), \underset{\sim}{\phi}_2(x), \cdots, \underset{\sim}{\phi}_n(x)\}$ at one point $x_0 \in I$ is ascertained, then the linear independence of these vectors is ensured over the whole interval I.

In example 9.14, to verify that the given matrix is a fundamental matrix of the given ode it suffices to show that determinant of that matrix computed at some point, say, $x = 0$ is non-zero.

(ii) If A be traceless, then $det[\Phi(x)] = det[\Phi(x_0)] \ \forall \ x \in I$, i,e, the determinant of the fundamental matrix is a constant function over I.

(iii) If $\Phi(x)$ be a fundamental matrix of the vector differential equation $y'(x) = A(x)\underset{\sim}{y}(x)$, then for any non-singular constant matrix C, $\Phi(x)C$ is again a fundamental matrix of the same system. The claim can be proved as follows :

$$\Phi(x) = [\phi_1(x) \ \phi_2(x) \ \cdots \ \phi_n(x)] \ ; \ C = [c_{ij}]_{n \times n}$$

$$\therefore \quad \Phi(x) \cdot C = \left[\sum_{k=1}^{n} c_{k1}\phi_k \quad \sum_{k=1}^{n} c_{k2}\phi_k \quad \cdots \sum_{k=1}^{n} c_{kn}\phi_k \right]$$

If one agrees to write $\Phi(x).C$ as $\Psi(x)$, then

$\Psi(x) \equiv [\underset{\sim}{\psi_1}(x) \ \underset{\sim}{\psi_2}(x) \cdots \underset{\sim}{\psi_n}(x)]$ is continuously differentiable,

where $\quad \psi_i(x) = \sum_{i=1}^{n} c_{ij} \ \phi_i \quad$ so that

$$\underset{\sim}{\psi_i'}(x) = \sum_{i=1}^{n} c_{ij} \ \underset{\sim}{\phi_i'}(x) = \sum_{i=1}^{n} c_{ij} A\underset{\sim}{\phi_i}(x) = A\left(\sum_{i=1}^{n} c_{ij} \ \underset{\sim}{\phi_i} \right) = A\psi_i(x)$$

Further due to product rule of determinants,

$$det[\underset{\sim}{\Psi}(x)] = det[\Psi(x).C] = det[\Psi(x)] \ det[C] \neq 0$$

Hence the proof.

[By uniqueness theorem it can be shown that any fundamental matrix of a vector differential equation is of this general form.]

(iv) A fundamental matrix $\Phi(x)$ of the vector differential equation $y'(x) = A(x)\underset{\sim}{y}(x)$ over the interval I is an invertible solution of the matrix differential equation $y'(x) = A(x)\underset{\sim}{y}(x)$ over the same interval. It follows from Liouville's theorem that an invertible solution of the above matrix differential equation over I can be had only if we ensure that at one point, say $x_0 \in I$, $\Phi(x)$ is invertible.

If $Y(x) = \Phi(x \; ; \; x_0)$ denotes the unique solution of the IVP :

$$\frac{dY}{dx} = A(x)Y \; ; \; Y(x_0) = I_n \; ; \; x_0 \in I,$$

then $\Psi(x \; ; \; x_0)$ is invertible for all $x \in I$ and moreover, $\underset{\sim}{y}(x) = \Phi(x \; ; \; x_0)\underset{\sim}{\xi}_0$ is the unique solution of the IVP :

$$\underset{\sim}{y}'(x) = A(x)\underset{\sim}{y} \; ; \; \underset{\sim}{y}(x_0) = \underset{\sim}{\xi}_0 \; ; \; \underset{\sim}{y}_0 \in \mathbb{R}^n$$

and the $n \times n$ matrix $Y(x) = \Phi(x \; ; \; x_0)C$ is the unique solution of the matrix IVP :

$$\frac{dY}{dx} = A(x)Y \; ; \; Y(x_0) = C \; ; \; C \in M_n(\mathbb{R})$$

Since $Y(x_0) = C$, it follows that $\Phi(x_0 \; ; \; x_0) = I_n$

If in particular $C \in GL(n \; ; \; \mathbb{R})$, then the general form of the fundamental matrix of the vector differential equation

$$\underset{\sim}{y}'(x) = A(x)\underset{\sim}{y}(x) \; ; \; x \in I$$

is given by

$$Y(x) = \Phi(x \; ; \; x_0)C \quad \forall \; x, x_0 \in I_n$$

In particular if $x_1 \in I$, then

$$\left.\begin{array}{l} Y(x_1) = \Phi(x_1 \; ; \; x_0)Y((x_0))^{-1} \\ Y(x) = \Phi(x \; ; \; x_1)Y(x_1) \end{array}\right\}$$

Therefore, $\Phi(x \; ; \; x_0) = Y(x)(Y(x_0))^{-1}$
$$= \Phi(x \; ; \; x_1)Y(x_1)(Y(x_0))^{-1}$$
$$= \Phi(x \; ; \; x_1)\Phi(x_1 \; ; \; x_0)Y(x_0) \cdot (Y(x_0))^{-1}$$
$$= \Phi(x \; ; \; x_1)\Phi(x_1 \; ; \; x_0) \quad \forall \; x, x_0, x_1 \in I$$

This result will be useful in finding the solution of the inhomogeneous linear vector differential equation

$$\underset{\sim}{y}'(x) = A(x)\underset{\sim}{y}(x) + \underset{\sim}{b}(x)$$

subject to the initial condition $\underset{\sim}{y}(x_0) = \underset{\sim}{\xi}_0 \; ; \; \underset{\sim}{\xi}_0 \in \mathbb{R}^n$.

First we apply the transformation

$$y(x) = \Phi(x \; ; \; x_0)z(x), \qquad\qquad (*)$$

where $\Phi(x \; ; \; x_0)$ is the unique solution of the I.V.P :

$$\frac{dY}{dx} = A(x)Y(x) \; ; \; Y(x_0) = I_n \; ; \; x_0 \in I$$

By differentiation of transformation rule $(*)$ we have :

$$
\begin{aligned}
\frac{d\underset{\sim}{y}}{dx} &= \Phi(x \; ; \; x_0)\frac{d\underset{\sim}{z}}{dx} + \frac{d}{dx}\Phi(x \; ; \; x_0)\underset{\sim}{z}(x) \\
&= \Phi(x \; ; \; x_0)\frac{d\underset{\sim}{z}}{dx} + A(x)\Phi(x \; ; \; x_0)\underset{\sim}{z}(x) \\
&= \Phi(x \; ; \; x_0)\frac{d\underset{\sim}{z}}{dx} + A(x)\underset{\sim}{y}(x) \\
&= \Phi(x \; ; \; x_0)\frac{d\underset{\sim}{z}}{dx} + \frac{d\underset{\sim}{y}}{dx} - \underset{\sim}{b}(x) \\
\therefore \quad \Phi(x \; ; \; x_0)\frac{d\underset{\sim}{z}}{dx} &= \underset{\sim}{b}(x)
\end{aligned}
$$

Equivalently this relation can be written as :

$$\frac{d\underset{\sim}{z}}{dx} = (\Phi(x \; ; \; x_0)^{-1} \underset{\sim}{b}(x) \; [\text{ since } \Phi(x \; ; \; x_0) \text{ is invertible }]$$

Integrating this relation w.r.to x between x_0 and x we get :

$$
\begin{aligned}
\underset{\sim}{z}(x) - \underset{\sim}{z}(x_0) &= \int_{x_0}^{x} (\Phi(x' \; ; \; x_0))^{-1} \underset{\sim}{b}(x') \, dx' \\
\text{i,e, } \underset{\sim}{z}(x) &= \underset{\sim}{z}(x_0) + \int_{x_0}^{x_1} (\Phi(x' \; ; \; x_0))^{-1} \underset{\sim}{b}(x') \, dx' \\
\therefore \quad \underset{\sim}{y}(x) &= \Phi(x \; , \; x_0)\underset{\sim}{\xi}_0 + \Phi(x \; ; \; x_0)\int_{x_0}^{x} (\Phi(x' \; ; \; x_0)^{-1} \underset{\sim}{b}(x') \, dx' \\
&= \Phi(x \; , \; x_0)\underset{\sim}{\xi}_0 + \int_{x_0}^{x} \Phi(x \; ; \; x_0)\Phi(x' \; , \; x_0)^{-1} \underset{\sim}{b}(x') \, dx' \\
&= \Phi(x \; , \; x_0)\underset{\sim}{\xi}_0 + \int_{x_0}^{x} \Phi(x' \; ; \; x_0) \underset{\sim}{b}(x') \, dx'
\end{aligned}
$$

Let's now solve the following IVP to illustrate how the above method works. (This method is also known as **Variation of Constant**)

Example 9.15 Solve the IVP :

$$\frac{dy}{dx} = Ay + \underset{\sim}{b}(x) \; ; \; \underset{\sim}{y}(0) = \underset{\sim}{\xi_0} \quad \text{where}$$

$$A = \begin{pmatrix} -2 & 1 & 0 \\ 0 & -2 & 0 \\ 3 & 2 & 1 \end{pmatrix} \; ; \; \underset{\sim}{b}(x) = \begin{pmatrix} 2 \\ 0 \\ x \end{pmatrix} \; ; \; \underset{\sim}{\xi_0} = \begin{pmatrix} 1 \\ 1 \\ 0 \end{pmatrix}$$

First let's solve for eigenvalues of A. As the characteristic equation of A is

$$(1 - \lambda)(2 + \lambda)^2 = 0,$$

it follows that $\lambda = 1 \; ; \; -2, -2$

For $\lambda = 1$, $\begin{pmatrix} 0 \\ 0 \\ 1 \end{pmatrix}$ is an eigenvector.

For $\lambda = -2 \; \exists$ only one linearly independent eigenvector $\begin{pmatrix} 1 \\ 0 \\ -1 \end{pmatrix}$. So

we have to find a generalised eigenvector of A corresponding to $\lambda = -2$. If $(x_1 \; x_2 \; x_3)^T$ be a generalised eigenvector of A corresponding to $\lambda = -2$, then

$$(A + 2I_3)^2 \begin{pmatrix} x_1 \\ x_2 \\ x_3 \end{pmatrix} = \begin{pmatrix} 0 \\ 0 \\ 0 \end{pmatrix}$$

and by computation it is found to be $\begin{pmatrix} 1 \\ 0 \\ -1 \end{pmatrix}$ or $\begin{pmatrix} 0 \\ 1 \\ -1 \end{pmatrix}$.

Having a set of three generalised eigenvectors of A we construct P as :

$$P = \begin{pmatrix} 0 & 1 & 0 \\ 0 & 0 & 1 \\ 1 & -1 & -1 \end{pmatrix} \; ; \; P^{-1} = \begin{pmatrix} 0 & 1 & 0 \\ 0 & 0 & 1 \\ 1 & 1 & -1 \end{pmatrix}$$

so that

$$S = P[\lambda_j \; \delta_{ij}]P^{-1} = \begin{pmatrix} 0 & 1 & 0 \\ 0 & 0 & 1 \\ 1 & -1 & -1 \end{pmatrix} \begin{pmatrix} 1 & 0 & 0 \\ 0 & -2 & 0 \\ 0 & 0 & -2 \end{pmatrix} \begin{pmatrix} 1 & 1 & 1 \\ 1 & 0 & 0 \\ 0 & 1 & 0 \end{pmatrix}$$

$$= \begin{pmatrix} -2 & 0 & 0 \\ 0 & -2 & 0 \\ 3 & 3 & 1 \end{pmatrix}$$

S is diagonalisable and by Decomposition theorem,

$$N = A - S = \begin{pmatrix} 0 & 1 & 0 \\ 0 & 0 & 0 \\ 0 & -1 & 0 \end{pmatrix}$$

It is a routine matter to check that $N^2 = 0$ (i,e, nilpotent of index 2) and N commutes with S

$$\left[SN = NS = \begin{pmatrix} 0 & -2 & 0 \\ 0 & 0 & 0 \\ 0 & 2 & 0 \end{pmatrix} \right]$$

Once all the pre-requisites being done, we construct the fundamental matrix $\Phi(x\ ;\ 0)$, where

$$\Phi(x\ ;\ 0) = P\left(e^{\lambda_j x} \cdot \delta_{ij}\right) P^{-1}(I_3 + Nx)$$

$$= \begin{pmatrix} 0 & 1 & 0 \\ 0 & 0 & 1 \\ 1 & -1 & -1 \end{pmatrix} \begin{pmatrix} e^x & 0 & 0 \\ 0 & e^{-2x} & 0 \\ 0 & 0 & e^{-2x} \end{pmatrix} \begin{pmatrix} 1 & 1 & 1 \\ 1 & 0 & 0 \\ 0 & 1 & 0 \end{pmatrix} \begin{pmatrix} 1 & x & 0 \\ 1 & 0 & 0 \\ 0 & 1 & 0 \end{pmatrix}$$

$$= \begin{pmatrix} e^{-2x} & xe^{-2x} & 0 \\ 0 & e^{-2x} & 0 \\ e^x - e^{-2x} & e^x - (x+1)^{-2x}e & e^x \end{pmatrix}$$

Now $\displaystyle\int_0^x [\Phi(s\ ;\ 0)]^{-1}\, \underset{\sim}{b}(s)ds$

$$= \int_0^x \left[\begin{array}{ccc} e^{2s} & -se^{2s} & 0 \\ 0 & e^{2s} & 0 \\ e^{-s} - e^{2s} & e^{-s} - (1-s)^{2s}e & e^{-s} \end{array} \right] \left[\begin{array}{c} 2 \\ 0 \\ s \end{array} \right] ds$$

$$= \int_0^x \left[\begin{array}{c} 2e^{2s} \\ 0 \\ 2e^{-s} - 2e^{2s} + se^{-s} \end{array} \right] ds = \left[\begin{array}{c} e^{2x} - 1 \\ 0 \\ -e^{2x} - (x+3)e^{-x} \end{array} \right]$$

$$\therefore \quad \underset{\sim}{\xi_0} + \int_0^x [\Phi(s\ ;\ 0)]^{-1}\, \underset{\sim}{b}(s)\, ds = \begin{pmatrix} e^{2x} \\ 1 \\ 4 - (x+3)e^{-x} - e^{2x} \end{pmatrix}$$

$$\therefore \quad y_{\sim}(x) = \Phi(x \; ; \; 0)\{\xi_0 + \int_0^x exp(-sA) \, b_{\sim}(x)\}dx$$

$$= \begin{pmatrix} e^{-2x} & xe^{-2x} & 0 \\ 0 & e^{-2x} & 0 \\ e^x - e^{-2x} & e^x - xe^{-2x} - e^{-2x} & e^x \end{pmatrix} \begin{pmatrix} e^{2x} \\ 1 \\ 4 - e^{2x} - (x+3)e^{-x} \end{pmatrix}$$

$$= \begin{pmatrix} 1 + xe^{-2x} \\ e^{-2x} \\ -4 - x - (x+1)e^{-2x} + 5e^x \end{pmatrix}$$

This is the desired solution of the inhomogeneous Initial value problem. From the above workout we claim that if for such a IVP, A is independent of x, i,e, A is a constant matrix, then the formula becomes simplified as:

$$y_{\sim}(x) = exp\,[\,(x - x_0)A\,]\,\xi_0 + \int_{x_0}^x exp\,[\,(s - x_0)A\,]\,b_{\sim}(s)\,dx$$

9.8 Alternative Ansatz for Computation of e^{Ax}

From the theoretical discussions and previous examples it is quite clear that for the initial value problems

$$y_{\sim}'(x) \quad = \quad Ay_{\sim}(x) \text{ subject to } y_{\sim}(0) = y_0$$

or,

$$y_{\sim}'(x) \quad = \quad Ay_{\sim}(x) + b_{\sim}(x) \text{ subject to } y_{\sim}(x_0) = \xi_0$$

with $y_0, \xi_0, b_{\sim}(x), y_{\sim}(x) \in \mathbb{R}^n$, if the co-efficient matrix A happens to be a constant matrix of order $n \times n$, then the computation of e^{Ax} is the prime job. We have so far made use of the S-N decomposition of A fruitfully to compute e^{Ax}. In this section (optional to the reader) we quote two alternative ansatz for computing the same − one being related to Laplace transform and the other being Putzer Algorithm that makes use of Cayley Hamilton theorem.

Laplace Transformation Technique with Illustrations

Recall that the Laplace transform of the function

$$f(x) = \begin{cases} e^{ax} & , \quad x > 0 \\ 0 & , \quad \text{otherwise} \end{cases}$$

is $F(s) = \frac{1}{s-a}$; $s \in \mathbb{R}$. One can prove an analogous result for matrices :

Laplace transform of e^{Ax} is $(sI_n - A)^{-1}$, where $A \in M_n(\mathbb{R})$ and $x \in \mathbb{R}$ One may therefore think of computing e^{Ax} as the inverse laplace transform of $(sI_n - A)^{-1}$. We skip the theoretical details of this result since it is based on contour integration but shall find it very handy for our computational purpose. In the following we shall illustrate the same for the matrices

$$A = \begin{pmatrix} 2 & 0 & 0 \\ 1 & 2 & 0 \\ 1 & 0 & 3 \end{pmatrix} \quad \text{appearing in Example 9.9}$$

$$\text{and} \quad A = \begin{pmatrix} -2 & 1 & 0 \\ 0 & -2 & 0 \\ 3 & 2 & 1 \end{pmatrix} \quad \text{appearing in Example 9.15}$$

For first case, $\quad (sI_3 - A) = \begin{pmatrix} s-2 & 0 & 0 \\ -1 & s-2 & 0 \\ -1 & 0 & s-3 \end{pmatrix}$

$$(sI_3 - A)^{-1} = \begin{pmatrix} \frac{1}{(s-2)} & 0 & 0 \\ \frac{1}{(s-2)^2} & \frac{1}{s-2} & 0 \\ \frac{1}{(s-2)(s-3)} & 0 & \frac{1}{s-3} \end{pmatrix}$$

Apply inverse Laplace transform to the elements of $(sI_3 - A)^{-1}$ to have :

$$e^{Ax} = \begin{bmatrix} e^{2x} & 0 & 0 \\ xe^{2x} & e^{2x} & 0 \\ e^{3x} - e^{2x} & 0 & e^{3x} \end{bmatrix}$$

This is exactly what we had via S-N decomposition method! The brevity is of course the reward !!

For the second problem, $\quad (sI_3 - A) = \begin{pmatrix} s+2 & -1 & 0 \\ 0 & s+2 & 0 \\ -3 & -2 & s-1 \end{pmatrix}$

$$(sI_3 - A)^{-1} = \begin{bmatrix} \frac{1}{s+2} & \frac{1}{(s+2)^2} & 0 \\ 0 & \frac{1}{s+2} & 0 \\ \frac{3}{(s+2)(s-1)} & \frac{2s+7}{(s+2)^2(s-1)} & \frac{1}{s-1} \end{bmatrix}$$

By method of partial fractions one can write

$$\frac{2s+7}{(s+2)^2(s-1)} = \frac{1}{s-1} - \frac{1}{(s-2)} - \frac{1}{(s+2)^2}$$

so that by Inverse Laplace transform applied to each of the elements of $(sI_3 - A)^{-1}$ we have :

$$e^{Ax} = \begin{bmatrix} e^{-2x} & xe^{-2x} & 0 \\ 0 & e^{-2x} & 0 \\ e^x - e^{-2x} & e^x - xe^{-2x} - e^{-2x} & e^x \end{bmatrix}$$

This is again same as that obtained via S-N decomposition.

The Laplace transform technique applies well to matrices where the eigenvalues are complex. For example consider the matrix $A = \begin{pmatrix} 0 & 1 \\ -1 & 0 \end{pmatrix}$, whose eigenvalues are $\pm\, i$.

We simply compute the inverse of $(sI_2 - A) = \begin{pmatrix} s & -1 \\ 1 & s \end{pmatrix}$.

$$(sI_2 - A)^{-1} = \begin{bmatrix} \frac{s}{s^2+1} & \frac{1}{s^2+1} \\ -\frac{1}{s^2+1} & \frac{s}{s^2+1} \end{bmatrix} \; ; \; s \in \mathbb{R}$$

Using Inverse Laplace transform of each element of $(sI_2 - A)^{-1}$ we get :

$$e^{Ax} = \begin{pmatrix} \cos x & \sin x \\ -\sin x & \cos x \end{pmatrix} \; ; \; x \in \mathbb{R}$$

Putzer Algorithm for Computing Matrix Exponentials

Putzer Algorithm is another alternative method to compute e^{At} when A is a constant matrix, no matter whether the eigenvalues of A are repeated or not.

According to this algorithm, for a $n \times n$ constant matrix A having eigenvalues $\lambda_1, \lambda_2, \cdots, \lambda_n$ (not necessarily distinct),

$$e^{Ax} = u_1(x)P_0 + u_2(x)P_1 + u_3(x)P_2 + \cdots + u_n(x)P_{n-1} \qquad (9.17)$$

where

$$P_0 = I_n$$
$$P_1 = (A - \lambda_1 I_n)$$
$$P_2 = (A - \lambda_1 I_n)(A - \lambda_2 I_n)$$
$$\cdots\cdots\cdots\cdots\cdots\cdots\cdots\cdots\cdots\cdots\cdots\cdots$$
$$\cdots\cdots\cdots\cdots\cdots\cdots\cdots\cdots\cdots\cdots\cdots\cdots$$
$$P_n = (A - \lambda_1 I_n)(A - \lambda_2 I_n)\ldots\ldots(A - \lambda_n I_n) \qquad (9.18)$$

and $u_j(x)(j = 1, 2, \cdots, n)$ are solutions of a system of n first order linear differential equations together with designed initial conditions :

$$u_1' = \lambda_1 u_1 \text{ with } u_1(0) = 1$$
$$u_2' = \lambda_2 u_2 + u_1 \text{ with } u_2(0) = 0$$
$$u_3' = \lambda_3 u_3 + u_2 \text{ with } u_3(0) = 0$$
$$\cdots\cdots\cdots\cdots\cdots\cdots\cdots\cdots\cdots\cdots\cdots\cdots\cdots$$
$$\cdots\cdots\cdots\cdots\cdots\cdots\cdots\cdots\cdots\cdots\cdots$$
$$u_n' = \lambda_n u_n + u_{n-1} \text{ with } u_n(0) = 0$$

i, e, $\quad u_1'(x) = \lambda_1 u_1 \; ; \; u_1(0) = 1$

$$\left.\begin{array}{ll} u_i'(x) = u_{i-1}(x) + \lambda_i u_i(x) & \forall \, i = 2, 3, \cdots, n \; ; \\ u_i(0) = 0 & \forall \, i = 2, 3, \cdots, n \end{array}\right\}$$

We also observe that

$$P_{k+1} = P_k(A - \lambda_{k+1} I_n) \; \forall \, k = 1, 2, \cdots, \overline{n-1} \qquad (9.19)$$

and each P_k commutes with A as they are matric polynomials in A. Finally we conclude that $P_n = 0$, the null matrix as Cayley Hamilton theorem ensures that every square matrix satisfies its own characteristic equation.

For establishing Putzer Algorithm it suffices to prove that the matrix function

$$M(x) \equiv u_1(x)P_0 + u_2(x)P_1 + u_3(x)P_2 + \cdots + u_n(x)P_{n-1} \qquad 9.20(a)$$

satisfies the matrix I.V.P

$$\frac{dY}{dx} = AY \; ; \; Y(0) = I_n \; ; \; 0 \in I, \text{ with } Y \in M_n(\mathbb{R}) \qquad 9.20(b)$$

because "Fundamental theorem for linear systems" (Theorem 9.7) ensures that $Y(x) = e^{Ax}$ is the unique solution of the above matrix I.V.P. Observe that

$$M(0) \; = \; u_1(0)P_0 + u_2(0)P_1 + u_3(0)P_2 + \cdots + u_n(0)P_{n-1} = I_n$$

$$
\begin{aligned}
M'(x) \; &= \; u_1'(x)P_0 + u_2'(x)P_1 + u_3'(x)P_2 + \cdots + u_n'(x)P_{n-1} \\[2mm]
&= \; \lambda_1\, u_1(x)P_0 + \sum_{k=1}^{n-1} u_{k+1}'(x)P_k \\
&= \; \lambda_1\, u_1(x)P_0 + \sum_{k=1}^{n-1} \{u_k(x) + \lambda_{k+1}\, u_{k+1}(x)\}P_k \\
&= \; \lambda_1\, u_1(x)P_0 + \sum_{k=1}^{n-1} u_k(x)P_k + \sum_{k=1}^{n-1} u_{k+1}(x)(P_k A - P_{k+1}) \\
&= \; \lambda_1\, u_1(x)P_0 + \sum_{k=1}^{n-1} u_{k+1}(x)P_k\, A + u_1(x)P_1 - u_n(x)P_n \\
&= \; u_1(x)(P_1 + \lambda_1\, P_0) + \sum_{k=1}^{n-1} u_{k+1}(x)P_k\, A \quad (\because\; P_n = 0) \\
&= \; u_1(x)A + \sum_{k=1}^{n-1} u_{k+1}(x)A\, P_k \quad [\because\; P_k's \text{ commute with } A] \\
&= \; A\left[u_1(x)P_0 + \sum_{k=1}^{n-1} u_{k+1}(x)P_k \right] \quad (\because\; P_0 = I_n) \\
&= \; AM(x)
\end{aligned}
$$

This proves the claim that $M(x)$ is a solution of the matrix I.V.P. quoted before and consequently by (9.19)

$$e^{Ax} = M(x) = u_1(x)I_n + u_2(x)P_1 + \cdots + u_n(x)P_{n-1}$$

Example 9.16 Consider matrix $A = \begin{pmatrix} 1 & 0 & 0 \\ -1 & 2 & 0 \\ 1 & 0 & 2 \end{pmatrix}$ and compute e^{Ax} by Putzer algorithm.

As the matrix A is lower triangular, are its eigenvalues $1, 2, 2$.

So $\lambda_1 = 1$; $\lambda_2 = 2$; $\lambda_3 = 2$

$P_0 = I_3$

$$P_1 = (A - \lambda_1 I_3) = (A - I_3) = \begin{pmatrix} 0 & 0 & 0 \\ -1 & 1 & 0 \\ 1 & 0 & 1 \end{pmatrix}$$

$$P_2 = (A - I_3)(A - 2I_3) = \begin{pmatrix} 0 & 0 & 0 \\ -1 & 1 & 0 \\ 1 & 0 & 1 \end{pmatrix} \begin{pmatrix} -1 & 0 & 0 \\ -1 & 0 & 0 \\ 1 & 0 & 0 \end{pmatrix} = \begin{pmatrix} 0 & 0 & 0 \\ 0 & 0 & 0 \\ 0 & 0 & 0 \end{pmatrix}$$

Consequently $P_3 = 0$

Again as per the algorithm, $u_1' = u_1$; $u_1(0) = 1$ so that $u_1(x) = e^x$

$$u_2' = u_1 + 2u_2 = e^x + 2u_2 \; ; \; u_2(0) = 0$$

i, e, $u_2' - 2u_2 = e^x$ subject to $u_2(0) = 0$

On solving, $u_2(x) = e^{2x} - e^x$

$\therefore \quad e^{Ax} = M(x) = u_1(x)I_3 + u_2(x)P_1$

$$= \begin{pmatrix} e^x & 0 & 0 \\ 0 & e^x & 0 \\ 0 & 0 & e^x \end{pmatrix} + \begin{pmatrix} 0 & 0 & 0 \\ e^x - e^{2x} & e^{2x} - e^x & 0 \\ e^{2x} - e^x & 0 & e^{2x} - e^x \end{pmatrix}$$

$$= \begin{pmatrix} e^x & 0 & 0 \\ e^x - e^{2x} & e^{2x} & 0 \\ e^{2x} - e^x & 0 & e^{2x} \end{pmatrix}$$

Note : (i) This algorithm of computing e^{Ax} is equally valid even if A has complex eigenvalues.

(ii) If the order of the matrix be large, then computation by e^{Ax} will be lengthy—nevertheless much better than S-N decomposition scheme depending on idea of generalised eigenvectors.

(iii) The method clicks irrespective of whether the matrix is diagonalisable or defective, possessing real or complex eigenvalues.

Exercise 9

1. Solve the initial value problem :

$$y'(x) = A\underset{\sim}{y}(x) \quad \text{subject to} \quad y(0) = y_0$$

where the matrix A and $\underset{\sim}{y}_0$ are given by

(a) $A = \begin{pmatrix} 1 & 0 & 0 \\ 2 & 1 & 0 \\ 3 & 2 & 1 \end{pmatrix}$; $\underset{\sim}{y}_0 = \begin{pmatrix} 1 \\ 1 \\ 1 \end{pmatrix}$

(b) $A = \begin{pmatrix} -1 & 1 & -2 \\ 0 & -1 & 4 \\ 0 & 0 & 1 \end{pmatrix}$; $\underset{\sim}{y}_0 = \begin{pmatrix} 1 \\ 0 \\ 2 \end{pmatrix}$

(c) $A = \begin{pmatrix} 0 & 1 & 0 & 0 \\ 1 & 0 & 0 & 0 \\ 0 & 0 & 1 & -1 \\ 0 & 0 & 1 & 1 \end{pmatrix}$; $\underset{\sim}{y}_0 = \begin{pmatrix} 1 \\ 0 \\ -1 \\ 1 \end{pmatrix}$

by computing the fundamental matrix for $(a), (b)$ and (c) by

(i) S-N decomposition

(ii) Laplace transform technique

2. Use Laplace transformation technique to solve the IVP

$$\underset{\sim}{y}'(x) = A\underset{\sim}{y}(x) ; \ \underset{\sim}{y}(x_0) = \begin{pmatrix} c_1 \\ c_2 \\ c_3 \end{pmatrix} \quad \text{and} \quad A = \begin{pmatrix} -3 & 0 & 0 \\ 0 & 3 & -2 \\ 0 & 1 & 1 \end{pmatrix}$$

(Observe that although the eigenvalues of A are complex, Laplace technique enables us to determine the fundamental matrix e^{Ax})

Try the same problem by the Method of diagonalisation (§ 9.3)

3. (a) Find a basis of $\mathbb{R}^3$ that will put the matrix A given by

$$A = \begin{pmatrix} 2 & -2 & 2 \\ 2 & 2 & 2 \\ 1 & 1 & 2 \end{pmatrix}$$

to Jordan form. Can you compute A^n for $n \in \mathbb{N}$?

(b) Using the methods discussed in § 9.6, transform the matrix

$$A = \begin{pmatrix} 5 & 4 & 2 & 1 \\ 0 & 1 & -1 & -1 \\ -1 & -1 & 3 & 0 \\ 1 & 1 & -1 & 2 \end{pmatrix}$$

to the Jordan canonical form and also verify that one of the possible Jordan forms of A is

$$J = \begin{pmatrix} 4 & 1 & 0 & 0 \\ 0 & 4 & 0 & 0 \\ 0 & 0 & 2 & 0 \\ 0 & 0 & 0 & 1 \end{pmatrix}$$

4. (a) Reduce to Jordan form the matrix $A = \begin{pmatrix} 2 & 1 \\ -1 & 4 \end{pmatrix}$ and

hence solve the system $\frac{dy}{dx} = Ay$, where $y \equiv (y_1 \ y_2)^T$.

(b) Reduce to Jordan form the matrix A defined by

$$A = \begin{pmatrix} 5 & 1 & -4 \\ 4 & 3 & -5 \\ 3 & 1 & -2 \end{pmatrix}$$

and hence solve the system $\frac{dy}{dx} = Ay$, where $y \equiv (y_1 \ y_2 \ y_3)^T$.

(c) Prove that the differentiation operator D on the linear space $P_n(\mathbb{R})$ of all polynomials of degree $\leq n$ and having real coefficients is strictly upper triangular and hence nilpotent of index $(n+1)$. Explicitly give the matrix representation of D operator w.r.to the canonical basis $\{1, x, x^2, \cdots, x^n\}$ of $P_n(\mathbb{R})$.

(d) Using the fact that every complex square matrix A can be Jordanised by means of a similarity transformation with a suitable base-changing matrix P, prove that any $n \times n$ complex matrix A is similar to its transpose.

5. Use the Variation of Constant method (§ 9.7) to solve the IVP's :

(a) $y'(x) = \begin{pmatrix} 0 & 1 \\ -1 & 0 \end{pmatrix} y + \begin{pmatrix} 0 \\ x \end{pmatrix} ; y(0) = \begin{pmatrix} 0 \\ 1 \end{pmatrix}$

(b) $\underset{\sim}{y}'(x) = \begin{pmatrix} -2 & 0 \\ 0 & 4 \end{pmatrix} + \begin{pmatrix} 1 \\ e^{-x} \end{pmatrix}$; $\underset{\sim}{y}(0) = \begin{pmatrix} -1 \\ 2 \end{pmatrix}$

6. Verify Cayley Hamilton Theorem for each of the following matrices and hence use **Putzer algorithm** (§ 9.8) to find e^{Ax} :

(a) $A = \begin{pmatrix} 2 & -2 & 2 \\ 0 & 1 & 1 \\ -4 & 8 & 3 \end{pmatrix}$ (b) $A = \begin{pmatrix} 1 & 0 & 2 \\ 2 & 1 & -1 \\ 1 & -1 & 2 \end{pmatrix}$

(c) $A = \begin{pmatrix} -3 & 0 & 0 \\ 0 & 3 & -2 \\ 0 & 1 & 1 \end{pmatrix}$ (d) $A = \begin{pmatrix} 1 & 1 & 0 \\ 0 & 1 & 0 \\ 0 & 1 & 1 \end{pmatrix}$

Also find solution of the inhomogeneous vector differential equation

$$\underset{\sim}{y}'(x) = A\underset{\sim}{y}(x) + \underset{\sim}{b}(x), \quad \text{where } \underset{\sim}{b}(x) = \begin{pmatrix} 1 \\ x \\ x^2 \end{pmatrix}$$

Appendix

A. Dirac Delta Function:

If $\{\phi_n\}$ be a sequence of functions that satisfies the properties

(i) $\phi_n(x) = \phi_n(-x) \; \forall x \in \mathbb{R}$ and for each n (evenness property)

(ii) $\displaystyle\int_{-\infty}^{+\infty} \phi_n(x)dx = 1$ for each n (normalisation),

then for any arbitrary continuous function $f(x)$ one can establish that

$$\lim_{n\to\infty} \int_{-\infty}^{+\infty} \phi_n(x)f(x)dx = f(0).$$

We formally express this result by the statement : 'Each set of functions $\{\phi_n\}$ forms a delta sequence.' Alternatively, one may state that the Dirac delta function $\delta(\cdot)$ is defined through the relation

$$\int_{-\infty}^{+\infty} \delta(x)f(x)dx = f(0) \qquad (A.1)$$

f being any continuous function of x, generalising it one gets

$$\int_{-\infty}^{+\infty} \delta(x - \xi)f(\xi)d\xi = f(x) \qquad (A.2)$$

Some authors like to put the definition (A.1) or (A.2) of Dirac delta function through the limiting relation

$$\lim_{\varepsilon\to 0+} \int_{-\infty}^{+\infty} \delta_\varepsilon(x)f(x)dx = f(0),$$

or more generally by the relation

$$\lim_{\varepsilon\to 0+} \int_{-\infty}^{+\infty} \delta_\varepsilon(x - \varepsilon)f(\xi)d\xi = f(x), \qquad (A.3)$$

From this viewpoint, the definition of Dirac delta function shows that one can generate any continuous function $f(x)$ by a succession of 'pulses' that are of short duration and which follow each other in a proper sequence.

Here the $\delta_\varepsilon(x)$'s have the generic properties :

(i) $\delta_\varepsilon(x)$ vanishes outside ε – neighbourhood of $x = 0$ and within the ε-neighbourhood of $x = 0$ it does never change sign, i, e, it is either positive or zero.

(ii) $\displaystyle\int_{|x|<\varepsilon} \delta_\varepsilon(x)dx = 1$

B.   Another way of classifying the odes:

The differential equations are also classified as 'non-autonomous' and 'autonomous' according as it involves the independent variable explicitly or not. A first order differential equation of the form $\frac{dy}{dx} = f(x, y, \lambda)$, λ being a parameter is said to be 'non-autonomous' while that appearing in the form $\frac{dy}{dx} = g(y, \lambda)$ is called autonomous. Since any non-autonomous first order differential equation can be cast into an equivalent system of the form

$$\left. \begin{array}{rcl} \frac{dy}{dx} &=& f(u, y, \lambda) \\ \frac{du}{dx} &=& 1 \end{array} \right\}$$

all ordinary differential equations are translatable to first order autonomous system of differential equations — u being a new variable.

Higher order linear differential equation (with variable co-efficients)

$$y^{(n)} + p_1(x)y^{(n-1)} + p_2(x)y^{(n-2)} + \cdots\cdots + p_{n-1}(x)y' + p_n(x)y = q(x)$$
$$(B.1)$$

can be written in the form

$$\frac{d}{dx} \begin{pmatrix} y \\ y_1 \\ \cdots \\ \cdots \\ y_{n-2} \\ y_{n-1} \end{pmatrix} - \begin{pmatrix} 0 \\ 0 \\ \cdots \\ \cdots \\ 0 \\ q(x) \end{pmatrix}$$

$$
= \begin{pmatrix} 0 & 1 & \cdots & 0 & 0 \\ 0 & 0 & \cdots & 0 & 0 \\ \cdots & \cdots & \cdots & \cdots & \cdots \\ \cdots & \cdots & \cdots & \cdots & \cdots \\ 0 & 0 & \cdots & 0 & 1 \\ -p_n(x) & -p_{n-1}(x) & \cdots & -p_2(x) & -p_1(x) \end{pmatrix} \begin{pmatrix} y \\ y_1 \\ \cdots \\ \cdots \\ y_{n-2} \\ y_{n-1} \end{pmatrix}
$$

In the compact matrix form it reads

$$
\frac{d\vec{y}}{dx} = A(x)\vec{y} + \vec{q}(x), \tag{B.1}
$$

where
$$
\left.\begin{array}{l} \vec{y} \equiv (y,\ y_1,\ y_2, \cdots\cdots y_{n-2},\ y_{n-1})^T \in \mathbb{R}^n \\ \vec{q}(x) \equiv (0, 0, \cdots\cdots, 0, q(x))^T \in \mathbb{R}^n \end{array}\right\}
$$

and $A(x)$ is the above matrix function of x.

This gives us an equivalent first order vector-valued differential equation in x and y. The equation (B.1) is non-autonomous.

In case $q(x)$ of (B.1) were zero and $p_k(x)$'s $(k = 1, 2, \cdots\cdots, n)$ were all constants, we see that it would become autonomous. Involvement of one or more parameters like λ etc. in the system is optional.

As a passing remark we note that an autonomous first order equation of the form $\frac{dy}{dx} = g(y, \lambda)$ is separable and hence is easily integrable to produce solutions in the implicit form.

C. Tabular Integration — an Auxiliary Method to Integration by Parts:

Integration by parts arises out of the product rule of differentiation and here the main objective is to swap the roles of integrand and integrator. For example, if we are asked to compute $\int u(x)d(v(x))$, then $u(x)$ is the integrand and $v(x)$ is the integrator. In this method, we try to express $\int u(x)d(v(x))$ in terms of $v(x)d(u(x))$, the working formula being

$$
\int u(x)d(v(x)) = u(x)v(x) - \int v(x)d(u(x))
$$

If now $v(x)$ be a C^1-function, then we could write $\int u(x)d(v(x))$ as $\int u(x)v'(x)dx$, where $v'(x) \equiv \frac{dv}{dx}$.

If we agree to write $v'(x) \equiv \omega(x)$, then

$$\int u(x)d(v(x)) = \int u(x)\omega(x)dx.$$

If $u(x)$ can be differentiated successively and $\omega(x)$ can be integrated successively, then integration by parts provides us a technique for simplifying integrals of the form $\int u(x)\omega(x)dx$.

In particular, if $u(x)$ be a polynomial of certain degree n, say,

$$u(x) = \sum_{k=0}^{n} a_k x^k,$$

then one observes that $\frac{d^{n+1}}{dx^{n+1}}(u(x)) = 0$ but $\frac{d^n}{dx^n}(u(x)) = a_n n! \neq 0$. However, in this special case we may organise the cumbersome calculations in a tabular form as illustrated schematically in the following :

$u(x)$ & its derivatives	$\omega(x)$ & its integrals
$u(x) = \sum_{k=0}^{n} a_k x^k$	$\omega(x)dx$
$u'(x) = \sum_{k=1}^{n} k a_k x^{k-1}$	$\int \omega(x)dx$
$u''(x) = \sum_{k=2}^{n} k(k-1)a_k x^{k-2}$	$\int dx \int dx\, \omega(x)$
$\cdots\cdots\cdots$	$\cdots\cdots\cdots$
$\cdots\cdots\cdots$	$\cdots\cdots\cdots$
$u^{(n)}(x) = a_n\, n!$	$\int\int\cdots\int(dx)^n\omega(x)$
$u^{(n+1)}(x) = 0$	$\int\int\cdots\int(dx)^{n+1}\omega(x)$
$u(x)$ & its derivatives	$\omega(x)$ & its integrals

$x^3 - 5x^2$	e^{2x}
$3x^2 - 10x$	$\frac{1}{2}e^{2x}$
$6x - 10$	$\frac{1}{4}e^{2x}$
6	$\frac{1}{8}e^{2x}$
0	$\frac{1}{16}e^{2x}$

We add the products of the functions connected by arrows, with middle sign changed, to obtain the desired result. As an illustration, consider $I = \int(x^3 - 5x^2)e^{2x}dx$, where $u(x) = x^3 - 5x^2$ and $v(x) = e^{2x}$

We add the products of functions connected by arrows, with middle sign changed to obtain

$$I = \left(\frac{1}{2}(x^3 - 5x^2) - \frac{1}{4}(3x^2 - 10x) + \frac{1}{8}(6x - 10) - \frac{3}{8} \right) e^{2x}$$

D. Bairstow's Method of Factorisation:

This method determines the quadratic factors of a given polynomial. Once the quadratic factor is extracted, it may give rise to a pair of real or a pair of complex roots. The method is based on the division algorithm. We seek a quadratic factor of the form $q(x) \equiv x^2 - ux - v$, (u and v being not yet determined) for the nth degree polynomial $p(x)$ given by

$$p(x) \equiv a_0 x^n + a_1 x^{n-1} + a_2 x^{n-2} + \cdots\cdots + a_{n-1}x + a_n, a_k \in \mathbb{R}$$
for $k = 0, 1, 2, \cdots\cdots, n$.

Had $q(x)$ been not a factor of $p(x)$, we would have

$$p(x) = q(x)s(x) + (r_0 x + r_1),$$

where $s(x)$ is a polynomial of degree $(n-2)$ and r_0, r_1 are constants, but strictly speaking, functions of u and v.

$$\therefore \ p(x) = (a_0 x^n + a_1 x^{n-1} + a_2 x^{n-2} + \cdots + a_{n-1}x + a_n)$$
$$= (x^2 - ux - v) \times (b_0 x^{n-2} + b_1 x^{n-3} + \cdots + b_{n-2}) + (r_0 x + r_1)$$

Comparing co-efficients of like powers of x, we have :

$$
\begin{aligned}
a_0 &= b_0 \\
a_1 &= b_1 - ub_0 \\
a_2 &= b_2 - ub_1 - vb_0 \\
&\ \cdots\cdots \quad \cdots\cdots \\
&\ \cdots\cdots \quad \cdots\cdots \\
a_{n-2} &= b_{n-2} - ub_{n-3} - vb_{n-4} \\
a_{n-1} &= r_0 - ub_{n-2} - vb_{n-3} \\
a_n &= r_1 - vb_{n-2}
\end{aligned}
$$

In order to fit r_0 and r_1 into the other's format, we write $r_0 \equiv b_{n-1}$ and $r_1 \equiv b_n - ub_{n-1}$

Hence compact form of the recurrence relation reads :

$$b_k = a_k + ub_{k-1} + vb_{k-2} \ \forall \ k \geq 2 \qquad (D.1)$$

We may extend it for $k = 0$ and $k = 1$ provided we set $b_{-1} = b_{-2} = 0$

i, e, $b_k = a_k + ub_{k-1} + vb_{k-2} \ \forall \ k \geq 0$ with $b_{-1} = b_{-2} = 0$

In case $q(x)$ is a factor of $p(x)$, $r_0 = 0$ and $r_1 = 0$ conversely.

This in turn, implies that

$$b_{n-1} = 0; \ b_n = 0 \qquad (D.2)$$

We should keep in mind that all $b_k (k = 0, 1, 2, \cdots, n)$ are functions of u and v. Our next job is to solve the pair of equations appearing in (C.2) by Newton-Raphson method.

From set of recurrence relations (C.1), $\frac{\partial b_0}{\partial u} = 0$.

We agree to write $c_k \equiv \frac{\partial b_{k+1}}{\partial u} \ \forall k \geq 1$

In this way, we have another set of recurrence relations

$$\left.\begin{array}{rl} c_k &= b_k + uc_{k-1} + v.c_{k-2} \quad \forall \ k = 0, 1, 2, \cdots, \overline{n-1} \\ c_{-1} &= c_{-2} = 0 \end{array}\right\} \quad (D.3)$$

Let's make Taylor-expansion of b_n and b_{n-1} upto first order:

$$0 = b_{n-1}(u,v) \approx b_{(n-1)}(u_0, v_0) + (u - u_0) \left.\frac{\partial b_{n-1}}{\partial u}\right|_{(u_0,v_0)}$$

$$+ (v - v_0) \left.\frac{\partial b_{n-1}}{\partial v}\right|_{(u_0, v_0)}$$

$$0 = b_n(u,v) \approx b_n(u_0, v_0) + (u - u_0) \left.\frac{\partial b_n}{\partial u}\right|_{(u_0, v_0)}$$

$$+ (v - v_0) \left.\frac{\partial b_n}{\partial v}\right|_{(u_0, v_0)}$$

Writing $\delta = (u - u_0)$ and $\varepsilon = (v - v_0)$, the above approximation reduces to the following set of linear equations in δ and ε:

$$\left.\begin{array}{rl} \delta.c_{n-2} + \varepsilon.c_{n-3} + b_{n-1} &= 0 \\ \delta.c_{n-1} + \varepsilon.c_{n-2} + b_n &= 0 \end{array}\right\}$$

so that $\delta = \dfrac{c_{n-3}b_n - b_{n-1}c_{n-2}}{c_{n-1}^2 - c_{n-1} \cdot c_{n-3}}$ and $\varepsilon = \dfrac{c_{n-1}b_{n-1} - b_n c_{n-2}}{c_{n-2}^2 - c_{n-1}c_{n-3}}$

Once these corrections are obtained, we replace u_0 and v_0 by its updated values–viz, $(u_0 + \delta)$ and $(v_0 + \varepsilon)$ respectively.

Denoting $u_1 \equiv u_0 + \delta$ and $v_1 \equiv v_0 + \varepsilon$ as the first iterates, we restart the above sequence of steps in order. After m iterations, we therefore reach a quadratic factor, viz, $(x^2 - u_m x - v_m)$ of $p(x)$. However, u_m and v_m are obtained through numerical calculations and so are prone to inherent error. The number of iterations that we have to undergo depends solely on tolerance level of accuracy.

Illustration: Solve the general solution of $(D^4 - 5D^2 + D + 4)y = 0$, a fourth order linear ode. We may use the above method of Bairstow to show that with suitable approximations the quadratic factors are (approx) $(D^2 - 3D + 2.172)$ and $(D^2 - 3D + 1.849)$.

Remark :

(i) If the co-efficients of polynomials are allowed to be complex, Bairstow's method is needless.

(ii) When the roots of the polynomial are very close, this method fails miserably. If the linear system appearing in (C.3) is ill-conditioned, Bairstow method fails.

(iii) Often this method is very fruitful in determining eigenvalues of a matrix associated with a linear system.

E.   Differential Operator as Matrix : Discretization

Consider the homogeneous second order differential equation

$$y'' + p(x)y' + q(x)y = 0,$$

where $p(x)$ and $q(x)$ are given functions of x. We 'atomise' this equation by breaking it up into a large number of algebraic equations. For this purpose we replace the continuum of x-values by a countable dense set. In fact this is always possible as $\mathbb{R}$ and any homeomorphic image, viz, an interval I is separable. Without loss of generality we assume $[0, 1]$ to be the domain of our problem. As a first step towards atomisation, we replace $[0, 1]$ by the discrete set $\mathcal{S} = \{k \in /k = 0, 1, 2, \cdots, n \,\&\, n\varepsilon = 1\}$. Once we atomise $y(x), p(x)$ and $q(x)$ over $[0,1]$, our interest gets restricted to the values of these functions over $\mathcal{S}$ only and hence the continua of function values gets replaced by $(n + 1)$ tuple vectors, viz, $y = (y_0, y_1, \cdots, y_n)$; $p = (p_0, p_1, \cdots, p_n)$ and $q = (q_0, q_1, \cdots, q_n)$ with $y_k = y(x_k)$; $p_k = p(x_k)$ and $q_k = q(x_k)$, $k = 0, 1, 2, \cdots, n$. We transcribe the given ode into a set of linear algebraic difference equations by replacing the first order derivative $y'(x_k)$ by the forward difference $\frac{1}{\varepsilon}(y_{k+1} - y_k)$ and the second order derivative $y''(x_k)$ by the central difference $\frac{1}{\varepsilon^2}(y_{k+1} - 2y_k + y_{k-1})$. This gives us the following set of linear algebraic equations:

$$\frac{1}{\varepsilon^2}(y_{k+1} - 2y_k + y_{k-1}) + \frac{1}{\varepsilon}p_k(y_{k+1} - y_k) + q_k y_k = 0,$$

i.e,
$$\left(\frac{1}{\varepsilon^2} + \frac{p_k}{\varepsilon}\right) y_{k+1} + \left(q_k - \frac{1}{\varepsilon}p_k - \frac{2}{\varepsilon^2}\right) y_k + \frac{1}{\varepsilon^2}y_{k-1} = 0,$$

$$\text{for } k = 1, 2, \cdots, \overline{n-1}. \qquad \text{(E.1)}$$

Note that we are compelled to keep out the end on cases, viz, $k = 0$ and $k = n$ as there is no knowledge of values of $y(x)$ outside the range $[0,1]$. Since the above system involves fewer equations than the total number of unknowns, it possesses infinitely many solutions. Some people call this system 'incomplete' as it fails to provide a unique solution unless it is supplemented by further data. Usually the boundary conditions provide this supplementary data. Enroute this algebraisation, we have an indirect evidence to the truth that a homogeneous differential equation alone has infinitely many solutions—arbitrary solutions appearing in the general solution of the given ode being the mantles.

The above system (D.1) is compactly written in the matrix form $A\underset{\sim}{y} = 0$, where A is an $(n-1) \times (n+1)$ matrix; $\underset{\sim}{y}$ is an $(n+1)$ - component column vector. To curb the arbitrariness of solution of system (D.1) we have three alternative ways of passing boundary conditions.

(a) To append two more linear equations to the system (1) and have a square matrix of order $(n + 1)$. If values of the endon derivatives $y'(0)$ and $y'(1)$ are preassigned, then due to the algebraic transcription,

$$\left.\begin{array}{ll} y'(0) & \approx \frac{1}{\varepsilon}(y_1 - y_0) \quad \text{(forward difference)} \\ y'(1) & \approx \frac{1}{\varepsilon}(y_n - y_{n-1}) \;\; \text{(backward difference)} \end{array}\right\}$$

(b) To cut down the number of unknowns by two and have a square matrix of order $(n-1)$. If values $y(0)$ and $y(1)$ are prescribed, then the demand case is satisfied.

(c) To cut down the number of unknowns by unity and append one or more linear equation to the above system and have a square matrix of order n. If we know $y(x)$ at one endpoint and $y'(x)$ at the same/other endpoint of $[0,1]$, then again on suitable algebraic transcription of the derivative by means of difference co-efficient, we fulfil the demand.

The all important point to be focussed through this discussion is that the differential equation alone, without the associated boundary conditions cannot give a unique solution. Hence the matrix which substitutes the differential operator involved in the given equation must comprise of the allied boundary conditions.

References

1. Differential Equations: D.A. Murray [Orient Longman]

2. Differential Equations: G.F.Simmons [Tata Mc Graw Hill]

3. Differential Equations: S.L. Ross [John Wiley & Sons]

4. Elements of Differential Equations: W. Kaplan [Addison Wesley Publishing]

5. Elementary Differential Equations & Boundary Value Problems: W.E. Boyce & R.C.Diprima [John Wiley & Sons]

6. Differential Equations : A Dynamical System Approach: J.H. Hubbard & B. West [Springer]

7. An Elementary Treatise on Differential Equations: H.T.H.Piaggio [G.Bell & Sons]

8. Differential Equations and the Calculus of Variations: L. Elsgolts [Mir Publishers]

9. Differential Equations and their Applications: M. Braun [Springer]

10. An Introduction to Ordinary Differential Equations: E. Coddington [Prentice Hall of India]

11. Linear Algebra: S. Fredberg, A. Insel & L.Spence [Prentice Hall of India]

12. Linear Algebra Done Right: S. Axler [Springer]

13. Matrix Theory: D.W. Lewis [Allied Publishers]

14. Mathematical Methods : M.C.Potter & Goldberg [Prentice Hall of India]

Index

Printed in the United States
by Baker & Taylor Publisher Services

Printed in the United States
by Baker & Taylor Publisher Services